Marine Nutraceuticals and Functional Foods

NUTRACEUTICAL SCIENCE AND TECHNOLOGY

Series Editor

FEREIDOON SHAHIDI, PH.D., FACS, FCIC, FCIFST, FIAFoST, FIFT, FRSC
University Research Professor
Department of Biochemistry
Memorial University of Newfoundland
St. John's, Newfoundland, Canada

1. Phytosterols as Functional Food Components and Nutraceuticals, *edited by Paresh C. Dutta*
2. Bioprocesses and Biotechnology for Functional Foods and Nutraceuticals, *edited by Jean-Richard Neeser and Bruce J. German*
3. Asian Functional Foods, *John Shi, Chi-Tang Ho, and Fereidoon Shahidi*
4. Nutraceutical Proteins and Peptides in Health and Disease, *edited by Yoshinori Mine and Fereidoon Shahidi*
5. Nutraceutical and Specialty Lipids and their Co-Products, *edited by Fereidoon Shahidi*
6. Anti-Angiogenic Functional and Medicinal Foods, *edited by Jack N. Losso, Fereidoon Shahidi, and Debasis Bagchi*
7. Marine Nutraceuticals and Functional Foods, *edited by Colin Barrow and Fereidoon Shahidi*

Marine Nutraceuticals and Functional Foods

Edited by

Colin Barrow
Fereidoon Shahidi

CRC Press
Taylor & Francis Group
Boca Raton London New York

CRC Press is an imprint of the
Taylor & Francis Group, an **informa** business

CRC Press
Taylor & Francis Group
6000 Broken Sound Parkway NW, Suite 300
Boca Raton, FL 33487-2742

Printed in the United States of America on acid-free paper
10 9 8 7 6 5 4 3 2 1

International Standard Book Number-13: 978-1-57444-487-2 (Hardcover)

Library of Congress Cataloging-in-Publication Data

Marine nutraceuticals and functional foods / [edited by] Colin Barrow and Fereidoon Shahidi
p. ; cm. -- (Nutraceutical science and technology ; 7) "A CRC title."
Includes bibliographical references and index.
ISBN 978-1-57444-487-2 (hardcover : alk. paper)
1. Functional foods. 2. Fisheries--By-products--Health aspects. I. Barrow, Colin J. II. Shahidi, Fereidoon, 1951- III. Series.

[DNLM: 1. Dietary Supplements. 2. Fish Products. QU 145.5 M338 2008]
QP144.F85M37 2008 612.3'97--dc22 2007012218

Visit the Taylor & Francis Web site at
http://www.taylorandfrancis.com

and the CRC Press Web site at
http://www.crcpress.com

Contents

Preface

With better appreciation of the health benefits of seafood, there has been an increase in the harvesting of wild fish, fish aquaculture, and seaweed harvesting and cultivation. Increased harvesting leads to a greater abundance of by-products. Thus, finfish heads, gut, and backbone as well as shellfish and crustacean shell/exoskeleton are often discarded. Because seaweeds are often harvested for the production of specific components, such as carrageenan, alginate, and agar, a large amount of waste is also generated. Many hundreds of tons of various marine by-products are annually available for developing value-added products, and thus a more efficient utilization of our natural resources is possible. This has led to the development of several marine nutraceuticals with potential health benefits. Although there are many definitions of *nutraceutical*, for the purposes of this book we define marine nutraceutical as "a marine-derived substance that can be used as a dietary supplement or a food ingredient that provides a medicinal or health benefit beyond basic nutrition."

This book discusses a variety of marine nutraceuticals, including both commercialized products and those in development. Chapter 1 provides an overview of current marine nutraceutical products. Chapters 2 through 5 deal with omega-3 oils, describing their origin, and their health benefits, especially with regard to brain health, and how they can be stabilized and delivered into functional foods. Most commercial omega-3 oils derived from fish come from South American fish meal processing, where the oil was originally a waste product. Several years ago this fish oil was hydrogenated and sold as an inexpensive hydrogenated fat, but with the recognition of the health benefit of highly polyunsaturated fatty acids, and those derived from fish in particular, this waste product from fish meal manufacturing is now one of the most widely consumed nutritional supplements and its use is growing rapidly as a functional food ingredient. Chapters 6 and 7 discuss chitin, chitosan, and partially hydrolyzed chitosan. These polymers are derived from shrimp and crab shell and have been sold for several years as dietary supplements, mostly for their fat- and cholesterol-absorbing properties. Chapter 8 discusses the health benefits and methods for the production of glucosamine. Glucosamine is fully hydrolyzed chitosan and is also derived mainly from shrimp and crab shell. Glucosamine and fish oil are the two most commercially successful marine nutraceuticals. Glucosamine is widely consumed as a nutritional supplement for joint health.

Chapters 9 and 10 provide an overview of the bioactive properties of hydrolyzed marine protein, in particular ACE-inhibitory and blood pressure–lowering properties of finfish protein hydrolyzate. Chapters 11 through 14 discuss the bioactive constituents of marine algae, including fucoxanthin, astaxanthin, and marine polysaccharides. Seaweeds are widely consumed as food products in Asia and highly regarded as healthy foods. However, their consumption is limited in most Western countries. The isolation of specific components from seaweeds and proof of specific health benefits through human clinical studies are important and growing areas of marine nutraceuticals. Chapter 15 provides an overview of nutraceutical development from marine microorganisms. This includes methods for controlled production of established products, like omega-3 oils, as well as the production of novel marine compounds via a renewable resource. Chapter 16 describes the potential of shark cartilage for the prevention of cancer. Chapter 17 discusses potential marine sources of calcium, one of the most widely consumed nutritional supplements and functional food ingredients, normally derived from terrestrial sources. Finally, Chapter 18 describes the discovery and development of a novel immunoenhancing polysaccharide complex derived from the microalgae, *Chlorella*.

The discovery and development of marine nutraceuticals is a relatively new area compared to that of nutraceuticals derived from terrestrial sources. However, the commercial success of glucosamine and omega-3 marine oils, in particular, together with the overall interest in novel healthy food ingredients and nutritional supplements is driving both research and commercialization of marine nutraceuticals. The next few decades will see continued growth and should be an exciting and challenging time for researchers in this field.

The book may be used as a text or reference for students in food science, nutrition, biochemistry, and health sciences at the senior undergraduate and graduate levels. Scientists in the academia, government laboratories, and industry will also find it of interest. The extensive bibliography provided would allow readers to further consult the source materials for further information.

We are indebted to the renowned contributors whose efforts led to this state-of-the-art publication on marine nutraceuticals. We are also grateful to Peggy-Ann Parsons for her efforts in formatting the manuscripts.

Fereidoon Shahidi
Colin J. Barrow

Editors

Colin J. Barrow is the vice president of research and development for Ocean Nutrition Canada and is responsible for a research team of 40 scientists developing marine-derived supplement and healthy food ingredients, with emphasis on omega-3 oils. Dr. Barrow has previously been on the faculty of the School of Chemistry at the University of Melbourne, Australia, and spent a decade in the Japanese and American biotech and pharmaceuticals industries. Dr. Barrow's areas of expertise include marine natural products and the chemistry of Alzheimer's disease. Dr. Barrow has over 80 peer-reviewed publications and has presented at numerous conferences and workshops.

Fereidoon Shahidi, PhD, FACS, FCIC, FCIFST, FIAFoST, FIFT, FRSC, is a university research professor at Memorial University of Newfoundland in Canada. Dr. Shahidi has authored some 600 research papers and book chapters, authored or edited 40 books, and given over 400 presentations at scientific conferences. His research contributions have led to several industrial developments around the globe. Dr. Shahidi's current research interests include different areas of nutraceuticals and functional foods, natural antioxidants, marine foods, and aquaculture. Dr. Shahidi serves as the editor-in-chief of the *Journal of Food Lipids*, an editor of *Food Chemistry*, an editorial board member of the *Journal of Food Science, Journal of Agricultural and Food Chemistry, Nutraceuticals and Food, International Journal of Food Properties*, and *Current Nutrition and Food Science*, and is on the editorial advisory board of *Inform*. Dr. Shahidi has received numerous awards, including the 1996 William J. Eva Award from the Canadian Institute of Food Science and Technology, the 1998 Earl P. McFee Award from the Atlantic Fisheries Technological Society, the 2002 ADM Award from the American Oil Chemists' Society, and the 2005 Stephen Chang Award from the Institute of Food Technologists, for his outstanding contributions to science and technology. Dr. Shahidi is the most published and a most highly cited scientist for the period 1996–2006 in the area of food, nutrition, and agricultural sciences as listed by the ISI. He was a founding member and chair of the Nutraceuticals and Functional Food Division of the Institute of Food Technologists and has served and chaired numerous committees including the Natural Health Product Advisory Board of Health Canada.

Contributors

Colin J. Barrow
Ocean Nutrition Canada
Dartmouth, Nova Scotia, Canada

Adam M. Burja
Metabolic Engineering and Fermentation
Ocean Nutrition Canada
Dartmouth, Nova Scotia, Canada

Julie Conquer
Human Biology and Nutritional Sciences
University of Guelph
Guelph, Ontario, Canada

Jonathan Curtis
Ocean Nutrition Canada
Dartmouth, Nova Scotia, Canada

J. Helen Fitton
Department of Medicine
University of Tasmania
Hobart, Australia

Hiroyuki Fujita
Research and Development Department
Nippon Supplement, Inc.
Osaka, Japan

Masashi Hosokawa
Laboratory of Biofunctional Material Chemistry
Faculty of Fisheries Science
Hokkaido University
Hakodate, Japan

Mohammad R. Irhimeh
Department of Medicine
University of Tasmania
Hobart, Australia

Chieko Itho
Department of Food Sciences and Nutritional Health
Kyoto Prefectural University
Shimogamo Kyoto, Japan

Yulai Jin
Ocean Nutrition Canada
Dartmouth, Nova Scotia, Canada

Won-Kyo Jung
Department of Chemistry
Pukyong National University
Busan, South Korea

Se-Kwon Kim
Department of Chemistry
Pukyong National University
Busan, South Korea

Tsukasa Kitahashi
Department of Food Sciences and Nutritional Health
Kyoto Prefectural University
Shimogamo Kyoto, Japan

Jaroslav A. Kralovec
Ocean Nutrition Canada
Dartmouth, Nova Scotia, Canada

Hordur G. Kristinsson
Laboratory of Aquatic Food Biomolecular Research
University of Florida
Gainesville, Florida

Eresha Mendis
Department of Chemistry
Pukyong National University
Busan, South Korea

Kazuo Miyashita
Laboratory of Biofunctional Material Chemistry
Faculty of Fisheries Science
Hokkaido University
Hakodate, Japan

Michael T. Morrissey
Department of Food Science and Technology
Oregon State University
Astoria, Oregon

Miguel Olaizola
Mera Pharmaceuticals, Inc.
Kailua-Kona, Hawaii

Chi Perrie
Ocean Nutrition Canada
Dartmouth, Nova Scotia, Canada

Helia Radianingtyas
Metabolic Engineering and Fermentation
Ocean Nutrition Canada
Dartmouth, Nova Scotia, Canada

Niranjan Rajapakse
Department of Chemistry
Pukyong National University
Busan, South Korea

Rosalee S. Rasmussen
Department of Food Science and Technology
Oregon State University
Astoria, Oregon

Kenji Sato
Department of Food Sciences and Nutritional Health
Kyoto Prefectural University
Shimogamo Kyoto, Japan

Fereidoon Shahidi
Department of Biochemistry
Memorial University of Newfoundland
St. John's, Newfoundland, Canada

Cai Song
Department of Biomedical Sciences
University of Prince Edward Island
Charlottetown, Prince Edward Island, Canada

Jane Teas
Department of Medicine
University of South Carolina
Columbia, South Carolina

Masahiro Tsutsumi
Department of Oncological Pathology
Nara Medical University
Nara, Japan

Corrine Van Diepen
Ocean Nutrition Canada
Dartmouth, Nova Scotia, Canada

Masaaki Yoshikawa
Division of Food Science and Biotechnology
Graduate School of Agriculture
Kyoto University
Kyoto, Japan

Genevieve Young
Human Biology and Nutritional Sciences
University of Guelph
Guelph, Ontario, Canada

Yvonne V. Yuan
School of Nutrition
Ryerson University
Toronto, Ontario, Canada

Wei Zhang
Ocean Nutrition Canada
Dartmouth, Nova Scotia, Canada

1 Marine Fisheries By-Products as Potential Nutraceuticals: An Overview

Se-Kwon Kim, Eresha Mendis, and Fereidoon Shahidi

CONTENTS

Maintenance of good health and sense of well-being are top priorities for many people. Dietary components play a vital role in accomplishing these needs. With the increasing knowledge of biofunctional properties associated with marine foods, utilization of these materials has accelerated. As a result, during the past decade, development of new technologies to identify and isolate bioactive materials has progressed. Proteins, peptides, and carbohydrates are the major contributors of the bioactivities in different marine food sources. A three-step recycle membrane reactor has been used for production of bioactive peptides and for exploration of new bioactivities. In addition, it is expected that the knowledge on isolation methods as well as specific properties of fish skin collagen and gelatin will provide a solution for the present or concern associated with the use of these materials from a variety of sources. Also chitin, chitosan, and their derivatives obtained from crustacean shell wastes have long been of interest as biomaterials in a variety of applications. The continuous enzymatic production of desirable molecular weight chitosan oligosaccharides has increased the possibilities of utilizing these materials for human use. The bioactivities such as antihypertensive, antioxidative, antitumor, and calcium absorption acceleration observed in different marine by-products could be utilized for the development of potential nutraceuticals that enhance and promote human health.

1.1 INTRODUCTION

There has been a growing interest in the health-promoting role of certain foods above their nutritional value. Thus, research efforts to identify functional and bioactive components from many natural sources including plants, animals, microorganisms, and marine organisms have intensified. As a result, two new groups of products, "functional foods" and "nutraceuticals" have been introduced to the market and currently some are produced in large scale. Functional foods are foods that enrich with functional components to offer medical and physiological benefits or to reduce the risk of chronic diseases, beyond their basic nutritional functions [1]. In contrast, nutraceuticals are bioactive materials isolated or purified from foods to be utilized in the medicinal form. As a whole, both physiological functional foods and nutraceuticals demonstrate many physiological and health benefits [2].

According to the Food and Agriculture Organization, the world marine fisheries production increases annually [3]. The increasing demand for seafood may be due to its health benefits and willingness to include marine foods into the daily diet. Specially, fish and shellfish are important dietary components in health-conscious ethnic groups. Southeast Asian countries including Korea have a well-known background in processing fish into various value-added products. In the year 2001, 59% of the total marine fish catch in Korea was utilized for production of processed foods [3]. In marine fish–processing industry, the main concern is food and nutritional value of the products. However, in addition to this main production stream, there is a potential to identify and develop new industries related to marine fisheries. As a result, identification of biological properties and development of nutraceuticals from these bioresources have yielded a considerable number of drug candidates in recent years. However, most of these compounds are still at the developmental stages. Fish protein hydrolysates and fish bone–derived functional materials have attracted the main focus among functional bioactive materials identified from marine fisheries by-products. Some bioactive peptides isolated from fish protein hydrolysates have shown to act as antioxidants [4,5] and angiotensin I converting enzyme (ACE)-inhibitors [6] that work toward lowering the blood pressure by inhibiting ACE. In addition to the health benefits of these materials, they do not exert any harmful influences on the human body even at higher doses. The safe nature of these biomaterials for use as nutraceuticals or physiological functional foods further increases their demand compared to other synthetic bioactive materials. In addition to the above bioactive materials, chitosan and their oligomers derived from chitin of crustacean exoskeletons have attracted much interest due to their wide range of applications in numerous fields including pharmacy and medicine.

For many years, identification of bioactive materials from marine sources including fish and shellfish has been a major effort in many research groups. To identify and isolate bioactive materials, a new continuous production method has been developed and the isolated compounds have shown promising bioactivities. However, to develop these bioactive materials as nutraceuticals, further research work is needed. This chapter focuses on current isolation efforts of bioactive materials from marine fisheries sources and their potential as nutraceuticals.

1.2 DEVELOPMENT OF MARINE BIOACTIVE MATERIALS

Initial identification of bioactive materials from marine sources paved the way for utilization of huge amounts of fish-processing waste, which had a considerable impact on environmental pollution. Following identification of the potential to isolate new functional biomaterials from fish and shellfish, fisheries processing waste acquired a commercial value as industrial raw material. In addition, various fish and shellfish source materials such as skin, muscle, frame, bone, and internal organs are presently utilized to isolate a number of bioactive materials. Fish bone is a good source of calcium. Skin and protein remaining on fish frame could be used as cheap materials for identification and isolation of bioactive peptides. In addition, attempts were successful in isolating crude enzymes from fish intestine. Further, they were utilized to develop some bioactive peptides from fish protein sources.

1.3 FISH SKIN

1.3.1 COLLAGEN

Fish processing by-products contain a considerable amount of skin, which has been identified as a potential source to isolate collagen and gelatin. Collagen is the major structural protein found in the skin and bones of all animals. Based on structural roles and better compatibility within human body, collagen is commonly used in medical and pharmaceutical industries especially as drug carriers [7]. In addition, it is commonly used in the cosmetics industry for the production of some skin lotions as it forms a superior protective film to soothe and hydrate the skin [8]. This is because it is chemically bound to water and so provides long-lasting moisturizing effects. Currently, there is increased interest to use collagen as a nutraceutical, especially in Southeast Asian countries such as Japan, China, and Korea. It is used in the production of skin rejuvenating lotions and taken orally as well. This use is the latest breakthrough in skin-care technology. Even though by-products from fish processing are a potential source for collagen, fewer studies have been performed to identify the potential uses of fish skin collagen than that derived from mammalian origin. The main sources of industrial collagen are limited to bovine and pig skins and bones. Owing to some religious reasons pig skin collagen has become comparatively unpopular. The use of bovine-derived collagen is in active discussion due to the mad cow disease and the risk it poses to humans. This disease has affected a number of countries, including England, Canada, and the United States. Mad cow disease, technically known as bovine spongiform encephalopathy, is a chronic degenerative disease affecting the central nervous system of cattle. While it is important to pay attention to the issue of mad cow disease in relation to eating beef products, the possibility of ingredients used in cosmetics that harbor the disease and cause health risks are issues of interest to women. In this context, the knowledge on fish skin collagen is expected to attract the interest of the industry.

Attempts have been made to examine some properties of fish skin and scale collagens [9–11]. Thus, collagens isolated from filefish (*Novoden modestus*) and cod (*Gadus macrocephalus*) were tested for their physicochemical properties [9]. The solubility of both collagens was lowest at pH 7 and increased with decreasing pH. The observed viscosities of filefish skin collagen and cod skin collagen were highest at pH 4.0 and 2.0, respectively. It was observed that cod skin collagen had a higher hydration capacity than that of filefish skin collagen. Further, acid-soluble collagen from filefish was modified by papain-catalyzed incorporation of L-leucine alkyl ester (Leu-Ocn) [10]. The functional properties of enzymatically modified collagen showed a better emulsifying ability and foamability and an ideal material as a low-fat proteinaceous surfactant.

1.3.2 Gelatin

Gelatin, a heterogeneous mixture of high-molecular-weight water-soluble proteins, is derived from collagen. Owing to its unique physical properties, gelatin is widely used in the pharmaceutical and food industries for encapsulation of drugs and as a food additive to enhance texture, water-holding capacity, and stability of several food products. Gelatin has a unique amino acid arrangement in its sequence and contains relatively high amounts of glycine, proline, and alanine. Despite their uses considering unique physical and structural properties, possible biological activities of gelatin and collagen have not been adequately discussed. However, preliminary reports on gelatin suggest that consumption of gelatin can help to improve the structure and health of hair and nails [12,13].

Fish skin gelatin and fish bone gelatin can be easily extracted by a simple hot-water treatment, and optimum extraction conditions vary only slightly with species [14–16]. The concentration of alkaline treatment, water to fish skin ratio, pH, temperature, and time of extraction are the main considerations in fish skin gelatin extraction. The intact form of the gelatin is invaluable with regard to the bioactivity. Many of the bioactivities of proteins are attributable to the presence of biologically active peptide sequences in their primary structure. Therefore, a number of methods are used to release bioactive peptide fractions from native proteins

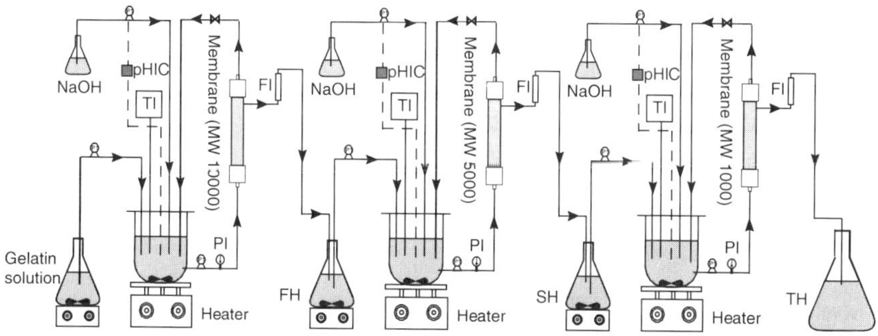

FIGURE 1.1 Schematic diagram of the three-step recycling membrane reactor for the production and separation of enzymic hydrolysates of Alaska pollack skin gelatin. TI, temperature indicator; PI, pressure indicator; FI, flow indicator; pHIC, pH indicator controller; FH, first hydrolysates; SH, second hydrolysates; TH, third hydrolysate. (From Byun, H.G. and Kim, S.K., *Process Biochem.*, 36, 1155–1162, 2001.)

and proteolytic digestion has become the most common method. A sequential digestion method utilizing three enzymes and different molecular weight cut-off (MWCO) ultrafiltration (UF) membranes was developed to obtain bioactive peptides [17–19]. A three-step recycling membrane reactor, as shown in Figure 1.1 was designed to ensure the continuous production of gelatin peptides [20,21]. In this system, UF was combined with three different enzyme reactors to separate hydrolyzed peptide products based on their molecular weights. In the first step of the membrane reactor, 1% (w/v) gelatin was digested with Alcalase® (pH 8, 50°C) and fractionated through a membrane with a 10,000 Da MWCO. After hydrolysis with Pronase-E (pH 8, 50°C), the resultant hydrolysate is fractionated through a membrane with a 5000 Da MWCO, and finally hydrolyzed with collagenase (pH 7.5, 37°C). The fraction that passes through the 5000 Da MWCO is further fractionated through a membrane with a 1000 Da MWCO and the fraction having molecular weights less than 1000 Da is obtained. Separated fractions clearly demonstrate different molecular distribution profiles and these peptide fractions were screened to identify different bioactivities. Interestingly, separated peptides could act as ACE inhibitors and antioxidants in lipid peroxidation systems.

1.3.2.1 ACE-Inhibitory Activity

Two peptides having high ACE-inhibitory potencies were purified from Alaska pollack skin gelatin extract using the above method, and their sequences were determined as Gly-Pro-Leu (IC_{50} = 2.65 μM) and Gly-Pro-Met (IC_{50} = 17.13 μM). Many peptides derived from different protein sources have been reported to act as ACE inhibitors [6,7]. These activities were comparable to those of potent peptides reported from other protein sources. Further, these peptide sequences acted as competitive inhibitors of ACE. Generally, ACE inhibitors lower blood pressure by inhibiting ACE, the key component of the renin angiotensin system, whose main function is to convert angiotensin I (Ang I) to angiotensin II (Ang II). Inhibition of Ang II synthesis with ACE inhibitors has been demonstrated to be beneficial in modifying human disease progression including hypertension. Binding of inhibitor to the enzyme takes place predominantly via carboxyl terminal tripeptide residues. These purified peptides had favorable amino acids in their sequences, which positively contributed to their strong activities.

1.3.2.2 Antioxidant Activity

Oxidation causes many unfavorable impacts on food and biological systems. In aerobic organisms, oxidation is associated with the occurrence of several disease conditions including atherosclerosis, inflammation, and cancer. Antioxidant compounds are generally utilized to overcome oxidation-mediated problems. Three fish skin gelatin peptide fractions (F1, F2, and F3) prepared in the three-step recycling membrane reactor (Figure 1.1), respectively from first, second, and third enzymatic hydrolysates exhibited different potencies in inhibiting lipid peroxidation over time (Figure 1.2) [21]. Antioxidative activity was measured in linoleic acid–oxidizing system and the activity was measured as the amount of malondialdehyde (MDA) produced. Relatively low-molecular-weight peptides derived after hydrolysis of two enzymes (P2 and P3) could strongly inhibit lipid peroxidation and these activities were higher than that of α-tocopherol and lower than that

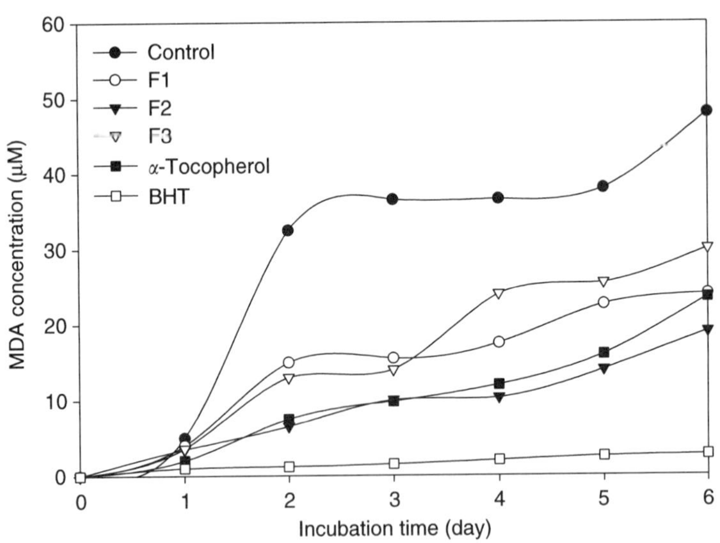

FIGURE 1.2 Antioxidative activity of fish-gelatin peptide fractions (F1, F2, and F3) obtained from the three-step recycling membrane reactor. Oxidation was performed in a linoleic acid system and antioxidative activity was measured by malondialdehyde (MDA) concentrations. Lower the MDA concentration, higher the antioxidative activity. BHT, butylated hydroxytoluene. (From Kim, S.K., Kim, Y.T., Byun, H.G., Nam, K.S., Joo, D.S. and Shahidi, F., *J. Agric. Food Chem.*, 49, 1984–1989, 2001.)

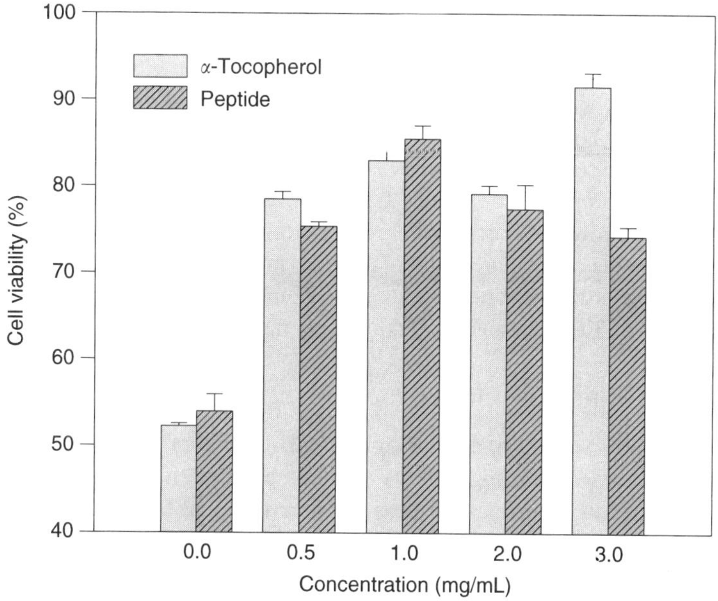

FIGURE 1.3 Comparison of cell viability in oxidation-induced cultured human hepatoma cells after treatment with antioxidative peptides derived from fish skin gelatin. (From Kim, S.K., Kim, Y.T., Byun, H.G., Nam, K.S., Joo, D.S. and Shahidi, F., *J. Agric. Food Chem.*, 49, 1984–1989, 2001.)

of butylated hydroxytoluene (BHT). In addition, purified peptides could enhance cell viability in oxidation-induced cultured human hepatocytes (Figure 1.3). To identify whether these antioxidative effects of fish skin gelatin were comparable to gelatin derived from other sources, a similar study was conducted using bovine skin gelatin and antioxidative activities were compared [22]. The activities observed with regard to lipid peroxidation inhibition as well as increment of oxidative stress–induced cell viability were stronger in fish skin gelatin than those of bovine-skin gelatin. These results suggest that the antioxidative activities of Alaskan pollack skin gelatin were superior to that of bovine. A number of natural substances including peptides have been reported to exhibit strong antioxidative properties in different oxidative systems [23,24]. Based on these activities it can be expected that fish gelatin–derived peptides have the potential to use as safe and potent natural antioxidants.

1.4 FISH FRAME PROTEIN

Utilization of fish protein extracts as sauce or flavor compounds has a long history in different cultures, and in the present world there is a renewed interest to incorporate some fish and shellfish protein hydrolysates in food formulations as additional supplements. Fish frames resulting from filleting operations contain considerable amount of muscle protein. Therefore, frame proteins can be utilized as an ideal protein source to obtain functional hydrolysates. Frames of several fish species have been utilized to obtain functional hydrolysates and to identify fish mussel–derived bioactive peptides. In a research dealt with cod frame, enzymatic hydrolysis was applied to efficiently recover the protein remaining on the fish frame [25]. The enzyme used for the hydrolysis process was a fish-derived crude proteinase, extracted in a previous research from tuna pyloric caeca [26]. The functional properties of peptide fractions were dependent on their molecular weights. To improve the functional properties, the resultant hydrolysate was separated based on the molecular weight of peptides and tested for several bioactivities. For this purpose the hydrolysate was passed through a series of UF membranes with MWCO of 30, 10, 5, and 3 kDa; four separate fractions were obtained.

1.4.1 ACE-Inhibitory Activity and Antioxidative Activity

Among the four isolated fractions, higher ACE-inhibitory activity was observed in the fraction passed through 3 kDa membrane and the IC_{50} values of all fractions are shown in Figure 1.4. Antioxidative activities of the four fractions were determined by 2-thiobarbituric acid (TBA) assay. Results were expressed as oxidation inhibition ratio compared to the control and α-tocopherol was used as the positive control. Among the four fractions obtained, fraction 5–10 kDa exhibited the highest antioxidative activity (Figure 1.5). To isolate and characterize antioxidative peptides from enzymatic hydrolysate of yellowfin sole (*Limanda aspera*) frame protein, it was fractionated using an UF membrane system into four fractions employing 30, 10, 5, 3, and 1 kDa MWCO [27]. Hydrolysis was

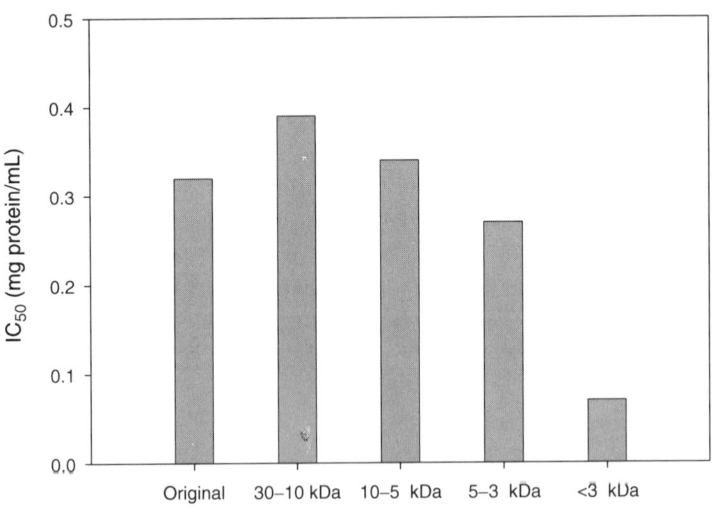

FIGURE 1.4 ACE-inhibitory activity of different molecular weight peptides obtained from cod frame protein hydrolysate. 30–10, 10–5, 5–3, and <3 kDa were different molecular weights of peptide fractions and original represents unfractionated peptide mixture. (From Jeon, Y.J., Byun, H.G. and Kim, S.K., *Process Biochem.*, 35, 471–478, 2000.)

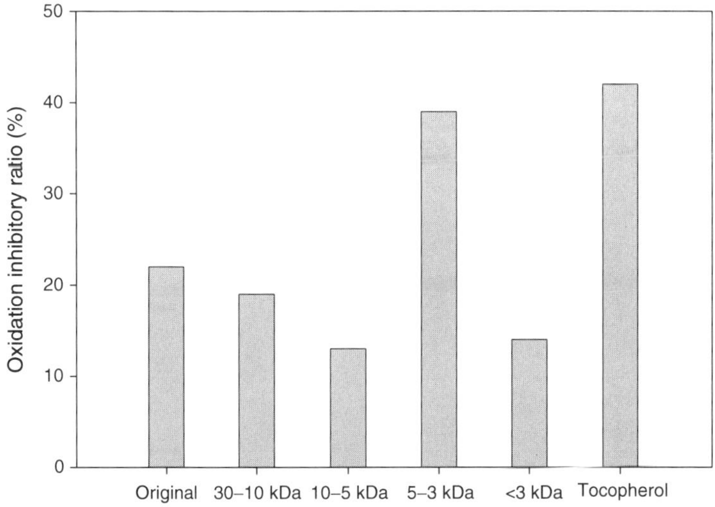

FIGURE 1.5 Antioxidative activity of different molecular weight peptides obtained from cod frame protein hydrolysate. 30–10, 10–5, 5–3, and <3 kDa were different molecular weights of peptide fractions and original represents unfractionated peptide mixture. (From Jeon, Y.J., Byun, H.G. and Kim, S.K., *Process Biochem.*, 35, 471–478, 2000.)

carried out using enzyme combination of pepsin and mackerel intestine crude enzyme (MICE) [28]. Antioxidative activities of the fractions were compared and a 13 kDa peptide was isolated from the highest active fraction. This activity was higher than that of α-tocopherol.

1.4.2 Calcium Absorption Acceleration Effect

Frame protein hydrolysate from hoki (*Johnius belengeri*) was utilized to recover phosphorylated and nonphosphorylated peptides and these were named as phosphorylated hoki frame protein (PHFP) and nonphosphorylated hoki frame protein (HFP). These two isolated peptides were tested for their calcium absorption accelerating effects compared to a positive control, casein phosphopeptide (CPP) [29]. In an *in vitro* experiment, HFP and PHFP inhibited calcium phosphate precipitation and this was dependent on the concentration of the peptides present. The effects of HFP, PHFP, and CPP on calcium absorption as a function of concentration are shown in Figure 1.6. In an *in vivo* experiment carried out using Sprague-Dawley (SD) rats, in groups fed with HFP and PHFP calcium content in the femur was significantly increased (Table 1.1). Under certain conditions, dietary calcium becomes unavailable for absorption due to the formation of insoluble forms. Developing methods to increase solubility can contribute significantly to calcium availability in biological systems. These results suggest that the use of protein hydrolysates for improvement of bioactivities in functional foods might be beneficial with regard to health promotion.

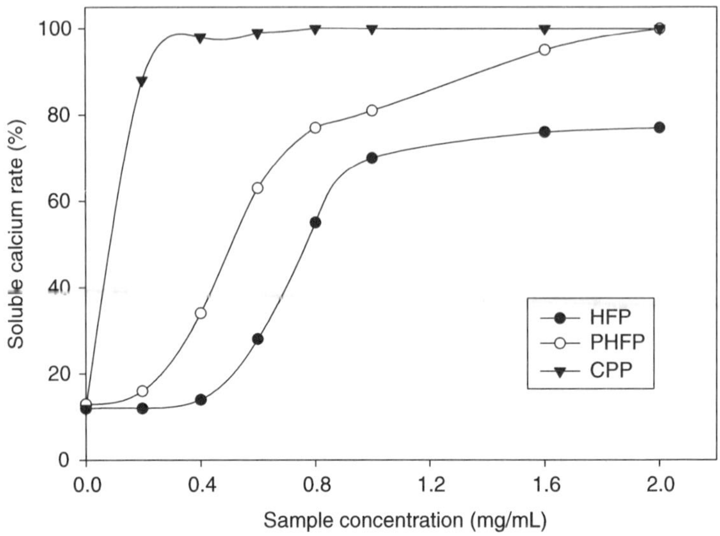

FIGURE 1.6 Effect of nonphosphorylated hoki frame protein (HFP), phosphorylated hoki frame protein (PHFP), and casein phosphopeptide (CPP) on calcium absorption tested in Sprague-Dawley rat models. (From Kim, S.K., Jeon, Y.J., Byun, H.G., Park, P.J., Kim, G.H., Choi, Y.R. and Lee, Y.S., *Korean Fish. Soc.*, 32, 713–717, 1999.)

TABLE 1.1
Changes in Weight, Length, and Ash and Calcium Content in the Femur of Rats Fed with HFP and PHFP

	Wet Weight[a] (g)	Lengthh[b] (mm)	Ash (mg/g)	Ca (mg/g)	Ca, % (Ca/Ash × 100)
Control	1.09 ± 0.04	28.2 ± 0.4	160.4 ± 7.5	64.2 ± 2.3	40.1 ± 0.7
HFP[c]	1.23 ± 0.09	27.9 ± 0.2	164.5 ± 2.7	70.8 ± 1.6	43.1 ± 0.7
PHFP[d]	1.24 ± 0.02	28.3 ± 0.2	166.0 ± 3.2	73.2 ± 3.1	44.0 ± 0.7

[a] Total weight of left and right femur.
[b] Mean length of left and right femur.
[c] Nonphosphorylated hoki frame protein.
[d] Phosphorylated hoki frame protein.

Source: Kim, S.K., Jeon, Y.J., Byun, H.G., Park, P.J., Kim, G.H., Choi, Y.R. and Lee, Y.S., *Korean Fish. Soc.*, 32, 713–717, 1999.

1.5 FISH BONE

Generally, bone substitution materials such as autografts, allografts, and xenografts are used to solve problems related to bone fracture and damage. However, none of these materials provide a perfect bone healing due to mechanical instability and incompatibility. Currently, calcium phosphate bioceramics such as tetracalcium phosphate, amorphous calcium phosphate, tricalcium phosphate, and hydroxyapatite are identified as most suitable bone substitution materials to serve the demand. Therefore, a potential exists to identify alternates from different sources and hence fish bone recovered from fish frame was used as a potential source to isolate hydroxyapatite. Approximately 65–70% of the fish bone is composed of inorganic substances. Almost all of these inorganic substances are hydroxyapatite, composed of calcium, phosphorus, oxygen, and hydrogen. A wet milling process was used to isolate hydroxyapatite compounds and their physical properties were tested [30]. Temperature effect for sintering was tested using various temperatures and 1300°C was found to be favored for a better sintering. The change of mean particle size distribution was changed as a function of temperature, and was clearly observed using x-ray measurements. The major phases were identified as hydroxyapatite at below 1300°C, but the whitlockite [$Ca_3(PO_4)_2$] phases appeared due to the decomposition of hydroxyapatite with the increment in temperature. The major phases for the glass-ceramic prepared by using hydroxyapatite were identified as pseudowollastonite and tricalcium phosphate, respectively. The measured maximum strength of the glass-ceramics (G-C) prepared at 900°C for 4 h in the air was 90 MPa and this value was in the strength range of cortical bone (Figure 1.7) [31].

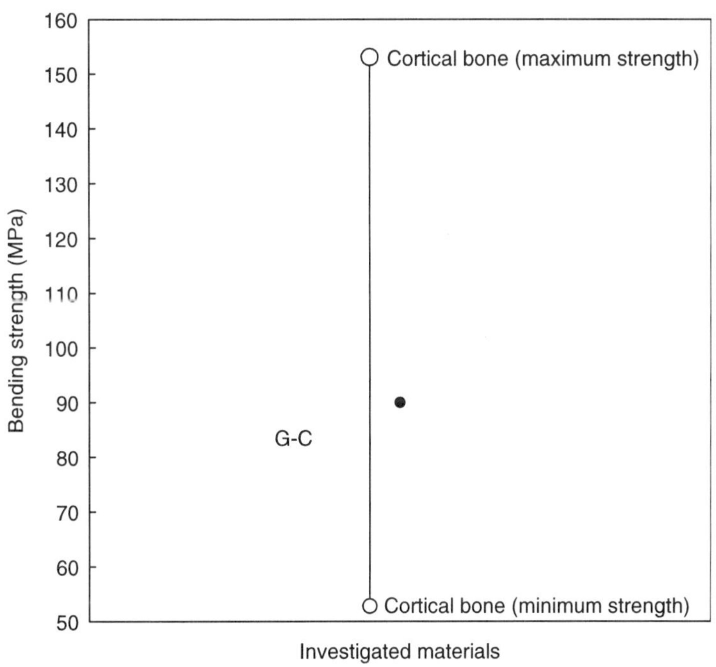

FIGURE 1.7 Comparison of bending strength between glass-ceramics (G-C) and cortical bone. (From Kim, S.K. and Park, P.J., *Korean J. Life Sci.*, 10, 605–609, 2000.)

Chemical bonding was also investigated in simulated body fluid for several selected hydroxyapatite-containing composites such as 5:1, 6:1, and 7:1 weight ratio of hydroxyapatite:wollastonite [32]. The hydroxyapatite-containing composites were chemically bonded to each other after 4 weeks. Their composite bodies were bonded by heterogeneous nucleation and a clear growth of the interface in the simulated body fluid was observed. Scanning electron microscopy (SEM) photographs of an interface of mutually bonded 5:1 composite specimen are shown in Figure 1.8. Bioglass bonding was strong in the simulated body fluid but

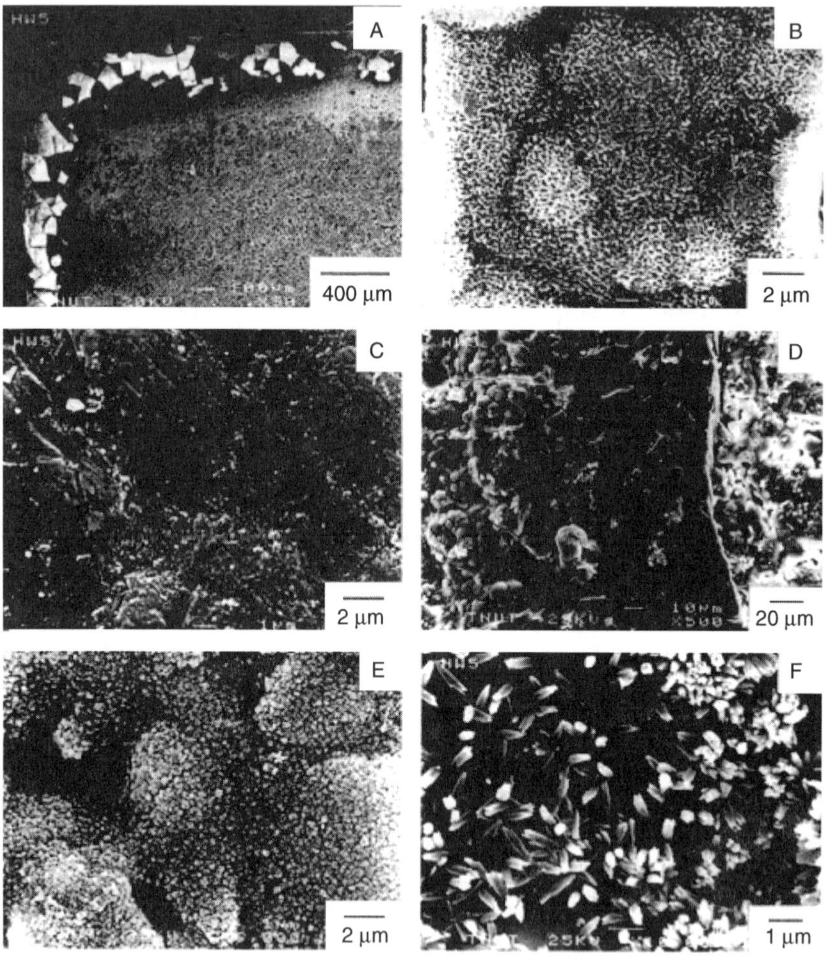

FIGURE 1.8 SEM photographs of an interface of mutually bonded 5:1,hydroxyapatite: wollastonite composite specimen. (A) Interface (4 weeks); (B) inner (4 weeks); (C) after 4 weeks; (D) after 6 months; (E) the ridge inner (6 months); (F) the central portion of the interface after 6 months. (From Kim, S.K., Choi, J.S., Lee, C.K., Byun, H.G., Jeon, Y.J. and Lee, E.H.J., *J. Korean Ind. Eng. Chem.*, 9, 322–329, 1998.)

bonding strength was not dependent on the composition. Unlike other calcium phosphates, hydroxyapatite did not break down under physiological conditions. In fact, it is thermodynamically stable at physiological pH and actively takes part in bone bonding, forming strong chemical bonds with the surrounding bone. This property has been exploited for rapid bone repair after major trauma or surgery. To evaluate the safety of hydroxyapatite sinter produced from tuna bone, oral mucous membrane irritation test was carried out in Syrian hamsters [33]. The study was carried out by injecting the hamsters a dose of 5 g/kg body weight hydroxyapatite sinter under pentobarbital sodium anesthesia. Abnormal clinical signs were not observed both in the control and treated groups of hamster cheeks observed for 14 days. No difference was observed between histopathological lesions of oral mucosa of treated and controlled groups. In addition, acute toxicity of hydroxyapatite was evaluated in SD rats [34]. Hydroxyapatite sinter was subcutaneously administered at different doses and rats were observed for 14 days. The hydroxyapatite sinter did not induce any toxic signs in mortalities, clinical findings, body weights, and gross findings of rats. Therefore, it was concluded that hydroxyapatite from tuna bone has no effect on acute toxicity and side effects in rats.

1.6 FISH INTERNAL ORGANS

Fish internal organs may also offer a potential source for bioactive materials, which could enhance the value of fish-processing waste. Two ACE-inhibitory peptides were isolated and characterized from enzymatic hydrolysate of cod-liver after fractionation through UF membrane reactor system. The amino acid sequences of the two purified peptides were Met-Ile-Pro-Pro-Tyr-Tyr (IC_{50} = 10.9 μM) and Gly-Leu-Arg-Asn-Gly-Ile (IC_{50} = 35.0 μM) [35]. In another study, it was observed that the crude hydrolysates of cod-liver and head exert high ACE-inhibitory activities and these activities were dependent on the proteolytic enzyme used for the hydrolysis process. Hydrolysates showed a similar potency as antioxidants in *in vitro* systems [36]. Further, a strong antioxidative peptide with the amino acid sequence of Ser-Asn-Pro-Glu-Trp-Ser-Trp-Asn was isolated from the cod teiset protein hydrolysate following consecutive chromatographic separations. The antioxidative activity observed was higher than that of α-tocopherol [37]. These results suggested that protein components of the internal organs may also be used to isolate bioactive peptides.

Fish proteinases in fish tissues have been reported to possess some distinctive properties such as higher catalytic efficiency at low temperatures, and effective catalytic activity and stability at neutral to alkaline pH [38]. However, the enzymes derived from marine sources have not been extensively utilized. Very few studies have been carried out to identify and characterize enzymes from marine fish sources, and hence limited information is available on their potential use. As a result, potential exists to isolate and characterize enzymes especially from fish internal organs. Thus, marine fish–processing waste also offers a better potential to isolate fish-derived enzymes that can be used to hydrolyze fish-derived materials and in developing methods to utilize fish products. Thus, several crude

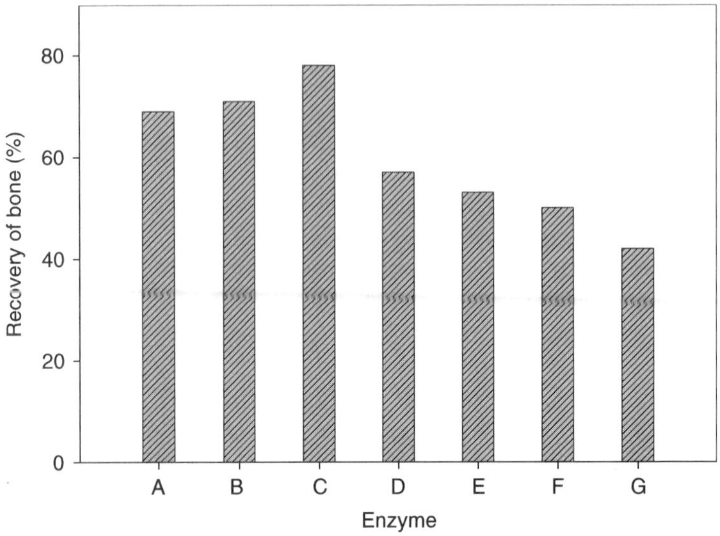

FIGURE 1.9 Comparison of the fish bone recovery rates of MICE and other commercial proteinases under optimal conditions. (A) MICE (pH 9.0, 40°C); (B) Pronase-E (pH 8.0, 40°C); (C) papain (pH 6.0, 40°C); (D) Alcalase (pH 7, 50°C); (E) trypsin (pH 8.0, 40°C); (F) α-chymotrypsin (pH 8.0, 40°C); (G) Neutrase (pH 8.0, 50°C). (From Kim, S.K., Park, P.J., Byun, H.G., Je, J.Y. and Moon, S.H., *J. Food Biochem.*, 27, 255–266, 2003.)

enzymes were isolated and subsequently used to recover fish materials from processing waste [26,28]. To efficiently recover cod proteins that remained on the frame after filleting in fish processing, a crude proteinase was partially purified from tuna *Thunnus thynnus* pyloric caeca [26]. Under optimum reaction conditions, the protein recovery was similar to that of α-chymotrypsin, Pronase-E, and papain. Another crude enzyme isolated from MICE was utilized to recover the fish bone from *Johnius belengeri* frame [39]. Under optimum conditions, the bone recovery using MICE was approximately 90% and this was comparatively higher than that of Alcalase, trypsin, α-chymotrypsin, and Neutrase® (Figure 1.9). In addition, serine collagenolytic proteases were isolated from internal organs of filefish (*Novoden modestrus*), mackerel (*Scomber japonicus*), and tuna (*Thunnus thynnus*) [40–42]. Serine collagenases are thought to be involved in the production of hormones and pharmacologically active peptides as well as other cellular functions [40]. The serine collagenase isolated from mackerel internal organs was further tested for its ability to cleave native collagen types I, II, III, and V; a higher cleavage was exerted on type I collagen (Figure 1.10).

1.7 SHELLFISH

The amount of fat and the proportions of saturated, monounsaturated, and polyunsaturated fat in shellfish contribute to a healthful diet. Shellfish also provide high-quality protein with all the dietary essential amino acids for maintenance

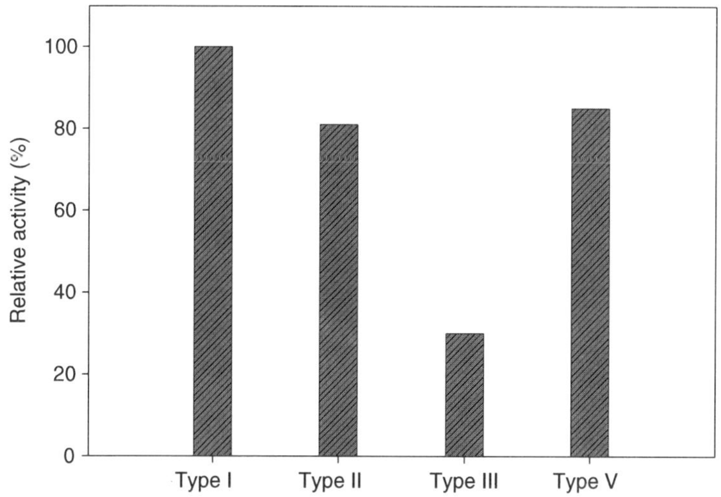

FIGURE 1.10 Substrate specificity of the purified collagenase on various collagens. The collagens were incubated with the purified enzyme:substrate ratio (1:200) for 1 h at 55°C. Type I, II, III, and V represent different native collagen types. (From Kim, S.K., Park, P.J., Kim, J.B. and Shahidi, F., *J. Biochem. Mol. Biol.*, 35, 165–171, 2002.)

and growth of the human body. With increasing knowledge about the nutrition of shellfish, it is becoming popular among the nutrition-conscious population. Recently, there is a growing interest in searching for biological activities of shellfish to be used them as functional foods. However, use of shellfish is mainly common among Asians and remains underutilized in the Western world. Therefore, alternate methods can be developed to improve the use of this underutilized valuable source. Several bioactivities associated with some shellfish species were investigated. The bioavailability of calcium compounds obtained from oyster shell was determined [43]. For that, calcium oxide was prepared by burning oyster shell at 1200°C. Its purity was approximately 98.5% and by means of chemical reactions calcium compounds $CaCl_2$ and $CaHPO_4$ were prepared for further testing. The effect of gelatin peptides on oyster shell–derived Ca absorption was tested *in vitro* as well as *in vivo* using calcium-deficient rats. For this purpose, fish skin gelatin peptides were prepared using enzymatic hydrolysis of skin gelatin for 4 h with tuna pyloric caeca crude enzyme (TPCCE). *In vitro* experiments revealed that calcium absorption was approximately 70% higher in fish gelatin–derived peptide treated group than that of the control. In the *in vivo* test conducted in calcium-deficient rats, the group fed with 3% peptides and $CaHPO_4$ significantly improved the amount of calcium and ash in femur and its strength. These results suggest that peptides derived from fish sources have a potential to increase calcium bioavailability. Moreover, calcium compounds obtained from Oyster shell and fish gelatin–derived peptides combination could be used as an effective dietary calcium source (Table 1.2).

TABLE 1.2
Effect of the Addition of Fish Skin Gelatin Peptides on Body Weight, Food Intake, and Food Efficiency Ratio in Calcium Deficiency–Induced Rats

	Final Body Weight (g)	Daily Food Intake (g)	Daily Weight Gain (g)	FER[a]
Ca deficient	195.0 ± 6.25	15.78 ± 0.25	3.79 ± 0.20	0.24 ± 0.15
Control	333.1 ± 7.55	18.14 ± 0.65	7.05 ± 0.35	0.39 ± 0.02
$CaCl_2$	323.6 ± 4.84	19.34 ± 0.52	6.30 ± 0.27	0.33 ± 0.01
$CaHPO_4$	336.1 ± 8.68	18.79 ± 0.30	6.85 ± 0.35	0.36 ± 0.02
$CaCl_2$+FGH(1%)	292.9 ± 9.92	17.83 ± 0.41	4.93 ± 0.43	0.28 ± 0.02
$CaHPO_4$+FGH(1%)	287.5 ± 12.59	16.56 ± 0.45	4.64 ± 0.69	0.28 ± 0.03
$CaCl_2$+FGH(3%)	323.4 ± 8.19	18.69 ± 0.54	6.34 ± 0.43	0.34 ± 0.02
$CaHPO_4$+FGH(3%)	348 ± 18.47	20.40 ± 0.23	7.47 ± 0.67	0.37 ± 0.03

[a] Food efficiency ratio (FER): weight gain (g)/food intake (g).

Source: Kim, G.H., Jeon, Y.J., Byun, H.G., Lee, Y.S., Lee, E.H. and Kim, S.K., *J. Korean Fish. Soc.*, 31, 149–159, 1998.

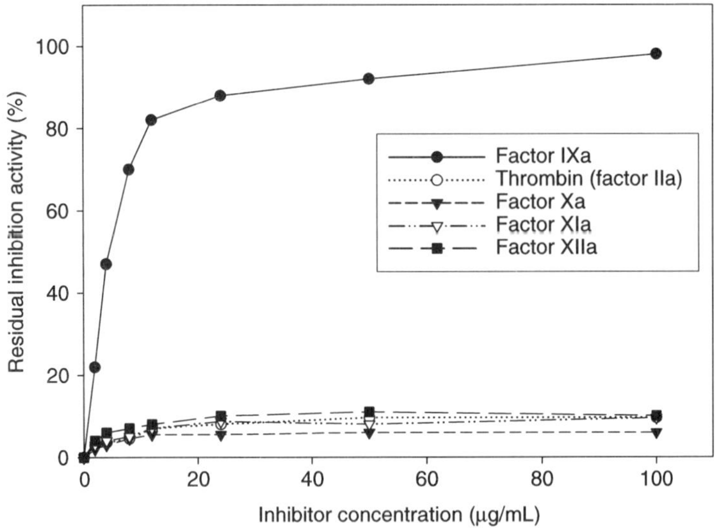

FIGURE 1.11 Inhibition of activated blood coagulation factors by the purified anticoagulant obtained from *Scapharca broughtonii*. (From Jung, W.K., Je, J.Y., Kim, H.J. and Kim, S.K., *J. Biochem. Mol. Biol.*, 35, 199–205, 2002.)

In another research, the edible portion of blood ark shell, *Scapharca broughtonii*, was used to isolate an anticoagulant protein. This protein prolonged the activated partial thromboplastin time and acted as an inhibitor of activated blood coagulation factor IXa (FIXa) in the intrinsic pathway of coagulation (Figure 1.11) [44]. In addition, many anticoagulant proteins have been isolated from different sources, and anticoagulants from marine organisms other than marine algae have rarely been discussed. These protein sequences can be used as safe anticoagulants in anticoagulant therapy to solve blood coagulation–related abnormalities.

Further, bioactive peptides from shellfish were isolated following microbial digestion. Oyster (*Crassostrea gigas*) was fermented with 25% NaCl (w/w) for 6 months [45]. Fermentation is one of the oldest food preservation techniques specifically practiced for long storage of fish, shrimp, croakers, and shellfish in Southeast Asian countries such as China, Japan, and Korea, upon which food is biochemically altered due to microorganism-led proteolysis. Fermentation of protein is expected to break down the substrate into a variety of peptides possessing different bioactivities. Recently, interest has been expressed in searching for biological activities of fermented fish and shellfish for use as potential nutraceuticals. An ACE-inhibitory peptide sequence was identified from the fermented shellfish sauce following purification using consecutive chromatographic techniques. The purified peptide was further evaluated for its antihypertensive effects in spontaneously hypertensive rats (SHR), following oral administration. The blood pressure was significantly decreased in the treated group of rats compared to the control group (Figure 1.12), suggesting strong antihypertensive activity of the fermented oyster-derived peptides.

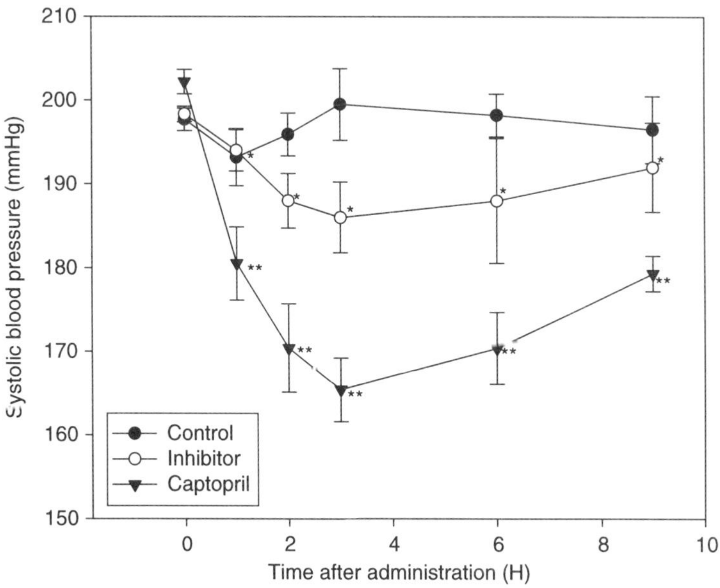

FIGURE 1.12 Changes in systolic blood pressure (SBP) of spontaneously hypertensive rats after administration of oyster-derived ACE-inhibitory peptide. Single oral administration was performed with the dose of 10 mg/kg body weight, and SBP was measured 0, 1, 2, 3, 6, and 9 h after the administration. Significance of the difference compared to the control at $^{*}P<0.05$, $^{*}P<0.01$. (From Je, J.Y., Park, J.Y., Jung, W.K., Park, P.J. and Kim, S.K., *Food Chem.*, in press.)

1.8 CRUSTACEAN EXOSKELETONS

Exoskeletons of crustaceans such as crabs, lobsters, and shrimps are used to obtain chitin and chitosan, well-known bioactive materials used in several industries. Chitin is a complex polymer of *N*-acetylglucosamine (β-1,4 linked 2-acetamido-D-glucose), whereas chitosan is the deacetylated form of chitin. Recently, low-molecular-weight chitin oligosaccharides and chitosan oligosaccharides (COSs) have received considerable attention as physiologically functional materials with the findings that they are responsible for biological activities such as antitumor activity [46], immunoenhancing activity [47], and antibacterial activity [48]. Annually, over 80,000 tons of chitin is produced from industrial waste [49]. Production of COSs from the hydrolysis of chitosan can be achieved either by chemical or enzymatic methods. The chemical method needs high energy and produces considerable amounts of harmful industrial chemicals. Therefore, the enzymatic method is preferred due to desirable and less harmful properties. However, the enzymatic method is comparatively expensive because of the cost associated with the enzyme. Therefore, we developed a new continuous low-cost production method to produce desired molecular size COSs [50].

Chitin, chitosan, and their oligomers are also shown to be responsible for varying bioactivities both *in vitro* and *in vivo*. However, high viscosity and insoluble nature at neutral pH restrict the use of chitin and chitosan in many instances. Therefore, compared to chitin and chitosan, their hydrolyzed low molecular forms are suitable materials for many biological applications. The activities of these materials are related to their structural properties. Especially, the degree of deacetylation and molecular weights were shown to be beneficial for their observed biological

TABLE 1.3
Some Biological Activities of Chitosan and Their Oligomers

Derivative	Molecular Weight Range (kDa)	Activity	Reference
COSs[a]	1.5–5.5	Antitumor	46
HMWCOSs[b]	5–10	Antibacterial	48
Heterochitosan[c]/Hetero-COSs[d]	1–5	Antibacterial	51
Hetero-COSs	1–5	Free radical scavenging	52
COSs	1–3	Free radical scavenging	53
Hetero-COSs	<10	ACE inhibitory	54
Sulfated hetero-COSs[e]	<10	Anticoagulant activity	55

[a] chitosan oligosaccharides prepared by enzymatic hydrolysis of chitosan.
[b] High-molecular-weight chitosan oligosaccharides prepared by enzymatic hydrolysis of chitosan.
[c] Differentially deacetylated chitosan.
[d] Differentially deacetylated chitosan oligosaccharides.
[e] Sulfated form of differentially deacetylated chitosan oligosaccharides.

activities. Generally, low-molecular-weight COSs are potential biomaterials to fight against bacterial infection and control growth of some bacteria [48,51]. Interestingly, for many biological activities, 75–90% deacetylation was found to be more effective than lower degrees of deacetylation. Heterochitosan oligosaccharides, those derived from partially deacetylated chitosans, can also be considered as promising biomaterials to positively affect some adverse biological events such as cancer [46], free radical generation [52,53] hypertension [54], and blood coagulation [55]. The activity differences observed in these structures could be due to their molecular weights, which facilitate their solubility and higher reaction rates. In addition, free amino groups may also contribute substantial effects on biological activities. Some observed bioactivities of chitosan and their oligomers are listed in Table 1.3.

REFERENCES

1. Schmidl, M.K. 1993. Food products for medical purposes. *Trends Food Sci. Technol.* 4, 163–168.
2. Defelice, S.L. 1995. The nutritional revolution: its impact on food industry R & D. *Trends Food Sci. Technol.* 6, 59–61.
3. FAOSTAT, FAO statistical databases, fisheries data, http://apps.fao.org/default.jsp.
4. Park, P.J., Jung, W.K., Nam, K.S., Shahidi, F. and Kim, S.K. 2001. Purification and characterization of antioxidative peptides from protein hydrolysate of lecithin-free egg yolk. *J. Am. Oil Chem. Soc.* 78, 651–656.
5. Saiga, A., Tanabe, S. and Nishimura, T. 2003. Antioxidant activity of peptides obtained from porcine myofibrillar proteins by protease treatment. *J. Agric. Food Chem.* 51, 3661–3667.
6. Meisel, H. 1997. Biochemical properties of bioactive peptides derived from milk proteins: potential nutraceuticals for food and pharmaceutical applications. *Livestock Prod. Sci.* 50, 125–138.
7. Ho, H.O., Lin, L.H. and Sheu, M.T. 1997. Characterization of collagen and application of collagen gel as a drug carrier. *J. Controlled Rel.* 44, 103–112.
8. Swatschek, D., Schatton, W., Kellermann, J., Muller, W.E.G. and Kreuter, J. 2002. Marine sponge collagen: isolation, characterization and effects on the skin parameters surface-pH, moisture and sebum. *Eur. J. Pharm. Biopharm.* 53, 107–113.
9. Kim, S.K., Kang, O.J. and Kwak, D.C. 1993. Physicochemical characteristics of filefish and cod skin collagen. *J. Korean Agric. Chem. Soc.* 36, 163–171.
10. Kim, S.K. and Kwak, D.C. 1991. The enzymatic modification and functionalities of filefish skin collagen. *J. Korean Agric. Chem. Soc.* 34, 265–272.
11. Jeon, Y.J., Kim, Y.T. and Kim, S.K. 1998. Analysis of compositions for effective utilization of fish scales. *Korean J. Life Sci.* 8, 589–597.
12. Morganti, P. and Randazzo, S.D. 1984. Nutrition and Hair. *J. Appl. Cosmetol.* 2, 41–49.
13. Tyson T.L. 1950. The effect of gelatin on fragile finger nails. *Invest. Dermatol.* 14, 323–325.
14. Kim, S.K., Byun, H.G. and Lee, E.H. 1994. Optimum extraction conditions of gelatin from fish skins and its physical properties. *J. Korean Ind. Eng. Chem.* 547–559.
15. Kang, T.J., Yang, H.P., Kim, S.K. and Song, D.J. 1992. Proteolytic conditions for the hydrolysate of flounder skin gelatin. *J. Koean Soc. Food Nutr.* 21, 398–406.

16. Kim, S.K., Jeon, Y.J., Lee, B.J. and Lee, C.K. 1996. Purification and characterization of the gelatin from the bone of cod, *Gadus macrocephalus*. *Korean J. Life Sci.* 6, 14–26.
17. Kim, S.K., Byun, H.G., Jeon, Y.J., Ahn, C.B., Cho, D.J. and Lee, E.H. 1995. Functional properties of produced fish skin gelatin hydrolysate in a recycle three-step membrane enzyme reactor. *J. Korean Ind. Eng. Chem.* 984–996.
18. Kim, S.K. and Byun, H.G. 1994. Development of optimum process for continuous hydrolysis of fish skin gelatin using a three-step recycle membrane reactor. *J. Korean Ind. Eng. Chem.* 5, 681–697.
19. Kim, S.K., Byun, H.G. and Cheryan, M. 1991. Continuous hydrolysis of cod skin gelatin in an ultrafiltration reactor. *Korean J. Biotechnol. Bioeng.* 6, 309–319.
20. Byun, H.G. and Kim, S.K. 2001. Purification and characterization of angiotensin I converting enzyme (ACE) inhibitory peptides from Allaska pollack (*Theragra chalcogramma*) skin. *Process Biochem.* 36, 1155–1162.
21. Kim, S.K., Kim, Y.T., Byun, H.G., Nam, K.S., Joo, D.S. and Shahidi, F. 2001. Isolation and characterization of antioxidative peptides from gelatin hydrolysate of Allaska pollack skin. *J. Agric. Food Chem.* 49, 1984–1989.
22. Kim, S.K., Park, P.J., Song, B.K. and Kim, J.B. 2000. Comparison of antioxidative activity on fish and bovine skin gelatin hydrolysates produced in a three-step membrane enzyme reactor. *Korean J. Biotechnol. Bioeng.* 15, 635–643.
23. Chen, H.M., Muramoto, K., Yamauchi, F., Fujimoto, K. and Nokihara, K. 1998. Antioxidative properties of histidine-containing peptides designed from peptide fragments found in the digests of a soybean protein. *J. Agric. Food Chem.* 46, 49–53.
24. Suetsuna, K., Ukeda, H. and Ochi, H. 2000. Isolation and characterization of free radical scavenging activities peptides derived from casein. *J. Nutr. Biochem.* 11, 128–131.
25. Jeon, Y.J., Byun, H.G. and Kim, S.K. 2000. Improvement of functional properties of cod frame protein hydrolysates using ultrafiltration membranes. *Process Biochem.* 35, 471–478.
26. Kim, S.K., Jeon, Y.J., Byun, H.G., Kim, Y.T. and Lee, C.K. 1997. Enzymatic recovery of cod frame proteins with crude proteinase from tuna pyloric caeca. *Fish. Sci.* 63, 421–427.
27. Jun, S.H., Park, P.J., Jung, W.K. and Kim, S.K. 2004. Purification and characterization of an antioxidative peptide from enzymatic hydrolysate of yellowfin sole (*Limanda aspera*) frame protein. *Eur. Food Res. Technol.* 219, 20–26.
28. Kim, S.K., Park, P.J., Byun, H.G., Je, J.Y. and Moon, S.H. 2003. Recovery of fish bone from hoki (*Johnius belengeri*) frame using a proteolytic enzyme isolated from mackerel intestine. *J. Food Biochem.* 27, 255–266.
29. Kim, S.K., Jeon, Y.J., Byun, H.G., Park, P.J., Kim, G.H., Choi, Y.R. and Lee, Y.S. 1999. Calcium absorption acceleration effect on phosphorylated and nonphosphorylated peptides from hoki (*Johnius belengeri*) frame. *Korean Fish. Soc.* 32, 713–717.
30. Kim, S.K., Choi, J.S., Lee, C.K., Byun, H.G., Jeon, Y.J. and Lee, E.H.J. 1997. Synthesis and biocompatibility of the hydroxyapatite ceramic composites from tuna bone (II)—The sintering properties of hydroxyapatite treated with wet milling process. *J. Korean Ind. Eng. Chem.* 8, 1000–1005.
31. Choi, J.S., Lee, C.K., Jeon, Y.J., Byun, H.G. and Kim, S.K. 1999. Properties of the ceramic composites and glass ceramics prepared by using the natural hydroxyapatite derived from tuna bone. *J. Korean Ind. Eng. Chem.* 10, 394–399.
32. Kim, S.K., Choi, J.S., Lee, C.K., Byun, H.G., Jeon, Y.J. and Lee, E.H.J. 1998. Synthesis and biocompatibility of the hydroxyapatite ceramic composites from tuna

bone (III)—SEM photographs of bonding properties hydroxyapatite ceramic composites in the simulated body fluid. *J. Korean Ind. Eng. Chem.* 9, 322–329.
33. Kim, S.K. and Park, P.J. 2000. Evaluation of mucous membrane irritation by hydroxyapatite sinter produced from tuna bone in Syrian hamsters. *Korean J. Life Sci.* 10, 605–609.
34. Kim, S.K., Park, P.J. and Kim, Y.T. 2001. Study on acute subcutaneous toxicity of hydroxyapatite sinter produced from tuna bone in Sprague-Dawly rats. *Korean J. Life Sci.* 11, 97–102.
35. Choi, Y.R., Park, P.J., Choi, J.H., Byun, H.G., Jeong, I.C., Moon, S.H. and Kim, S.K. 2000. Purification and characterization of angiotensin I converting enzyme inhibitory peptides from enzymatic hydrolysate of cod liver protein. *Korean J. Life Sci.* 2, 140–149.
36. Kim, S.K., Choi, Y.R., Park, P.J., Choi, J.H. and Moon, S.H. 2000. Screening of biofunctional peptides from cod processing wastes. *J. Korean Soc. Agric. Chem. Biotechnol.* 43, 225–227.
37. Kim, S.K., Choi, Y.R., Park, P.J., Choi, J.H. and Monn, S.H. 2000. Purification and characterization of antioxidative peptides from enzymatic hydrolysate of cod teiset protein. *J. Korean Fish. Soc.* 33, 198–204.
38. Haard, N.F. and Simpson, B.K. Protease from aquatic organisms and their uses in the sea food industry, in *Fisheries Processing Biotechnological Applications*, Martin, A.M., Ed., Chapman & Hill, London, 1994, pp.132–154.
39. Kim, S.K., Park, P.J., Byun, H.G., Je, J.Y. and Moon, S.H. 2003. Recovery of fish bone from hoki (*Johnius belengeri*) frame using a proteolytic enzyme isolated from mackerel intestine. *J. Food Biochem.* 27, 255–266.
40. Kim, S.K., Park, P.J., Kim, J.B. and Shahidi, F. 2002. Purification and characterization of collagenolytic protease from the file fish, *Novoden modestus. J. Biochem. Mol. Biol.* 35, 165–171.
41. Park, P.J., Lee, S.H., Byun, H.G., Kim, S.H. and Kim, S.K. 2002. Purification and characterization of a collagenase from the mackerel, *Scomber japonicus. J. Biochem. Mol. Biol.* 35, 576–582.
42. Byun, H.G., Park, P.J., Sung, N.J. and Kim, S.K. 2003. Purification and characterization of serine proteinase from the tuna pyloric caeca. *J. Food Biochem.* 26, 479–494.
43. Kim, G.H., Jeon, Y.J., Byun, H.G., Lee, Y.S., Lee, E.H. and Kim, S.K. 1998, Effect of calcium compounds from oyster shell bound fish skin gelatin peptide in calcium deficient rats. *J. Korean Fish. Soc.* 31, 149–159.
44. Jung, W.K., Je, J.Y., Kim, H.J. and Kim, S.K. 2002. A novel anticoagulant protein from *Scapharca broughtonii. J. Biochem. Mol. Biol.* 35, 199–205.
45. Je, J.Y., Park, J.Y., Jung, W.K., Park, P.J. and Kim, S.K. 2005. Isolation of angiotensin I converting enzyme (ACE) inhibitor from fermented oyster sauce, *Crassostrea gigas. Food Chem.* 90, 809–814.
46. Jeon, Y.J. and Kim, S.K. 2002. Antitumor activity of chisan oligosaccharides produced in an ultra filtration membrane reactor system. *J. Microbiol. Biotechnol.* 12, 503–507.
47. Tsukada, K., Matsumoto, T., Aizawa, K., Tokoro, A., Naruse, R., Suzuki, S. and Suzuki, M. 1990. Antimetastatic and growth inhibitory effects of N-acetyl chitohexaose in mice bearing Lwis lung carcinoma. *Jpn. J. Cancer Res.* 81, 259–265.
48. Jeon, Y.J., Park, P.J. and Kim, S.K. 2001. Antimicrobial effect of chitooligosaccharides produced by bioreactor. *Carbohyd. Polym.* 44, 71–76.
49. Subasinghe, S. The development of crustacean and mollusc industries for chitin and chitosan resource. In *Chitin and Chitosan*, Zakaria, M.B.; Wan Muda, W.M.;

Abdullah, M.P., Eds., Penerbit University Kebangsaan, Malaysia, 1995, pp. 27–34.
50. Jeon, Y.J. and Kim, S.K. 2000. Continuous production of chitooligosaccharides using a dual reactor system. *Process Biochem.* 35, 623–632.
51. Park, P.J., Je, J.Y., Byun, H.G., Moon, S.H. and Kim, S.K. 2004. Antimicrobial activity of hetero-chitosans and their oligosaccharides with different molecular weights. *J. Microbiol. Biotechnol.* 14, 317–323.
52. Je, J.Y., Park, P.J. and Kim, S.K. 2004. Radical scavenging activity of hetero-chitooligosaccharides. *Eur. Food Res. Technol.* 219, 60–65.
53. Park, P.J., Je, J.Y. and Kim, S.K. 2003. Free radical scavenging activity of chito-oligosaccharides by electron spin resonance spectrometry. *J. Agric. Food Chem.* 51, 4624–4627.
54. Park, P.J., Je, J.Y. and Kim, S.K. 2003. Angiotensin I converting enzyme (ACE) inhibitory activity of hetero-chitooligosaccharides prepared from partially different deacetylated chitosans. *J. Agric. Food Chem.* 51, 4930–4934.
55. Park, P.J., Je, J.Y., Jung, W.K., Ahn, C.B. and Kim, S.K. 2004. Anticoagulant activity of hetero-chitosan and their oligosaccharide sulfates. *Eur. Food Res. Tech.* 219, 529–533.

2 Omega-3 Oils: Sources, Applications, and Health Effects

Fereidoon Shahidi

CONTENTS

2.1 INTRODUCTION

Marine oils serve as a rich source of long-chain omega-3 polyunsaturated fatty acids (PUFA), which have attracted much attention in recent years. They originate primarily from the body of fatty fish such as mackerel and herring, the liver of white lean fish, and the blubber of marine mammals such as seals and whales. They are also present in high amounts in certain algal and fungal oils. The main sources of fish oils are pelagic species caught in large quantities, particularly those with oily flesh, such as salmon, tuna, mackerel, and herring or small fish such as anchovy and capelin. The oily flesh is often used for the purpose of fish meal and oil production, but fish oil can also be produced from fish processing by-products. The latter may include those from aquaculture industries.

Table 2.1 summarizes the content of omega-3 fatty acids, namely eicosapentaenoic acid (EPA, C20:5n-3) and docosahexaenoic acid (DHA, C22:6n-3) in representative fish of commercial significance. Both omega-3 and omega-6 fatty acids are essential PUFA that cannot be made in the human body [1]. The Western diet is abundant in omega-6 fatty acids, mainly from vegetable oils rich in linoleic acid (LA, C18:2n-6). However, humans lack the necessary enzymes to convert omega-6 fatty acids into their omega-3 counterparts, and the latter must be obtained from appropriate dietary sources [2]. The n-6 PUFA include LA and its long-chain product arachidonic acid (AA, C20:4n-6) and intermediate γ-linolenic acid (GLA, C18:3n-6). The n-3 PUFA include the essential fatty acid γ-linolenic acid (ALA, C18:3n-3) and its long-chain metabolites, EPA, docosapentaenoic acid (DPA, C22:5n-3), and DHA. ALA is available from certain plants such as the seeds and oils of flax or linseed, and to a lesser extent perilla, soybean, and canola as well as walnuts [3,4]. EPA and DHA, however, are derived from marine resources (i.e., fish, shellfish and algal species) [3,5–11], but DPA is less abundant and found in 0–2% in most fish oils. Humans can synthesize up to approximately 5% EPA and DHA through desaturation and elongation from dietary ALA [12]. This pathway is an important source of these long-chain n-3 PUFA in strict vegetarians, who do not consume fish. Nonvegetarians can also obtain PUFA from a variety of food products [13].

Table 2.2 summarizes fatty acid composition of lipids from several marine species. Marine lipids contain mainly monounsaturated fatty acids along with PUFA and some saturated fatty acids in different proportions [5–11]. Among these, the ratio of DHA to EPA and possible presence of DPA in modest amounts are most important [14]. Although EPA and DHA are found abundantly in different marine oils, DPA is present in significant amounts only in marine mammal oils such as seal blubber oil [15–18].

TABLE 2.1
The Content of Omega-3 Polyunsaturated Fatty Acids (% of Total Fatty Acids) in Various Fish

Fish	Eicosapentaenoic Acid (20:5 n-3) (%)	Docosahexaenoic Acid (22:6 n-3) (%)
Sardine	3	9–13
Pacific anchovy	18	11
Mackerel	8	8
Capelin	9	3
Herring	3–5	2–3
Freshwater fish	5–13	1–5

Source: Newton, I. and Snyder, D., *Cereal Foods World*, 42, 126–131, 1997.

TABLE 2.2
Distribution of Selected FA (% Wet Weight) in Selected Marine Oils

Component	Atlantic Menhaden[a]	Northern Krill (*Meganyctiphanes norvegica*)[b]	Atlantic Salmon (*Salmo gairdneri*)[c]	Sardine (*Sardine pilchardus*)[d]	Anchovy (*Engraulis encrasicholus*)[e]	Skipjack Tuna[f]
14:0	7.3	4.6	5.5	5.7	10.2	8.2
C16:0	19.0	18.4	10.2	17.8	23.5	26.5
C18:0	4.2	2.2	2.7	3.1	3.4	11.9
SFA	30.5	25.2	18.4	27.2	40.4	38.4
C16:1	9.1	3.5	8.1	6.8	10.6	8.2
C18:1	13.2	14.7	16.9	7.8	10.3	6.8
C20:1	2.0	2.3	15.1	–	5.0	0.5
C22:1	0.6	0.9	14.4	5.1	0.7	–
MUFA	24.9	21.4	54.5	19.7	28.4	7.3
C18:2n-6	1.3	1.8	4.5	1.9	0.8	0
C18:3n-3	1.3	1.3	0.9	–	–	1.3
C18:4n-3	2.8	3.1	1.7	8.4	8.2	11.1
C20:4n-6	0.2	1.0	0.6	–	0.9	–
C20:5n-3	11.0	10.8	6.2	15.5	15.2	29.1
C22:5n-3	1.9	1.0	1.8	2.6	1.6	1.1
C22:6n-3	9.1	28.6	9.1	–	0.7	3.1
PUFA	27.6	45.8	24.8	28.4	29.4	45.7
Others	17.0	7.6	2.3	25.0	1.8	–

[a] Ackman, R.G. in *Nutritional Evaluation of Long-Chain Fatty Acids in Fish Oils*, Academic Press, London, UK, 1982, 25–88.
[b] Mayzaud, P., Virtue, P. and Albessard, E., *Mar. Ecol. Prog. Ser.*, 186, 199–210, 1999.
[c] Oliveira, A.C.M. and Bechtel, P.J., *J. Aquat. Food Prod. Technol.*, 14(1), 73–89, 2005.
[d] Newton, I. and Snyder, D., *Cereal Foods World*, 42, 126–131, 1997.
[e] Kalogeropoulos, N., Andrikopoulos, N.K. and Hassapidou, M., *J. Sci. Food Agric.*, 84, 1750–1758, 2004.
[f] Copeman, L. and Parrish, C.C., *J. Agric. Food Chem.*, 52, 4872–4881, 2004.

The distribution of fatty acids in triacylglycerols (TAG) in fish and marine mammal oils differ; while fish oils contain long-chain PUFA mainly in the sn-2 position of TAG, in marine mammal lipids they are predominantly in the sn-1 and sn-3 positions. These factors greatly influence the metabolism, deposition, and potential health benefits of marine lipids [19].

2.1.1 Oil from By-Products

There is a considerable amount of bycatch in targeted fishery and there are several fish species that are primarily harvested for fish meal production such as menhaden. Menhaden (*Brevoortin tyrannus*) oil, which contains 18% EPA and nearly 10% DHA is a by-product of menhaden meal production. The major fatty acids of menhaden oil are given in Table 2.3. Meanwhile, capelin (*Mytilus edulis*) is a small marine fish that is also used for fish meal and oil production. It is a major prey species that is often used as a bait in some fisheries. Capelin contains 7–10% oil, mainly TAG, and about 20% PUFA as shown in Table 2.3. The oil from capelin is used in aquaculture feed formulation as well as other applications.

By-products from gutting, filleting, and other processing operations are also good raw materials for fish meal and oil production. One can obtain oil from different parts of fish with diverse nutritional composition. Composition, lipid content, and fatty acid profile of individual by-products are of increasing importance, as different by-products are being segregated and used for different end products.

Shark liver is the principal site of lipid storage. The oil content and fatty acid composition in shark liver are influenced by the gender, season, and species of shark. Table 2.4 presents the fatty acid profile of liver oil from male and female blue shark. In addition to fatty acids, shark liver oil contains high amounts of squalene (90–92.8%), low-density lipids (diacylglyceryl ethers), and vitamin A [18,23]. In recent years, scientific evidence has emerged in support of therapeutic value of shark products, particularly shark fins, cartilage, and liver oil as a good source of n-3 PUFA. At present, cod-liver oil is dominant in the market and contains high levels of vitamins A and D [24]. Here it should be noted that cod- as well as halibut-liver oils, although rich in n-3 PUFA, are used primarily as a source of vitamins A and D and hence may be diluted with vegetable oils.

Salmon deposits lipid mainly in its head, at approximately 15–18%, as well as its flesh. However, salmon oil is generally produced from viscera, whole fish (down-graded), and filleting by-products (heads, trimmings such as belly-flaps as well as skin, and frame bones). The lipid and fatty acid content and composition of salmon by-products were found to be quite different. Pink salmon heads had the highest lipid content and viscera the lowest. The n-3/n-6 ratios for pink salmon samples ranged from 7.7 to 10.5 and only values for viscera were statistically different [15–18]. The fatty acid composition of salmon oil also depends on the composition of the raw material used. The average content of n-3 PUFA in salmon oil is in the medium range compared with other fish oils.

TABLE 2.3
Major Fatty Acids of Menhaden and Capelin Oil

Fatty Acid	Menhaden (%)	Capelin (%)
14:0	7.30	5.9
15:0	0.65	
16:0	19.45	8.7
16:1	9.05	10.5
16:2 n-7	0.50	
16:2 n-4	1.55	
16:3 n-4	1.70	
16:4 n-1	2.60	
17:0	1.05	
18:0	4.45	0.6
18:1	10.40	6.0
18:2 n-6	1.30	0.5
18:2 n-4	0.50	
18:3 n-3	0.65	0.2
18:4 n-3	2.65	1.2
20:1	1.45	17.6
20:2 n-6	0.30	
20:3 n-3	1.00	
20:4 n-3	0.80	
20:5 n-3	18.30	9.3
21:5 n-3	0.90	
22:1	1.55	27.8
22:4 n-3	0.60	
22:5 n-3	1.80	0.7
22:6 n-3	9.60	4.1

Source: Kalogeropoulos, N., Andrikopoulos, N.K. and Hassapidou, M., *J. Sci. Food Agric.*, 84, 1750–1758, 2004; Copeman, L. and Parrish, C.C., *J. Agric. Food Chem.*, 52, 4872–4881, 2004.

Even though the by-products from most other fish processing industries tend to vary with season, both in quality and quantity, salmon slaughterhouses generate high-quality offal at a relatively constant rate [25,26].

The by-products of catfish processing consist of heads, frames, skin, and viscera, which often end up in landfills or rendering plants. Producing edible oil from viscera may add value to catfish viscera. The total unsaturated fatty acids in the purified oil from catfish viscera was 67.7%. The combined n-3 fatty acids of the purified catfish viscera oil was only 4.6 mg/g of oil [17].

Herring oil is produced from three different types of by-products; only heads, mixed, and headless by-products are of interest [17]. Even though by-products from heads and their oil have the highest oxidation levels and the lowest α-tocopherol content, heads contain the lowest PUFA and the highest amount of

TABLE 2.4
Fatty Acid Composition in Total Lipids from Blue Shark (*P. glauca*) Liver (w/w%)

Fatty Acid	Male	Female
14:0	2.3	2.8
16:0	22.0	16.8
18:0	4.9	3.9
Total SFA	36.0	30.3
16:1	3.9	5.3
18:1	16	27.6
20:1	3.0	4.9
22:1	0.0	0.0
24:1	0.8	0.9
Total MUFA	23.6	38.7
18:2n-6	0.8	0.7
18:3	0.2	0.2
18:4n-3	0.4	0.4
20:5n-3	4.5	2.7
22:5n-3	2.1	2.5
22:6n-3	23.2	18.4
Total PUFA	39.2	30.2

Source: Jayasinghe, C., Gotoh, N., Tokairin, S., Ehara, H. and Wada, S., *Fish. Sci.*, 69, 644–653, 2003.

saturated fatty acids. No significant differences were found between the fatty acid composition of the mixed and the headless by-products or their oil [16].

2.1.2 Oil from Blubbers of Marine Mammals

Marine mammals are unique in that they are protected from cold by a layer of insulating fat under their skin, known as blubber, which also helps in their movement and buoyancy [27]. Blubber may vary in thickness, depending on a number of variables, but is on average about 5 cm thick for seals. Seal blubber oil is a by-product of seal meat and seal skin industries. The oils from marine mammals contain various lipid classes, including TAG, diacylglycerols, monoacylglycerols, free fatty acids, wax esters, cholesterol, cholesterol esters, hydrocarbons, vitamins, and ether lipids. TAG of seal blubber oils are the main component of neutral lipids, which contain a variety of lipid classes. Neutral lipids account for 98.9% of blubber in contrast to intramuscular lipids (78.8% neutral and 21.1% polar lipids) [28].

In addition to TAG, wax esters (long-chain alcohols esterified to fatty acids) are another important group of neutral lipids found in marine mammals. Most species of marine mammals have C32, C34, C36, and C38 (total of alcohol plus acid) as major components [29]. Whale oils are especially interesting because some contain fatty acids that are largely in the form of wax esters [30]. The oils from the blubber of the Physeteridae may consist mainly of wax esters. Sperm whale blubber oil consists of a mixture of about 79% wax esters and 21% TAG [31]. Dwarf sperm whale (*K. simus*) blubber oil consists of 42% wax esters and 58% TAG [32]. The blubber fat of beaked whales (*Berardius, Hyperoodon, and Ziphius*) is composed almost entirely of wax esters (94–99%) along with low levels of TAG (2–6%) [33]. A number of possible functions for wax esters in marine mammals has been proposed; these include their role as a reserve energy store, buoyancy, metabolic water, thermal insulation, and biosonar [34–36].

Among unsaponifiable matters, hydrocarbons, especially long-chain hydrocarbons, are found in detectable amounts in marine mammal oils. Some marine oils contain less than 0.1% hydrocarbons, while others contain as much as about 90% [23]. In the liver of the seal, *Arctocephalus* (Pinnipedia), squalene was present at 0.50% of the oil [37]. High squalene contents (90, 91, and 92.8%) occur in shark liver oils [23,37]. Total hydrocarbons were present at 0.3% of dry matter weight of the blubber, 1.6% in liver, and 1.3% in the muscle [38]. Among cetaceans, limited data for two dolphins have been published: in *Delphinus longirostris* liver, very long-chain hydrocarbons (C44) were detected and zamene was present in *Langenorynchus acutus* [39].

The fatty acid composition of marine lipids varies significantly, but all contain a large proportion of long-chain highly unsaturated fatty acids, similar to fish oils. However, the proportion of fatty acids in fish and marine mammals varies considerably [19]. A marine oil typically contains some 40 different fatty acids with carbon numbers varying from 10 to 24, resulting in a large number of different TAG with the same carbon number, but with different levels of unsaturation [28,40]. Even-numbered carbon fatty acids make up about 97% of the total fatty acids, with a few notable exceptions [31]. Some fatty acids with odd-numbered carbon chain such as C15:0 and C17:0, along with traces of C13:0 and C19:0 have also been found in marine oils [41]. Besides, monomethyl branched fatty acids have been isolated from marine oils, such as 3-methyldodecanoic acid from blubber of the sperm whale *Physeter catodon* [41].

In contrast to relatively small amounts of saturated fatty acids, marine mammal oils have been characterized by high amounts of monounsaturated fatty acids (MUFA) and n-3 PUFA [42,43]. For instance, the contents of MUFA in neutral and polar lipids in seal blubber are more than 60 and 46%, respectively [44]. Most of these fatty acids are long-chain with 20–22 carbon atoms and are of the n-3 (omega-3) type. Ackman et al. [45] have pointed out that the total C20 and C22 MUFA and PUFA in each layer of whale blubber is nearly constant, but the ratios of the MUFA to PUFA change very significantly. The most common long-chain PUFA in marine lipids are EPA and DHA as well as a smaller amount of DPA, all of which belong to the omega-3 family [46]. The high content of omega-3 fatty acids in marine lipids is suggested to be a consequence of cold temperature adaptation, because at lower habitat temperatures, omega-3 PUFA remain liquid and resist to crystallization [41]. Most of the long-chain PUFA are formed in unicellular phytoplankton and multicellular sea algae and eventually pass through the food web and become incorporated into the body of fish and other higher marine species, including marine mammals which often eat fish [47]. The fatty acid composition of oils from most species of marine mammals has been summarized [48]. Seal oils, owing to the increasing interest in seal fishery and product development, have been in focus and frequently studied by researchers. The fatty acid composition of oils from different species of seal has been reviewed [19]. Table 2.5 shows the fatty acids profiles of blubber lipid from the main species of seals.

The fatty acid composition of blubber in marine mammals such as seals is regulated by their diet [51], location [52,53], season, as well as physiological conditions such as age [24] and sex [52,53] of the animal. In some marine mammals, the depot fats are largely dietary fatty acids laid down with a minimum change, but the fatty acids of the lipids of the essential organs have terrestrial characteristics [48]. Fatty acid composition also depends on tissue and species of the animal. However, differences are most apparent among tissues. Seal blubber, for example, had a high content of MUFA but was low in arachidonic acid, dimethyl acetals, and DHA. Lung tissue lipids were high in palmitic acid and heart tissue lipids had a higher content of linoleic acid. The proportions and fatty acid constituents in different tissues are different, most probably due to their varying functional requirements [49,50]. The lipids of vital organs of seals and whales contain high proportions of fatty acids of the ω6 family, similar to those of terrestrial animals. The distinction between the fatty acids of functional organs such as liver, heart, and other organs with depot fat has been discussed in the literature [49,54].

As explained earlier, the fatty acid distribution in the TAG molecules in blubber oil are different from fish oil and the omega-3 fatty acids are located primarily in the sn-1 and sn-3 positions of TAG (Table 2.6), while in fish oils they are located abundantly in the sn-2 position of TAG [55]. Mag [56] has reported that the different distribution of fatty acids may influence the metabolism and potential health benefits of marine lipids, and moreover, may account for the better oxidative stability of marine mammal oils compared to fish oils.

2.1.3 Algal Oils

Algal oils, in contrast to other marine oils, have a dominance of one particular fatty acid, for example, EPA or DHA. Table 2.7 summarizes the fatty acid profile of selected algal oils [57]. As can be seen, dominance of EPA or DHA is clearly noticeable in the oils tested and assembled in Table 2.7.

2.2 NUTRACEUTICAL AND FOOD APPLICATIONS OF OMEGA-3 FATTY ACIDS

Omega-3 fatty acids may be used as dietary supplements in the liquid or capsule form. Salmon oil, cod-liver oil, halibut-liver oil, omega-3 concentrated oils, flax oil, and omega 3, 6, 9 combination oils are readily available in the drug and supplement stores. The liver oils are often consumed for not only their omega-3 components, but perhaps primarily as a source of fat-soluble vitamins, primarily vitamin A.

TABLE 2.5
Fatty Acid Composition (g/100 g) of Blubber of Various Species of Seal

Fatty Acid	Bearded	Gray	Harbor	Harp	Hooded	Ringed
14:0	3.05	3.83 ± 0.03	4.52 ± 0.13	4.66 ± 0.49	4.40 ± 0.38	3.36 ± 0.66
16:0 DMA	ND	ND	ND	ND	ND	ND
16:0	10.14	6.61 ± 0.08	8.03 ± 0.38	6.24 ± 0.44	9.81 ± 1.57	4.82 ± 2.07
16:1 n-7	17.77	12.77 ± 0.09	19.26 ± 0.53	14.93 ± 0.46	10.09 ± 0.35	23.12 ± 0.18
18:0 DMA	ND	ND	ND	ND	ND	ND
18:1 n-9 DMA	ND	ND	ND	ND	ND	ND
18:1 n-7 DMA	ND	0.45 ± 0.01	ND	0.46 ± 0.00	ND	ND
18:0	2.15	0.94 ± 0.02	0.85 ± 0.02	0.95 ± 0.03	1.83 ± 0.31	0.42 ± 0.19
18:1 n-9	16.76	24.50 ± 0.44	18.61 ± 0.55	18.59 ± 1.01	22.77 ± 2.66	19.72 ± 1.33
18:1 n-7	9.49	4.95 ± 0.09	5.16 ± 0.44	3.57 ± 0.36	3.75 ± 0.47	5.03 ± 0.46
18:2 n-6	2.30	1.28 ± 0.00	1.27 ± 0.04	1.36 ± 0.20	1.63 ± 0.20	2.58 ± 0.02
20:1 n-9	5.08	12.50 ± 0.43	9.06 ± 0.33	12.56 ± 2.92	13.00 ± 1.86	6.71 ± 2.17
20:4 n-6	0.94	0.51 ± 0.00	0.44 ± 0.00	0.36 ± 0.96	0.31 ± 0.03	0.30 ± 0.02
20:5 n-3	8.28	4.85 ± 0.13	9.31 ± 0.21	6.82 ± 0.69	5.21 ± 1.65	8.72 ± 1.06
22:0	0.63	<0.3	1.19 ± 0.02	<0.3	<0.3	0.75 ± 0.67
22:1 n-11	0.27	0.62 ± 0.03	0.31 ± 0.01	0.77 ± 0.61	0.86 ± 0.33	0.34 ± 0.01
22:5 n-3	4.26	5.06 ± 0.05	4.22 ± 0.14	4.78 ± 0.25	2.29 ± 0.08	5.46 ± 0.47
22:6 n-3	7.22	8.91 ± 0.29	7.76 ± 0.98	10.48 ± 1.98	9.56 ± 2.36	9.45 ± 1.74

Note: DMA, dimethyl acetal; ND, not detected.

Source: Durnford, E. and Shahidi, F., *J. Am. Oil Chem. Soc.*, 79, 1095–1102, 2002; Durnford, E., Shahidi, F. and Ackman, R.G., *J. Am. Oil Chem. Soc.*, 80, 405–406, 2003.

TABLE 2.6
Fatty Acid Distribution (%) in Different Positions of Triacylglycerols of Harp Seal Blubber Oil and Menhaden Oil

Fatty Acid	sn-1	sn-2	sn-3
Seal Blubber Oil			
EPA	8.36	1.60	11.2
DPA	3.99	0.79	8.21
DHA	10.5	2.27	17.9
Menhaden Oil			
EPA	3.12	17.5	16.3
DPA	1.21	3.11	2.31
DHA	4.11	17.2	6.12

Source: Shahidi, F. in *Seal Fishery and Product Development*, ScienceTech, NL, Canada, 1998, 99–146; Ackman, R.G., Hooper, S.N. and Hingley, J., *Can. J. Biochem.*, 50, 833–838, 1972.

Food application of marine oils is also popular and recent advances in the microencapsulation and refining technologies have allowed production of oils that do not render any off-flavor to foods. Foods in which omega-3 oils are included are bakery products, pastas, dairy products such as milk, yogurt, and juice, as well as nutrition bars and a myriad of other products. A novel microencapsulated oil prepared by Ocean Nutrition Canada, known as MEG 3, remains protected from the environment until it reaches the gastrointestinal tract. The availability of this technology has opened the way for inclusion of marine oils in different foods without concerns about their flavor reversion.

2.2.1 Health Effects of Marine Oils and Polyunsaturated Fatty Acids

Recognition of the health benefits associated with consumption of seafoods (n-3 fatty acids) is one of the most promising developments in human nutrition and disease prevention research in the past three decades. According to the current knowledge, long-chain n-3 PUFA play an important role in the prevention and treatment of coronary artery disease [58], hypertension [59], diabetes [60], arthritis and other inflammatory [61], and autoimmune disorders [62], as well as cancer [63,64] and are essential for normal growth and development, especially for the brain and retina [65]. The most direct and complete source of n-3 oils is found in fish oils and the blubber of certain marine mammals, especially harp seal. Among its advantages is that the body's absorption of n-3 fatty acids from marine mammal blubber may be faster and more thorough than is the case with flaxseed and fish oils [56]. Since marine mammal oils contain a high concentration of MUFA, it is possible that some of their beneficial effects may be ascribed to their MUFA or to the combined effect of MUFA and n-3 PUFA [66]. A pilot study

TABLE 2.7
Fatty Acid Profile of Selected Algal Oils

Organism	14:0	14:1	16:0	16:1	18:0	18:1	18:2	18:3	18:4	20:4 n-6	20:5 n-3	22:5	22:6 n-3
Thraustochytrium aureum (H,76)	3		8			16	2	2		3			52
Schizochytrium sp. (H,77)	18		38	6	1	5		1		1	1	9	18
Crypthecodinium cohnii (H,33)	15		15	2	2	15	1			6			37
Amphidinium carterae (H,78)	8	30	15	5	3	5	6	17			4	4	2
Isochrysis galbana (P,79)	12		15	11	1	3	2		11		25		11
Skeletonema costatum (P,80)	17		10	11		2	1		6		41		7
Amphidinium sp. (P,81)	5		17		18	17	2	2			8		17
Pavlova lutheri (P,82)	14		27	10	1	3			4		12		7

Note: H, heterotrophic growth; P, phototrophic growth.
Source: Mag, T. 2000. Patent WO00/44862A1, PCT International Patent Application.

indicated that a low dose of seal oil supplementation can reduce atherogenic risk indices in young healthy individuals, and the effects are strongly dependent on the integrated n-3 fatty acids dose [67,68]. The essential fatty acids found in seal oil include a high level of DPA (up to 10 times that of fish oils). There is growing evidence that DPA is the most important fatty acid that keeps artery walls soft and plaque-free [56]. Marine oils are also attractive from a nutritional point of view because they are thought to provide specific physiological functions against thrombosis, cholesterol build-up, and allergies [69]. Oils from the blubbers of seal and whale have beneficial effects on selected parameters that play a role in cardiovascular disease; it has been hypothesized that the effect of whale oil is not mediated by its n-3 fatty acids alone [70]. The difference in the beneficial effects of whale and seal oils on cardiovascular disease may argue against the distribution of n-3 fatty acids in TAG as being relevant to the superiority of whale oil, since the n-3 fatty acids are mainly in the sn-1 and sn-3 positions of both of these oils. The effect of whale oil is probably not mediated by n-3 fatty acids alone as the content of these fatty acids is relatively low in whale oil. Thus, in addition to n-3 fatty acids, other dietary factors may play a role in the protective effects against atherosclerosis and thrombosis in Greenland Eskimos [70].

The following sections of this chapter provide a cursory account of the specific health benefits of n-3 fatty acids in cardiovascular disease, cancer, autoimmune diseases, and mental health. Effects on gene expression are also discussed.

2.2.2 Omega-3 Fatty Acids and Cardiovascular Disease

Cardiovascular disease is the common term for all diseases that affect the heart and the circulatory system including ischemic heart disease, nonischemic myocardial heart disease, hypertensive heart disease, and valvular heart disease. It is the leading cause of death in the Western societies [71] and has been linked to the high fat intake, particularly saturated fat, common in Western diets [72]. Besides high saturated and trans fat intake, other risk factors for cardiovascular disease include diabetes mellitus, smoking, stress, physical inactivity, high sodium intake, and genetic predisposition. The hallmark of cardiovascular disease is cardiac dysfunction, which in most cases is caused by hypertension due to the narrowing of large arteries with atheromatous plaques, or the total occlusion of coronary arteries (thrombus) caused by atheromatous blockages leading to myocardial tissue necrosis. Both conditions reduce the heart's ability to pump blood and can result in either chronic or sudden heart failure.

It is becoming apparent that regular consumption of fatty fish or fish oils containing n-3 long-chain PUFA lowers the rate of incidence and death from cardiovascular heart disease [73,74]. The cardioprotective effects of marine oils were first postulated in the 1950s based on cross-cultural studies done on Greenland Inuits and Danish settlers of Greenland [75]. These studies revealed that the Greenland Inuits had a significantly lower incidence of heart disease compared to the Danish settlers, despite comparable fat intakes (40% of caloric intake) and a higher intake of dietary cholesterol. This anomaly was referred to as the "Eskimo

Paradox" [76]. Epidemiological studies done in the 1970s suggested a strong correlation between the low incidence of coronary heart disease in Greenland Inuits and their high consumption of fish and marine mammals, both being rich in long-chain n-3 fatty acids [77]. Other cross-cultural epidemiological studies among coastal Japanese and Alaskan populations have resulted in similar findings, showing inverse relationships between long-chain n-3 PUFA intake and cardiovascular disease [78,79].

The biochemical basis for cardioprotective effects of n-3 fatty acids are unknown but are probably multifactorial and may collectively result in increased heart rate variability (antiarrhythmic), reduced atheroma development (antiatherogenic), and decreased platelet reactivity/aggregation (antithrombotic). Investigations on the link between fish oils and cardiovascular disease in both animal and human models have concluded that this effect may be mediated by substrate competition between n-3 fatty acids and arachidonic acid (AA, 20:4n-6) for cyclooxygenase (COX) enzymes that produce prostaglandins and thromboxanes. Competition between n-3 fatty acids and AA could result in positive health benefits because (a) n-3 fatty acids inhibit the production of AA through substrate competition for the Δ^6 desaturase; (b) long-chain n-3 fatty acids compete with AA for incorporation into the sn-2 position of membrane phospholipids thereby reducing membrane AA levels [80]; and (c) eicosanoids produced from EPA have anti-inflammatory and antiaggregatory effects, for example, increasing the membrane EPA/AA ratio shifts eicosanoid production from the proaggregatory eicosanoids PGI_2 and TXA_2 toward the antiaggregatory TXA_3 in platelets [81] and PGI_3 in endothelial cells [77]. These actions would result in vasodilation and decreased platelet aggregation, both having antithrombotic effects.

The potential antiarrhythmic properties of n-3 fatty acids (ALA, EPA, and DHA) have been examined in animal models. In one such experiment, intravenous infusion of either fish oil or pure n-3 fatty acids in exercising dogs before an experimentally induced coronary artery obstruction reduced sudden cardiac death by preventing ventricular fibrillation [82]. The mechanisms of the antiarrhythmic effects of n-3 fatty acids have been further explored using spontaneously contracting cultured cardiac myocytes isolated from rats [83]. The cultured rat myocytes were induced to fibrillation using various toxic agents such as ouabain, β-adrenergic agonists and high Ca^{2+} concentrations added to the bathing medium. These researchers [84] have shown that EPA and DHA added at low doses of 5–15 μmol/L in the bathing medium inhibited the expected fibrillation when the toxic agents were added to the media. Interestingly, when the added EPA and DPA were removed from the culture media using delipidated bovine serum albumin, the cultured myocytes returned to fibrillation. Thus, n-3 fatty acids at low concentrations are able to modulate the activity of specific ion channels in myocardial sarcolemma.

One of the most effective ways to protect the myocardium from ischemic/reperfusion injury is by inhibition of the transmembrane Na^+/H^+ antiport exchanger. This transmembrane antiport system maintains the myocardial cell's pH, but during ischemia this system paradoxically participates in cell necrosis. The importance of transmembrane Na^+/H^+ antiport exchanger in ischemic heart disease was shown in a clinical trial using a specific Na^+/H^+ exchange inhibitor caporide [85]. The inhibitor showed a potential benefit in reducing the risk of cardiac death as long as it was taken prior to the ischemic event [86]. It has recently been shown that EPA and DHA at concentrations of 25–100 μM inhibited the Na^+/H^+antiport exchanger in isolated cardiomyocytes and thus could protect the myocardium from arrhythmias and cell death during ischemic events. This effect was limited to long-chain n-3 fatty acids, as LA and ALA showed no significant effects on the Na^+/H^+ exchanger [87].

The role of ALA in the prevention of arrhythmia caused by ischemia or during reperfusion remains unclear. The metabolic conversion of ALA to EPA is thought to mediate any of the cardioprotective effects of ALA acid, but at least one study using a canine model revealed cardioprotective effects with ALA [88]. More basic studies with α-linolenic acid are needed to explore the cardioprotective effects of this fatty acid, but first a better understanding of the conversion processes of ALA to EPA and DHA is needed.

The effect of dietary interventions, carried out by the Diet and Reinfarction Trial (DART), of 2033 Welsh men who had recovered from a previous heart attack, was the earliest controlled trial to examine the effects of dietary intervention in the secondary prevention of myocardial infarction [89]. The results of this study strongly suggest that marine n-3 fatty acids have a specific antiarrhythmic effect rather than antiatherogenic or antithombotic effects [90].

The *Gruppo Italiano per lo Studio della Sopravvienenza nell'Infarto Miocardio* (GISSI)-Prevenzione study was initiated in 1993, and was carried out for 3.5 years [91]. It was a multicentered trial conducted in Italy (172 centers) and included 11,324 patients who had suffered a heart attack less than 3 months prior to recruitment. Just as in the DART study [90], the GISSI-Prevenzione study revealed that marine n-3 fatty acid intake conferred early and progressive risk reductions for cardiovascular disease.

The Lyon Diet Heart Study [84] conducted in France was one of the earliest intervention trials (with 204 control subjects and 219 experimental subjects), making the hypothesis that a Mediterranean diet high in ALA could reduce the relative risk of cardiovascular events and death in previous heart attack victims. After 1 year of study, the total cardiovascular events were 24.5% in the fish oil group, 28% in the mustard oil group, and 34.7% in the control group ($p < 0.01$), suggesting that dietary modifications can improve the 1-year risk of recurrent cardiovascular events [92].

2.2.3 Omega-3 Fatty Acids and Cancer

Cancer is a general term for the more than 100 diseases that are characterized by uncontrolled and abnormal growth of cells (neoplasia) that are derived from normal tissues. The first description of these symptoms was in relation to breast carcinoma. Experimental and epidemiological studies have demonstrated that the composition of dietary fat affects the incidence and progression of some cancers; n-3 fatty acids have been shown to have anticarcinogenic effects while saturated and n-6 fatty acids may promote cancer development [93]. Several cancerous cell lines have been developed from animal and human tumors; these are established cell lines (e.g., Hep-G2 cells, caco-2 cells, LNCaP cells, PC-3 cells) derived from malignant tumors that proliferate indefinitely in culture under the appropriate conditions. Cell lines serve as excellent *in vitro* models for cancer studies because the biochemical processes that occur within these cells are remarkably similar to those within their parent tumors.

Early evidence from epidemiological studies indicated that n-3 fatty acids might be protective against prostate cancer. Cross-cultural studies among the Inuit and non-Inuit people of Canada, Alaska, and Greenland from 1969 to 1988 showed that the incidence rate of prostate cancer among the Inuit populations was 70–80% less than the non-Inuit populations [94]. This observation was attributed to dietary differences between the two populations, in particular the traditional seafood diet of the Inuit people that are exceptionally rich in n-3 fatty acids were speculated as having anticarcinogenic effects [94]. Terry et al. [95] studied the association between fatty fish consumption and prostate cancer in a long-term prospective cohort of 6272 Swedish men. An inverse association between fatty fish consumption and prostate cancer was observed. After adjustment for other dietary and lifestyle habits (multivariate analysis), significant inverse associations were observed between fatty fish consumption and prostate cancer incidence as well as prostate cancer death ($p = 0.05$ and $p = 0.01$, respectively, using COX proportional hazard models). Mamalakis et al. [96] examined adipose tissue and prostate tissue fatty acid composition in 71 prostate cancer and benign hyperplasia patients from the island of Crete. Relative to benign hyperplasia patients, cancer patients had elevated adipose tissue levels of saturated fatty acids and reduced adipose tissue levels of MUFA ($p < 0.05$, *t*-tests). Compared to hyperplasia patients, cancer patients had reduced prostate tissue stearic acid to oleic acid ratios and total stearic acid levels ($p < 0.0005$). Relative to benign hyperplasia patients, cancer patients had reduced prostate tissue levels of AA, DHA ($p < 0.0005$), EPA ($p < 0.05$), total n-3 fatty acids ($p < 0.0005$), and n-3/n-6 fatty acid ratios. The pronounced elevation in adipose tissue saturated fatty acid levels in cancer patients highlights a possible role of dietary saturated fats in neoplastic processes, since adipose tissue fatty acid composition mimics the composition of dietary fats ingested [86]. The decreased prostate tissue level of C_{20} and C_{22} PUFA in cancer patients possibly stems from enhanced metabolism of these fatty acids via lipoxygenase and COX pathways. Augustsson et al. [97] analyzed data from the 12-year Health Professionals Follow-Up Study to investigate whether high dietary intake of fish and long-chain n-3 fatty acids reduced the risk of prostate cancer in 47,882 male American participants. Consumption of three or more fish meals per week was associated with a reduced risk of prostate cancer compared to infrequent fish consumption (less than two fish meals per month), and the strongest association was for metastatic cancer (RR = 0.56, 95% CI: 0.37–0.86). Intake of n-3 fatty acids from foods other than fish showed a similar but weaker association. Each additional daily n-3 fatty acid intake of 0.5 g from food was associated with a 24% decreased risk of metastatic prostate cancer. The results of this study show dietary n-3 fatty acids from fish and other sources reduce the risk of prostate cancer, especially advanced forms of prostatic carcinomas. These results imply that long-term consumption of fish meat and n-3 fatty acids may slow the progression of prostate cancer toward metastasis, as evidenced by the significantly lowered relative risk for metastatic prostate cancer among the participants of this study [98].

Experimental and epidemiological studies suggest that n-3 fatty acids have antitumor effects during the initiation and postinitiation stages of colon carcinoma [99]. Western populations exhibit significantly higher colon cancer incidence and mortality rates compared to Asian populations, which experts have long associated with high dietary fat and animal fat consumption by Western populations [100]. Caygill et al. [101] examined colon cancer mortality data from 24 European countries, showing a significant inverse correlation between colon cancer mortality and fish meat or fish oil consumption. This inverse correlation was significant for both men and women who consumed fish or fish oil for 1, 10, or 23 years before cancer mortality. This study strongly suggests that fish oil consumption can significantly reduce colorectal cancer mortality. Unfortunately, dietary amounts of fish or fish fatty acids were not adequately assessed in this study, making it impossible to critically assess these findings. Anti et al. [102] studied the effects of n-3 fatty acids on colonic cell proliferation in subjects at high risk for colon cancer. The results of this short-term study show that n-3 fatty acids reduce the proliferation of early-stage colonic cancers, which may reduce the progression colorectal polyps to colorectal carcinoma and may protect high-risk individuals from colon cancer.

The vast majority of research on n-3 fatty acids and colorectal cancer has been carried out using animal models and *in vitro* studies. Takahashi et al. [103] studied the effects of DHA supplementation on colon cancer using a rat model. The results of this study do not strongly support the premise that n-3 fatty acids (DHA) protect against colon carcinoma, but the aberrant results may be due to the low amount of DHA supplement. It is also possible that the protocol used to supplement DHA, specifically intragastric injection rather than dietary supplementation, may have caused additional stress in the animals and compromised their immune system, which could have enhanced the tumor-promoting effects of AOM regardless of DHA supplementation. Dwivedi et al. [104] used a similar protocol to Takahashi et al. [103] to study the effects of dietary n-3 and n-6 fatty acid supplementation on colon cancer development in male Wistar rats. This study also shows that oils rich in EPA and DHA are not as chemoprotective against colon carcinogenesis when compared to oils rich in ALA. The corn oil supplementation had the least chemoprotective effects against colon carcinogenesis. Since no control group or saturated fatty acid group were employed in this study, it is impossible to assess the promotional or inhibitory effects of corn oil and n-6 fatty acids on colon cancer development, except to say that n-6 fatty acids are less inhibitory than their n-3 counterparts.

Several studies have investigated the effects of n-3 and n-6 fatty acids on colon carcinoma cell line development and COX expression or activity. In one such study [105], the effects of n-3 and n-6 fatty acids on the proliferation of two COX overexpressing colon cancer cell lines in culture (Caco-2 cell line and

HT-29 cell line) were investigated; this group also assessed COX activity by measuring PGE_2 production in response to fatty acid treatments. Narayanan et al. [106] studied the effects of DHA on Caco-2 cell growth, proliferation, and transcription rates of 3800 genes belonging to 156 functional categories using DNA microarrays. Functional gene groups examined included several tumor suppressors, apoptosis factors, growth factors, chemokines, lipooxygenases, cyclooxygenases, transcription factors, cellular receptors, and nuclear receptors. Compared to control cultures, the DHA treatment resulted in a 30% decrease in Caco-2 cell proliferation after 48 h. Hybridization occurred in only 13% of the genes present on the microarray, implying that the array used in this study was improperly designed. To confirm the microarray results of several representative genes, Narayanan et al. [106] amplified RNA extracts using RT–PCR followed by separation using a denaturing polyacrylamide gel and Northern blotting. Their results showed that DHA down-regulated the expression rate of several genes encoding transcription factors, transcriptional enhancers, RNA polymerases, lipooxygenases, COX-2, and the inducible nitric oxide synthase. DHA treatments enhanced the expression of peroxisome proliferator activated receptors α and γ by over twofold. These changes seem to indicate that DHA promotes Caco-2 cell apoptosis through modulation of several biological activities, and suggests that DHA is an effective chemopreventive agent against colon carcinogenesis [106]. A recent study using the same protocol and cells showed that DHA enhances the expression of several cell cycle inhibitors, which further illustrates the antitumor effects of DHA in colon carcinogenesis [107].

Very few epidemiological studies have assessed the effect of dietary n-3 fatty acids and breast cancer. Holmes et al. [108] analyzed data from 1982 breast cancer patients (mean age 54 years) registered in the 18-year Nurses Health Study. The results of this assessment showed that n-3 fatty acid intake significantly reduced breast cancer mortality by 48% (RR = 0.52; 95% CI:0.30–0.93). Recently, Holmes et al. [109] re-examined the data of the 121,700 female nurses registered in the Nurses Health Study to find associations between breast cancer and dietary intake of meat, fish, and eggs. After the 18-year follow-up period, 4107 cases of breast cancer were diagnosed. Women in the highest quintile for meat, egg, and fish intake showed no difference in breast cancer risk; secondary analyses did not affect these results. Similar results have been observed in the 8-cohort international pooling project involving 350,000 women who were followed up for 15 years [110]. Maillard et al. [111] evaluated the fatty acid composition of adipose tissue from 241 women patients from central France with nonmetastatic breast cancer and 88 patients with benign breast tumors to assess the protective effects of dietary n-3 fatty acid intake against breast cancer, using adipose fatty acid composition as a biomarker of past dietary fatty acid composition. This study showed that adipose n-3 fatty acid levels and breast cancer risk were inversely associated. Women with the highest adipose levels of ALA were 6% less likely to develop breast cancer compared to women with the lowest adipose ALA levels (RR = 0.39; 95% CI:0.19–0.78). Similarly, adipose DHA level and n-3 fatty acid to n-6 fatty acid ratio were inversely associated with breast cancer risk (RR = 0.31 and 0.33, respectively; 95% CI:0.13–0.75 and 0.17–0.66, respectively).

These results suggest that n-3 fatty acids have protective effects against breast cancer risk, and also show that n-3 and n-6 fatty acids affect breast cancer risk.

The proposed antitumorigenic effects of n-3 fatty acids in breast cancer have been studied using *in vitro* and animal models of this disease; several studies show that n-3 fatty acids are able to modulate second messenger systems and cell signaling cascades in cancerous breast cells. The ability of n-3 fatty acids to inhibit breast tumor development has been shown in several tumor transplant studies [112]. Kort et al. [113] showed that female rats with transplanted mammary carcinoma tumors (BN472 cells) who were fed with 25% fish oil after tumor transplantation for 6 weeks exhibited significantly less tumor development compared to rats fed 25% cacao butter. Rose et al. [114] showed a similar suppressive effect of n-3 fatty acids on transplanted human mammary tumors (MDA-MB-435 cells). Nude mice given 20% fish oil diets after tumor transplantation showed less metastatic growth of the implanted tumors into the lungs and overall suppression of tumor development. Recently, Robinson et al. [115] showed that dietary fish oil supplementation (50 mg/g chow) for 21 days after tumor transplantation did not significantly affect tumor development.

Several nonhuman studies support the premise that n-3 fatty acids inhibit breast carcinoma development by influencing the biochemical events that follow tumor initiation. Unfortunately, these findings do not correlate well with human breast cancer studies. This may imply that n-3 fatty acids at attainable human dietary levels (1–3% of total calories) do not affect breast cancer development.

2.2.4 Omega-3 Fatty Acids and Inflammatory Diseases

Arachidonic acid–derived cytokines have proinflammatory actions *in vivo*, whereas those derived from EPA are significantly less proinflammatory [116,117]. Studies investigating the effects of n-3 fatty acids on *ex vivo* cytokine production by leukocytes have produced inconsistent results. Mantzioris et al. [118] showed a 20% decrease in *ex vivo* IL-1β production in healthy men after 4 weeks of supplementation with 1.8 g of fish oil per day, demonstrating that n-3 fatty acids affect cytokine production by leukocytes, but this study was not a controlled trial. Results of a recent placebo-based, double-blind, parallel study involving 150 healthy men and women aged 25–72 years, who were supplemented with ALA or fish oil, revealed no significant differences in *ex vivo* cytokine production (TNF-α, IL-6, IL-1β, and IL10) between the placebo and intervention groups after 6 months of supplementation [119]. However, they did show that monocytes had significantly increased levels of ALA in participants supplemented with α-linolenic acid, and increased monocyte EPA and DHA levels in those given fish oil. Both n-3 groups also had lowered monocyte AA levels compared to the control group, which might lead to decreased synthesis of the proinflammatory LTB_4 *in vivo* [120]. Although the mechanisms by which n-3 fatty acids suppress the production of inflammatory cytokines are unknown, the suppression of inflammatory eicosanoid production by EPA is likely to be involved.

Inflammatory bowel disease (IBD) is a general term for chronic inflammatory diseases of the gastrointestinal tract and mainly includes ulcerative colitis and

Crohn disease. The incidence of ulcerative colitis and Crohn disease is higher and rising in Western countries than in Asian countries, and epidemiological studies have attributed this trend to high intakes of saturated and n-6 PUFA in typical Western diets [121].

Many animal models have been used to study the effects of n-3 fatty acid supplementation in IBD. Shoda et al. [122] studied the effects of perilla oil (n-3, α-linolenic acid rich), fish oil (n-3 long-chain fatty acid rich), and safflower oil (n-6 fatty acid rich) supplementation on ulcer formation and proinflammatory cytokine production in rats. These results suggest that ALA may be superior to EPA and DHA for controlling intestinal inflammation in experimentally induced Crohn disease, but the authors could not rule out the possibility of synergistic effects between n-3 fatty acids and other bioactives in perilla oil. Nieto et al. [123] recently used the trinitrobenzenesulfonic acid model to study the effects of n-3 fatty acid supplementation on ultrastructural and histological changes during experimentally induced ulcerative colitis in rats and the production of biological inflammatory markers. The histological results of this study revealed that experimental rats given an n-3 fatty acid–rich diet had significantly less macroscopic and microscopic colonic damage when compared to both the n-6 group and the n-6 + n-3 group; also, the n-3 group had significantly lower inflammatory marker levels when compared to both other groups, both of which strongly suggest that n-3 fatty acids are therapeutic, whereas n-6 fatty acids exacerbate experimentally induced ulcerative colitis. Another model used to study IBD is the acetic acid protocol in which rats are fed 4% acetic acid to induce IBD. Using this model, Empey et al. [124] showed that rats given a fish oil (EPA)–enriched diet for 6 weeks after treatment had improved intestinal function and considerably less histologic injury compared to rats given low n-3 fatty acid diets after treatment, demonstrating that n-3 fatty acids, especially EPA, have protective effects against acetic acid–induced colitis.

Although studies using animal models provide strong evidence for the protective effects of n-3 fatty acids against induced IBD, animal models may not accurately portray the human etiology of this disease since it is induced using noxious chemicals. Several epidemiological studies have shown an inverse relationship between n-3 fatty acid intake and the risk of IBD. In addition, some intervention studies have shown that n-3 fatty acid supplementation is an effective therapeutic approach for management of these diseases [125]. A 24-year study showed that the Greenlandic people exhibited a significantly lower incidence of IBD when compared to Western populations, which was attributed to a diet rich in marine-derived n-3 fatty acids. Later, Shoda et al. [121] examined the incidence of Crohn disease and dietary habits among Japanese men and women over a 19-year period. This study showed that individuals with lower dietary n-6/n-3 ratios were 21% less likely to suffer from Crohn disease (RR 0.79). IBD sometimes exhibits alternating relapses and remissions, and some clinical studies have investigated the potential of n-3 fatty acids to prolong periods of remission. Belluzzi et al. [126] carried out a double-blind, placebo-based study to investigate the effects of 2.7 g/day of fish oil supplements in 78 patients with Crohn disease who were at high risk for relapse as assessed by the Crohn

Disease Activity Index. After 1 year, 59% of patients in the fish oil group remained in remission (23 out of 39) compared to 26% in the placebo group (10 out of 39). Further analysis revealed the difference in relapse rate between the two groups to be due to fish oil supplementation only; cigarette smoking, gender, previous surgery, age, and duration of the disease did not affect the likelihood of relapse. Also, examination of blood for indicators of inflammation (serum α_1-acid glycoprotein, serum α_2-globulin) revealed that the fish oil group had significant decreases in all inflammatory markers assayed compared to the control group after 1 year.

Not all studies have supported the therapeutic effects of n-3 fatty acids in IBD sufferers. For example, Lorenz-Meyer et al. [127] performed a double-blind, placebo-based trial on 204 patients with Crohn disease in remission to study the effects of highly concentrated n-3 PUFA on the maintenance of remission over a 12-month period. At the end of this trial there was no difference between the n-3 and control groups; specifically, 30% of patients in both groups remained in remission. However, at the end of this study it was noticed that the n-3 group required less drug therapy (prednisolone) to manage the disease compared to the control group. This result implies that n-3 fatty acid supplementation may be somewhat helpful in the treatment of Crohn disease. A recent clinical trial by Middleton et al. [128] of 63 ulcerative colitis patients studied the effects of a combination of fish- and plant-derived n-3 fatty acids on disease remission. After 12 months, the duration of remission was not significantly different ($p > 0.05$) between groups (n-3 group: 55% remained in remission, control group: 38% remained in remission). Based on these results, Middleton et al. [128] were not able to support the postulated therapeutic benefits of n-3 fatty acid supplementation, although there was a 17% increase in disease remission in the n-3 group. The insignificant effects of n-3 fatty acids may be due to the relatively low doses of n-3 fatty acids used in the study (1.9 g total).

There is a wealth of evidence both supporting and refuting the therapeutic potential of n-3 fatty acids for IFD. The conflicting results are most likely due to differences in study size, duration, source of n-3 fatty acids, and the amount of n-3 fatty acids provided. More animal studies are needed to develop a comprehensive biochemical basis for the theorized effects of n-3 fatty acid supplementation in the treatment of IBD.

The effects of n-3 fatty acid supplementation in patients with arthritis, particularly rheumatoid arthritis have been investigated. Kremer et al. [129] examined the effects of manipulating dietary fat intake on clinical outcomes in patients with rheumatoid arthritis. The n-3 group reported noticeable improvement (reduced morning stiffness and number of tender joints). The beneficial results were attributed to the intervention regimen conducted on the n-3 supplemented group by this group. Volker et al. [130] performed a randomized, placebo-based, double-blind clinical study to determine the effects of fish oil supplementation on clinical variables. After 15 weeks of supplementation, there was a significant improvement ($p < 0.02$) in the clinical status of patients in the n-3 group compared to the placebo group. Although trials by Kremer et al. [129] and Volker et al. [130] do provide evidence about therapeutic benefits for n-3 fatty acid supplementation in rheumatoid arthritis, neither of these clinical trials were long-term studies lasting 12 and 15 weeks, respectively. Geusens et al. [131] studied the long-term effects of n-3 fatty acid supplementation in patients with active rheumatoid arthritis in a 12-month double-blind, randomized study that included 90 subjects who were supplemented daily with either 2.6 g of fish oil, or 1.3 g of fish oil and 3 g of olive oil, or 6 g of olive oil. No dietary interventions were made. After a 12 month supplementation period only the fish oil group exhibited significant clinical improvements. Specifically, significant improvements in both the patient's evaluation of pain and the physician's assessment of pain occurred in the group receiving 2.6 g/day of fish oil. In addition, a significant number of patients in this group had reduced antirheumatic medication use throughout the 12-month trial. No significant improvements occurred in the combined fish and olive oil group or in the olive oil only group, implying that the observed therapeutic benefits of fish oil supplementation in patients with rheumatoid arthritis were dose dependent, with doses less than 2.6 g/day being ineffective.

Considerable evidence from *in vitro* and human studies suggest that n-3 fatty acids serve as effective therapeutic agents for the management of inflammatory arthritic diseases, but the biochemical basis for these observations are not well understood. However, it is likely that n-3 fatty acids exert their antiarthritic affects through modulation of inflammatory cytokine production. More in-depth knowledge of the roles of cytokines in inflammatory arthritic diseases is needed to understand how n-3 fatty acids influence this disease. Also, longer-term and large-scale intervention studies investigating the effects of n-3 fatty acid supplementation on arthritis symptoms are needed to strengthen the proposed inverse relationship between n-3 fatty acids and inflammatory arthritic diseases.

Several human studies have investigated the immunosuppressive effects of n-3 fatty acids in transplant patients. Homan van der Heide et al. [132] studied the effects of fish oil supplementation on kidney transplant acceptance and renal function. After 1 year there was an overall improvement of renal function in the fish oil supplemented group. The total number of rejection episodes was lower in the fish oil group compared to the control, as was mean arterial blood pressure. The authors of this study speculated that the observed hemodynamic and immunomodulatory effects of fish oil were due to a shift away from the vasoactive and proinflammatory AA eicosanoids to EPA-derived eicosanoids. However, a similar but more sophisticated study by Hernández et al. [133] investigated the effects of fish oil supplementation (6 g/day) on kidney function and kidney rejection rate as well as on proinflammatory cytokine production in 86 kidney transplant patients, and after 12 months no differences existed in the above parameters between the n-3 group (fish oil, experimental group) and control group (6 g/day of soybean oil). However, this study may have been complicated by the choice of soybean oil as the placebo fatty acid source. Soybean oil contains approximately 8% ALA, which may have benefited the soybean oil group and reduced the significance of differences between the experimental and control groups.

Studies on animals and humans investigating the potential immunosuppressive effects of postoperative n-3 fatty acid supplementation in organ transplant patients have been inconsistent. While some studies report significant therapeutic benefits, others do not. These conflicting results may be due to differences in study design such as amounts and sources of fatty acid supplements, the duration of study, and the type of organ transplant surgery studied. More clinical trials are needed to clearly support the beneficial effects of n-3 fatty acid immunonutrition in organ transplant patients.

The frequency of inflammatory lung diseases is increasing in Western societies [134]. Some have speculated that this trend may be due to high n-6 to n-3 fatty acid ratios in typical Western diets, which may cause increased proinflammatory cytokine production and lead to bronchial inflammation in those prone to inflammatory lung diseases. Case control studies indicate that children who do not consume fish early in life are three times more likely to have asthma than those who do [135]. Fish fatty acids, EPA and DHA, tend to reduce the incorporation of arachidonic acid into membrane phospholipids, and have been shown to decrease the production of proinflammatory arachidonic acid–derived eicosanoids. Thus, fish oil may have therapeutic effects on inflammatory lung disease symptoms.

Koch et al. [136] studied the effects of short-term infusions with PUFA emulsions on the pulmonary response to inflammatory stimulation (increased vascular resistance and permeability) in perfused rabbit lungs. Results for pulmonary artery pressure and lung weight gain (indicating edema formation) were significantly lower in the n-3 fatty acid group than in the control and n-6 fatty acid groups.

Several clinical and prospective studies have investigated the therapeutic potential of n-3 fatty acids on bronchial inflammation in asthmatics, most studied asthmatic children, or children at high risk of developing asthma. Hodge et al. [137] investigated the association between consumption of oily fish and recurrence of pulmonary wheeze in 584 previously diagnosed asthmatic children (8–11 years old) living in Sydney, Australia. There was a small but significantly decreased ($P < 0.05$) risk for current wheeze in children who consumed any amount of fresh fish or fatty fish (one or more fresh or fatty fish meals per week). These results remained significant after adjustment for other possible risk factors such as parental asthma, parental smoking, ethnicity, early respiratory illness, and sex. Although Hodge et al. [137] reported that oily fish consumption reduced the risk of asthma, this study only assessed asthma risk when questionnaires were distributed, it did not assess the prevalence or extent of asthma symptoms in subjects throughout the 5-month study period. Troisi et al. [138] examined the association between several dietary factors and adult onset of asthma in the 10-year Nurses Health Study (77,866 women). Their results revealed no association between n-3 or n-6 fatty acids and asthma, but positive associations were observed between asthma and antioxidant vitamin intake (vitamins C and E, β-carotene). These data suggest that n-3 fatty acid intake during adulthood is not an important determinant of asthma.

The symptoms of asthma are quite variable; questionnaires cannot adequately assess the prevalence and severity of asthma. Therefore, epidemiological studies are less reliable than clinical trials. Nagakura et al. [139] performed a 10-month placebo-based, randomized trial to evaluate the effects of fish oil capsules in 29 young patients (4–17 years old) with asthma who were receiving long-term treatment at the Department of Pediatrics, Higashi-Saitama Hospital, Japan. The results indicate that EPA reduces the bronchoconstrictive effects of acetylcholine but not the frequency or severity of asthma attacks. The authors of this study speculate that their results may be due to the ability of n-3 fatty acids to reduce the production and release of proinflammatory eicosanoids but not of histamine, which may explain the decreased response to inhaled allergens but unchanged overall asthma symptoms. Emelyanov et al. [140] performed an 8-week, placebo-based, randomized clinical trial to examine the effects of a lipid extract from New Zealand green-lipped mussel (rich in EPA and DHA) on the symptoms and biochemical markers of asthma in adults (18–56 years old). There were no significant differences in forced expiratory volume between the n-3 and placebo groups after 8 weeks of supplementation, but significant decreases in mean expired H_2O_2 ($p = 0.0001$) and mean daytime wheeze ($p = 0.026$) were observed after 8 weeks in the Lyprinol® group compared to placebo. Based on these results, Emelyanov et al. [140] concluded that Lyprinol supplementation (100 mg of n-3 fatty acids daily) improved symptom management in adult asthma sufferers. It was also noted that throughout this study no significant changes in blood pressure, serum creatinine, bilirubin, liver transaminase, or alkaline phosphatase occurred in either group, which indicates no ill effects of Lyprinol supplementation.

Thus, there is a considerable body of evidence that both support and refute the potential therapeutic benefits of n-3 fatty acid supplementation in asthma. To date, only one large-scale intervention study investigating n-3 fatty acid supplementation in asthmatics has been initiated; the results of this 5-year study are highly anticipated because most previous trials investigating n-3 fatty acid supplementation in asthmatics have been on a small scale. Table 2.8 provides a summary of some research findings on the effects of n-3 fatty acids on inflammatory immune responses.

TABLE 2.8
Research Findings on the Effects of n-3 Fatty Acids on Inflammatory Immune Responses

Condition/Disease Studied	Reference	Overall Findings
Cytokine production	1–5	The n-3 fatty acids may reduce proinflammatory cytokine production
Inflammatory bowel disease	7–15	Current findings show little consensus. More large-scale trials are needed
Arthritis	16–22	*In vitro* studies are promising, however, few large-scale studies have been conducted
Organ transplantation	23–26	Animal and human findings do not always agree
Bronchial inflammation	27	Results from some human studies are promising

2.2.5 Omega-3 Fatty Acids in Mental Health and Neural Function

Excluding adipose tissue, the human nervous system has the highest lipid content compared to all other tissues; 50–60% of the total dry weight of the adult human brain is lipid [141] and approximately one-third of these lipids are n-3 fatty acids, mostly DHA [142]. DHA is especially important during prenatal human brain development; incorporation of DHA into growing neurons is a prerequisite for synaptogenesis (formation of synapses) [143]. The period of greatest brain development occurs from the third trimester of pregnancy until 18 months after birth; this period correlates well with the accumulation of DHA in this organ [144]. The importance of n-3 fatty acids during prenatal development is best indicated by the observation that deficiency of these fatty acids during development greatly increases the likelihood of diminished visual acuity, cerebellar dysfunction and several cognitive impairments and neurological disorders [145]. The importance of n-3 fatty acids during human development is also evident by the fact that both the placenta and mammary tissues supply large amounts of DHA to the developing fetus and infants [146].

The effects of n-3 fatty acids on the clinical symptoms of depression and schizophrenia have received considerable attention. Depression is the most prevalent psychiatric disorder in North America; in the United States 1 in 20 people suffer from unipolar depression and 1 in 100 experience bipolar or manic depression [147]. Omega-3 fatty acid supplementation is receiving much attention as a possible adjunct therapy for depression; epidemiological and clinical studies suggest inverse association between n-3 fatty acid consumption and depression. Recent studies among Inuit populations show an overall decline in mental health characterized by increased rates of depression as well as other mental illnesses, which may be linked to the rapid changes to a westernized culture including a shift from traditional seafoods to processed foods [148]. Tanskanen et al. [149] performed a large survey to assess depression symptoms and frequency of fish intake among a cohort of 3204 Finnish adults aged 25–64 years. The results of this survey showed that mild to severe depression symptoms were 31% more prevalent among infrequent fish consumers (<3 fish meals per month); gender-based assessments reached significance among female participants but not in men ($p<0.01$, chi-square tests). Thus, results from this large cohort of adults report strong correlations between infrequent fish consumption and depression, but unfortunately this study could not investigate the effects of n-3 fatty acids since information of the type of fish consumed was not obtained. Recently, Marangell et al. [150] evaluated the effectiveness of DHA supplementation for the treatment of depression. The difference between groups did not reach statistical significance (two-way t-tests), however, these results may be due to the relatively low level of n-3 fatty acids supplemented. A similar but smaller clinical trial by Su et al. [151] showed that daily supplementation with 9.6 g fish oil for 8 weeks significantly reduced depressive symptoms compared to a placebo group ($p<0.05$, Wilcoxon signed rank test).

Although the etiology of depression is not completely understood, several pathophysiological features of this disease have been identified, including overproduction of inflammatory cytokines [152]. The beneficial effects of n-3 fatty acids in depression may be due to modulation of eicosanoid production. The fact that EPA-derived eicosanoids are the least proinflammatory ones provides a possible explanation for the beneficial effects of n-3 fatty acid supplementation in depression, since depression has been linked to proinflammatory cytokine production. Because the eicosanoid products of n-3 fatty acids do not activate macrophages to any extent compared to those derived from n-6 fatty acids, replacement of membrane n-6 fatty acids with n-3 fatty acids would reduce proinflammatory cytokine production, especially if cyclooxygenase activity is enhanced in depressive patients. Several lines of evidence support beneficial effects of n-3 fatty acids on depressive disorders, but this evidence is far from conclusive and it is premature to recommend n-3 fatty acid supplementation for the treatment of depressive symptoms.

Many research groups have evaluated the effects of n-3 fatty acid supplementation in schizophrenia. Schizophrenia is a mental disorder that affects 1% of all people regardless of race or nationality. Previous family history of schizophrenia is the major risk factor for this disease; however, oxidative injury to neuronal cells and abnormal neuronal membrane phospholipid composition have been observed in schizophrenic patients postmortem. Reduced DHA levels have been observed in neurons of schizophrenic patients that may be the result of phospholipase A_2 overexpression [153]. Many schizophrenic patients show signs of excessive *in vivo* lipid peroxidation; these include increased plasma thiobarbituric acid reactive substances (TBARS) [154] and breath pentane [155], which suggests that reduced DHA in schizophrenics may be due to increased oxidative stress. The potential role of dietary n-3 fatty acids on antioxidant enzymes and parameters of oxidative stress was recently studied in rat neurons [156]. Results showed that n-3 fatty acid–treated rats had significantly lower TBARS in corpus striatum neurons ($p<0.001$, ANOVA) and significantly less corpus striatum nitric oxide levels compared to controls ($p<0.002$, ANOVA). Omega-3 fatty acid–treated rats had significantly lower corpus striatum xanthine oxidase activity ($p<0.005$, ANOVA). These results indicate that n-3 fatty acids can improve oxidant parameters in normal neural tissue, and thus may reduce the prooxidative symptoms observed in schizophrenia. Hibbeln et al. [157] quantified the erythrocyte fatty acid compositions of 76 medicated schizophrenic patients before and after 16 weeks of EPA (3 g/day) or placebo supplementation. Several schizophrenic indices were performed on each patient before and after the supplementation period. Although plasma EPA levels were increased in the n-3 fatty acid group ($p<0.05$, Mann–Whitney tests), these differences did not correlate with reduced schizophrenia symptoms. Table 2.9 provides a summary of some research findings on the effects of n-3 fatty acids on mental health and neural development/function.

2.2.6 Omega-3 Fatty Acids and Gene Expression

Several reports indicate that the genomic effects of long-chain PUFA are mediated through specific interactions with hydrophobic binding sites on transcription factors; the earliest of such studies began with the discovery that peroxisome

TABLE 2.9
Summary of Research Findings on the Effects of n-3 Fatty Acids on Mental/Neural Health

Condition/Disease Studied	Reference	Overall Findings
Visual development	69–76	The n-3 fatty acids may improve visual function scores in some indices of visual development
Depression	79–88	The n-3 fatty acids exert beneficial actions on some depressive symptoms
Schizophrenia	89–93	Some evidence exists showing therapeutic effects of n-3 fatty acids on symptoms of schizophrenia

proliferator activated receptors (PPARs) are regulated by long-chain PUFA [158]. Several other fatty acid–regulated transcription factors have since been identified, including the hepatic nuclear factor 4α (HNF4α) [159], retinoid X receptor (RXR) [160], and liver X receptors (LXR) [161]. Experimental studies show that the level of sterol regulatory element binding protein-1 (SREBP1) synthesis and activation is modulated by fatty acids.

2.2.6.1 Omega-3 Fatty Acids and Peroxisome Proliferator Activated Receptors

The peroxisome proliferator activated receptors (PPARs isoforms: α, β, γ1, γ2, δ) are the best understood fatty acid–specific transcription factors; currently they are regarded as intracellular monitors of nonesterified fatty acid levels [162]. The PPARs regulate genes involved in fatty acid and glucose oxidation, fatty acid uptake, fatty acid activation, triacylglycerol biosynthesis, and lipoprotein metabolism; recent studies suggest that PPARs may be involved in other cellular functions including growth, differentiation, and proliferation [163]. The PPAR family are nuclear receptors whose DNA-binding affinities are enhanced when in complex with n-3 long-chain PUFA. Activated PPARs bind with peroxisome proliferator response elements (PPREs), which are promoter proximal regulatory elements located near initiator sequences of many eukaryotic genes. Activated PPARs bind to DNA as heterodimers with retinoic X receptors (RXRs); n-3 fatty acids have been shown to promote the dimerization of PPARs and RXRs [164].

All PPAR isoforms possess a fatty acid–binding activity; C_{18} and C_{20} fatty acids bind with greatest affinity [165]. Xu et al. [165] examined the three-dimensional structure of activated PPARγ and showed that fatty acids occupy a 1300 cubic Angstrom hydrophobic binding site. The binding of EPA changes the three-dimensional conformation of PPARγ and enhances its DNA binding affinity. The cellular alterations mediated by EPA would lead to decreased expression of genes involved in fatty acid and triacylglycerol biosynthesis while inducing genes involved in fatty acid oxidation. Ren et al. [166] used an animal model to examine the role of PPARα on fatty acid regulation of hepatic lipid metabolism by assessing hepatic mRNA transcript levels for fatty acid synthase, acyl-CoA oxidase, and CYP4A2 (CYP4A2 is one enzyme in the biosynthetic pathway of bile acids from cholesterol). Results indicated that PPARα is required for the n-3 fatty acid mediated upregulation of the hepatic CYP4A2 gene and the acyl-CoA oxidase gene, and that fish oil-mediated suppression of lipogenic genes do not involve PPARα. To further assess the effects of fatty acids on hepatic PPARα activity, Ren et al. [166] isolated primary hepatocytes from rat liver samples by digesting the livers with collagenase. Ren et al. [166] then treated the hepatocytes with pure fatty acids for 48 h and assessed the acyl-CoA oxidase mRNA levels. After the incubation period, cultures treated with oleic acid, LA, ALA, GLA (18:3, n-6), and AA showed no significant changes in acyl-CoA oxidase mRNA levels when compared to control cultures. Primary hepatocytes treated with EPA showed a twofold increase in acyl-CoA oxidase mRNA levels compared to control cultures, implying that only long-chain PUFA are able to induce acyl-CoA oxidase gene transcription. The results of Ren et al. [166] collectively imply that long-chain n-3 PUFA in fish oils promote hepatic fatty acid oxidation and bile acid synthesis through activation of PPARα transcription factor. Over time, these genomic effects of n-3 fatty acids may lead to reduced blood lipid levels and could affect the symptoms of hyperlipidemia and hypercholesterolemia.

Although several studies show that EPA enhances the transcriptional activity of the PPAR family of transcription factors, most of these studies are performed using *in vitro* models. Very few studies examine the genomic effects of dietary n-3 fatty acids, thus more live animal studies are needed to lend further support to the existing studies that show n-3 fatty acids, particularly EPA, activate PPARs. It still remains to be clarified whether the upregulation of PPARs by n-3 fatty acids operate via transcriptional or posttranscriptional mechanisms.

2.2.6.2 Omega-3 Fatty Acids and Sterol Regulatory Element Binding Proteins

Sterol regulatory elements binding proteins (SREBPs) are transcription factors that regulate the transcription of several genes involved in lipid, cholesterol, bile acid, and lipoprotein biosynthesis. Evidence from animal and cell culture studies indicate that SREBP1 isoforms regulate fatty acid and triacylglycerol biosynthesis while SREBP2 regulates cholesterol biosynthesis. Animal studies show that transgenic mice overexpressing SREBP1a or SREBP1b have higher transcription rates of hepatic genes involved in lipogenesis, triacylglycerol biosynthesis, and very low density lipoprotein secretion, and develop fatty liver [167]. Unlike PPARs, unsaturated fatty acids have not been shown to directly bind with SREBPs, but unsaturated fatty acids do modulate the activity and abundance of SREBP1, which in turn affects lipogenic gene expression.

Several animal studies show that dietary n-3 fatty acids suppress hepatic lipogenesis by inhibiting SREBP1 gene transcription and proteolytic activation, as well as increasing $SREBP1_{mRNA}$ decay [168]. In one study comparing hepatic levels of mature SREBP1 (nuclear SREBP or nSREBP) and precursor SREBP1 (pSREBP1, inactive membrane bound form) in mice fed either normal diets or diets rich in fish oils, it was shown that mice in the fish oil group had 90% lower hepatic levels of nSREBP1 (active transcription factor) and 75% lower hepatic levels of pSREBP1. The changes in the fish oil group were accompanied by a decrease in hepatic fatty acid synthase mRNA [169]. Unlike mice in the fish oil group, mice on diets rich in saturated and MUFA had normal hepatic SREBP1 levels and activity. Other animal studies have shown that n-3 fatty acids inhibit the transcription of several hepatic genes involved in glucose metabolism and lipogenesis, including glucokinase, acetyl-CoA carboxylase, and the Δ^5 and Δ^6 desaturases, which may be explained by n-3 fatty acid–mediated reductions in hepatic SREPB1 levels [170]. *In vitro* studies show that n-3 fatty acids reduce the nuclear content of activated SREBP1 in two separate ways. The first response is an inhibition of the proteolytic cleavage of pSREBP, which occurs within the first hour of n-3 fatty acid treatment. The second response to n-3 fatty acids is a reduction in SREBP1 gene transcription, which is accompanied by decreased nuclear or endoplasmic reticulum membrane levels of pSREBP1. The underlying molecular mechanisms by which n-3 fatty acids reduce cellular SREBP1 levels are not understood but it is believed that n-3 fatty acids somehow enhance the rate of $SREBP1_{mRNA}$ decay; studies using cultured liver cells have shown that n-3 fatty acids reduce the half-life of $SREBP1c_{mRNA}$ from 11 h to less than 6 h [171]. Omega-3 fatty acids have also been shown to inhibit lipogenic gene expression in cultured adipocytes; however this inhibition is SREBP1 independent [172].

2.2.6.3 Omega-3 Fatty Acids and Liver X Receptors

Liver X receptors (LXRα and LXRβ) are members of the nuclear hormone receptor superfamily; LXRα was recently identified as fatty acid–regulated transcription factor. LXRs are important regulators of cholesterol, bile acid, fatty acid, and triacylglycerol biosynthesis in tissues such as the liver, brain, and gonads [173]. Studies using human embryonic kidney (HEK), 293 cells have shown that unsaturated fatty acids bind to LXRα, antagonizing oxysterol activation of LXRα. However, a similar study by Pawar and Jump [162] showed that EPA had no effect on LXRα activity in two different cell lines or in rat primary hepatocytes. Animal model studies have not shown any change in LXRα activity or abundance in response to changes in dietary fatty acid composition.

Evidently, there is no consensus among the existing literature assessing unsaturated fatty acid regulation of LXRs; only a few research groups have investigated this recent observation. More studies are needed to assess the effects of several fatty acid types to clarify the possible regulatory actions of n-3 and n-6 fatty acids on LXRs.

2.2.6.4 Omega-3 Fatty Acids and Hepatic Nuclear Factor 4α (HNF4α)

Hepatic nuclear factor 4α (HNF4α) is a member of the steroid receptor superfamily; other members include the glucocorticoid receptor and the androgen receptor. HNF 4α is a transcription factor that enhances expression of several hepatic genes including CYP4A2, transferrin, apolipoprotein-CII, -CIII, -AII, -AIV, and pyruvate kinase [174]. Hertz et al. [174] were the first to report fatty acid regulation of HNF4α. Using cultured hepatocytes, Hertz et al. [174] showed that long-chain acyl-CoAs (C_{14} and up) at *in vivo* concentrations (approximately 2.6 μm) were ligands for HNF4α. Binding of saturated fatty acids enhanced the transcriptional activity of HNF4α, as evidenced by increased transferrin and apolipoprotein CIII expression. However, ALA, EPA, and DHA as their CoA thioesters inhibited HNF4α activity. Long-chain n-3 PUFA exert antilipogenic effects and promote lipid oxidation in hepatocytes.

Clearly, n-3 fatty acids modulate the activity and abundance of at least three transcription factor families (PPARs, SREBP, and HNF4α), which play important roles in hepatic fatty acid, cholesterol, apolipoprotein, and carbohydrate metabolism. Omega-3 fatty acid regulation of PPARs and SREBP1 isoforms has been well established through a number of animal and cell culture studies. The significance of fatty acid regulation of HNF4α *in vivo* is debatable, but the evidence that highly unsaturated acyl-CoAs may indeed influence HNF4α activity and is a topic that deserves further investigation. Based on the current literature, the genomic effects of n-3 fatty acids on hepatic metabolism involve a shift from TAG synthesis, storage, and apolipoprotein secretion toward hepatic oxidation of lipids. In turn, this response may reduce blood levels of TAG, cholesterol, and lipoproteins such as LDL, all of which are important risk factors for several chronic diseases. Further *in vivo* studies are needed to understand the mechanisms by which n-3 fatty acids enhance hepatic lipid oxidation while concurrently decreasing hepatic lipid storage. Elucidation of these pathways may provide evidence for novel therapeutic strategies for blood lipid and cholesterol disorders.

REFERENCES

1. Din, J.N., Newby, D.E. and Flapan, A.D. 2004. Omega 3 fatty acids and cardiovascular disease—fishing for a natural treatment. *Br. Med. J.* 328, 30–35.
2. Hulshof, K.F.A.M., van Erp-Baart, M.A., Anttolainen, M., Becker, W., Church, S.M., Couet, C., Hermann-Kunz, E., Kesteloot, H., Leth, T., Martins, I., Moreiras, O., Moschandreas, J., Pizzoferrato, L., Rimestad, A.H., Thorgeirsdottir, van Amelsvoort, J.M.M., Aro, A., Kafatos, A.G., Lanzmann-Petithory, D. and van Poppel, G. 1999. Intake of fatty acids in Western Europe with emphasis on trans fatty acids: the transfair study. *Eur. J. Clin. Nutr.* 53, 143–157.
3. Newton, I. and Snyder, D. 1997. Nutritional aspects of long-chain omega-3 fatty acids and their use in bread enrichment. *Cereal Foods World* 42, 126–131.
4. Kamal-Eldina, A. and Yanishlievab, N.V. 2002. N-3 fatty acids for human nutrition: stability considerations. *Eur. J. Lipid Sci. Technol.* 104, 825–836.
5. Watanabe, T. and Ackman, R.G. 1974. Lipids and fatty acids of the American (*Crassostrea virginica*) and European flat (*Ostrea edulis*) oysters from a common habitat and after one feeding with *Dicrateria inornata* or *Isochrysis galbana*. *J. Fish. Res. Board Can.* 31, 403–409.
6. Ackman, R.G. 1982. Fatty acid composition of fish oils, in *Nutritional Evaluation of Long-Chain Fatty Acids in Fish Oils*, Barlow, S.M. and Stansby, M.E., Eds., Academic Press, London, UK, pp. 25–88.

7. Myher, J.J., Kuksis, A., Geher, K., Park, P.W. and Diersen-Schade, D.A. 1996. Stereospecific analysis of triacylglycerols rich in long-chain polyunsaturated fatty acids. *Lipids* 31, 207–215.
8. Mayzaud, P., Virtue, P. and Albessard, E. 1999. Seasonal variations in the lipid and fatty acid composition of the euphausiid *Meganyctiphanes norvegica* from the Ligurian Sea. *Mar. Ecol. Prog. Ser.* 186, 199–210.
9. Tanabe, T., Suzuki, T., Ogura, M. and Watanabe, Y. 1999. High proportion of docosahexaenoic acid in the lipid of juvenile and young skipjack tuna, *Katsuwonus pelamis* from the tropical western Pacific. *Fish. Sci.* 65, 806–807.
10. Arts M.T., Ackman R.G. and Holub, B.J. 2001. Essential fatty acids in aquatic ecosystems: a crucial link between diet and human health and evolution. *Can. J. Fish Aquat. Sci.* 581, 122–137.
11. Oliveira, A.C.M. and Bechtel, P.J. 2005. Lipid composition of Alaska Pink Salmon (*Oncorhynchus gorbuscha*) and Alaska Walleye Pollock (*Theragra chalcogramma*) byproducts. *J. Aquat. Food Prod. Technol.* 14 (1), 73–89.
12. Aliam, S.S.M. 2003. Long chain polyunsaturated fatty acids, nutritional and healthy aspects. Review article. *Rivista Italiana delle sostanze grasse*. 80, 85–92.
13. Gebhardt, S.E. and Thomas, R.G. 2002. Nutritive value of foods. *U.S. Dep. Agric. Home Garden Bull.* 72, 1–103.
14. Shahidi, F. 2002. Marine nutraceuticals. *Inform* 13, 57–62.
15. Wanasundara, U.N., Shahidi, F. and Amarowicz, R. 1998. Effect of processing on constituents and oxidative stability of marine oils. *J. Food Lipids* 5, 29–41.
16. Aidos, I., Masbernat-Martinez, S., Luten, J.B., Boom, R.M. and Padt, A.V. 2002. Composition and stability of herring oil recovered from sorted byproducts as compared to oil from mixed byproducts. *J. Agric. Food Chem.* 50, 2818–2824.
17. Sathivel, S., Prinyawiwatkul, W., King, J.M., Grimm, C.C. and Lloyd, S. 2003. Oil production from catfish viscera. *J. Am. Oil Chem. Soc.* 80, 377–382.
18. Jayasinghe, C., Gotoh, N., Tokairin, S., Ehara, H. and Wada, S. 2003. Inter species changes of lipid compositions in liver of shallow-water sharks from the Indian Ocean. *Fish. Sci.* 69, 644–653.
19. Shahidi, F. 1998. Seal blubber, in *Seal Fishery and Product Development,* Shahidi, F., Ed., ScienceTech, NL, Canada, pp. 99–146.
20. Kalogeropoulos, N., Andrikopoulos, N.K. and Hassapidou, M. 2004. Dietary evaluation of Mediterranean fish and mollusks pan-fried in virgin olive oil. *J. Sci. Food Agric.* 84, 1750–1758.
21. Copeman, L. and Parrish, C.C. 2004. Lipid classes, fatty acids, and sterols in seafood from Gilbert Bay, Southern Labrador. *J. Agric. Food Chem.* 52, 4872–4881.
22. Ackman, R.G. 2005. Fish oils, in *Bailey's Industrial Oil and Fat Products*, Vol. 3, Shahidi, F., Ed., John Wiley, New York, pp. 279–317.
23. Heller, J.H., Heller, M.S., Springer, S. and Clark, E. 1957. Squalene content of various shark livers. *Nature* 179, 919–920.
24. Engelhardt, F.R. and Walker, B.L. 1974. Fatty acid composition of the harp seal, *Pagophilus groenlandicus* (*Phoca groenlandica*). *Comp. Biochem. Physiol. B.* 47, 169–179.
25. Park, Y., Kelleher, S.D., Mcclements, D.J. and Decker, E.A. 2004. Incorporation and stabilization of omega-3 fatty acids in surimi made from cod, *Gadus morhua*. *J. Agric. Food Chem.* 52, 597–601.
26. Rora, A.M.B., Birkelandb, S., Hultmannc, L., Rustadc, H., Rab, S.T. and Bjerkeng, B. 2005. Quality characteristics of farmed Atlantic salmon (*Salmo salar*) fed diets high in soybean or fish oil as affected by cold-smoking temperature. *L.W.T.* 38, 201–211.

27. Holmer, G.K. 1989. Triglycerides, in *Marine Biogenic Lipids, Fats, and Oils,* Vol. 1, Ackman, R.G., Ed., CRC Press, Boca Raton, FL, pp. 139–174.
28. Shahidi, F., Synowiecki, J., Amarowicz, R. and Wanasundara, U. 1994. Omega-3 fatty acid composition and stability of seal lipids, in *Lipids in Food Flavors,* Ho, C.T. and Hartaman, T.G., Eds., ACS Symposium Series 558, American Chemical Society, Washington, DC, pp. 233–243.
29. Lee, R.F. and Patton, J.S. 1989. Alcohol and waxes, in *Marine Biogenic Lipids, Fats, and Oils,* Vol. 1, Ackman, R.G., Ed., CRC Press, Boca Raton, FL, pp. 73–102.
30. Gruger, E.H. Jr. 1967. Fatty acid composition, in *Fish Oils*, Stansby, M.E., Ed., The AVI Publishing Company, Westport, CT, pp. 3–30.
31. Hansen, I.A. and Cheah, C.C. 1969. Related dietary and tissue lipids of the sperm whale. *Comp. Biochem. Physiol.* 31, 757–761.
32. Litchfield, C., Greenberg, A.J., Caldwell, D.K., Caldwell, M.C., Sipos, J. and Ackman, R.G. 1975. Comparative lipid patterns in acoustical and nonacoustical fatty tissues of dolphins, porpoises and toothed whales. *Comp. Biochem. Physiol. B,* 50, 591–597.
33. Litchfield, C., Greenberg, A.J. and Mead, J.G. 1976. Distinctive character of zyphiidae head and blubber fats. *Cetology* 23, 1–10.
34. Nevenzel, J.C. 1970. Occurrence, function and biosynthesis of wax esters in marine organisms. *Lipids* 5, 308–319.
35. Sargent, J.R., Lee, R.F. and Nevenzel, J.C. 1976. Marine waxes, in *Chemistry and Biochemistry of Natural Waxes,* Kolattukudy, P., Ed., Elsevier, Amsterdam, pp. 49–91.
36. Sargent, J.R. 1978. Marine wax esters. *Sci. Prog. Oxford* 65, 437–458.
37. Karnovsky, M.L. and Rapson, W.S. 1947. Application of the Fuelson method of "Squalene" determination to some marine oils. *J. Soc. Chem. Ind. London* 66, 124–125.
38. Bottino, N.R. 1978. Lipids of the antarctic sei whale, *Balaenoptera borealis. Lipids* 13, 18.
39. Blumer, M. and Thomas, D.W. 1965. "Zameane," isomeric C19 monoolefins from marine zooplankton, fishes, and mammals. *Science* 148, 370–371.
40. Borch-Jensen, C. and Mollerup, J. 1996. Supercritical fluid chromatography of fish, shark and seal oils. *Chromatographia* 42, 252–258.
41. Ackman, R.G. 1989. Fatty acids, in *Marine Biogenic Lipids, Fats, and Oils,* Vol. 1, Ackman, R.G., Ed., CRC Press, Boca Raton, FL, pp. 103–138.
42. Bang, H.O., Dyerberg, J. and Hjoorne, N. 1976. Composition of food consumed by Greenland Eskimos. *Acta Med. Scand.* 200, 69–73.
43. Bang, H.O., Dyerberg, J. and Sinclair, H.M. 1980. Composition of the Eskimo food in north western Greenland. *Am. J. Clin. Nutr.* 33, 2657–2661.
44. Shahidi, F., Wanasundara, U.N. and Amarowicz, R. 1997. A novel edible marine oil and its stabilization, in *Chemistry and Novel Foods,* Spa nier, A.M., Tamura, M., Hideo, O., and Mills, D., Eds., Allured Publishing, Carol Stream, IL, pp. 111–124.
45. Ackman, R.G., Eaton, C.A. and Jangaard, P.M. 1965. Lipids of the finwhale (*Balaenoptera physalus*) from North Atlantic waters. *Can. J. Biochem.* 43, 1513–1520.
46. Wanasundara, U.N. 1997. Marine Oil: Stabilization, Structural Characterization and Omega-3 Fatty Acid Concentration. Ph.D. thesis, Memorial University of Newfoundland, St. John's, NL, Canada.
47. Yongmanichai, W. and Ward, O.P. 1989. Omega-3 fatty acids: Alternative sources of production. *Prog. Biochem.* 24, 117–125.
48. Ackman, R.G. and Lamothe, F. 1989. Marine mammals, in *Marine Biogenic Lipids, Fats, and Oils,* Vol. 2, Ackman, R.G., Ed., CRC Press, Boca Raton, FL, pp. 179–382.

49. Durnford, E. and Shahidi, F. 2002. Analytical and physical chemistry—comparison of FA compositions of selected tissues of phocid seals of Eastern Canada using one-way and multivariate techniques. *J. Am. Oil Chem. Soc.* 79, 1095–1102.
50. Durnford, E., Shahidi, F. and Ackman, R.G. 2003. Processing and engineering technology—letters to the editor—phthalates and the overestimation of docosanoic acid in seal lipids. *J. Am. Oil Chem. Soc.* 80, 405–406.
51. Grahl-Nielsen, O. and Mjaavatten, O. 1991. Dietary influence on fatty acid composition of blubber fat of seals as determined by biopsy: a multivariate approach. *Marine Biol.* 110, 59–64.
52. West, G.C., Burns, J.J. and Modafferi, M. 1979. Fatty acid composition of Pacific walrus skin and blubber fats. *Can. J. Zool.* 57, 1249–1255.
53. West, G.C., Burns, J.J. and Modafferi, M. 1979. Fatty acid composition of blubber from the four species of Bering Sea phocid. Seals, *Can. J. Zool.* 57, 189–195.
54. Ackman, R.G., Hooper, S.N. and Hingley, J. 1972. The harbor seal *Phoca vitulina* concolor De Kay: comparative details of fatty acids in lung and heart phospholipids and triglycerides. *Can. J. Biochem.* 50, 833–838.
55. Wanasundara, U.N. and Shahidi, F. 1997. Structural characteristics of marine lipids and preparation of omega 3 concentrates, in *Flavor and Lipid Chemistry of Seafoods*, Shahidi, F. and Cadwallader, K.K., Eds., American Chemical Society Symposium Series 674, Washington DC, pp. 240–254.
56. Mag, T. 2000. Patent WO00/44862A1, PCT International Patent Application.
57. Sijtsma, L. and de Swaaf, M.E. 2004. Biotechnological production and applications of the ω-3 polyunsaturated fatty acid docosahexaenoic acid. *Appl. Microbiol. Biotechnol.* 64, 146–153.
58. Alexander, J.W. 1998. Immunonutrition: the role of omega-3 fatty acids. *Nutrition* 1, 627–633.
59. Howe, P.R.C. 1997. Dietary fats and hypertension: focus on fish oil. *Ann. NY Acad. Sci.* 827, 339–352.
60. Krishna Mohan, I. and Das, U.N. 2001. Prevention of chemically induced diabetes mellitus in experimental animals by polyunsaturated fatty acids. *Nutrition* 17, 126–151.
61. Babcock, T., Helton, W.S. and Espat, N.J. 2000. Eicosapentaenoic acid (EPA): an antiinflammatory ω-3 fat with potential clinical applications. *Nutrition* 16, 1116–1118.
62. Kelly, D.S. 2001. Modulation of human immune and inflammatory responses by dietary fatty acids. *Nutrition* 17, 669–673.
63. Rose, D.P. and Connolly, J.M. 1999. Omega-3 fatty acids as cancer chemopreventive agents. *Pharmcol. Ther.* 83, 217–244.
64. Akihisa, T., Tokuda, H., Ogata, M., Ukiya, M., Iizuka, M., Suzuki, T., Metori, K., Shimizu, N. and Nishino, H. 2004. Cancer chemopreventive effects of polyunsaturated fatty acids. *Cancer Lett.* 2–5, 9–13.
65. Anderson, G.J., Connor, W.E. and Corliss, J.D. 1990. Docosahexaenoic acid is the preferred dietary n-3 fatty acid for the development of the brain and retina. *Pediatr. Res.* 27, 89–97.
66. Hansen, J.C., Sloth Pedersen, H. and Mulvad, G. 1994. Fatty acids and antioxidants in the Inuit diet. Their role in ischemic heart disease (IHD) and possible interactions with other dietary factors. A review. *Arctic. Med. Res.* 53, 4–17.
67. Deutch, B., Bonefeld Jorgensen, E.C. and Hansen, J.C. 2000. N-3 PUFA from fish or seal oil reduce atherogenic risk indicators in Danish women. *Nutr. Res.* 20, 1065–1077.

68. Bonefeld Jorgensen, E.C., Moller, S.M. and Hansen, J.C. 2001. Modulation of atherosclerotic risk factors by seal oil: a preliminary assessment. *Int. J. Circumpolar Health* 60, 25–33.
69. Kimoto, H., Endo, Y. and Fujimoto, K. 1994. Influence of interesterification on the oxidative stability of marine oil triacylglycerols. *J. Am. Oil Chem. Soc.* 71, 69–473.
70. Osterud, B., Elvevoll, E., Barstad, H., Brox, J., Halvorsen, H., Lia, K., Olsen, J.O., Olsen, R.L., Sissener, C., Rekdal, O. and Vognild, E. 1995. Effect of marine oils supplementation on coagulation and cellular activation in whole blood. *Lipids* 30, 1111–1118.
71. de Lorgeril, M., Salen, P., Laporte, F. and de Leiris, J. 2002. Alpha-linolenic acid in the prevention and treatment of coronary heart disease. *Eur. Heart J.* 3, D26–D32.
72. Dolocek, T.A. and Granditis, G. 1991. Dietary polyunsaturated fatty acids and mortality in multiple risk factor intervention trial (MRFIT). *World Rev. Nutr. Diet.* 66, 205–216.
73. Albert, C.M., Campos, H., Stampfer, M.J., Ridker, P.M., Manson, J.E., Willett, W.C. and Ma, J. 2002. Blood levels of long-chain n-3 fatty acids and the risk of sudden death. *Engl. J. Med.* 346, 1113–1118.
74. Hu, F.B., Bronner, L., Willet, W.C., Stampfer, M.J., Rexrode, K.M., Albert, C.M., Hunter, D. and Manson, J.E. 2002. Fish and omega-3 fatty acid intake and risk of coronary heart disease in women. *J. Am. Med. Assoc.* 287, 1815–1821.
75. Sinclair, H.M. 1956. Deficiency of essential fatty acids and atherosclerosis, etcetera. *Lancet* 267, 381–383.
76. Bang, H.O., Dyerberg, J. and Sinclair, H.M. 1980. The composition of the Eskimo food in north western Greenland. *Am. J. Clin. Nutr.* 33, 2657–2661.
77. Fischer, S. and P.C. 1984. Webber, prostaglandin I_3 is formed in-vivo in man after dietary eicosapentaenoic acid. *Nature* 307, 165–168.
78. Hirai, A., Terano, T., Tamura, Y. and Yoshida, S. 1989. Eicosapentaenoic acid and adult diseases in Japan: epidemiological and clinical aspects. *Int. Med. Suppl.* 225, 69–75.
79. Davidson. M., Bulkow, L.R. and Gellin, B.G. 1993. Cardiac mortality in Alaska's indigenous and non-native residents. *Int. J. Epidemiol.* 22, 62–71.
80. Siess, W., Roth, P., Scherer, B., Kurzmann, I., Bohlig, B. and Weber, P.C. 1980. Platelet membrane fatty acids, platelet aggregation, and thromboxane formation during a mackerel diet. *Lancet* 1, 441–444.
81. Coker, S.J. and Parratt, J.R. 1985. AH23848, A thromboxane antagonist, suppresses ischemia and reperfusion induced in anaesthetized greyhounds. *Br. J. Pharmacol.* 86, 259–264.
82. Billman, G.E., Kang, J.X. and Leaf, A. 2000. Prevention of ischemia-induced cardiac sudden death by pure omega-3 polyunsaturated fatty acids in dogs. *Circulation* 99, 2452–2457.
83. Kang, J.X. and Leaf, A. 2000. Prevention of fatal cardiac arrhythmias by polyunsaturated fatty acids. *Am. J. Clin. Nutr.* 71, 202–207.
84. de Lorgeril, M., Salen, P., Martin, J.L., Monjaud, I., Boucher, P. and Mamelle, N. 1998. Mediterranean dietary pattern in a randomized trial prolonged survival and possible reduced cancer rate. *Arch. Int. Med.* 158, 1181–1187.
85. Theroux, P., Chaitman, B.R., Danchin, N., Erhardt, L., Meinertz, T., Schroeder, J.S., Togoni, G., White, H.D., Willerson, J.T. and Jessel, A. 2000. Inhibition of the sodium/hydrogen exchanger with cariporide to prevent myocardial infraction in high risk ischemic situations. Main results of the GUARDIAN trial. *Circulation* 102, 3032–3038.

86. Amusquivar, E., Ruperez, F.J., Barbas, C. and Herrera, E. 2000. Low arachidonic acid rather than alpha-tocopherol is responsible for the delayed postnatal development in offspring of rats fed fish oil instead of olive oil during pregnancy and lactation. *J. Nutr.* 130, 2855–2865.
87. Goel, D.P., Maddaford, T.G. and Pierce, G.N. 2002. Effects of omega-3 polyunsaturated fatty acids on cardiac sarcolemmal Na(+)/(H+) exchange. *Am. J. Physiol. Heart Circul. Physiol.* 283, H1688–H1694.
88. Chaudry, A.A., Wahle, K.W., McClinton, S. and Moffat, S.E. 1994. Arachidonic acid metabolism in benign and malignant prostatic tissue *in vitro*: effects of fatty acids and cyclooxygenase inhibitors. *Int. J. Cancer Res.* 57, 176–180.
89. Burr, M.L., Fehily, A.M., Gilbert, J.M., Rodgers, S., Holliday, R.M., Sweetnam, P.M., Elwood, P.C. and Deadman, N.M. 1989. Effects of changes in fat, fish, and fibre intakes on death and myocardial reinfarction: diet and reinfarction trial (DART). *Lancet* 2, 757–761.
90. Thun, M.J., Namboodiri, M.M. and Heath, C.W. Jr. 1991. Aspirin use and reduced risk of fatal colon cancer. *Engl. J. Med.* 325, 1593–1596.
91. GISSI-Prevenzione Study Investigators 1999. Dietary supplementation with n-3 polyunsaturated fatty acids and vitamin E in 11,324 patients with myocardial infraction: results of the GISSI-prevenzione trial. *Lancet* 354, 447–455.
92. Narayanan, B.A., Narayanan, N.K. and Reddy, B.S. 2001. Docosahexaenoic acid regulated genes and transcription factors inducing apoptosis in human colon cancer cells. *Int. J. Oncol.* 19, 1255–1262.
93. Prener, A., Storm, H.H. and Nielsen, N.H. 1996. Cancer of the male genital tract in Circumpolar Inuit. *Acta Oncologica Stockholm*, Sweden 35, 589–593.
94. Bhagavathi, A., Narayanan, B., Narayanan, K., Simi, B. and Reddy, B.S. 2003. Modulation of inducible nitric oxide synthase and related proinflammatory genes by the omega-3 fatty acid docosahexaenoic. *Acid. Cancer Res.* 63, 972–979.
95. Terry, P., Lichtenstein, P., Feychting, M., Ahlbom, A. and Wolk, A. 2001. Fatty fish consumption and risk of prostate cancer. *Lancet* 357, 1764–1766.
96. Mamalakis, G., Kafatos, A., Kalogeropoulos, N., Andrikopoulos, N., Daskalopulos, G. and Kranidis, A. 2002. Prostate cancer vs hyperplasia: relationships with prostatic and adipose tissue fatty acid composition. prostaglandins, leukot. *Essent. Fatty Acids* 66, 467–477.
97. Augustsson, K., Michaud, D.S., Rimm, E.B., Leitzmann, M.F., Stampfer, M.J., Willett, W.C. and Giovannucci, E.A 2003. Prospective study of intake of fish fatty acids and prostate cancer. *Cancer Epidemiol. Biomarkers Prev.* 12, 64–67.
98. Rose, D.P. and Connolly, J.M. 1999. Omega-3 fatty acids as cancer chemopreventative agents. *J. Natl. Cancer Inst.* 83, 217–244.
99. Reddy, B.S. 1994. Chemoprevention of colon cancer by dietary fatty acids. *Cancer Metastasis Rev.*13, 285–302.
100. Wynder, E.L., Kajitani, T., Ishikawa, S., Dodo, H. and Takano, A. 1969. Environmental factors of cancer of the colon and rectum. II. Japanese epidemiological data. *Cancer* 12, 1210–1220.
101. Caygill, C.P., Charlett, A. and Hill, M.J. 1996. Fat, fish oil and cancer. *Br. J. Cancer* 74, 159–164.
102. Anti, M., Marra, G., Amelao, F., Bartoli, G.M., Ficarelli, R., Percepce, A. De Vetis, I., Maria, G., Sofo, L. and Rapaccini, G.L. 1992. Effect of omega-3 fatty acids on rectal mucosa cell proliferation in subjects at high risk for colon cancer. *Gastroenterology* 103, 883–891.

103. Takahashi, M., Fukutake, M., Isoi, T., Fukuda, K., Sato, H., Yazawa, K., Sugimura, T. and Wakabayashi, K. 1997. Suppression of azomethane-induced rat colon carcinoma development by a fish oil component, docosahexaenoic acid (DHA). *Carcinogenesis* 18, 1337–1342.
104. Dwivedi, C., Muller, L.A., Goetz-Parten, D.E., Kasperson, K. and Mistry, V.V. 2003. Chemopreventitive effects of dietary mustard oil on Colon Tumour development. *Cancer Lett.* 196, 29–34.
105. Dommels, Y.E.M., Haring, M.G.M., Keestra, N.G.M., Alink, G.M., van Balderen, P.J. and van Ommen, B. 2003. The role of cyclooxygenase in n-6 and n-3 polyunsaturated fatty acid mediated effects on cell proliferation, PGE_2 synthesis and cytotoxicity in human colorectal carcinoma cell lines. *Carcinogenesis* 24, 385–392.
106. Narayanan, B.A., Narayanan, N.K. and Reddy, B.S. 2001. Docosahexaenoic acid regulated genes and transcription factors inducing apoptosis in human colon cancer cells. *Int. J. Oncol.* 19, 1255–1262.
107. Bhagavathi, A., Narayanan, B., Narayanan, K., Simi, B. and Reddy, B.S. 2003. Modulation of inducible nitric oxide synthase and related proinflammatory genes by the omega-3 fatty acid docosahexaenoic acid. *Cancer Res.* 63, 972–979.
108. Holmes, M.D., Stampfer, M.J., Colditz, G.A., Rosner, B., Hunter, D.J. and Willet, W.C. 1999. Dietary factors and the survival of women with breast carcinoma. *Cancer* 86, 826–835.
109. Holmes, M.D., Colditz, G.A., Hunter, D.J., Hakinson, S.E., Rosner, B., Speizer, F.E. and Willet, W.C. 2003. Meat, fish and egg intake and risk of breast cancer. *Int. J. Cancer Res.* 104, 221–227.
110. Bougnoux, P. 1999. N-3 Fatty acids and cancer. *Curr. Opin. Clin. Nutr. Metab. Care* 2, 121–126.
111. Maillard, V., Bougnoux, P., Ferrari, P., Jourdain, J.L., Pinault, M., Lavillonniere, F., Body, G., Le Floch, O. and Chajes, V. 2002. N-3 and N-6 fatty acids in breast adipose tissue and relative risk of breast cancer in a case control study in Tours, France. *Int. J. Cancer Res.* 98, 78–83.
112. Ip, C. 1997. Review of the effects of trans fatty acids, oleic acid, n-3 polyunsaturated fatty acids and conjugated linoleic acid on mammary carcinogenesis in animals. *Am. J. Clin. Nutr.* 66, 1523S–1529S.
113. Kort, W.J., Weijma, I.M., Vergroesen, A.J. and Westbroek, D.L. 1987. Conversion of diets at tumour induction shows the pattern of tumour growth and metastasis of the first given diet. *Carcinogenesis* 8, 611–614.
114. Rose, D.P., Connolly, J.M., Rayburn, J. and Coleman, M. 1995. Influence of diets containing eicosapentaenoic acid or docosahexaenoic acid on growth and metastasis in breast cancer cells in nude mice. *J. Natl. Cancer Inst.* 87, 587–592.
115. Robinson, L.E., Clandin, T. and Field, C.J. 2002. The role of dietary long-chain n-3 fatty acids in anti-cancer immune defence and R3230AC mammary tumour growth in rats: influence of diet fat composition. *Breast Cancer Res. Treat.* 73, 145–160.
116. Shoda, R., Matsueda, K., Yamato, S. and Umeda, N. 1995. Therapeutic efficacy of N-3 polyunsaturated fatty acid in experimental Crohn's disease. *J. Gastroenterol.* 30, 98–101.
117. Bagga, D., Wang, L., Farias-Eisner, R., Glaspy, J.A. and Reddy, T.A. 2003. Differential effects of prostaglandin derived from ω-6 and ω-3 polyunsaturated fatty acids on COX-2 expression and IL-6 secretion. *Proc. Natl. Acad. Sci. U.S.A* 100, 1751–1756.
118. Mantzioris, E., Cleland, L.G., Gibson, R.A., Neumann, M.A., Demasi, M. and James, M.J. 2000. Biochemical effects of a diet containing foods enriched with n-3 fatty acids. *Am. J. Clin. Nutr.* 72, 42–48.

119. Kew, S., Banerjee, T., Minihane, A.M., Finnegan, Y.E., Muggli, R., Albers, R., Williams, C.M. and Calder, P.C. 2003. Lack of effect of foods enriched with plant- or marine-derived n-3 fatty acids on human immune function. *Am. J. Clin. Nutr.* 77, 1287–1295.
120. Tilley, D.G. and Maurice, D.H. 2002. Vascular smooth muscle cell phosphodiesterase (PDE) 3 and PDE4 activities and levels are regulated by cyclic AMP *in vivo. Mol. Pharmacol.* 62, 497–506.
121. Shoda, R., Matsueda, K., Yamato, S. and Umeda, N. 1996. Epidemiologic analysis of Crohn's disease in Japan: increased dietary intake of n-6 polyunsaturated fatty acids and animal protein relates to the increased incidence of Crohn's disease in Japan. *Am. J. Clin. Nutr.* 63, 741–745.
122. Shoda, R., Matsueda, K., Yamato, S. and Umeda, N. 1995. Therapeutic efficacy of n-3 polyunsaturated fatty acid in experimental Crohn's disease. *J. Gastroenterol.* 30(8), 98–101.
123. Nieto, N., Torres, M.I., Rios, A. and Gil, A. 2002. Dietary polyunsaturated fatty acids improve histological and biochemical alterations in rats with experimental ulcerative colitis. *J. Nutr.* 132, 11–19.
124. Empey, L.R., Jewell, L.D., Garg, M.L., Thomson, A.B., Clandinin, M.T. and Fedorac, R.N. 1991. Fish oil-enriched diet is mucosal protective against acetic-acid induced colitis in rats. *Can. J. Physiol. Pharmacol.* 69, 480–487.
125. Alsan, A. and Triadafilopoulos, G. 1993. Fish oil fatty acid supplementation in active ulcerative colitis: a double blind, placebo controlled, crossover study. *Gut* 35, 345–357.
126. Belluzzi, A., Brignola, C., Campierim M., Pera, A., Boschi, S. and Miglioli, M. 1996. Effect of an enteric-coated fish oil preparation on relapses in Crohn's disease. *Engl. J. Med.* 334, 1557–1560.
127. Lorenz-Meyer, H., Bauer, P., Nicolay, C., Schultz, B., Purrmann, J., Fleig, W.E., Scheurlen, C., Koop, I., Pudel, V. and Carr, L. 1996. Omega-3 fatty acids and low carbohydrate diet for the maintainance of remission in Crohn's disease. A randomized controlled multicenter trial (German Crohn's disease study group). *Scand. J. Gastroenterol.* 31, 778–785.
128. Middleton, S.J., Naylor, S., Woolner, J. and Hunter, J.O. 2002. A double-blind, randomized, placebo-controlled trial of essential fatty acid supplementation in the maintenance of remission of ulcerative colitis. *Aliment. Pharmacol. Therapeut.* 16, 1131–1135.
129. Kremer, J.M., Bigauoette, J., Michalek, A.V., Timchalk, M.A., Lininger, L., Rynes, R.I., Huyck, C. and Zieminski, J. 1985. Effects of manipulation of dietary fatty acids on clinical manifestations of rheumatoid arthritis. *Lancet* 1, 184–187.
130. Volker, D., Fitzgerald, P., Major, G. and Garg, M. 2000. Efficacy of fish oil concentrate in the treatment of rheumatoid arthritis. *J. Rheumatol.* 27, 2305–2307.
131. Geusens, P., Wouters, C., Nijs, J., Jiang, Y. and Dequeker, J. 1994. Long-term effect of omega-3 fatty acid supplementation in active rheumatoid arthritis. A 12-month, double-blind, controlled study. *Arthrit. Rheumat.* 37, 824–829.
132. Homan van der Heide, J.J., Bilo, H., Donker, J.M., Wilmink, J.M. and Tegzess, A.M. 1993. Effect of dietary fish oil on renal function and rejection in cyclosporine treated recipients of renal transplants. *Engl. J. Med.* 329, 769–773.
133. Hernández, D., Guerra, R., Milena, A., Torres, A., García, G., García, C., Abreu, P., González, A., Gómez, M.A., Rufino, M., González-Posada, J., Lorenzo, V. and Salido, E. 2002. Dietary fish oil does not influence acute rejection rate and graft survival after renal transplantation: a randomized placebo-controlled study. *Nephrol. Dial. Transplant.* 17, 897–904.

134. Burney, P., Chinn, S. and Rona, R.J. 1990. Has the prevalence of asthma increased in children? Evidence from the national study of health and growth 1973–86. *Br. Med. J.* 300, 1306–1310.
135. Peat, J.K., Salome, C.M. and Woolcock, A.J. 1992. Factors associated with bronchial hyperresponsiveness in Australian adults and children. *Eur. Respir. J.* 5, 921–929.
136. Koch, T., Duncker, H.P., Klein, A., Schlotzer, E., Peskar, B.M., Van-Ackern, K. and Neuhof., H. 1993. Modulation of pulmonary vascular resistance and edema formation by short-term infusion of a 10% fish oil emulsion. *Infusionsther. Transfusions med.* 20, 291–300.
137. Hodge, L., Salome, C.M., Peat, J.K., Haby, M.M., Xuan, W. and Woolcock, A.J. 1996. Consumption of oily fish and childhood asthma risk. *Med. J. Aust.* 164, 137–140.
138. Troisi, R.J., Willet, W.C., Weiss, S.T., Trichopoulos, D., Rosner, B. and Speizer, F.E.A. 1995. Prospective study of diet and adult-onset asthma. *Am. J. Respir. Crit. Care Med.* 151, 1401–1408.
139. Nagakura, T., Matsuda, S., Shichijo, K., Sugimoto, H. and Hata, K. 2000. Dietary supplementation with fish oil rich in n-3 polyunsaturated fatty acids in children with bronchial asthma. *Eur. Respir. J.* 16, 861–865.
140. Emelyanov, A., Fedoseev, G., Kranoschekova, O., Abulimity, A., Trendeleva, T. and Barnes, P.J. 2002. Treatment of asthma with lipid extract of New Zealand green-lipped mussel: A randomized clinical trial. *Eur. Respir. J.* 20, 596–600.
141. Mahadik, S.P., Evans, D. and Lal, H. 2001. Oxidative stress and role of antioxidant and n-3 essential fatty acid supplementation in schizophrenia. *Prog. Neuropsychopharmacol. Biol. Psychiatr.* 25, 463–493.
142. Bourre, J.M. and Dumont, O. 1991. Essentiality of n-3 fatty acids for brain structure and function. *World Rev. Nutr. Diet.* 66, 103–117.
143. Martin, R.E. and Bazan, N.G. 1992. Changing fatty acid content of growth cone lipids prior to synaptogenesis. *J. Neurochem.* 59, 318–325.
144. Martinez, M. 1992. Abnormal fatty acid profiles of polyunsaturated fatty acids in the brain, liver, kidney and retina of patients with peroxisomal disorders. *Brain Res.* 583, 171–182.
145. Chamberlain, J.G. 1996. Fatty acids in human brain phylogeny. *Persp. Biol. Med.* 39, 436–445.
146. Neuringer, M. 1993. Cerebral cortex docosahexaenoic acid is lower in formula fed than in breast fed infants. *Nutr. Rev.* 51, 238–241.
147. Keller, J.R. 2002. Omega-3 fatty acids may be effective in the treatment of depression. *Top. Clin. Nutr.* 17, 21–27.
148. McGrath-Hanna, N.K., Greene, D.M., Tavernier, R.J. and Bult-Ito, A. 2003. Diet and mental health in the Arctic: is diet an important risk factor for mental health in circumpolar peoples? A review. *Int. J. Circumpolar Health* 62, 228–241.
149. Tanskanen, A., Hibbelin, J.R., Tuomilehto, J., Uutela, A., Haukkala, A., Viinamaki, H., Lehtonen, J. and Vartianen, E. 2001. Fish consumption and depressive symptoms in the general population in Finland. *Psychiatr. Serv.* 52, 529–531.
150. Marangell, L.B., Martinez, J.M., Zboyan, H.A., Kertz, B., Kim, H.F.S. and Puryear, L.J. 2003. A double-blind, placebo-controlled study of the omega-3 fatty acid docosahexaenoic acid in the treatment of major depression. *Am. J. Psychiatr.* 160, 996–998.
151. Su, K.P., Huang, S.Y., Chiu, C.C. and Shen, W.W. 2003. Omega-3 fatty acids in major depressive disorder. A preliminary double-blind, placebo-controlled trial. *Euro. Neuropsychopharmacol.* 13, 267–271.
152. Smith, R.S. 1991. The macrophage theory of depression. *Med. Hypotheses* 35, 298–306.

153. Horrobin, 1992. D.F. The relationship between schizophrenia and essential fatty acids and eicosanoid metabolism. *Prostagl. Leukotr. Essent. Fatty Acids* 46, 71–77.
154. McCreadie, R.G., Macdonald, E., Wiles, D., Campell, G. and Patterson, J.R. 1995. Plasma lipid peroxide and serum vitamin E levels in patients with and without tardive dyskinesia and normal subjects. *Br. J. Psychiatr.* 167, 1–8.
155. Phillips, M., Erickson, G.A., Sabas, N., Smith, J.P. and Greenberg, J. 1995. Volatile organic compounds in the breath of schizophrenic patients. *J. Clin. Pathol.* 48, 466–469.
156. Sarsilmaz, M., Songur, A., Ozyurt, H., Kus, I., Ozen, O.A., Ozyurt, B., Sogut, S. and Akyol, O. 2003. Potential role of dietary omega-3 fatty acids on some oxidant/antioxidant parameters in rats' corpus striatum. *Prostagl. Leukotr. Essen. Fatty Acids* 69, 253–259.
157. Hibbeln, J.R., Makino, K.K., Martin, C.E., Dickerson, F., Boronow, J. and Fenton, W.S. 2003. Smoking, gender, and dietary influences on erythrocyte essential fatty acid composition among patients with schizophrenia or schizoaffective disorder. *Biol. Psychiatr.* 53, 431–441.
158. Gottlicher, M., Widmark, E., Li, Q. and Gustafsson, J.A. 1992. Fatty acids activate a chimera of the clofibric acid-activated receptor and the glucocorticoid receptor. *Proc. Natl. Acad. Sci.* 89, 4653–4657.
159. Hertz, R., Magenheim, J., Berman, I. and Bar-Tana, J. 1998. Fatty acid CoA thioesters are ligands of hepatic nuclear factor-4α. *Nature* 392, 512–516.
160. de Urquiza, A.M., Liu, S., Sjoberg, M., Zetterstrom, R.H., Griffiths, W., Sjovall, J. and Perlmann, T. 2000. Docosahexaenoic acid, a ligand for the retinoid X receptor in mouse brain. *Science* 290, 2140–2144.
161. Ou, J., Tu, H., Shan, B., Luk, A., DeBose-Boyd, R.A., Bashmakov, Y., Goldstein, J.L. and Brown, M.S. 2001. Unsaturated fatty acids inhibit transcription of the sterol regulatory element binding protein-1c (SREBP-1c) gene by antagonizing ligand-dependent activation of the LXR. *Proc. Natl. Acad. Sci.* 98, 6027–6032.
162. Pawar, A. and Jump, D.B. 2003. Unsaturated fatty acid regulation of peroxisome proliferator activated receptor α activity in rat primary hepatocytes. *J. Biol. Chem.* 278, 35931–35939.
163. Desvergne, B. and Wahli, W. 2000. Peroxisome proliferator activated receptors: nuclear control of metabolism. *Endocr. Rev.* 20, 649–688.
164. Muerhoff, A.S., Griffin, K.J. and Johnson, E.F. 1992. The peroxisome proliferator activated receptor mediates the induction of CYP4A6, a cytochrome P450 fatty acid ω hydroxylase, by clofibric acid. *J. Biol. Chem.* 267, 19051–19053.
165. Xu, H.E., Lambert, M.H., Montata, V.G., Parks, D.J., Blanchard, S.G., Brown, P.J., Sternbach, D.D. and Lehmann, J.M. 1999. Molecular recognition of fatty acids by peroxisome proliferator activated receptors. *Mol. Cell* 3, 397–403.
166. Ren, B., Thelen, A.P., Peters, J.M., Gonzales, F.J. and Jump, D.B. 1997. Polyunsaturated fatty acid suppression of hepatic fatty acid synthase and S14 gene expression does not require peroxisome proliferator-activated receptor α. *J. Biol. Chem.* 272, 26827–26832.
167. Shimano, H., Horton, J.D., Shimomura, I., Hammer, R.E., Brown, M.S. and Goldstein, J.L. 1997. Isoform 1c of sterol regulatory element binding protein is less active than isoform 1a in livers of transgenic mice and in cultured cells. *J. Clin. Invest.* 99, 846–854.
168. Pawar, A., Botolin, D., Mangelsdorf, D.J. and Jump, D.B. 2003. The role of liver X receptor α in the fatty acid regulation of hepatic gene expression. *J. Biol. Chem.* 278, 40736–40743.
169. Xu, M.T., Nakamura, M.T., Cho, H.P. and Clarke, S.D. 1999. Sterol regulatory element binding protein-1 expression is suppressed by dietary polyunsaturated

fatty acids: a mechanism for the coordinate suppression of lipogenic genes by polyunsaturated fats. *J. Biol. Chem.* 274, 23577–23583.
170. Sessler, A.M. and Ntambi, J.M. 1998. Polyunsaturated fatty acid regulation of gene expression. *J. Nutr.* 128, 923–926.
171. Xu, J., Teran-Garcia, M. and Park, J.H. 2001. Polyunsaturated fatty acids suppress hepatic sterol regulatory element binding protein-1 expression by accelerating transcript decay. *J. Biol. Chem.* 276, 9800–9807.
172. Mater, M.K., Thelen, A.P., Pan, D.A. and Jump, D.B. 2001. Sterol response element binding protein-1c (SREBP1c) is involved in the polyunsaturated fatty acid suppression of hepatic S14 gene transcription. *J. Biol. Chem.* 274, 32725–32744.
173. Lu, T.T., Repa, J.J. and Mangelsdorf, D.J. 2001. Orphan nuclear receptors as elixirs and fixers of sterol metabolism. *J. Biol. Chem.* 276, 37735–37738.
174. Hertz, R., Sheena, V. and Kalderon, B. 2001. Suppression of hepatic nuclear factor-4α by acyl-CoA thioesters of hypolipidemic peroxisome proliferators. *Biochem. Pharmacol.* 61, 1057–1062.

3 Omega-3s and Their Impact on Brain Health

Genevieve Young and Julie Conquer

CONTENTS

3.1 INTRODUCTION

In the past decade, interest has surged in the area of omega-3 fatty acids and their role in normal brain functioning and neurological disease prevention. This chapter will summarize the evidence pointing to a role of omega-3 fatty acids in brain health. Although omega-3 fatty acids are present in plant-based sources such as alpha-linolenic acid (ALA; 18:3n-3), this chapter will focus mainly on the animal-derived long-chain (eicosapentaenoic acid; EPA; 20:5n-3 and docosahexaenoic acid; DHA; 22:6n-3) omega-3 fatty acids. Chemical structures and biosynthetic pathway for the formation of long-chain omega-3 fatty acids are given in Figures 3.1 and 3.2, respectively.

$$\mathrm{(CH_3)\text{-}CH_2\text{-}CH{=}CH\text{-}CH_2\text{-}CH{=}CH\text{-}CH_2\text{-}CH{=}CH\text{-}CH_2\text{-}CH_2\text{-}CH_2\text{-}CH_2\text{-}CH_2\text{-}CH_2\text{-}CH_2\text{-}(HO\text{-}C{=}O)}$$

Alpha-linolenic acid (ALA; 18:3n-3)

$$\mathrm{(CH_3)\text{-}CH_2\text{-}CH{=}CH\text{-}CH_2\text{-}CH{=}CH\text{-}CH_2\text{-}CH{=}CH\text{-}CH_2\text{-}CH{=}CH\text{-}CH_2\text{-}CH{=}CH\text{-}CH_2\text{-}CH_2\text{-}CH_2\text{-}(HO\text{-}C{=}O)}$$

Eicosapentaenoic acid (EPA; 20:5n-3)

$$\mathrm{(CH_3)\text{-}CH_2\text{-}CH{=}CH\text{-}CH_2\text{-}CH{=}CH\text{-}CH_2\text{-}CH{=}CH\text{-}CH_2\text{-}CH{=}CH\text{-}CH_2\text{-}CH{=}CH\text{-}CH_2\text{-}CH{=}CH\text{-}CH_2\text{-}CH_2\text{-}(HO\text{-}C{=}O)}$$

Docosahexaenoic acid (DHA; 22:6n-3)

FIGURE 3.1 Molecular formulas for the n-3 polyunsaturated fatty acids ALA, EPA, and DHA.

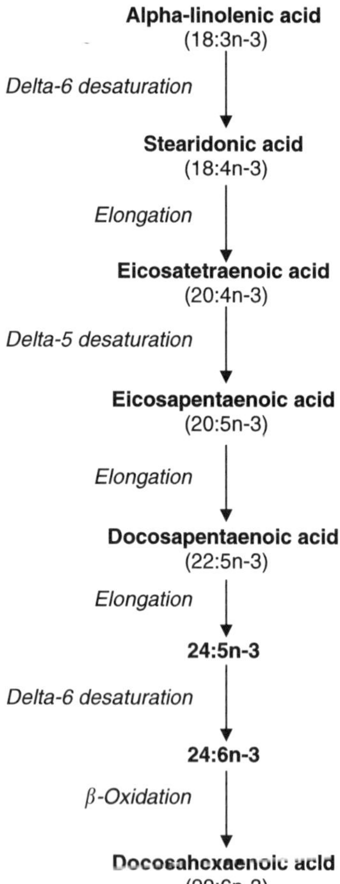

FIGURE 3.2 Pathway by which unsaturated omega-3 fatty acids are converted to long-chain polyunsaturated fatty acids in animals.

3.2 DIETARY CONSUMPTION OF n-3 FATTY ACIDS

Historical evidence suggests that human beings evolved consuming a diet that contained n-6 and n-3 fatty acids in the ratio of 1–2:1 respectively [1]. However, the current Western diet contains a ratio of up to 20–30:1, which means that the present diet is deficient in n-3 fatty acids compared to that on which our genetic patterns were established [2]. The evolution of change in dietary fat intake is illustrated in Figure 3.3. Today's intake of n-3 fatty acids is lower because of the decrease in consumption of fish and wild game, and because modern agriculture emphasizes consumption of cereal grains by animals destined for meat production [1]. The predominant sources of n-3 fatty acids in the modern diet are vegetable oils and fish, with fish being the major source of EPA and DHA, while vegetable oils are the major sources of ALA [3].

An analysis of the consumption of n-3 fatty acids in various populations shows that modern societies consume low levels of these dietary lipids, and has led to the establishment of guidelines concerning their recommended daily intake. In the United States, it has been recommended that EPA and DHA be consumed at an intake of 0.65 g/day, which is a fourfold increase from the current level of consumption of 0.1–0.2 g/day [3]. The adequate intake (AI) for LNA has been set at 1.6 and 1.1 g/day (adult men/women), and the target intake for EPA and DHA has been set at 160 or 110 mg/day (adult men/women) [4]. In Britain, the British Nutrition Foundation Task Force on Unsaturated Fatty Acids recommends a daily intake of 0.5–1.0 g of long-chain polyunsaturated n-3 fatty acids, which they suggest can be achieved through the consumption of an intake

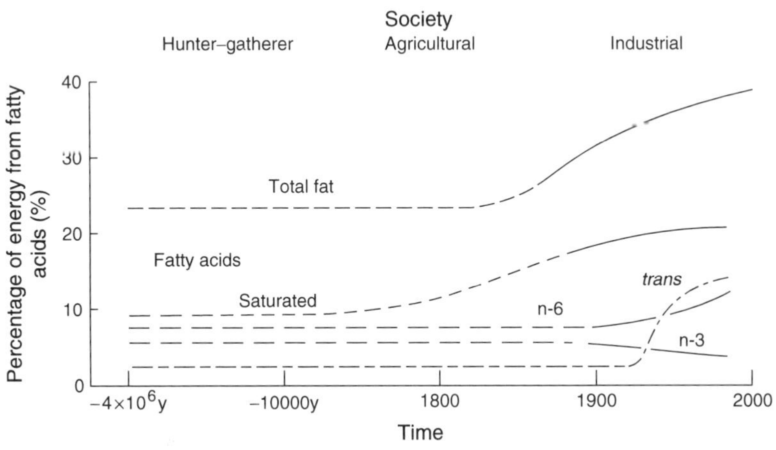

FIGURE 3.3 A hypothetical proposal of the percentage of energy from fat and fatty acids in human nutrition across time. (Reproduced from Simopoulos, *A.P., Am. J. Clin. Nutro.*, 70, 560S–569S, 1999 © American Journal of Clinical Nutrition. American Society for Clinical Nutrition. With permission)

equivalent to 1–2 portions of oily fish per week [5]. Even in Japan, where seafood has traditionally been consumed at very high levels, the ratio of n-6 to n-3 fatty acids is increasing as diets become more westernized, leading some authors to suggest that fish consumption be increased, particularly among young people [6]. There are several new products available in North America that have been supplemented with n-3 fatty acids (both short and long chain) including eggs, milk, and bread. Tables 3.1 and 3.2 show the fatty acid content of a number of n-3-containing food sources [7–10].

TABLE 3.1
The Omega-3 Fatty Acid Content, in Grams per 100 g Food Serving, of a Representative Sample of Commonly Consumed Fish, Shellfish, Fish Oils, Nuts and Seeds, and Plant Oils That Contain at Least 5 g of Omega-3 Fatty Acids per 100 g

Food Item	EPA	DHA	ALA
Fish (raw)			
Anchovy, European	0.6	0.9	–
Cod, Atlantic and Pacific	Trace	0.1	Trace
Haddock	Trace	0.1	Trace
Halibut, Atlantic and Pacific	Trace	0.3	Trace
Mackerel, Pacific and Jack	0.6	0.9	Trace
Ocean Perch, Atlantic	Trace	0.2	Trace
Pike, Walleye	Trace	0.2	Trace
Roughy, Orange	Trace	–	Trace
Salmon, Atlantic, farmed	0.6	1.3	Trace
Salmon, Atlantic, wild	0.3	1.1	0.3
Seabass, mixed species	0.2	0.4	–
Swordfish	0.1	0.5	0.2
Trout, Rainbow, armed	0.3	0.7	Trace
Trout, Rainbow, wild	0.2	0.4	0.1
Tuna, Bluefin, fresh	0.3	0.9	–
Tuna, Yellowfin, fresh	Trace	0.2	Trace
Whitefish, mixed species	0.3	0.9	0.2
Wolffish, Atlantic	0.4	0.3	trace
Shellfish (raw)			
Clam, mixed species	Trace	Trace	Trace
Crab, Blue	0.2	0.2	–
Lobster, Northern	–	–	–
Mussel, Blue	0.2	0.3	Trace
Oyster, Pacific	0.4	0.3	Trace
Scallop, mixed species	Trace	0.1	–
Shrimp, mixed species	0.3	0.2	Trace

(Continued)

TABLE 3.1
(Continued)

Food Item	EPA	DHA	ALA
Fish oils			
Cod-liver oil	6.9	11.0	0.9
Herring oil	6.3	4.2	0.8
Menhaden oil	13.2	8.6	1.5
Salmon oil	13.0	18.2	1.1
Sardine oil	10.1	10.7	1.3
Nuts and seeds			
Butternuts, dried	–	–	8.7
Flaxseed	–	–	18.1
Walnuts, English	–	–	9.1
Plant oils			
Canola (rapeseed)	–	–	9.3
Flaxseed	–	–	
Soybean	–	–	6.8
Walnut	–	–	10.4
Wheatgerm	–	–	6.9

Note: Trace = <0.1.

Source: Adapted from http://www.ahrq.gov/downloads/pub/evidence/pdf/o3asthma/tbls.pdf (accessed 13 September 2004).

TABLE 3.2
Omega-3 Fatty Acid Content of Commercially Available Omega-3 Enriched Foods

Product	Total Omega-3 Content	DHA Content
Omega-3 enriched eggs Gray Ridge® egg farms [8]	0.4 g/egg	0.085 g/egg
Omega-3 enriched milk Neilson Dairy® Oh! [9]	0.02 mg/cup (homogenized) 0.01 mg/cup (2%)	0.02 mg/cup (homogenized) 0.01 mg/cup (2%)
Omega-3-enriched bread Tip Top® UP [10]	0.27 g/2 slices	0.27 g/2 slices

3.3 OMEGA-3 FATTY ACIDS AND THE BRAIN

In the human body, EPA is found primarily in cholesterol esters (CE), triacylglycerols (TG), and phospholipids (PL); and DHA is found primarily in PL, and is highly concentrated in the cerebral cortex, retina, testes, and sperm [2]. In fact, DHA makes up a large proportion of the brain's lipids, and is the predominant n-3 fatty acid found in this organ [11]. The structural predominance of DHA in the brain suggests functional significance, and as will be demonstrated, both DHA and its long-chain counterpart EPA can be linked with several aspects of neural function, including, but not limited to, phospholipase A_2 (PLA_2) activity, inflammation, neurotransmission, membrane fluidity, oxidation, ion channel and enzyme regulation, and gene expression. Each of these functions will be considered in terms of their relationship with the n-3 polyunsaturated fatty acids (PUFA), and when available, evidence linking them with various neuropsychiatric disorders will be presented.

3.3.1 PHOSPHOLIPASE A_2

PLA_2 is an enzyme that acts on the *sn*-2 position of phospholipids, thereby generating a free fatty acid and a lysophospholipid [12]. DHA, along with the n-6 PUFA arachidonic acid (AA; 20:4n-6), predominates at this position in cerebral phospholipids [13]. Several classes of PLA_2 exist in the brain [14], and PLA_2 is expressed in many brain regions with the highest expression in the hippocampus [15]. While the function of PLA_2 in the human nervous system has not been fully elucidated, it has been implicated in the processes of phospholipid turnover, neurotransmitter release, detoxification, exocytosis, and membrane remodeling [16], and the free fatty acid and lysophospholipid produced by its action are known to be highly active cell-signaling molecules [17]. It has been suggested that under pathological conditions, seen in neuropsychiatric disorders such as schizophrenia, an increase in the activity of PLA_2 can result in a decrease in neuronal membrane phospholipid biosynthesis and an increase in phospholipid breakdown [18]. Owing to the frequent occupation of the *sn*-2 position by DHA in the brain, an increase in the activity of PLA_2 could potentially cause a decrease in levels of this fatty acid with functionally significant consequences. Figure 3.4 illustrates the phospholipid structure and the site of action of the main phospholipase enzymes.

EPA has been shown to inhibit PLA_2 activity [19], and administration of EPA has shown considerable success in the treatment of schizophrenia and bipolar disorder. Furthermore, nutritional analysis has revealed that a lower intake of EPA is associated with more severe psychopathology and tardive dyskinesia [20].

3.3.2 INFLAMMATION

Many mood and neurodevelopmental disorders appear to be linked to immune system activation, as evidenced by overactivity of the inflammatory response [21–24]. Omega-3 fatty acids, when consumed in adequate amounts, can exert anti-inflammatory actions *in vivo*. This is primarily accomplished through

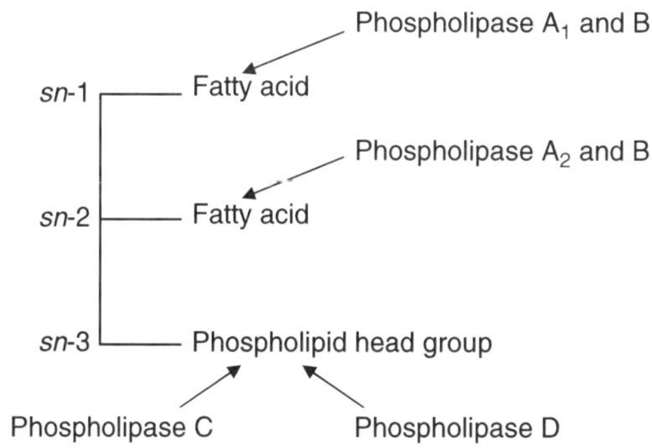

FIGURE 3.4 Phospholipid structure and sites of action of the phospholipase enzymes.

modification of the production of cytokines and eicosanoids. IL-1 has been shown to decrease [25], while tumor necrosis factor (TNF) has been shown to increase as a consequence of fish oil feeding [26], thereby reducing inflammation. Similarly, increasing the consumption of n-3 fatty acids, particularly the long-chain polyunsaturated n-3s, tends to shift the balance of eicosanoid production from pro- to anti-inflammatory mediators [27]. Figure 3.5 illustrates the effect of dietary n-3 fatty acids on the immune system via modulation of eicosanoids, such as the prostaglandins and thromboxanes. For example, increasing the amount of EPA in the diet causes a shift in the production of the inflammatory eicosanoid leukotriene B4 (LTB4) to the production of the anti-inflammatory leukotriene B5 (LTB5), thereby attenuating the inflammatory response [28,29]. Because of the relationship between inflammation and the pathology of many neuropsychiatric diseases, the influence of n-3 fatty acids on this physiological process is important to consider.

3.3.3 Neurotransmitters

Neurotransmitters are molecules that mediate intercellular communication, and include dopamine and serotonin. Levels of neurotransmitters have been shown to be affected by diet, which is not surprising considering that many, including serotonin and dopamine, are derived from nutrient precursors. Serotonin is derived from the amino acid tryptophan while dopamine is derived from the amino acid tyrosine [30]. Modulation of neurotransmitter levels has long been viewed as a causative factor in both unipolar and bipolar depression [31–35].

Although they do not serve directly as substrates for the formation of serotonin and dopamine, n-3 PUFA have been shown to influence levels of these molecules in the brain. When piglets are fed a diet deficient in AA and DHA, there is a decrease in both dopamine and serotonin concentration in the frontal cortex [36]. Conversely, when their diet was supplemented with AA and DHA, piglets showed an increase in the frontal cortex concentration of serotonin, possibly due to a decrease in degradation [37]. A similar situation has been found in rats, who

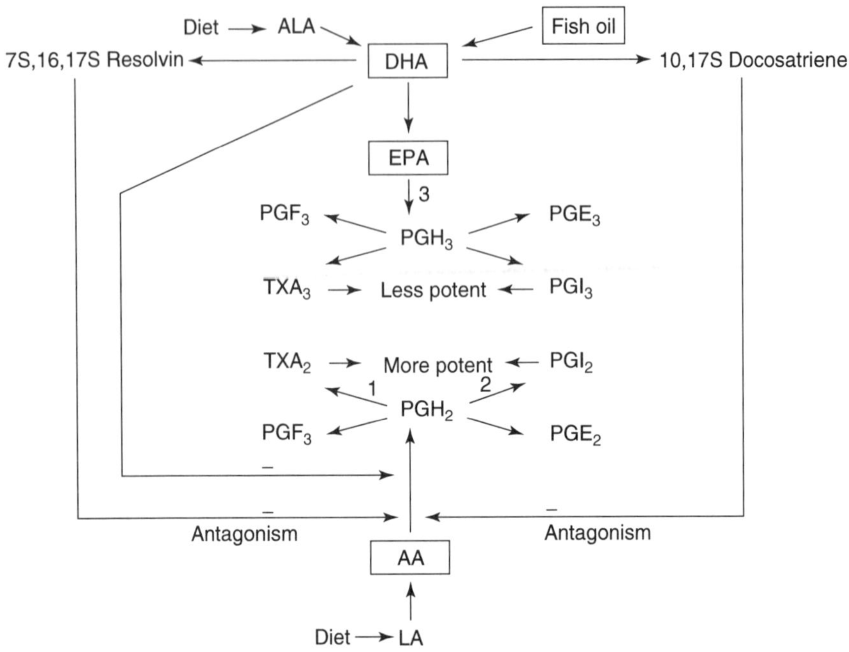

FIGURE 3.5 Interactions between the n-3 fatty acids DHA EPA, and the n-6 fatty acid AA, in the immune system. DHA antagonizes (-) the effects of prostaglandins and thromboxanes and increases the production of the less potent series-3 prostaglandins. AA increases the production of the more potent series-2 prostaglandins. Linoleic acid (LA; 18:2n-6); (1) thromboxane synthase; (2) prostaglandin I synthase; and (3) cyclooxygenase-1 and -2. (Reprinted from Horrocks, L.A. Docosahexaenoic Acid in the Diet: Its Importance in Maintenance and Restoration of Neural Membrance Function, Elsevier, Amsterdam, 2004. With permission.)

when fed a diet deficient in n-3 fatty acids, displayed inadequate storage of newly synthesized dopamine [38], as well as an overall reduction in the dopaminergic vesicle pool [39]. Alternatively, when rats are fed fish oil, there is a 40% increase in frontal cortex dopamine concentrations as well as a greater binding to dopamine D_2 receptors [40]. Thus, n-3 fatty acids may modulate concentrations of these neurotransmitters through an effect on storage and availability.

3.3.4 Membrane Fluidity

Membrane fluidity is believed to be greatly influenced by the fatty acid composition of the membrane PL [41]. In the brain, membrane lipids comprise 50–60% of the solid matter [42], the majority of which are found within PL. The presence of double bonds in these fatty acids has a significant effect on the physical properties of the membrane [43]. Furthermore, it is generally assumed that as the unsaturation of the fatty acid is increased, there is an increase in membrane fluidity as reflected in properties such as molecular area and phase transition temperature. In terms of molecular area, it is believed that more highly unsaturated fatty acyl chains will occupy more space. However, this assumption only holds true for the addition of the first and second double bonds; addition of a third, fourth, and sixth double bond does not cause a significant further increase [44], and DHA, while more expanded than the monounsaturated oleic acid at low lateral pressures, occupies approximately the same space at the higher pressures that are characteristic of lipid bilayers [45]. Similarly, phase transition temperature has been found to decrease upon addition of a first and second double bond but then increases upon addition of further double bonds [46]. Membrane fluidity has also been quite thoroughly investigated using fluorescent probes, and there are several negative reports in the literature of DHA-associated changes in membrane fluidity using such techniques [47–49]. However, there are also reports of DHA-induced increases in fluidity [50], leading Youdim et al. [28] to suggest that the relationship between the number of double bonds and fluidity is not a simple linear one.

3.3.5 Oxidative Stress

Free radicals are generated under normal physiological conditions, and play important roles in a variety of biological processes. However, when these molecules are generated in excess, they can initiate spontaneous chain reactions that may have negative consequences, such as abnormal neurodevelopment and neuronal function [51]. Free radicals are considered unstable because they carry one or more unpaired electrons, which make them highly reactive. Examples of free radicals are peroxyl radical (LOO·), alkoxyl radical (LO·), and hydroxyl radical (≅OH), all of which are oxygen-containing species and are therefore referred to as oxyradicals. These oxyradicals can react with PUFA, and cell membranes of tissues exposed to high concentrations of oxygen, such as the brain, are susceptible to oxidation because of the presence of unsaturated fatty acids in their phospholipids [52]. Oxyradicals are eliminated by enzymes such as superoxide dismutase, glutathione peroxidase, and catalase; nonenzymatic mechanisms such as glutathione and uric acid; and dietary antioxidants such as vitamins A, E, and C [53]. Free radical reactions involving the peroxyl, alkoxyl, and hydroxyl radicals are illustrated in Figure 3.6.

$$\text{(a) } Cu^{2+} + LOOH \rightarrow Cu^{+} + LOO^{\cdot} + H^{+} + Cu^{+}$$

$$\text{(b) } Cu^{+} + HOOH \rightarrow Cu^{2+} + OH^{\cdot} + OH^{-}$$

$$\text{(c) } Cu^{+} + LOOH \rightarrow Cu^{+} + LO^{\cdot} + H^{+} + OH^{-}$$

FIGURE 3.6 An example of free radical oxidation by the Cu^{2+} ion. This ion has been shown to react with any LOOH (lipid hydroperoxide) present to produce Cu^{+} and a peroxyl radical (a). The Cu^{+} product can go on to react with HOOH (hydrogen peroxide) to form a hydroxyl radical (b), or with LOOH forming an alkoxyl radical (c).

The influence of n-3 fatty acids on oxidative pathology is likely via replacement of lost membrane phospholipid PUFA following attack by oxyradicals [52]. In animals, consumption of dietary n-3 fatty acids has been shown to modulate levels of these in the brain [54], and human serum and erythrocyte phospholipids are very responsive to dietary n-3 modulation [55–57]. Because of the vulnerability of lipids to attack by oxyradicals, supplementation with n-3 PUFA should be accompanied by cotreatment with an antioxidant [51].

3.3.6 Ion Channel and Enzyme Regulation

Proper physiological function requires coordinated integration of a number of different cellular components, including ion channels and enzymes. Sodium channels, which are glycoproteins that form pores in the cell membrane, open and close in response to changes in membrane potential thereby regulating the generation of action potentials. A similar process occurs with potassium channels, which are also found in the cell membrane. Enzymes such as the Na + K + ATPase and Ca-ATPase perform the functions of ion transport, allowing for maintenance of proper intracellular ion concentration and cellular homeostasis, and the regulation of ion channels and enzymes is accomplished by molecules such as neurotransmitters and G proteins. Therefore, there is a multitude of levels at which neuronal functioning might be compromised, potentially resulting in pathology that could give rise to neuropsychiatric disorders.

Ion channel, enzyme, and regulatory molecule function may be influenced by polyunsaturated n-3 fatty acids. EPA has been found to inhibit voltage-activated Na^+ currents [58,59], as has DHA [59,60]. It appears that these n-3 PUFA modify the function of the Na^+ channel by binding directly to channel proteins [59]. DHA [61,62] and EPA [61] have also been shown to inhibit voltage-activated K^+ current, and DHA has been observed to do this via binding to an external site on the channel structure [63]. Importantly, the opening of the K^+ channel TREK-1 by DHA appears to exert a neuroprotective effect against ischemia and epileptic damage in the brain [64,65]. DHA [60,66] and EPA [66] have further been shown to inhibit voltage-activated Ca^{2+} currents. The ion-regulating enzyme Na + K + ATPase appears to be strongly influenced by the presence of DHA in the surrounding cell membrane, in that high concentrations of DHA have been associated with high Na + K + ATPase activity [67], and both Ca-ATPase and Na + K + ATPase activity have been shown to be inhibited by both EPA and DHA [68]. Other cellular functions, such as the rate of glutamate uptake [69], the responsiveness of the NMDA receptor [70], and the activation of protein kinase C [71,72] have also been shown to be affected by polyunsaturated n-3 fatty acids.

3.3.7 Gene Expression

The regulation of genetic expression dictates the rate at which genes are transcribed to effect changes in the production of various gene products. Genes can either be up- or down-regulated, resulting either in an increase or decrease in transcription. Up-regulation of a gene may lead to an increase in the synthesis of a particular protein, while down-regulation may have the opposite effect. Modulation of gene expression at the transcription level can be mediated by n-3 PUFA, and Wahle et al. [73] recently summarized the generally accepted mechanisms of this regulation. The first reported mechanism is activation of cell signal cascades that results in covalent modification of specific transcription factors, which can in turn then bind to promoter regions of a gene causing an up- or down-regulation of transcription. The second reported mechanism is by direct binding of the fatty acid (or its derivative) to specific transcription factors, which consequently has a positive or negative effect on its promoter-binding capacity. The third reported mechanism is modification of transcription factor mRNA, or alteration of the stability of such mRNA and possibly its DNA-binding capacity. There are also likely indirect mechanisms of regulation of gene expression, such as modulation of the redox state of the cell. Among the transcription factors that are known to be activated by n-3 PUFA are peroxisome proliferated activated receptors (PPARs), liver X receptors α and β, hepatic nuclear factor-4, and sterol regulatory element binding proteins (SREBPs) [29].

3.4 OMEGA-3 FATTY ACID STATUS OF BLOOD AND CELLS OF INDIVIDUALS WITH VARIOUS NEUROPSYCHIATRIC DISORDERS

Omega-3 fatty acid deficiencies are associated with a wide range of neuropsychiatric disorders, including, but not limited to, attention deficit hyperactivity disorder (ADHD), neurodevelopmental disorders such as dyslexia and autism, depression, aggression, and dementia. This chapter will present available information on blood levels of omega-3 fatty acids in individuals with these, and other neuropsychiatric disorders, as compared with healthy controls. We will also discuss available evidence as to the efficacy of omega-3 fatty acid supplementation in alleviating the symptoms of these conditions.

3.4.1 ATTENTION DEFICIT DISORDER AND HYPERACTIVITY

ADHD, also known as attention deficit disorder (ADD), is a condition characterized by disabling levels of inattention, impulsivity, or hyperactivity, which are inappropriate for the individual's level of development [74]. Although previously thought to be a condition of childhood, it is now recognized that in up to 60% of sufferers, ADHD persists into adulthood [75].

Both n-3 and n-6 long-chain polyunsaturated fatty acids (LCPUFAs) have been suspected of being associated with ADHD, since the 1980s. Stevens et al. [76] found that plasma and red blood cell (RBC) levels of AA, EPA, and DHA were significantly lower in ADHD patients as compared to controls, and that a subgroup of ADHD patients exhibiting symptoms of LCPUFA deficiency had even lower plasma concentrations of AA and DHA than did ADHD subjects with few LCPUFA-deficiency symptoms. Recently, Young et al. [77] demonstrated that adults with ADHD also have an altered phospholipid fatty acid status, specifically having lower levels of omega-6 fatty acids and DHA in serum and lower levels of omega-3 fatty acids, including DHA in RBC.

Very few clinical trials have been conducted in terms of omega-3 fatty acid supplementation and ADHD. In 2001, Voigt et al. [78] supplemented 63 children with ADHD with either placebo or 345 mg/day DHA for 4 months. DHA levels in blood increased but there were no significant improvements in any measure of ADHD symptoms. However, Richardson and Puri [79] showed that supplementation with a mixture of EPA, DHA, γ-linolenic acid (GLA, 18:3n-6), vitamin E, AA, LA (linoleic acid), and thyme oil for 12 weeks in children with specific learning disabilities, improved symptoms in 7 out of 14 symptoms of ADHD (although only 3 were significant) compared to none for placebo. Recently, Stevens et al. [80] supplemented children with ADHD with (per day) 480 mg DHA, 80 mg EPA, 40 mg AA, and 96 mg GLA for 4 months. There was an increase in both EPA and DHA in plasma as well as improvement in parent-rated conduct, teacher-rated attention, and oppositional defiant behavior. Furthermore, there was a significant correlation between increased RBC EPA and DHA and a decrease in disruptive behavior. Finally, Hirayama et al. [81] examined the effect of DHA supplementation in food sources for 2 months on symptoms of ADHD. On average, children received 0.5 g DHA/day versus control foods. There was no improvement of ADHD symptoms in this study. These findings seem to suggest that a combination of LCPUFAs is more likely to exert a positive effect on ADHD symptoms than omega-3 fatty acids alone.

3.4.2 Alzheimer's Disease and Dementia

Alzheimer's disease (AD) is a progressive neurodegenerative disorder characterized by memory loss, intellectual decline, and eventual global cognitive impairment. It has long been suggested that AD is associated with brain lipid defects [82–87]. More recently, epidemiological studies [88–93] have suggested that high fish and omega-3 fatty acid consumption is inversely associated with cognitive impairment, cognitive decline, and development of dementia or AD. Furthermore, at least one study suggests an inverse association between cognitive decline and ratio of n-3/n-6 fatty acids in RBC membranes [94]. Furthermore, serum cholesteryl ester (CE) EPA and DHA levels have been shown to be lower in AD patients [95]. This study also suggested that CE-DHA is an important determinant of mini-mental state examination (MMSE) score across the population. Another study by ourselves [96] suggests that phospholipid (PL) and phosphatidylcholine (PC) EPA, DHA, and total omega-3 fatty acid levels are decreased in cognitively impaired and demented (including AD) individuals. Only one study [97] found no significant difference in plasma PL omega-3 fatty acid levels between controls and cognitively impaired or demented individuals.

There are only two known studies in which individuals with AD or dementia have been supplemented with long-chain omega-3 fatty acids. In the first, by Terano et al. [98], individuals with dementia were supplemented with 0.72 g DHA/day for 1 year. Blood levels of omega-3 fatty acids increased and scores on dementia rating scales improved. In the second study, by Otsuka [99], individuals with AD were supplemented with 900 mg EPA/day for 6 months. MMSE scores increased maximally by 3 months and remained higher for 6 months.

3.4.3 Schizophrenia

Schizophrenia is a severe mental illness characterized by positive (hallucinations and delusions) and negative (lack of emotional responsiveness and drive) symptoms. Schizophrenic outcome has been suggested to be inversely related to the consumption of saturated fats and directly related to consumption of omega-3 fatty acids [100,101]. Furthermore, blood omega-3 fatty acid levels have been shown to be correlated with positive schizotypal trait measures [102] and this suggests that these fatty acids may offer protection against psychotic breakdown. Decreased RBC omega-3 fatty acids levels have been shown in first-episode psychotic patients (medication-free) [103–105], as well as in medicated schizophrenic patients [106–109], as compared with control subjects. Furthermore, levels of total omega-3 fatty acids and DHA are decreased in cultured skin fibroblasts from schizophrenic patients as compared with controls [110]. Some studies also suggest decreased levels of certain omega-6 fatty acids, mainly AA, accompanies the modification of omega-3 fatty acid levels in these individuals [107]. Medication itself may influence the levels of omega-3 fatty acids in RBC of schizophrenic patients in either a positive [104,106] or negative [111] manner. Interestingly, Hibbeln et al. [112], suggest that schizophrenic smokers have decreased blood levels of DHA and EPA as compared with schizophrenic nonsmokers.

The findings involving supplementation of schizophrenic individuals with omega-3 fatty acids have been reviewed by various authors [113–115]. There is evidence to suggest positive benefits for omega-3 fatty acid supplementation in schizophrenic individuals but more well-designed studies are still needed before definite conclusions can be reached. Most studies have investigated the ability of ethyl-EPA (E-EPA) to modify positive and negative schizophrenic symptoms in individuals with residual symptoms despite medication. Most of these studies have shown improvements in Positive and Negative Syndrome Scale (PANSS) scores after at least 12 weeks of supplementation [111,115–117]. Levels of E-EPA used range from 1 to 4 g/day with 2–3 g/day being the most common, and this level appears to have the most benefit. Benefits have also been noted in dyskinesia scores after 12 weeks of supplementation [116]. One study in the United States [118] found no difference in symptoms, mood, or cognition after supplementation with 3 g E-EPA for 16 weeks.

At least two studies have been conducted with EPA and DHA combinations in fish oil, investigating changes in omega-3 fatty acid levels of RBC, and improvement in symptoms. One of these studies investigated the combined supplementation of omega-3 fatty acids (EPA plus DHA) and antioxidant vitamins (E and C) [119], and the other [116], investigated the effect of 10 g/day MaxEPA (EPA plus DHA) in an open study. Both studies had positive results.

At least two authors have presented results of findings in individual patients. In one, a 30-year-old woman with exacerbation of symptoms during pregnancy, was supplemented with omega-3 fatty acids [120]. This resulted in an increase in omega-3 levels of RBC and improvement in positive and negative symptoms. In a second study, a drug-naïve patient was supplemented with 2 g/day EPA for 6 months [121–123]. Improvements were noted in PANSS scores, RBC omega-3 levels, cerebral atrophy, and hemispheric imbalance.

3.4.4 Depression and Postpartum Depression

Major depression is defined as ≥2 weeks of predominantly low mood or diminished interests in one's usual activities in combination with four or more of the following: increased or decreased sleep patterns, inappropriate guilt or loss of self-esteem, increased or decreased appetite, low energy, difficulty in concentrating, agitation or retardation, and suicidal thoughts [124]. Several observational studies have provided evidence that supports the theory that decreased fish or omega-3 fatty acid consumption may be involved with increased incidence of depression and this has been reviewed by various authors [125,126]. Societies consuming large amounts of fish and n-3 fatty acids appear to have lower rates of major depression and bipolar disorders [127,128] and the likelihood of having depressive symptoms increases among infrequent fish consumers versus frequent fish consumers [129]. Studies have also suggested that depression associated with disease diagnosis is also associated with decreased dietary intake of total omega-3 fatty acids [130,131]. Seasonal variation of serum LCPUFA, including EPA and DHA, correlates negatively with the number of violent suicidal deaths in Belgium [132]. However, one study found no associations between dietary intakes of omega-3 fatty acids and depressed mood or major depressive episodes [133].

Total n-3s, EPA, and DHA are depleted in RBC membranes of depressive patients or individuals at risk for the recurrent form of the major depressive disorder [134–136], and there is a negative correlation between levels of blood and adipose tissue n-3 PUFA and depressive symptoms [135,137–139]. n-3 Fatty acids are also depleted in the serum PL and CE of depressed and bipolar patients [140–142], and an increase in the AA to EPA ratio has repeatedly been positively correlated with depression [140,141,143].

Postpartum depression (PPD) also appears to be associated with omega-3 fatty acid levels. Higher concentrations of DHA in breast milk and higher seafood consumption predicts lower prevalence of PPD in society [144]. Higher plasma DHA of the mother is associated with a reduction in depressive symptom reporting in the immediate postpartum period [145]. Plasma DHA was influenced by maternal education and smoking, so these results should be interpreted with caution. Another study suggested that the postpartum decrease of DHA status was greater in individuals who were "possibly depressed" versus the nondepressed group

[146]. In mothers who developed PPD, DHA and total omega-3 fatty acids (PL and CE) were decreased and the omega-6:omega-3 ratio was increased as compared to mothers who did not develop PPD [147].

As with the other neurological disorders mentioned in this review, there are a few studies investigating the effect of omega-3 fatty acid supplementation on symptoms of depression. At least four studies have investigated the potential efficacy of omega-3 fatty acid supplementation in depressive individuals. Most of these studies involved individuals who were on standard medications. Two of these studies investigated the effects of E-EPA and found improvements in scores on depressive rating scales as well as suicidal thoughts and social phobia [148,149], with maximum benefits observed with the 1 g/day dose [148]. In another study, individuals were supplemented with 6.6 g of omega-3 fatty acids or placebo for 8 weeks [150]. A decreased score in the Hamilton rating scale was noted in the omega-3 supplemented group. One study investigated the effect of supplementation with DHA alone (2 g/day) versus placebo for 6 weeks and found no difference in rating scale scores between the groups [151]. In bipolar patients on medication, Stoll et al. [152] determined that 9.6 g/day omega-3 fatty acids for 4 months, as compared with placebo, resulted in improved outcome and longer remission.

At least two studies have been conducted on individuals with PPD. In the first study, individuals with a history of PPD were supplemented with approximately 3 g fish oil from the 34–36 week of pregnancy to 12 weeks postpartum [153]. There were no dropouts, but there was also no evidence of benefit based on the number of individuals who had depressive episodes during the study. In the second study, individuals were supplemented with 200 mg/day DHA or placebo for 4 months following delivery [154]. Increased plasma DHA was noted in the supplemented group, whereas there was a decrease in the placebo group. There was no difference in self-rating or diagnostic measures of depression. It is still unclear whether there are potential benefits of omega-3 fatty acid supplementation from earlier on in pregnancy or prior to conception.

3.4.5 Other Neurological Disorders

The role of omega-3 fatty acids has also been investigated in other neurological disorders although the information is scarce. These include dyslexia, autistic spectrum disorders, dyspraxia, borderline personality disorder (BPD) and obsessive compulsive disorder (OCD), as well as aggression and hostility.

In adults and children with dyslexia, supplementation with HUFA (186 mg EPA, 480 mg DHA, 96 mg GLA, 864 mg LA, and 42 mg AA per day) improved ADHD symptoms in children with learning disabilities (mainly dyslexia) [79] and supplementation with DHA for 1 month improved dark adaptation in dyslexic young adults [155].

Decreased DHA and total omega-3 fatty acids have been shown in blood of autistic children versus mentally retarded children [156]. Furthermore, there is also decreased HUFA in RBC membranes of autistic children, which has been shown to break down faster than control samples when stored at –20°C versus –80°C [157]. There do not appear to be any published trials in which individuals with autistic spectrum disorder are supplemented with omega-3 fatty acids.

Two supplementation trials have been conducted in children with dyspraxia. These trials suggest improvement in movement skills with high DHA fish oil plus EPO for 4 months [155], and improvement in reading and spelling, with a decrease in ADHD symptoms after supplementation with fish oil and EPO for 12 weeks [158].

Supplementation with 1 g of E-EPA per day for 8 weeks was investigated in individuals with BPD. Improvement was noted in aggression and severity of symptoms [159].

Omega-3 fatty acid supplementation has also been investigated in individuals with OCD on traditional selective serotonin reuptake inhibitors (SSRIs). Individuals were supplemented with 2 g EPA or placebo for 6 weeks. Scores on the Yale Brown Obsessive Compulsive Scale decreased in both groups [160].

Cross-nationally, it has been observed that rates of death from homicide are lower in countries with high n-3 consumption [161]. In a 5-year prospective interventional study, an increase in dietary n-3 fatty acids caused a significant decrease in hostility [162].

Omega-3 supplementation has been investigated in terms of aggression and hostility. Supplementation with 1.5 g DHA versus placebo for 2 months resulted in decreased aggression in educated university workers but not uneducated villagers after a videotape stressor [163]. When aggression was measured during exam time, after 3 months of supplementation with 1.5 g DHA/day, there was an increase in aggression in the control group, but no change in the DHA group [164]. This same dose appeared to decrease the level of norepinephrine (NE), but not other catecholamines, in medical students during exams [165]. Interestingly, when aggression was determined during a nonstressful time, there was no change in aggression in the DHA group and a slight decrease in the control group [166]. It is possible that DHA supplementation is offering protection during stressful conditions.

In cocaine addicts admitted to hospital, aggressive patients were shown to have decreased levels of total omega-3 fatty acids and DHA as well as an increased n-6:n-3 ratio in blood [167].

3.5 CONCLUSIONS

It is obvious that there is a limited amount of work in the field of omega-3 fatty acids and brain health, and thus there are exciting opportunities for researchers. Evidence suggests decreased blood levels of omega-3 fatty acids in individuals with various psychiatric conditions. The reasons for this are not clear; decreased synthesis or increased breakdown are the only two possibilities. Both would increase dietary requirements in this population. Epidemiological evidence suggests that decreased plasma or RBC omega-3 fatty acids are a risk factor for the development of at least some of these conditions. This suggests that they may play a role in the actual development of the disorder as opposed to being a consequence of the disorder. So far, the benefits of supplementation, in terms of decreasing disease risk or aiding in symptom management are not clear and more research is needed. The timing of dietary changes and supplementation use as well as levels and specific types of omega-3 fatty acids are still in the process of being investigated.

REFERENCES

1. Simopolous, A.P. 1999. Essential fatty acids in health and chronic disease. *Am. J. Clin. Nutr.* 70, S560–S569.
2. Simopolous, A.P. 1991. Omega-3 fatty acids in health and disease and in growth and development. *Am. J. Clin. Nutr.* 54, 438–463.
3. Kris-Etherton, P.M., Shaffer, D., Yu-Poth, S., Huth, P., Moriarty, K., Fishell, V., Hargrove, R.L., Zhao, G. and Etherton, T.D. 2000. Polyunsaturated fatty acids in the food chain in the United States. *Am. J. Clin. Nutr.* 71, S179–S188.
4. Institute of Medicine (IOM). *Dietary References Intakes for Energy and Macronutrients.* National Academy Press, Washington. 2002.
5. British Nutrition Foundation. *Task Force on Unsaturated Fatty Acids.* Chapman and Hall, London. 1992.
6. Sugano, M. and Hirahara, F. 2000. Polyunsaturated fatty acids in the food chain in Japan. *Am. J. Clin. Nutr.* 71, 189S–196S.
7. http://www.ahrq.gov/downloads/pub/evidence/pdf/o3asthma/tbls.pdf (accessed 13 Sept 2004).
8. Nutrition Facts "Omega 3" eggs, Gray Ridge Egg Farms, RR#7, Strathroy, ON N7G 3H8
9. http://www.dairy-oh.com/nutrition.htm (accessed 13 Sept 2004).
10. http://www.tiptop.com.au/driver.asp?page=main/brands/bread/up+omega+3+dha (accessed 13 Sept 2004).
11. Sastry, P.S. 1985. Lipids of nervous tissue: composition and metabolism. *Prog. Lipid Res.* 24, 69–176.
12. Dennis, E.A. 1994. Diversity of group types, regulation, and function of phospholipase A2. *J. Biol. Chem.* 269, 13057–13060.
13. Horrobin, D.F. 1996. Schizophrenia as a membrane lipid disorder which is expressed throughout the body. *Prostaglandins Leukot. Essent. Fatty Acids* 55, 3–7.
14. Balboa, M.A., Varela-Nieto, I., Killermann Lucas, K. and Dennis, E.A. 2002. Expression and function of phospholipase A(2) in brain. *FEBS Lett.* 531, 12–17.
15. Molloy, G.Y., Rattray, M. and Williams, R.J. 1998. Genes encoding multiple forms of phospholipase A2 are expressed in rat brain. *Neurosci Lett.* 258, 139–142.
16. Farooqui, A.A., Yang, H.C., Rosenberger, T.A. and Horrocks, L.A. 1997. Phospholipase A2 and its role in brain tissue. *J. Neurochem.* 69, 889–901.
17. Farooqui, A.A. and Horrocks, L.A. 2004. Brain phospholipases A2: a perspective on the history. *Prostaglandins Leukot. Essent. Fatty Acids* 71, 161–169.
18. Horrobin, D.F., Glen, A.I. and Vaddadi, K. 1994. The membrane hypothesis of schizophrenia. *Schizophr. Res.* 13, 195–207.
19. Finnen, M.J. and Lovell, C.R. 1991. Purification and characterization of phospholipase A2 from human epidermis. *Biochem. Soc. Trans.* 19, 91S.
20. Mellor, J.E., Laugharne, J.D. and Peet, M. 1996. Omega-3 fatty acid supplementation in schizophrenia patients. *Hum. Psychopharmacol.* 11, 39–46.

21. Maes, M., Stevens, W.J., Declerck, L.S., Bridts, C.H., Peeters, D., Schotte, C. and Cosyns, P. 1993. Significantly increased expression of T-cell activation markers (interleukin-2 and HLA-DR) in depression: further evidence for an inflammatory process during that illness. *Prog. Neuropsychopharmacol. Biol. Psychiatr.* 17, 241–255.
22. Maes, M., Bosmans, E., Calabrese, J., Smith, R. and Meltzer, H.Y. 1995. Interleukin-2 and interleukin-6 in schizophrenia and mania: effects of neuroleptics and mood stabilizers. *J. Psychiatr. Res.* 29, 141–152.
23. Maes, M., Delange, J., Ranjan, R., Meltzer, H.Y., Desnyder, R., Cooremans, W. and Scharpe, S. 1997. Acute phase proteins in schizophrenia, mania and major depression: modulation by psychotropic drugs. *Psychiatr. Res.* 66, 1–11.
24. Segman, R.H., Meltzer, A., Gross-Tsur, V., Kosov, A., Frisch, A., Inbar, E., Darvasi, A., Levy, S., Goltser, T., Weizman, A. and Galili-Weisstub, E. 2002. Preferential transmission of interleukin-1 receptor antagonist alleles in attention deficit hyperactivity disorder. *Mol. Psychiatr.* 7, 72–74.
25. James, M.J., Gibson, R.A. and Cleland, L.G. 2000. Dietary polyunsaturated fatty acids and inflammatory mediator production. *Am. J. Clin. Nutr.* 71(Suppl 1), 343S–348S.
26. Hardardottir, I. and Kinsella, J.E. 1991. Tumor necrosis factor production by murine resident peritoneal macrophages is enhanced by dietary n-3 polyunsaturated fatty acids. *Biochim. Biophys. Acta* 1095, 187–195.
27. Kinsella, J.E., Broughton, K.S. and Whelan, J.W. 1990. Dietary unsaturated fatty acids: interactions and possible needs in relation to eicosanoid synthesis. *J. Nutr. Biochem.* 1, 123–141.
28. Youdim, K.A., Martin, A. and Joseph, J.A. 2000. Essential fatty acids and the brain: possible health implications. *Int. J. Devl. Neurosci.* 28, 383–399.
29. Horrocks, L.A. and Farooqui, A.A. 2004. Docosahexaenoic acid in the diet: its importance in maintenance and restoration of neural membrane function. *Prostaglandins Leukot. Essent. Fatty Acids* 70, 361–372.
30. Wainright, P.E. 2002. Dietary essential fatty acids and brain function: a developmental perspective on mechanisms. *Proc. Nutr. Soc.* 61, 61–69.
31. Paez, X. and Hernandez, L. 1996. Simultaneous brain and blood microdialysis study with a new removable venous probe. Serotonin and 5-hydroxyindolacetic acid changes after D-norfenfluramine or fluoxetine. *Life Sci.* 58, 1209–1221.
32. Post, R.M., Jimerson, D.C., Bunney, W.E. Jr. and Goodwin, F.K. 1980. Dopamine and mania: behavioral and biochemical effects of the dopamine receptor blocker pimozide. *Psychopharmacology (Berl)* 67, 297–305.
33. Malison, R.T., Price, L.H., Berman, R., van Dyck, C.H., Pelton, G.H., Carpenter, L., Sanacora, G., Owens, M.J., Nemeroff, C.B., Rajeevan, N., Baldwin, R.M., Seibyl, J.P., Innis, R.B. and Charney, D.S. 1998. Reduced brain serotonin transporter availability in major depression as measured by [123I]-2 beta-carbomethoxy-3 beta-(4-iodophenyl)tropane and single photon emission computed tomography. *Biol. Psychiatr.*. 44, 1090–1098.
34. Ogilvie, A.D., Battersby, S., Bubb, V.J., Fink, G., Harmar, A.J., Goodwim, G.M. and Smith, C.A. 1996. Polymorphism in serotonin transporter gene associated with susceptibility to major depression. *Lancet* 347, 731–733.
35. Mundo, E., Walker, M., Cate, T., Macciardi, F. and Kennedy, J.L. 2001. The role of serotonin transporter protein gene in antidepressant-induced mania in bipolar disorder: preliminary findings. *Arch. Gen. Psychiatr.* 58, 539–544.
36. De la Presa Owens, S. and Innis, S.M. 1999. Docosahexaenoic acid and arachidonic acid prevent a decrease in dopaminergic and serotoninergic neurotransmitters in

frontal cortex caused by a linoleic and α-linolenic acid deficient diet in formula-fed piglets. *J. Nutr.* 129, 2088–2093.

37. Austead, N., Innis, S.M. and de la Presa Owens, S. 2000. Auditory evoked response and brain phospholipid fatty acids and monoamines in rats fed formula with and without arachidonic acid (AA) and/or docosahexaenoic acid (DHA). In: Watkins P, Spector A, Hamilton J, and Katz R (Eds), *Brain Uptake and Utilization of Fatty Acids: Applications to Peroxisomal Biogenesis Disorders*. National Institutes of Health Conference, Maryland, p. 3.
38. Zimmer, L., Durand, G., Guilloteau, D. and Chalon, S. 1999. n-3 Polyunsaturated fatty acid deficiency and dopamine metabolism in the rat frontal cortex. *Lipids* 34, S251.
39. Zimmer, L., Delpal, S., Guilloteau, D., Aioun, J., Durand, G. and Chalon, S. 2000. Chronic n-3 polyunsaturated fatty acid deficiency alters dopamine vesicle density in the rat frontal cortex. *Neurosci. Lett.* 284, 25–28.
40. Chalon, S., Delion-Vancassel, S., Belzung, C., Guilloteau, D., Leguisquet, A., Besnard, J.C. and Durand, G. 1998. Dietary fish oil affects monoaminergic neurotransmission and behavior in rats. *J. Nutr.* 128, 2512–2519.
41. Stubbs, C.D. and Smith, A.D. 1984. The modification of mammalian membrane polyunsaturated fatty acid composition in relation to membrane fluidity and function. *Biochim. Biophys. Acta.* 779, 89–137.
42. O'Brien, J.S. 1986. Stability of the myselin membrane. *Science* 147, 1099–1107.
43. Yao, J., Stanley, J.A., Reddy, R.D., Keshavan, M.S. and Pettegrew, J.W. 2002. Correlations between peripheral polyunsaturated fatty acid content and in vivo membrane phospholipid metabolites. *Biol. Psychiatr.* 15, 823–830.
44. Demel, R.A., Geuts van Kessel, W.S.M. and Van Deene, L.L.M. 1972. The properties of polyunsaturated lecithins in monolayers and liposomes and the interactions of these lecithins with cholesterol. *Biochim. Biophys. Acta* 266, 26–40.
45. Urquhart, R., Chan, R.Y., Li, Q.T., Tilley, L., Grieser, F. and Sawyer, W.H. 1992. Omega-6 and omega-3 fatty acids: monolayer packing and effects on bilayer permeability and cholesterol exchange. *Biochem. Int.* 26, 831–841.
46. Coolbear, K.P., Berde, C.B. and Keough, K.M. 1983. Gel to liquid-crystalline phase transitions of aqueous dispersions of polyunsaturated mixed-acid phosphatidylcholines. *Biochemistry* 22, 1466–1473.
47. Gibney, M.J. and Bolton-Smith, C. 1988. The effect of a dietary supplement of n-3 polyunsaturated fat on platelet lipid composition, platelet function and platelet plasma membrane fluidity in healthy volunteers. *Br. J. Nutr.* 60, 5–12.
48. Wahnon, R., Cogan, U. and Mokady, S. 1992. Dietary fish oil modulates the alkaline phosphatase activity and not the fluidity of rat intestinal microvillus membrane. *J. Nutr.* 122, 1077–1084.
49. Clamp, A.G., Ladha, S., Clark, D.C., Grimble, R.F. and Lund, E.K. 1997. The influence of dietary lipids on the composition and membrane fluidity of rat hepatocyte plasma membrane. *Lipids* 32, 179–184.
50. Salem, N.J., Kim, H.Y. and Yergey, J.A. 1986. Docosahexaenoic acid: membrane function and metabolism. In: Simopolous AP, Kifer RR, and Martin RE (Eds), *Health Effects of Polyunsaturated Fatty Acids in Seafoods*. Academic, New York, pp. 319–351.
51. Mahadik, S.P. and Mukherjee, S. 1996. Free radical pathology and antioxidant defense in schizophrenia: a review. *Schizophr. Res.* 19, 1–17.
52. Groff, J.L., Gropper, S.S. and Hunt, S.M. 1995. *Advanced Nutrition and Human Metabolism*, 2nd Edition. West Publishing Co, MN.

53. Mahadik, S.P., Evans, D. and Lal, H. 2001. Oxidative stress and role of antioxidant and ω-3 essential fatty acid supplementation in schizophrenia. *Prog. Neuro. Psychoparmacol. Biol. Psychiatr.* 25, 463–493.
54. Hargreaves, K.M. and Clandinin, M.T. 1998. Dietary control of diacylphosphatidylethanolamine species in brain. *Biochim. Biophys. Acta* 962, 98–104.
55. Bjerve, K.S., Brubakk, A.M., Fougner, K.J., Johnsen, H., Midthjell, K. and Vik, T. 1993. Omega-3 fatty acids: essential fatty acids with important biological effects, and serum phospholipid fatty acids as markers of dietary omega 3-fatty acid intake. *Am. J. Clin. Nutr.* 57, 801S–805S.
56. Godley, P.A., Campbell, M.K., Miller, C., Gallagher, P., Martinson, F.E., Mohler, J.L. and Sandler, R.S. 1996. Correlation between biomarkers of omega-3 fatty acid consumption and questionnaire data in African American and Caucasian United States males with and without prostatic carcinoma. *Cancer Epidemiol. Biomarkers Prev.* 5, 115–119.
57. Hjartaker, J., Lund, E. and Bjerve, K.S. 1997. Serum phospholipid fatty acid composition and habitual intake of marine foods registered by a semi-quantitative food frequency questionnaire. *Eur. J. Clin. Nutr.* 51, 736–742.
58. Xiao, Y.F., Kang, J.X., Morgan, J.P. and Leaf, A. 1995. Blocking effects of polyunsaturated fatty acids on Na^+ channels of neonatal rat ventricular myocytes. *Proc. Natl. Acad. Sci.* 92, 11000–11004.
59. Kang, J.X. and Leaf, A. 1996. Evidence that free polyunsaturated fatty acids modify Na^+ channels by directly binding to the channel proteins. *Proc. Natl. Acad. Sci.* 93, 3542–3546.
60. Vreugdenhil, M., Bruehl, C., Voskuyl, R.A., Kang, J.X., Leaf, A. and Wadman, W.J. 1996. Polyunsaturated fatty acids modulate sodium and calcium currents in CA1 neurons. *Proc. Natl. Acad. Sci.* 93, 12559–12563.
61. Bogdanov, K.Y., Spurgeon, H.A., Vinogradova, T.M. and Lakatta, E.G. 1998. Modulation of the transient outward current in adult rat ventricular myocytes by polyunsaturated fatty acids. *Am. J. Physiol.* 274, H571–H579.
62. Seebungkert, B. and Lynch, J.W. 2002. Effects of polyunsaturated fatty acids on voltage-gated K^+ and Na^+ channels in rat olfactory receptor neurons. *Eur. J. Neurosci.* 16, 2085–2094.
63. Honore, E., Barhanin, J., Attali, B., Lesage, F. and Lazdunski, M. 1994. External blockade of the major cardiac delayed-rectifier K^+ channel (Kv1.5) by polyunsaturated fatty acids. *Proc. Natl. Acad. Sci.* 91, 1937–1944.
64. Lauritzen, I., Blondeau, N., Heurteaux, C., Widmann, C., Romey, G. and Lazdunski, M. 2000. Polyunsaturated fatty acids are potent neuroprotectors. *EMBO*, 19, 1784–1793.
65. Heurteaux, C., Guy, N., Laigle, C., Blondeau, N., Duprat, F., Mazzuca, M., Lang-Lazdunski, L., Widmann, C., Zanzouri, M., Romey, G. and Lazdunski, M. 2004. TREK-1, a K^+ channel involved in neuroprotection and general anesthesia. *EMBO* 23, 2684–2695.
66. Xiao, Y.F., Gomez, A.M., Morgan, J.P., Lederer, W.J. and Leaf, A. 1997. Suppression of voltage-gated L-type Ca^{2+} currents by polyunsaturated fatty acids in adult and neonatal rat ventricular myocytes. *Proc. Natl. Acad. Sci.* 94, 4182–4187.
67. Turner, N., Else, P.L. and Hulbert, A.J. 2003. Docosahexaenoic acid (DHA) content of membranes determines molecular activity of the sodium pump: implications for disease states and metabolism. *Naturwissenschaften* 90, 521–523.
68. Kearns, S.D. and Haag, M. 2002. The effect of omega-3 fatty acids on Ca-ATPase in rat cerebral cortex. *Prostaglandins Leukot. Essent. Fatty Acids* 67, 303–308.

69. Gegelashvili, G. and Schousboe, A. 1997. High affinity glutamate transporters: regulation of expression and activity. *Mol. Pharmacol.* 52, 6–15.
70. Nishikawa, M., Kimura, S. and Akaike, N. 1994. Facilitatory effect of docosahexaenoic acid on *N*-methyl-D-aspartate response in pyramidal neurons of rat cerebral cortex. *J. Physiol.* 475, 83–93.
71. Graber, R., Sumida, C. and Nunez, E.A. 1994. Fatty acids and cell signal transduction. *J. Lipid Mediat. Cell Signal.* 9, 91–116.
72. Kogteva, G.S. and Bezuglov, V.V. 1998. Unsaturated fatty acids as endogenous bioregulators. *Biochemistry (Moscow)* 63, 6–15.
73. Wahle, K.W., Rotondo, D. and Heys, S.D. 2003. Polyunsaturated fatty acids and gene expression in mammalian systems. *Proc. Nutr. Soc.* 62, 349–360.
74. Pary, R., Lewis, S., Matuschka, P.R. and Lippmann, S. 2002. Attention-deficit/hyperactivity disorder: an update. *South Med.* 95, 743–749.
75. Spencer, T., Biederman, J., Wilens, T.E. and Faraone, S.V. 1994. Is attention deficit hyperactivity disorder in adults a valid disorder? *Harv. Rev. Psychiatr.* 1, 326–335.
76. Stevens, L.J., Zentall, S.S., Deck, J.L., Abate, M.L., Watkins, B.A., Lipp, S.R. and Burgess, J.R. 1995. Essential fatty acid metabolism in boys with attention-deficit hyperactivity disorder. *Am. J. Clin. Nutr.* 62, 761–768.
77. Young, G.S., Maharaj, N.J. and Conquer, J.A. 2004. Blood phospholipid fatty acid analysis of adults with and without attention deficit/hyperactivity disorder. *Lipids* 39, 117–123.
78. Voigt, R.G., Llorente, A.M., Jensen, C.L., Fraley, J.K., Berretta, M.C. and Heird, W.C. 2001. A randomized, double-blind, placebo-controlled trial of docosahexaenoic acid supplementation in children with attention-deficit/hyperactivity disorder. *J. Pediatr.* 139, 189–196.
79. Richardson, A.J. and Puri, B.K. 2002. A randomized double-blind, placebo-controlled study of the effects of supplementation with highly unsaturated fatty acids on ADHD-related symptoms in children with specific learning difficulties. *Prog. Neuro-Psychopharmacol. Biol. Psychiatr.* 26, 233–239.
80. Stevens, L., Zhang, W., Peck, L., Kuczek, T., Grevsted, N., Mahon, A., Zentall, S.S., Arnold, E. and Burgess, J.R. 2003. EFA supplementation in children with inattention, hyperactivity, and other disruptive behaviours. *Lipids* 38, 1007–1021.
81. Hirayama, S., Hamazaki, T. and Terasawa, K. 2004. Effect of docosahexaenoic acid-containing food administration on symptoms of attention-deficit/hyperactivity disorder—a placebo-controlled double-blind study. *Eur. J. Clin. Nutr.* 58, 467–473.
82. Nitsch, R., Pittas, A., Blusztajn, J.K., Slock, B.E., and Growdon, J. 1991. Alterations of phospholipid metabolites in post-mortem brains from patients with Alzheimer's disease. *Ann. N.Y. Acad. Sci.* 640, 110–113.
83. Soderburg, M., Edlund, C., Kristensson, K. and Dallner, G. 1991. Fatty acid composition of brain phospholipids in agin and Alzheimer's disease. *Lipids* 26, 421–428.
84. Soderburg, M., Edlund, C., Kristensson, K., Alafuzoff, I. and Dallner, G. 1992. Lipid composition in different regions of the brain in Alzheimer's disease/senile dementia of Alzheimer's type. *J. Neurochem.* 59, 1646–1653.
85. Coorigan, F.M., Horrobin, D.F., Skinner, E.R., Besson, J.A. and Cooper, M.B. 1998. Abnormal content of n-6 and n-3 long-chain unsaturated fatty acids in the phosphoglycerides and cholesterol esters of parahippocampus cortex from Alzheimer's disease patients and its relationship to acetyl CoA content. *Int. J. Biochem. Cell. Biol.* 30, 197–207.

86. Mulder, M., Ravid, R., Swaab, D.F., deLoet, E.R., Haasdijk, E.D., Julk, J., van der Bloom, J. and Havekes, L.M. 1998. Reduced levels of cholesterol, phospholipids, and fatty acids in CSF of Alzheimer's disease patients are not related to Apo E4. *Alz. Dis. Assoc. Dis.* 12, 198–203.
87. Prasad, M.R., Lovell, M.A., Yatin, M., Dhillon, H. and Markesbery, W.R. 1998. Regional membrane phospholipid alterations in Alzheimer's disease. *Neurochem. Res.* 23, 81–88.
88. Grant, W.B. 1997. Dietary links to Alzheimer's disease. *Alzheimer Dis. Rev.* 2, 42–55.
89. Kalmijn, S., Feskens, E.J., Launer, L.J. and Kromhout, D. 1997. Polyunsaturated fatty acids, antioxidants, and cognitive function in very old men. *Am. J. Epidemiol.* 145, 33–41.
90. Grant, W.B. 2003. Diet and risk of dementia: does fat matter? The Rotterdam study. *Neurology* 60, 2020–2021.
91. Morris, M.C., Evans, D.A., Bienias, J.L., Tangney, C.C., Bennett, D.A., Wilson, R.S., Aggarwal, N. and Schneider, J. 2003. Consumption of fish and n-3 fatty acids and risk of incident Alzheimer disease. *Arch. Neurol.* 60, 940–946.
92. Kalmijn, S., van Boxtel, M.P.J., Ocke, M., Verschuren, W.M.M., Kromhout, D. and Launer, L.J. 2004. Dietary intake of fatty acids and fish in relation to cognitive performance at middle age. *Neurology* 62, 275–280.
93. Kyle, D.J., Schaefer, E., Patton, G. and Beiser, A. 1999. Low serum docosahexaenoic acid is a significant risk factor for Alzheimer's dementia. *Lipids* 34, S245.
94. Heude, B., Ducimetiere, P., Berr, C. and EVA Study. 2003. Cognitive decline and fatty acid composition of erythrocyte membranes—The EVA study. *Am. J. Clin. Nutr.* 77, 803–808.
95. Tully, A.M., Roche, H.M., Doyle, R., Fallon, C., Bruce, I., Lawlor, B., Coakley, D. and Gibney, M.J. 2003. Low serum cholesteryl ester-docosahexaenoic acid levels in Alzheimer's disease: a case-control study. *Br. J. Nutr.* 89, 483–489.
96. Conquer, J.A., Tierney, M.C., Zecevic, J., Bettger, W.J. and Fisher, R.H. 2000. Fatty acid analysis of blood plasma of patients with Alzheimer's disease, other types of dementia and cognitive impairment. *Lipids* 35, 1305–1312.
97. Laurin, D., Verreault, R., Lindsay, J., Dewailly, E. and Holub, B.J. 2003. Omega-3 fatty acids and risk of cognitive impairment and dementia. *J. Alz. Dis.* 5, 315–322.
98. Terano, T., Fujishiro, S., Ban, T., Yamamoto, K., Tanaka, T., Noguchi, Y., Tamura, Y., Yazawa, K. and Hirayama, T. 1999. Docosahexaenoic acid supplementation improves the moderately severe dementai from thrombotic cerebrovascular diseases. *Lipids* 34, S345–S346.
99. Otsuka, M. 2000. Analysis of dietary factors in Alzheimer's disease: clinical use of nutritional intervention for prevention and treatment of dementia. *Nippon Ronen Igakkai Zasshi* 37, 970–973.
100. Christensen, O. and Christenesen, E. 1988. Fat consumption and schizophrenia. *Acta Psychiatr. Scand.* 78, 587–591.
101. Mellor, J.E., Laugharne, J.D.E. and Peet, M. 1996. Omega-3 fatty acid supplementation in schizophrenia patients. *Hum. Psychopharmacol.* 11, 39–46.
102. Richardson, A.J., Cyhlarova, E. and Ross, M.A. 2003. Omega 3 and omega 6 fatty acid concentrations in red blood cell membranes relate to schizotypal traits in healthy adults. *Prostaglandins Leukotrienes Essent. Fatty Acids* 69, 461–466.
103. Evans, D.R., Parikh, V.V., Khan, M.M., Coussons, C., Buckley, P.F. and Mahadik, S.P. 2003. Red blood cell membrane essential fatty acid metabolism in early psychotic patients following antipsychotic drug treatment. *Prostaglandins Leukotrienes Essent. Fatty Acids* 69, 393–399.

104. Arvindakshan, M., Sitasawad, S., Debsikdar, V., Ghate, M., Evans, D., Horrobin, D.F., Bennett, C., Ranjekar, P.K. and Mahadik, S.P. 2003. Essential polyunsaturated fatty acid and lipid peroxide levels in never-medicated and medicated schizophrenia patients. *Biol. Psychiatry* 53, 56–64.
105. Khan, M.M., Evands, D.R., Gunna, V., Scheffer, R.E., Parikh, V.V. and Mahadik, S.P. 2002. Reduced erythrocyte membrane essential fatty acids and increased lipid peroxides in schizophrenia at the never-medicated first-episode of psychosis and after years of treatment with antipsychotics. *Schizophr. Res.* 58, 1–10.
106. Assies, J., Lieverse, R., Vreken, P., Wanders, R.J., Dingemans, P.M. and Linszen, D.H. 2001. Significantly reduced docosahexaenoic and docosapentaenoic acid concentrations in erythrocyte membranes from schizophrenic patients compared with a carefully matched control group. *Biol. Psychiatr.* 49, 510–522.
107. Peet, M., Laugharne, J., Rangarajan, N., Horrobin, D. and Reynolds, G. 1995. Depleted red cell membrane essential fatty acids in drug-treated schizophrenic patients. *J. Psychiatr. Res.* 29, 227–232.
108. Ranjekar, P.K., Hinge, A., Hegde, M.V., Ghate, M., Kale, A., Sitasawad, S., Wagh, U.V., Debsikdar, V.B. and Mahadik, S.P. 2003. Decreased antioxidant enzymes and membrane essential polyunsaturated fatty acids in schizophrenic and bipolar mood disorder patients. *Psychiatr. Res.* 121, 109–122.
109. Laugharne, J.D., Mellor, J.E. and Peet, M. 1996. Fatty acids and schizophrenia. *Lipids* 31, S163–S165.
110. Mahadik, S.P., Mukherjee, S., Horrobin, D.F., Jenkins, K., Correnti, E.E. and Scheffer, R.E. 1996. Plasma membrane phospholipid fatty acid composition of cultured skin fibroblasts from schizophrenic patients: comparison with bipolar patients and normal subjects. *Psychiatr. Res.* 63, 133–142.
111. Fischer, S., Kissling, W. and Kuss, H.J. 1992. Schizophrenic patients treated with high dose phenothiazine or thioxanthene become deficient in polyunsaturated fatty acids in their thrombocytes. *Biochem. Pharmacol.* 44, 317–323.
112. Hibbeln, J.R., Makino, K.K., Martin, C.E., Dickerson, F., Boronow, J. and Fenton, W.S. 2003. Smoking, gender, and dietary influences on erythrocyte essential fatty acid composition among patients with schizophrenia or schizoaffective disorder. *Biol. Psychiatr.* 53, 431–441.
113. Joy, C.B., Mumby-Croft, R. and Joy, L.A. 2003. Polyunsaturated fatty acid supplementation for schizophrenia. *Cochrane Database Syst. Rev.* CD001257.
114. Peet, M. 2003. Eicosapentaenoic acid in the treatment of schizophrenia and depression: rationale and preliminary double-blind clinical trial results. *Prostaglandins Leukotrienes Essent. Fatty Acids* 69, 477–485.
115. Peet, M. 2004. Nutrition and schizophrenia: beyond omega-3 fatty acids. *Prostaglandins Leukotrienes Essent. Fatty Acids* 70, 417–422.
116. Emsley, R., Myburgh, C., Oosthiuzen, P. and van Rensburg, S.J. 2002. Randomized, placebo-controlled study of ethyl-eicosapentaenoic acid as supplemental treatment in schizophrenia. *Am. J. Psychiatr.* 159, 1596–1598.
117. Peet, M., Horrobin, D.F. and E-E Muticentre group. 2002. A dose-ranging exploratory study of the effects of ethyl-eicosapentaenoate in patients with persistent schizophrenic symptoms. *J. Psychiatr. Res.* 36, 7–18.
118. Fenton, W.S., Dickerson, F., Boronow, J., Hibbeln, J.R. and Knable, M. 2001. A placebo-controlled trial of omega-3 fatty acid (ethyl eicosapentaenoic acid) supplementation for residual symptoms and cognitive impairment in schizophrenia. *Am. J. Psychiatr.* 158, 2071–2074.
119. Arvindakshan, M., Ghate, M., Ranjekar, P.K., Evans, D.R. and Mahadik, S.P. 2003. Supplementation with a combination of omega-3 fatty acids and antioxidants

(vitamins E and C) improves the outcome of schizophrenia. *Schizophr. Res.* 62, 195–204.
120. Su, K.P., Shen, W.W. and Huang, S.Y. 2001. Omega-3 fatty acids as a psychotherapeutic agent for a pregnant schizophrenic patient. *Eur. Neuropsychopharmacol.* 11, 295–299.
121. Richardson, A.J., Easton, T. and Puri, B.K. 2000. Red cell and plasma fatty acid changes accompanying symptom remission in a patient with schizophrenia treated with eicosapentaenoic acid. *Eur. Neuropsychopharmacol.* 10, 189–193.
122. Puri, B.K., Richardson, A.J., Horrobin, D.F., Easton, T., Saeed, N., Oatridge, A., Hajnal, J.V. and Bydder, G.M. 2000. Eicosapentaenoic acid treatment in schizophrenia associated with symptom remission, normalisation of blood fatty acids, reduced neuronal membrane phospholipid turnover and structural brain changes. *Int. J. Clin. Pract.* 54, 57–63.
123. Richardson, A.J., Easton, T., Gruzelier, J.H. and Puri, B.K. 1999. Laterality changes accompanying symptom remission in schizophrenia following treatment with eicosapentaenoic acid. *Int. J. Psychophysiol.* 34, 333–339.
124. Stoll, A.L., Damico, K.E., Daly, B.P., Severus, E. and Marangell, L.B. 2001. Methodological considerations in clinical studies of ω3 fatty acids in major depression and bipolar disorder. *World Rev. Nutr. Diet.* 88, 58–67.
125. Logan, A.C. 2003. Neurobehavioural aspects of omega-3 fatty acids: possible mechanisms and therapeutic value in major depression. *Altern. Med. Rev.* 8, 410–425.
126. Mischoulon, D. and Fava, M. 2000. Docosahexaenoic acid and omega-3 fatty acids in depression. *Psychiatr. Clin. North Am.* 23, 785–794.
127. Hibbeln, J.R. and Salem, N. 1995. Dietary polyunsaturated fatty acids and depression: when cholesterol does not satisfy. *Am. J. Clin. Nutr.* 62, 1–9.
128. Noaghiul, S. and Hibbeln, J.R. 2003. Cross national relationship of seafood consumption and rates of bipolar disorders. *Am. J. Psychiatr.* 160, 2222–2227.
129. Tanskanen, A., Hibbeln, J.R., Tuomilehto, J., Uutela, A., Haukkala, A., Viinamaki, H., Lehtonen, J. and Vartiainen, E. 2001. Fish consumption and depressive symptoms in the general population in Finland. *Psychiatr. Serv.* 52, 529–531.
130. Suzuki, S., Akechi, T., Kobayashi, M., Taniguchi, K., Goto, K., Sasaki, S., Tsugane, S., Nishiwaki, Y., Miyaoka, H. and Uchitomi, Y. 2004. Daily omega-3 fatty acid intake and depression in Japanese patients with newly diagnosed lung cancer. *Br. J. Cancer,* 90, 787–793.
131. Frasure-Smith, N., Lesperance, F. and Julien, P. 2004. Major depression is associated with lower omega-3 fatty acid levels in patients with recent acute coronary syndromes. *Biol. Psychiatr.* 55, 891–896.
132. De Vriese, S.R., Christophe, A.B. and Maes, M. 2004. In humans, the seasonal variation in poly-unsaturated fatty acids is related to the seasonal variation in violent suicide and serotonergic markers of violent suicide. *Prostaglandins Leukotrienes Essent. Fatty Acids* 71, 13–18.
133. Hakkarainen, R., Partonen, T., Haukka, J., Virtamo, J., Albanes, D. and Lonnqvist, J. 2004. Is low dietary intake of omega-3 fatty acids associated with depression? *Am. J. Psychiatr.,* 161, 567–569.
134. Peet, M., Murphy, B., Shay, J. and Horrobin, D. 1998. Depletion of omega-3 fatty acid levels in red blood cell membranes of depressive patients. *Biol. Psychiatr.* 43, 315–319.
135. Edwards, R., Peet, M., Shay, J. and Horrobin, D. 1998. Omega-3 polyunsaturated fatty acid levels in the diet and in red blood cell membranes of depressed patients. *J. Affect. Disord.* 48, 149–155.

136. Assies, J., Lok, A., Bockting, C.L., Weverling, G.J., Lieverse, R., Visser, I., Abeling, N.G.G.M., Duran, M. and Schene, A.H. 2004. Fatty acids and homocysteine levels in patients with recurrent depression: an explorative pilot study. *Prostaglandins Leukotrienes Essent. Fatty Acids* 70, 349–356.
137. Mamalakis, G., Kiriakakis, M., Tsibinos, G. and Kafatos, A. 2004. Depression and adipose polyunsaturate d fatty acids in the survivors of the Seven Countries Study population of Crete. *Prostaglandins Leukotrienes Essent. Fatty Acids* 70, 495–501.
138. Tiemeier, H., van Tuijl, H.R., Hofman, A., Kiliaan, A.J. and Breteler, M.M. 2003. Plasma fatty acid composition and depression are associated in the elderly: the Rotterdam study. *Am. J. Clin. Nutr.* 78, 40–46.
139 Mamalakis, G., Tornaritis, M. and Kafatos, A. 2002. Depression and adipose essential polyunsaturated fatty acids. *Prostaglandins Leukotrienes Essent. Fatty Acids* 67, 311–318.
140. Maes, M., Smith, R., Christophe, A., Cosyns, P., Desnyder, R. and Meltzer, H. 1996. Fatty acid composition in major depression: decreased ω3 fractions in cholesteryl esters and increased C20:4 ω6/C20:5 ω3 ration in cholesteryl esters and phospholipids. *J. Affect. Disord.* 38, 35–46.
141. Maes, M., Christophe, A., Delanghe, J., Altamura, C., Neels, H., and Meltzer, H.Y. 1999. Lowered omega3 polyunsaturated fatty acids in serum phospholipids and cholesteryl esters of depressed patients. *Psychiatr. Res.* 85, 275–291.
142. Ranjekar, P.K., Hinge, A., Hegde, M.V., Ghate, M., Kale, A., Sitasawad, S., Wagh, U.V., Debsikdar, V.B. and Mahadik, S.P. 2003. Decreased antioxidant enzymes and membrane essential polyunsaturated fatty acids in schizophrenic and bipolar mood disorder patients. *Psychiatr. Res.* 121, 109–122.
143. Adams, P.B., Lawson, S., Sanigorski, A. and Sinclair, A.J. 1996. Arachidonic acid to eicosapentaenoic acid ratio in blood correlates positively with clinical symptoms of depression. *Lipids* 31, S157–S161.
144. Hibbeln, J.R. 2002. Seafood consumption, the DHA content of mothers' milk and prevalance rates of postpartum depression: a cross-national, ecological analysis. *J. Affect. Disord.* 69, 15–29.
145. Makrides, M., Crowther, C.A., Gibson, R.A., Gibson, R.S. and Skeaff, C.M. 2003. Docosahexaenoic acid and post-partum depression—is there a link? *Asia Pac. J. Clin. Nutr.* 12, S37.
146. Otto, S.J., de Groot, R.H. and Hornstra, G. 2003. Increased risk of postpartum depressive symptoms is associated with slower normalization after pregnancy of the functional docosahexaenoic status. *Prostaglandins Leukotrienes Essent. Fatty Acids* 69, 237–243.
147. De Vriese, S.R., Christophe, A.B. and Maes, M. 2003. Lowered serum n-3 polyunsaturated fatty acid (PUFA) levels predict the occurrence of postpartum depression: further evidence that lowered n-3 PUFAs are related to major depression. *Life Sci.* 73, 3181–3187.
148. Peet, M. and Horrobin, D.F. 2002. A dose-ranging study of the effects ethyl-eicosapentaenoate in patients with ongoing depression despite apparently adequate treatment with standard drugs. *Arch. Gen. Psychiatr.* 59, 913–919.
149. Puri, B.K., Counsell, S.J., Hamilton, G., Richardson, A.J. and Horrobin, D.F. 2001. Eicosapentaenoic acid in treatment resistant depression associated with symptom remission, structural brain changes and reduced neuronal phospholipid turnover. *Int. J. Clin. Pract.* 55, 560–563.
150. Su, K.P., Huang, S.Y., Chiu, C.C. and Shen, W.W. 2003. Omega-3 fatty acids in major depressive disorder. A preliminary double-blind, placebo-controlled trial. *Eur. Neuropsychopharmacol.* 13, 267–271.

151. Marangell, L.B., Martinez, J.M., Zboyan, H.A., Kertz, B., Kim, H.F. and Puryear, L.J. 2003. A double-blind, placebo-controlled study of the omega-3 fatty acid docosahexaenoic acid in the treatment of major depression. *Am. J. Psychiatr.* 160, 996–998.
152. Stoll, A.L., Severus, W.E., Freeman, M.P., Rueter, S., Zboyan, H.A., Diamond, E., Cress, K.K. and Marangell, L.B. 1999. Omega-3 fatty acids in bipolar disorder: a preliminary double-blind, placebo-controlled trial. *Arch. Gen. Psychiatr.* 56, 407–412.
153. Marangell, L.B., Martinez, J.M., Zboyan, H.A., Chong, H. and Puryear, L.J. 2004. Omega-3 fatty acids for prevention of postpartum depression: negative data from a preliminary, open-label pilot study. *Depress. Anxiety* 19, 20–23.
154. Llorente, A.M., Jensen, C.L., Voigt, R.G., Fraley, J.K., Berretta, M.C. and Heird, W.C. 2003. Effect of maternal docosahexaenoic acid supplementation on postpartum depression and information processing. *Am. J. Obstet. Gynecol* 188, 1348–1353.
155. Stordy, B.Y. 2000. Dark adaptation, motor skills, docosahexaenoic acid, and dyslexia. *Am. J. Clin. Nutr.* 71(Suppl 1), 323S–326S.
156. Vancassel, S., Durand, G., Barthelemy, C., Lejeune, B., Martineau, J., Guillooteau, D., Andres, C. and Chalon, S. 2001. Plasma fatty acid levels in autistic children. *Prostaglandins Leukotrienes Essent. Fatty Acids* 65, 1–7.
157. Bell, J.G., Sargent, J.R., Tocher, D.R. and Dick, J.R. 2000. Red blood cell fatty acid compositions in a patient with autistic spectrum disorder: a characteristic abnormality in neurodevelopmental disorders? *Prostaglandins Leukotrienes Essent. Fatty Acids* 63, 21–25.
158. Richardson, A.J. 2004. Clinical trials of fatty acid treatment in ADHD, dyslexia, dyxpraxia and the autistic spectrum. *Prostaglandins Leukotrienes Essent. Fatty Acids* 70, 383–390.
159. Zanarini, M.C. and Frankenburg, F.R. 2003. Omega-3 fatty acid treatment of women with borderline personality disorder: a double-blind, placebo-controlled pilot study. *Am. J. Psychiatr.* 160, 167–169.
160. Fux, M., Benjamin, J. and Nemets, B. 2004. A placebo-controlled cross-over trial of adjunctive EPA in OCD. *J. Psychiatr. Res.* 38, 323–325.
161. Hibbeln, J.R. 2001. Seafood consumption and homicide mortality. *World Rev. Nutr. Diet.* 88, 41–46.
162. Iribarren, C., Markovitz, J.H., Jacobs, D.R. Jr., Schreiner, P.J., Daviglus, M. and Hibbeln, J.R. 2004. Dietary intake of n-3, n-6 fatty acids and fish: relationship with hostility in young adults—the CARDIA study. *Eur. J. Clin. Nutr.* 58, 24–31.
163. Hamazak, T., Thienprasert, A., Kheovichai, K., Samuhaseneetoo, S., Nagasawa, T. and Watanabe, S. 2002. The effect of docosahexaenoic acid on aggression in elderly Thai subjects—a placebo controlled double blind study. *Nutr. Neurosci.* 5, 37–41.
164. Hamazaki, T., Sawazaki, S., Itomura, M., Asaoka, E., Nagao, Y., Nishimura, N., Yazawa, K., Kuwamori, T. and Kobayashi, M. 1996. The effect of docosahexaenoic acid on aggression in young adults. A placebo-controlled double-blind study. *J. Clin. Invest.* 97, 1129–1133.
165. Sawazaki, S., Hamazaki, T., Yazawa, K. and Kobayashi, M. 1999. The effect of docosahexaenoic acid on plasma catecholamine concentrations and glucose tolerance during long-lasting psychological stress: a double-blind placebo-controlled study. *J. Nutr. Sci. Vitaminol.* 45, 655–665.

166. Hamazaki, T., Sawazaki, S., Nagao, Y., Kuwamori, T., Yazawa, K., Mizushima, Y. and Kobayashi, M. 1998. Docosahexaenoic acid does not affect aggression of normal volunteers under nonstressful conditions. A randomized, placebo-controlled, double-blind study. *Lipids* 33, 663–667.
167. Buydens-Branchey, L., Branchey, M., McMakin, D.L. and Hibbeln, J.R. 2003. Polyunsaturated fatty acid status and aggression in cocaine addicts. *Drug Alcohol Depend.* 71, 319–323.

4 Omega-3 Fatty Acids in the Treatment of Neurodegenerative Diseases through Anti-Inflammation and Neuroprotection: A Review of Studies in Animal Models

Cai Song

CONTENTS

4.1 INTRODUCTION

In the past decade, several epidemiologic investigations have reported that the consumption of fish (rich in unsaturated omega-3 fatty acids) is associated with the lower onset of neurodegenerative diseases, such as Alzheimer's disease (AD), depression, and Parkinson's disease (PD). For example, intake of fatty fish more than twice per week has been associated with a reduction in the risk of dementia by 28% and AD by 41% in comparison to those who ate fish less than once per month [1]. A recent longitudinal investigation has shown that high omega (n)-3 fatty acid intake is associated with a lower risk of PD [2]. In addition, significant decreases in n-3 fatty acids have been found in the blood of AD patients. These findings suggest that high intake of n-3 fatty acids may protect against AD or PD [3,4]. Indeed, clinical and experimental studies have demonstrated that n-3 fatty acids benefit psychiatric and neurodegenerative diseases. For example, unsaturated fatty acids and pure ethyl eicosapentaenoate (EPA) successfully treated depression and Huntington's disease [5–7], and docosahexaenoic acid (DHA) or EPA can significantly improve memory [8,9]. Recently, many studies have been carried out on animal models of neurodegeneration or on animal tissues or cells. These studies from behavioral, neuroendocrine, neurochemical, and immune aspects have revealed important mechanisms by which n-3 fatty acids in the treatments of neurodegenerative diseases, which provide information for further developments of n-3 fatty acids as a new generation of drugs. This chapter provides a review of the therapeutic mechanisms and recent research progress in this field.

4.2 THE FUNCTION OF n-3 FATTY ACIDS IN THE CENTRAL NERVOUS AND IMMUNE SYSTEMS

Long-chain polyunsaturated fatty acids (PUFA) are synthesized from dietary precursors such as α-linolenic (n-3) and linoleic (n-6) fatty acids. The n-3 fatty acids include EPA and DHA, while the n-6 fatty acids include dihomo-γ-linolenic acid (DGLA) and arachidonic acid (AA). These fatty acids are important components of membrane phospholipids in neurons, glial, and immune cells, and are involved in many functions of the immune and central nervous system (CNS) for the following reasons [10]. First, changes in membrane fatty acid components may change the function of receptors, enzymes, and peptides in the CNS and immune system. It is known that the quaternary structures of proteins and the final modeling and folding often depend on the precise nature of the lipid environment of the proteins because a high proportion of proteins in the cell is actually embedded in the membrane. Second, fatty acids can influence signal transduction. Neurotransmitters, hormones, and cytokines hit a target and induce functional changes by activating phospholipases that then generate a wide range of cell signaling or signal transduction. Third, fatty acids and other lipids can switch on and off many different genes. In particular, by binding to peroxisome proliferator activated receptors, fatty acids can switch on and off the whole genetic programs. Peroxsomal enzymes are essential for the synthesis of plasmalogen, which is used for myelin formation [11]. Lipids or carbohydrates also modulate heat shock proteins that aid the expression of mRNA and the synthesis of proteins. Fourth, fatty acids influence ongoing metabolic regulation. Several studies have revealed that phospholipids undergo constant remodeling, with key fatty acid components having half-lives of a few minutes. Therefore, changes in membrane fatty acid structure and concentration may directly affect neuronal and immune cell functions.

In the immune system, the onset of autoimmune and inflammatory diseases has been related to an imbalanced intake of n-3 and n-6 fatty acids [12]. Inflammation is an important component of the early immunological responses, while inappropriate or dysfunctional immune responses underlie chronic inflammatory and autoimmune diseases. AA, an n-6 fatty acid, is the precursor of eicosanoids that produce prostaglandins (PG) E2, leukotrienes, thromboxanes (TXs), and related compounds that activate macrophages, produce proinflammatory cytokines, and shift the response of T helper type (Th) 1 and Th2. Th1 cells trigger proinflammatory responses, while Th2 cells suppress Th1 responses. In contrast to AA, high intake of long-chain n-3 fatty acids, such as EPA (in fish oils) inhibit certain immune functions, for example, antigen presentation, adhesion molecule expression, Th1 responses, and the production of eicosanoids and proinflammatory cytokines. Clinical studies have also reported that oral fish oil supplementation has beneficial effects on rheumatoid arthritis and asthma [12], supporting the idea that the n-3 fatty acids are anti-inflammatory.

It is known that 60% of the brain's weight is made up of lipids in which essential fatty acids (n-3 and n-6) are important membrane components. Free fatty acids that are released into the blood, and then across the blood–brain barrier, can act at specific binding sites in the brain. Changes in the phospholipid content of neuronal membranes directly affect membrane viscosity and fluidity (unsaturated fatty acids reduce cholesterols and increase fluidity), which may cause abnormalities in basic physiological functions such as neuronal function, neurotransmitter binding and reuptake, membrane enzyme binding, receptor density and affinity, ion channels, and hormone secretion. In the CNS, n-3 and n-6 fatty acids have been shown to fulfill different roles. AA enhances the release of glutamate neurotransmitter, inhibits neurotransmitter uptake, stimulates stress hormone secretion, and enhances synaptic transmission [10,13]. In the brain, AA may trigger microglia activity and induce inflammatory responses. As a consequence of glial activation and inflammatory responses, oxidants will be produced, which are a major cause for neurodegeneration. Therefore, AA may contribute to inflammatory and oxidative toxicity in neurodegenerative diseases. In contrast, n-3 fatty acids have been found to compete with n-6 fatty acids. Both EPA and DHA have been found to protect neurons from inflammation and oxidants. However, n-3 and n-6 fatty acids also interact with and synergize each other. For example, it has been reported that combinations of n-3 and n-6 fatty acids at a ratio of 1:1 or 1:4 yield stronger anti-inflammatory or antistress effects [14,15].

4.3 INFLAMMATION, CYTOKINES, AND NEURODEGENERATION

Recent evidence shows that inflammatory abnormalities are involved in the onset and progression of a number of neurodegenerative disease, including AD, amyotrophic lateral sclerosis (ALS), and PD [16,17]. In the blood of AD patients, increased autoimmune and inflammatory responses have been found, which include increased lymphocyte function, and production of proinflammatory cytokines and its complement [17,18]. Proinflammatory cytokines, such as interleukin (IL)-1β, can activate glial cells in the brain. Glial cells (microglia and astrocytes) play a major role in the neuroinflammation [19,20] because (1) the activation of glial cells leads to enhancements in the production of inflammatory mediators, such as tumor necrosis factor-α (TNF-α), IL-1β, and prostaglandin (PG)E2 [117]; (2) these cytokines stimulate the hypothalamic–pituitary–adrenal (HPA) axis to release glucocorticoids (GCs). Blunted HPA axis function has been associated with autoimmune/inflammatory diseases and neurodegeneration [21,22]. Excessive production of GCs may result in hippocampal atrophy, suppress long-term potentiation (LTP, for memory), and suppress acetylcholine release [23–25]; (3) microglia- and proinflammatory cytokines-induced inflammation also rapidly release toxic reactive oxygen species (ROS) and turn on inducible nitric oxide (iNOS) to produce NO, which is mediated by phospholipases (PL)A_2 enzymes [26]. These oxidants are highly toxic to cells and cause apoptosis; and (4) IL-1β can induce the expression of amyloid precursor proteins (APP). Indeed, in the brains of patients with AD, activated microglia, increased expression of PLA_2 and production of proinflammatory cytokines, and increased production of NO and other oxidants have been consistently reported [27,28].

In the brain of PD patients, proinflammatory cytokines and oxidants are also regulated by microglia and astrocytes. The substantia nigra (SN) is the brain region with the highest density of microglial cells [29]. Robust microglial activation and reduced antioxidant capacity of dopamine (DA) neurons have been reported in the brain of PD patients [30]. This implies that dopamine neurons are particularly susceptible to microglia-mediated inflammation and oxidative toxicity in PD because both inflammation and oxidation can directly damage DA neurons. Recent evidence shows that GC issue may also be associated with PD neuropathology [31]. Glucocorticoid receptors (GRs) are widely distributed in the striatum, frontal cortex, and subtantia nigra. Owing to stimulation by inflammation, the overproduction of GCs may downregulate GRs. It has been reported that the GR deficiency in transgenic (Tg) mice expressing GR antisense RNA from early embryonic life has a dramatic impact on "programming" the vulnerability of DA neurons to 1-methyl-4-phenyl-1,2,3,6-tetrahydropyridine (MPTP), a model of PD [32]. Dexamethasone protects against DA neuron damage in a mouse model of PD [33].

Brain inflammation might be an essential cofactor in HD. A microarray investigation has revealed that TH1-mediated cytokines may be important regulators of affected genes. It is known that inflammation increases indoleamine dioxygenase activity, which may cause abnormalities in the kynurenine pathway in HD. Increased production of oxidants such as NO and NOS have been also reported in HD [34,35].

It is well known that ALS is also a neuroinflammatory disease. Large accumulations of microglia and macrophages have been found in the motor cortex and spinal cord of patients with ALS [16]. An increase in the mRNA expression and synthesis of cyclooxygenase 2 (COX2) mRNA, a trigger for inflammation, in the spinal cord has been reported [16]. Tumor necrosis factor (TNF)-α and other proinflammatory cytokines may also play important roles in ALS [36].

Except for inflammatory disorders, the other common change in neurodegenerative disease is the dysfunction of the neurotrophic system. In contrast with the neurotoxic effects, a neuroprotective effect of astrocytes in the process of neurodegeneration has been observed [37]. Some investigations also support that both microglia and astrocytes, while responding to proinflammatory agents, such as lipopolysaccharides (LPS) and IL-1β, can protect neuronal cells by secreting neurotrophic factors (NTFs) [38,39]. NTFs, such as nerve growth factor (NGF), brain-derived neurotrophic factor (BDNF), and glial cell line–derived neurotrophic factor (GDNF), act as growth factors for the phenotypic development and maintenance of specific neuronal populations in developing and adult vertebrate nervous system [40]. These diffusible proteins that act via retrograde signaling from target neurons and by paracrine and autocrine mechanisms regulate many aspects of neuronal and glial structure and function [41]. When neurons fail to obtain a sufficient quantity of NTFs, programmed cell-death is triggered [42]. NTF signals are processed via specific multicomponent receptor complexes. Among them, the NGF-super family receptors consist of $p75^{NTR}$ and the receptor protein tyrosine kinases (Trk), including TrkA, TrkB, and TrkC. The $p75^{NTR}$ receptor has been found to induce apoptosis [43], whereas the Trk family, after binding to specific NTFs, is able to promote neuron survival and function [44]. Changes in NTF levels have been suggested to contribute to the pathogenesis of AD [45,46]. Alzheimer's-like neurodegeneration was demonstrated in anti-NGF transgenic mice [47].

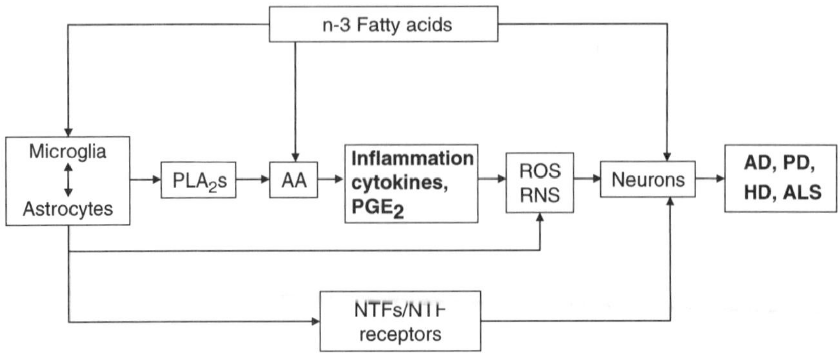

FIGURE 4.1 Interrelationship among glial activation, PLA_2, inflammation, oxidants, and neurodegeneration, and the possible target of n-3 fatty acids.

Upon the activation of astrocytes (maybe microglia as well), phospholipase A_2 (PLA_2) activity and expression will be triggered [28]. Many PLA_2 enzymes are expressed in both immune and brain cells [48]. Accumulating evidence has shown that these PLA_2s are involved in the pathology of neurodegenerative diseases. More recently, beta-amyloid (Aβ) stimulated ROS production through an extracellular signal-regulated kinase-PLA_2-dependent pathway has been reported, strongly indicating that PLA_2 activation is positively associated with Aβ-induced neurotoxicity [49]. Among three major PLA_2s, $cPLA_2$ is the key enzyme releasing n-6 fatty acid AA from membrane phospholipids. As explained earlier, AA is the precursor of eicosanoids that produce PGE_2 and proinflammatory cytokines. Overintake of n-6 fatty acids has been correlated with a high risk of inflammatory or autoimmune disorders and neurodegenerative diseases [50,51]. The possible pathway and interrelationship among glial activation, PLA_2, inflammation, oxidants, and neurodegeneration, and the possible target of n-3 fatty acids are summarized in Figure 4.1.

4.4 ANIMAL MODELS OF ALZHEIMER'S DISEASE AND n-3 FATTY ACID TREATMENTS

To explore the mechanisms involved in different neurodenegerative diseases, many rodent models have been popularly used. Since inflammation is characteristic of a broad spectrum of neurodegenerative diseases, several major animal models related to neurodegeneration and neuroinflammation are reviewed and reported here.

4.4.1 Aβ-Induced Models

4.4.1.1 *In Vivo* Model

Studies in animal models have revealed that the formation of Aβ and other derivatives of the APP are key factors in cellular changes in the AD brain, including the generation of inflammation and oxidative damage. Intracerebraventricular (i.c.v.) administration of Aβ 25–35 causes the loss of axons, dendrites, and synapses in the brain [52]. Aβ peptide 1–40 can suppress LTP in ageing rats, and induce memory deficiency [53]. Aβ 25–35 can also induce memory impairment in mice or rats [54,55]. After chronic Aβ administration, in the CA1 and the dentate gynus of the hippocampus, increases in PLA_2 expressions and inflammatory responses, increased immunostainings of NK-κB, OX6 (a marker for microglial activation), PGE_2 and COX2, have been reported [26,56].

4.4.1.2 *In Vitro* Models

To understand the cellular and molecular abnormalities in AD, neurons or neuronlike cell lines and glial cells have been cocultured with Aβ peptides. With rat astrocyte cultures, oligomeric and fibrillar Aβ induced IL-1β release. In addition, oligomeric Aβ increased the production of iNOS, NO, and TNF-α [57]. In a similar *in vitro* model, Aβ also activated microglial activity, which produced proinflammatory cytokines, and triggered inflammatory response by releasing leukotriene B4 and producing oxidants [58]. On differentiated PC12 cells, a reliable model of neuronal cells, Aβ (1–42) was incubated for 72 h. A markedly reduced cell viability was found by measuring MTT conversion [59], which indicates Aβ causing cell apoptosis or death. With regards to the GC function, an interesting experiment studied whether the GCs altered Aβ uptake in glialike N9 cell line *in vitro*. GCs dose-dependently enhanced these cells to accumulate Aβ. GCs also dose-dependently potentiated Aβ-induction of nitric oxide [60]. These data further provide evidence that GC release in response to inflammation or stress can induce neurodegeneration.

4.4.2 Transgenic Models

4.4.2.1 The Tg2576 Transgenic Mouse Model

Since Aβ plaque deposition is the major pathological hallmark of AD, Tg2576 mice, which overexpress the human APP, are widely used as an AD model [61]. In these animals, amyloid plaques are distributed in cortical and hippocampal regions after 10 months of aging in a pattern similar to that observed in humans [62]. Levels of Aβ 1–40 and 1–42 also increase prior to plaque deposition, with measurable levels apparent by 6–8 months of age [62]. In aged Tg2576 mice, memory impairment in attentional set-shifting test, spatial memory, and working memory in different testing apparatuses have been reported [26]. In this model, increased inflammatory responses, such as the microglia activity, brain concentrations of IL-1, TNF-α and PGE_2, and oxidative production have been reported [63,64]. Furthermore, Th1-produced proinflammatory cytokine interferon (IFN)-γ is increased, while anti-inflammatory cytokine IL-4 is decreased in aged Tg2576 mice [65]. The imbalance between Th1- and Th2-produced cytokines may also contribute to neuroinflammation and the formation of the amyloid plaque.

4.4.2.2 Other Transgenic Models

Other Aβ-related transgenic models include amyloid APP23 transgenic mice, transgenic APPswe/PS1dE9, TgCRND8 and BACE/APPV717F single- or double-transgenic mice. The common character of these models is the deposition of Aβ or expressed human APP in the brain [66–68]. In these models, activated microglia and astrocytes, and increased production of proinflammatory cytokines and oxidants, have also been found [67,69]. Animals also show a deficiency in spatial memory. In addition, impaired acetylcholine receptors and a deficiency in noradrenaline neurotransmission (which occurs in early stages of AD) have been reported in APPswe/PS1dE9 mice and APP23 transgenic mice, respectively [68,70,71].

4.4.3 Inflammation-Induced Model

4.4.3.1 IL-1-Induced Model

Experimental and clinical data have suggested IL-1β as being the most important mediator of neurodegenerative diseases. First a genetic study has shown that the inheritance of a specific IL-1 gene polymorphism is associated with an earlier age of AD onset and increases the risk for AD development by as much as sixfolds [72]. Second, IL-1 cocultured with purified cortical neurons has been found to dose-dependently increase the mRNA expression and the concentration of APP [19,72]. Third, IL-1 can activate and mediate glial functions. When cocultured with microglia, IL-1β-induced APP expression was exacerbated [72], and IL-1 receptor antagonist (RA) can block microglia-induced elevation of α-synuclein, total tau and p-tau, synaptophysin, and MAPK-p38 expression [72]. Fourth, IL-1β injected into the nucleus basalis of rats activates microglia and astrocytes, induces nitric oxide (NO) production, and increases glutamate release from the cortex [73]. In the past decade, we have demonstrated that IL-1-induced changes in memory, some neurotransmitter functions, and GC concentrations are similar to those observed in patients with AD. These could be summarized as (1) subchronic and i.c.v. administration of IL-1β significantly impairs animal spatial and working memory in the Morris water maze and the 8-arms radial maze. The memory deficiency can be blocked by IL-1RA [74,75]; (2) IL-1β administration induces an imbalance between prostaglandin (PG)E2 and IL-10 in the limbic system [75]; (3) IL-1β administration reduces noradrenalin but increases the turnover of serotonergic and dopaminergic systems in the limbic areas [42,74,75], and reduces ACh release from the hippocampus [25]; (4) IL-1β stimulates the HPA axis to secret GCs, and induces stress and anxietylike behavior [8,42]. Moreover, we have reported that IL-1-induced deficiency in working memory can be attenuated by an antagonist of GCs [8]; and (5) IL-1β reduces the expression of NGF in the hippocampus [76].

4.4.3.2 Inflammation-Induced *in Vitro* Models

In an *in vitro* model, an inflammatory reaction in the brain was studied by coculturing microglial and neuronal cell lines. Upon stimulation with IFN-γ and lipopolysaccharide (LPS, an endotoxin from bacterial walls), the microglial cells were activated and secreted TNF-α, PGE_2, and nitric oxide (NO). Neuronal degeneration was quantified by measuring the concentrations of the microtubule-associated protein, tau, and neuron-specific enolase, which are also used as diagnostic tools in AD, in supernatants. In activated contact cocultures, the levels of these neuronal markers were significantly raised compared to nonactivated cocultures. NO-synthase inhibitors significantly diminished the rise of tau in activated cocultures, while indomethacin, superoxide dismutase, or a neutralizing TNF-α antibody did not. When a chemical NO-donor or TNF-α was added to pure neuronal cultures, cell viability was significantly reduced. TNF-α increased neuronal sensitivity toward NO, and a part of the cells died by apoptosis [77]. These findings further support that inflammation is a major contributor to AD.

4.4.4 Effects of n-3 Fatty Acid Treatments

Recently, the effect of n-3 fatty acids on Aβ-induced changes have been reported. For example, the effect of DHA on the levels of Aβ peptide (1–40) and cholesterol in membrane fractions of the cerebral cortex were studied. Animal learning-related memory in the eight-arm radial-maze (for both spatial and working memory) was tested after chronically infusing Aβ (1–40) into the cerebral ventricle [78]. The infusion increased the levels of Aβ peptide and cholesterol, and increased reference memory errors (measured by tasks) compared with those of vehicle rats. After DHA feeding for 7 weeks to AD model rats, a decrease in the level of Aβ peptides and cholesterol in the membrane fraction were observed. Regression analysis revealed a significant and positive correlation between Aβ peptide and cholesterol, and between the number of reference memory errors and cholesterol. Moreover, a significant negative correlation was found between the number of reference memory errors and the mole ratio of DHA to cholesterol. These results suggest that the DHA-induced protection against memory deficits in AD model rats is related to the modulation of the interaction between cholesterol and Aβ peptides [78]. In another experiment, Aβ central administration also impaired rat-conditioned memory in a passive avoidant apparatus. Preadministration of DHA had a profound beneficial effect on the decline in avoidance learning ability in the AD model rats, associated with an increase in the cortico-hippocampal DHA/AA mole ratio, and a decrease in neuronal apoptotic products. Furthermore, DHA preadministration increased cortico-hippocampal glutathione levels and glutathione reductase activity (both are antioxidants), and suppressed the increase in lipid peroxide and reactive oxygen species levels in the cerebral cortex and hippocampus of the AD model rats, suggesting an increase in antioxidative defence [9]. In addition, in the aging rats, the deficit in LTP was accompanied by an increased expression of cell surface markers of activated microglia (major histocompatibility complex II and CD40) and increased IL-1β production. Treatment of aged rats with EPA attenuated these changes, and EPA up-regulated IL-4 production in the brain, indicating that EPA increased the antiinflammatory system. After cerebral administration of Aβ peptide 1–40, the LTP was decreased in aged, but not in young, rats. EPA treatment protects the aged brain against Aβ-induced inflammation and decreased LTP [53].

Only in the last 2–3 years, n-3 fatty acids have been used to treat some gene transgenic models. For example, APPsw (Tg2576) transgenic mouse model was fed with dietary DHA, and APP processing and amyloid burden were studied. Inclusion of 0.6% DHA in the diet significantly reduced cortical total Aβ by 70% when compared with control diets; DHA also decreased Aβ 42 levels. Image analysis of brain sections with an antibody against Aβ (amino acids 1–13) revealed that the overall plaque burden was reduced by 40% in the hippocampus and the parietal cortex. DHA modulated APP processing by decreasing both α- and β-APP C-terminal fragment products and full-length APP. BACE1 (beta-secretase activity of the beta-site APP-cleaving enzyme), ApoE (apolipoprotein E), and transthyretin gene expression were unchanged in the high-DHA diet. Together, these results suggest that dietary DHA could be protective against Aβ production, accumulation, and potential downstream toxicity [79].

To further understand the functions of different fatty acids, seven experimental diets with varying n-6/n-3 ratios, saturated and PUFA and cholesterol contents were fed to transgenic APPswe/PS1dE9 mice for 3–4 months beginning at a young adult age (6 months). A typical Western diet with 40% saturated fatty acids and 1% of cholesterol increased, while diets supplemented with DHA decreased, Aβ levels in the hippocampus compared to the regular (soy oil–based) diet. DHA diet also decreased the number of activated microglia and increased exploratory activity in these transgenic mice, but did not improve their spatial learning in the water maze [80]. Furthermore, a diet low in DHA depleted postsynaptic proteins and exacerbated behavioral alterations in the Tg 2576 transgenic mouse model of AD [81]. However, DHA was found to have no significant effects on neuroinflammation in AD [82]. Therefore, other fatty acids and combinations of fatty acids need to be further studied in the future.

Because EPA has been found to have anti-inflammatory effects in IL-1-induced model, the mechanism by which it improves stress or anxiety-like behaviors and memory impairment have been studied. Chronic treatment with EPA for 7 weeks balanced pre- and anti-inflammatory mediators, reducing blood and brain PGE_2 concentrations and increasing anti-inflammatory cytokine IL-10 lever. EPA also increased NA synthesis, and increased ACh release during maze training and learning in rats that received central IL-1 administration for 7 days [25,75]. EPA also blocked IL-1-induced elevation in the glucocorticoids, and reversed the IL-1-induced decrease in the expression of NGF [8]. In *in vitro* experiments, the anti-inflammatory effect of n-3 fatty acids has been demonstrated by coculturing cortical neurons with different doses of LPS. At doses of 20–200 nM, LPS caused a significant decrease in cell viability. After incubating EPA with cortical neurons for 72 h, the LPS-induced decrease in cell viability was largely reversed [83].

The mechanism by which EPA protects neurons may result from its interaction with NTFs and receptors. We have previously reported that EPA feeding for 7 weeks markedly increases nerve growth factor expressions in the hippocampus [76].

In an *in vitro* model, we reported that EPA significantly increased the cell viability of fully differentiated SH-SY5Y cells in a dose-dependent manner, but only in the presence of brain-derived neurotrophic factor (BDNF). Neurotrophic receptor TrkB expression (which indicates promoting cell survival) was increased in cells treated with different concentrations of EPA (0.1, 1.0, 10.0 μM), while the receptor $P75^{NTF}$ (indicates cell apoptosis) was decreased by EPA treatment (at 10 μM) [84].

Studies on DHA effects have shown that DHA pretreatment significantly increases neuronal survival upon Aβ treatment by preventing cytoskeleton perturbations, caspase activation and apoptosis, as well as by promoting extracellular signal-related kinase (ERK). These data suggest that DHA enrichment probably induces changes in neuronal membrane properties with functional outcomes, thereby increasing protection from Aβ [85].

4.5 PARKINSON'S DISEASE MODELS

4.5.1 *In Vivo* Model

4.5.1.1 MPTP-Induced Model

The most popular PD model is MPTP intoxication in the mouse. MPTP is metabolized to MPP^+ that is selectively taken by the DA transporter (DAT) in DA neurons into the mitochondria and therein selectively inhibits complex-1 of the mitochondrial respiratory chain. Inhibition of complex-1 has also been observed in PD [86]. Interestingly, MPP^+ toxic effects are present exclusively in dopaminergic neurons in the SN pars compacta (pc) and not in other catecholamine and noncatecholamine neurons [87]. Neuropathology induced by MPTP is strikingly similar to those observed in PD, with loss of DA neurons in the SNpc [88] and neurodegeneration in the putamen [89]. Animals may show some behavioral abnormalities, such as rigidity, slowness, or absence of voluntary movement (bradykinesia and akinesia, respectively), reduction of movement amplitude (hypokinesia), postural instability, and the inability to initiate a voluntary movement (freezing). Growing evidence suggests a pivotal role of inflammation in this model. MPTP injection can lead to the activation of microglia and astrocytes, and increase IL-1β and IL-6 in both the striatum and SN [37]. Genetic deletion or pharmacological inhibition of COX-2 was found to attenuate MPTP toxicity to DA neurons [90,91], indicating that inflammation plays a role in MPTP toxicity. Furthermore, it was found that $cPLA_2$ knockout mice were resistant to MPTP toxicity, suggesting a role of $cPLA_2$ in MPTP-mediated inflammation [92]. Moreover, MPP^+ was found to enhance the release of AA following activation of $cPLA_2$ [93], which may be responsible for the increased inflammation in this model, since AA is a precursor of several inflammatory mediators. The consequence of inflammation and mitochondrial toxicity is that free radicals and oxidants are overproduced, which further cause DA neuron apoptosis and degeneration. The other oxidative effect of MPP^+ is produced by its interaction with synaptic vesicles through its binding to vesicular monoamine transporter-2, and so MPP^+ translocates into synaptic vesicles, where it stimulates the extrusion of synaptic DA. This causes the excess of cytosolic DA to undergo autooxidation, thus generating a huge burst of aggressive ROS [94].

4.5.1.2 6-OHDA-Induced Model

The compound 6-hydroxydopamine (6-OHDA) is another neurotoxin, which has been used to develop PD models for many years. However, 6-OHDA deletes DA in a wide range of areas of the brain with effects on other catecholamine systems. Recent evidence has shown that nigrostriatal degeneration induced by 6-OHDA is accompanied by early microglial and astroglial activation, which precedes the onset of dopaminergic cell loss, in the SNc, without significant changes in cytokine levels. However, another investigation demonstrated that IL-1β and TNF-α were increased in the brain of this model [95]. The COX inhibitor ibuprofen protected dopaminergic neurons against 6-OHDA [96], suggesting that inflammation is also involved in this model.

4.5.2 *In Vitro* Model

The SH-SY5Y neuroblastoma cell line has been suggested to be a good *in vitro* model of the dopaminergic cell and model of PD after MPP^+ treatment [97], expressing cell death markers and other physiological changes similar to PD [98]. Although free radical formation in MPP^+ treatment has indeed been observed *in vitro*, it occurs usually at concentrations higher than the physiological concentrations of MPP^+ in animals after MPTP treatment. *In vitro* MPTP-induced dopaminergic toxicity could be suppressed by an inhibitor of NADPH, but only in the presence of microglia [99]. We have reported that MPP^+ markedly reduces SH-SY5Y cell viability, and increases PLA_2 and Bax expression (a proapoptosis marker) [100]. These *in vitro* studies further demonstrate the role of inflammation and oxidants in PD.

4.5.3 Inflammation-Induced Model

Intranigral injection of LPS can activate microglia and cause a dose-dependent selective loss of DA neurons. RNase protection assays revealed that mRNA for Bax, Fas, and the proinflammatory cytokines IL-1, IL-6, and TNF-α were increased in the LPS-injected side of SN, while expression of the antiapoptotic gene Bcl-2 was decreased. IL-10 injection protected against LPS-induced cell death of DA neurons, with a corresponding decrease in the number of activated microglia, suggesting that the reduction in microglia-mediated release of inflammatory mediators may contribute to the anti-inflammatory effect of IL-10 [101]. LPS exposure also largely increased production of NO in a time- and dose-dependent manner as well as increased subsequent generation of other reactive nitrogen species such as peroxynitrite anions in both *in vivo* and *in vitro* models [102]. However, chronic LPS treatment for 30 days did not further result in dopamine neuron death and oxidant production, but adaptation to this endotoxin was found [103].

4.5.4 Eicosapentaenoate Treatment in Parkinson's Disease Models

Even though less fish intake has been associated with increased risk of PD onset, n-3 fatty acids have not been tried in PD models except in some findings from our laboratory. We found that 1 week after subchronic MPTP injection, mice locomotor activity in "open field" was decreased, which was attenuated by EPA pretreatment for 7 weeks [100]. In an *in vitro* model, MPP^+ reduced SH-SY5Y cell viability in dose- and time-dependent manners, which could be partially or fully reversed by EPA treatment. $cPLA_2$ mRNA expression was downregulated significantly by EPA treatment as well in SH-SY5Y cell model.

Our previous study may indirectly support the finding that EPA has beneficial effects in treatment of PD. Using an *in vivo* microdialysis method, systemic IL-1 administration significantly increased extracellular levels of DA, its metabolites 3,4-dihydroxyphenylacetic acid (DOPAC) and homovanillic acid (HVA) in the shell of the nucleus accumbens (NAc), an important part of the mesolimbic system that is involved in the MPTP-induced model [104]. IL-1 also increased the blood concentration of corticosterone, increased accumbal PGE_2 level, and the activities of cytosolic and sectory PLA_2. These IL-1-induced changes were significantly attenuated by EPA treatment. Furthermore, glucocoticoid receptor antagonist mifepristone (RU486) pretreatment significantly blocked IL-1-induced changes in DA and metabolites. Quinacrine, a PLA_2 antagonist, significantly blocked IL-1-induced increases in PGE_2 and corticosterone concentrations [83]. These results demonstrated that EPA-attenuated IL-1 effects may be through the suppression of PLA_2 expression that reduced PGE_2 and corticosterone secretion.

4.6 ANIMAL MODELS OF HUNTINGTON'S DISEASE AND EFFECTS OF UNSATURATED FATTY ACID TREATMENTS

4.6.1 Huntington's Disease Models

HD is a progressive neurodegenerative disorder caused by CAG repeat expansion in the gene that codes for the protein huntingtin. The neuronal loss selectively occurs in the striatum. The main symptoms are choreiform, involuntary movement, and dementia. In the brain of HD patients or transgenic models of HD, the level of quinolinic acid (QA) was increased, and striatal injection of QA can induce most symptoms and neuropathological changes. QA administration also resulted in similar behavioral symptoms [105], and inflammation characterized by microgliosis, astrogliosis, and enhanced expressions of proinflammatory enzymes inducible nitric oxide synthase COX2 [106]. Seven days after QA injections, extensive oxidative stress evident as lipid peroxidation, oxidative DNA damage, and reactive oxygen species formation was observed. These changes can be attenuated by antibiotic treatments [106].

Mice transgenic for a human genomic fragment containing promoter elements exon 1 and a portion of intron 2 of the *Huntingtin* gene responsible for HD have also been developed. These mice display a progressive neurological phenotype and carry the following approximate CAG repeat expansions: R6/1, 115; R6/2, 150; R6/5, 128–156 [107]. In R6/2 mice, striatal gene expressions were analyzed by DNA microarray. Results revealed that neuroinflammation is associated with HD [108]. In addition, abnormal expression of a number of genes involved in cholesterol and fatty acid metabolism have been found in clonal striata-derived cells expressing huntingtin fragments (a cellular model of early HD), suggesting that these metabolic pathways may play a role in HD pathogenesis [109].

In R6/1 and normal mice, a mixture of essential fatty acids containing 54% of n-6 and 5% n-3 fatty acids fed throughout life has been reported to reverse or attenuate progressive reductions in locomotion, elements of rearing, sniffing, sifting and chewing, and an increase in grooming. However, the diet did not improve the reductions in body weight, and in brain dopamine D_1-like and D_2-like quantitative receptors [110]. These findings suggest that essential fatty acids may have therapeutic potential in HD. Therefore, higher doses of n-3 fatty acids or pure n-3 fatty acids should be used.

4.7 AMYOTROPHIC LATERAL SCLEROSIS

The most used model for studying ALS is G93A SOD1 transgenic mice that develop progressive loss of spinal cord motor neurons. Activated NADPH oxidase, the main reactive oxygen species–producing enzyme during inflammation, was reported in both the spinal cords of ALS patients and in the spinal cords of these genetic animals [111]. In this ALS model, proinflammatory cytokines IL-1β, IL-6, IFN-γ and TNF-α, and Th 3, which produced cytokine TGF-β, were all increased [112]. An abnormal kynurenine pathway and increased QA have also been found in ASL patients or animal models, which indicate an interaction between inflammation and tryptophan catabolism in these neurodegenerative diseases [113]. However, COX2 inhibitor rofecoxib treatment only delayed the onset of locomotor impairment in ALS mice, but did not affect animal survival [114].

The high intake of PUFA and vitamin E have only very recently been associated with a reduced risk of developing ALS [115]. However, no evidence shows whether n-3 fatty acids can significantly delay the onset of the disease, or prolong survival. In our pilot experiment, we found that EPA treatment for 7 weeks did not delay the onset of ALS symptoms in G93A SOD1 transgenic mice, but about 50% of the animals lived longer with better cognitive performance (unpublished results).

4.8 A SUMMARY OF MECHANISMS BY WHICH n-3 FATTY ACIDS BENEFIT NEURODEGENERATIVE DISEASES

As described above, these neurodegenerative diseases seem to have similar inflammatory changes; activated microglia and astrocytes, which overproduce proinflammatory cytokines and oxidants. Both inflammation and oxidants can cause dysfunctions in neuroendocrine and neurotransmitter systems, and induce neuron apoptosis and neurodegeneration. The mechanisms by which n-3 fatty acids improve neurodegenerative diseases and protect neurons may result from several aspects. First, ageing, Aβ, inflammation and oxidants can reduce membrane fluidity, while n-3 fatty acids increase membrane fluidity [13,116]. As mentioned in the introduction, changes in membrane fluidity and structure can change the function of receptors, neurotransmitters, and ion channels. Second, n-3 fatty acids can modulate and compete with n-6 fatty acid concentrations and functions. As n-6 fatty acids are precursors of inflammatory mediators, replacing n-6 fatty acids with n-3 fatty acids should reduce inflammatory effects. Third, because inflammatory cytokines and PGE_2 can stimulate the HPA axis to release GCs that may damage hippocampal neurons, the suppression of inflammation should reduce the HPA axis activity. Fourth, upon the activation of glial cells, the activity of PLA_2 enzymes is increased, which induces inflammation from membrane AA and increases oxidants from mitochondria. The effects of n-3 fatty acids on PLA_2 and oxidants may occur through modulating membrane functions of glial cells and mitochondria. Fifth, n-3 fatty acids can reduce cholesterol. Cholesterol can reduce membrane fluidity, and derived glucocorticoids. Even though both n-3 and n-6 fatty acids can reduce blood cholesterol, n-6 fatty acids redistribute, while n-3 fatty acid reduce cholesterol in the brain [13] and thereby protect neurons.

4.9 LIMITATION AND FUTURE RESEARCH DIRECTIONS

From the information provided above, several limitations in n-3 fatty acid treatments should be emphasized: (1) it is unclear whether different fatty acids have different benefits. For example, EPA may have more anti-inflammatory effects than DHA. However, DHA has been studied to treat AD but not EPA; (2) previously, some research groups, including our team, have reported that n-3 and n-6 combination may produce stronger anti-inflammatory effects. However, almost no efforts have been put into neurodegeneration studies yet; (3) even though neurodegenerative diseases show many similar changes in inflammation and oxidative productions, the causes and affected areas are still different. For example, the hippocampus is the most vulnerable and affected region in the brain in AD, and striatum and SN are the degenerative areas in PD. How diets enriched with n-3 fatty acids can specifically improve the affected brain area, and produce specific effects, are unknown. At this stage, we can only assume that n-3 fatty acids can produce general anti-inflammatory and neuroprotective effects; (4) the relationship between membrane structure and function in neurodegeneration and effects of n-3 fatty acid treatments have not been understood enough; and (5) ALS, HD, MS, and other neurodegenerative diseases have not been studied with n-3 fatty acid treatments. These limitations could be addressed in future research.

REFERENCES

1. Huang, T.L., Zandi, P.P., Tucker, K.L., Fitzpatrick, A.L., Kuller, L.H., Fried, L.P., Burke, G.L. and Carlson, M.C. 2005. Benefits of fatty fish on dementia risk are stronger for those without APOE epsilon4. *Neurology* 65, 1409–1414.
2. de Lau, L.M., Bornebroek, M., Witteman, J.C., Hofman, A., Koudstaal, P.J. and Breteler, M.M. 2005. Dietary fatty acids and the risk of Parkinson disease: the Rotterdam study. *Neurology* 64, 2040–2045.
3. Tully, A.M., Roche, H.M., Doyle, R., Fallon, C., Bruce, I., Lawlor, B., Coakley, D. and Gibney, M.J. 2003. Low serum cholesteryl ester-docosahexaenoic acid levels in Alzheimer's disease: a case-control study. *Br. J. Nutr.* 89, 483–489.
4. Conquer, J.A., Tierney, M.C., Zecevic, J., Bettger, W.J. and Fisher, R.H. 2000. Fatty acid analysis of blood plasma of patients with Alzheimer's disease, other types of dementia, and cognitive impairment. *Lipids* 35, 1305–1312.
5. Peet, M. and Stockes, C. 2005. Omega-3 fatty acids in the treatment of psychiatric disorders. *Drugs* 65, 1051–1059.
6. Puri, B.K. 2001. Impaired phospholipid-related signal transduction in advanced Huntington's disease. *Exp. Physiol.* 86, 683–685.
7. Vaddadi, K.S., Soosai, E., Chiu, E. and Dingjan, P. 2002. A randomised, placebo-controlled, double blind study of treatment of Huntington's disease with unsaturated fatty acids. *Neuroreport* 13, 29–33. Erratum in: *Neuroreport* 2002. 13, inside back cover.
8. Song, C., Phillips, A.G., Leonard, B.E. and Horrobin, D.F. 2004. Ethyl-eicosapentaenoic acid ingestion prevents corticosterone-mediated memory impairment induced by central administration of interleukin-1beta in rats. *Mol. Psychiatr.* 9, 630–638.
9. Hashimoto, M., Hossain, S., Shimada, T., Sugioka, K., Yamasaki, H., Fujii, Y., Ishibashi, Y., Oka, J. and Shido, O. 2002. Docosahexaenoic acid provides protection from impairment of learning ability in Alzheimer's disease model rats. *J. Neurochem.* 81, 1084–1091.
10. Peet, M., Glen, I. and Horrobin, D.F., Ed., 2003. *Phospholipid Spectrum Disorders in Psychiatry and Neurology.* Second edition. Marius Press, Lancashire, UK.
11. Mazza, M., Pomponi, M., Janiri, L., Bria, P. and Mazza, S. 2007. Omega-3 fatty acids and antioxidants in neurological and psychiatric diseases: an overview. *Prog. Neuropsychopharmacol. Biol. Psychiatr.* (Epub ahead of print).
12. James, M.J., Gibson, R.A. and Cleland, L.G. 2000. Dietary polyunsaturated fatty acids and inflammatory mediator production. *Am. J. Clin. Nutr.* 71(Suppl 1), 343S–348S.
13. Yehuda, S., Rabinovitz, S., Carasso, R.L. and Mostofsky, D.I. 2002. The role of polyunsaturated fatty acids in restoring the aging neuronal membrane. *Neurobiol. Aging* 23, 843–853.
14. Song, C., Li, X., Leonard, B.E. and Horrobin, D.F. 2003. Effects of dietary n-3 or n-6 fatty acids on interleukin-1beta-induced anxiety, stress, and inflammatory responses in rats. *J. Lipid. Res.* 44, 1984–1991.
15. Yehuda, S., Rabinovitz, S., Carasso, R.L. and Mostofsky, D.I. 2000. Fatty acid mixture counters stress changes in cortisol, cholesterol, and impair learning. *Int. J. Neurosci.* 1, 73–87.
16. McGeer, P.L. and McGeer, E.G. 2002. Inflammatory processes in amyotrophic lateral sclerosis. *Muscle Nerve* 26, 459–470.

17. Ringheim, G.E. and Conant, K. 2004. Neurodegenerative disease and the neuroimmune axis (Alzheimer's and Parkinson's disease, and viral infections). *J. Neuroimmunol.* 147, 43–49.
18. Song, C., Vandewoude, M., Stevens, W., De Clerck, L., Van der Planken, M., Whelan, A., Anisman, H., Dossche, A. and Maes, M. 1999. Alterations in immune functions during normal aging and Alzheimer's disease. *Psychiatr. Res.* 85, 71–80.
19. Griffin, W.S., Liu, L., Li, Y., Mrak, R.E. and Barger, S.W. 2006. Interleukin-1 mediates Alzheimer and Lewy body pathologies. *J. Neuroinflammation* 3, 5.
20. Blasko, I., Stampfer-Kountchev, M., Robatscher, P., Veerhuis, R., Eikelenboom, P. and Grubeck-Loebenstein, B. 2004. How chronic inflammation can affect the brain and support the development of Alzheimer's disease in old age: the role of microglia and astrocytes. *Aging Cell* 3, 169–176.
21. Eskandari, F., Webster, J.I. and Sternberg, E.M. 2003. Neural immune pathways and their connection to inflammatory diseases. *Arthritis Res. Ther.* 5, 251–265.
22. Andreini, I., Getuli, C., Pacelli, V., Manno, R., Ragazzoni, E., Nunziata, A. and Navarra, P. 2002. Function of the hypothalamo-pituitary-adrenal axis and humoral immune mechanisms during experimental allergic encephalomyelitis in SJL/J mice. *Neuroimmunomodulation* 10, 9–16.
23. Pavlides, C., Nivon, L.G. and McEwen, B.S. 2002. Effects of chronic stress on hippocampal long-term potentiation. *Hippocampus* 12, 245–257.
24. Elgh, E., Lindqvist Astot, A., Fagerlund, M., Eriksson, S., Olsson, T. and Nasman, B. 2006. Cognitive dysfunction, hippocampal atrophy and glucocorticoid feedback in Alzheimer's disease. *Biol. Psychiatr.* 59, 155–161.
25. Taepavarapruk, P. and Song, C. 2007. Effects of omega-3 fatty acid, eicosapentaenoic acid, on IL-1beta-induced memory impairment and acetylcholine efflux in rat hippocampus. *CINP Conference*, Bangkok, Thailand.
26. Zhu, D., Lai, Y., Shelat, P.B., Hu, C., Sun, G.Y. and Lee, J.C. 2006. Phospholipases A2 mediate amyloid-beta peptide-induced mitochondrial dysfunction. *J. Neurosci.* 26, 11111–11119.
27. Gebicke-Haerter, P.J. 2001. Microglia in neurodegeneration: molecular aspects. *Microsc. Res. Tech.* 54, 47–58.
28. Sun, G.Y., Xu, J., Jensen, M.D. and Simonyi, A. 2004. Phospholipase A2 in the central nervous system: implications for neurodegenerative diseases. *J. Lipid Res.* 45, 205–213.
29. Lawson, L.J., Perry, V.H., Dri, P. and Gordon, S. 1990. Heterogeneity in the distribution and morphology of microglia in the normal adult mouse brain. *Neuroscience* 39, 151–170.
30. Czlonkowska, A., Kurkowska-Jastrzebska, I., Czlonkowski, A., Peter, D. and Stefano, G.B. 2002. Immune processes in the pathogenesis of Parkinson's disease—a potential role for microglia and nitric oxide. *Med. Sci. Monit.* 8, 165–177.
31. van Craenenbroeck, K., De Bosscher, K., Vanden Berghe, W., Vanhoenacker, P. and Haegeman, G. 2005. Role of glucocorticoids in dopamine-related neuropsychiatric disorders. *Mol. Cell Endocrinol.* 245, 10–22.
32. Morale, M.C., Serra, P.A., Delogu, M.R., Migheli, R., Rocchitta, G., Tirolo, C., Caniglia, S., Testa, N., L'Episcopo, F., Gennuso, F., Scoto, G.M., Barden, N., Miele, E., Desole, M.S. and Marchetti, B. 2004. Glucocorticoid receptor deficiency increases vulnerability of the nigrostriatal dopaminergic system: critical role of glial nitric oxide. *FASEB J.* 18, 164–166.
33. Kurkowska-Jastrzebska, I., Litwin, T., Joniec, I., Ciesielska, A., Przybylkowski, A., Czlonkowski, A. and Czlonkowska, A. 2004. Dexamethasone protects against

dopaminergic neurons damage in a mouse model of Parkinson's disease. *Int. Immunopharmacol.* 4, 1307–1318.

34. Stoy, N., Mackay, G.M., Forrest, C.M., Christofides, J., Egerton, M., Stone, T.W. and Darlington, L.G. 2005. Tryptophan metabolism and oxidative stress in patients with Huntington's disease. *J. Neurochem.* 93, 611–623.
35. Deckel, A.W. 2001. Nitric oxide and nitric oxide synthase in Huntington's disease. *J. Neurosci. Res.* 64, 99–107.
36. Ghezzi, P. and Mennini, T. 2001. Tumor necrosis factor and motoneuronal degeneration: an open problem. *Neuroimmunomodulation* 9, 178–182.
37. Teismann, P. and Schulz, J.B. 2004. Cellular pathology of Parkinson's disease: astrocytes, microglia and inflammation. *Cell Tissue Res.* 318, 149–161.
38. Nakajima, K. and Kohsaka, S. 2001. Microglia: activation and their significance in the central nervous system. *J. Biochem. (Tokyo)* 130, 169–175.
39. Toyomoto, M., Inoue, S., Ohta, K., Kuno, S., Ohta, M., Hayashi, K. and Ikeda, K. 2005. Production of NGF, BDNF and GDNF in mouse astrocyte cultures is strongly enhanced by a cerebral vasodilator, ifenprodil. *Neurosci. Lett.* 379, 185–189.
40. Blesch, A., Grill, R.J. and Tuszynski, M.H. 1998. Neurotrophin gene therapy in CNS models of trauma and degeneration. *Prog. Brain Res.* 117, 473–484.
41. Yuen, E.C. and Mobley, W.C. 1996. Therapeutic potential of neurotrophic factors for neurological disorders. *Ann. Neurol.* 40, 346–354
42. Connor, B. and Dragunow, M. 1998. The role of neuronal growth factors in neurodegenerative disorders of the human brain. *Brain Res. Brain Res. Rev.* 27, 1–39.
43. Kaplan, D.R. and Miller, F.D. 1997. Signal transduction by the neurotrophin receptors. *Curr. Opin. Cell Biol.* 9, 213–221.
44. Huang, E.J. and Reichardt, L.F. 2003. Trk receptors: roles in neuronal signal transduction. *Annu. Rev. Biochem.* 72, 609–642.
45. Siegel, G.J. and Chauhan, N.B. 2000. Neurotrophic factors in Alzheimer's and Parkinson's disease brain. *Brain Res. Rev.* 33, 199–227.
46. Peng, S., Wuu, J., Mufson, E.J. and Fahnestock, M. 2005. Precursor form of brain-derived neurotrophic factor and mature brain-derived neurotrophic factor are decreased in the pre-clinical stages of Alzheimer's disease. *J. Neurochem.* 93, 1412–1421.
47. Capsoni, S., Ugolini, G., Comparini, A., Ruberti, F., Berardi, N. and Cattaneo, A. 2000. Alzheimer-like neurodegeneration in aged antinerve growth factor transgenic mice. *Proc. Natl. Acad. Sci. USA* 97, 6826–6831.
48. Zanassi, P., Paolillo, M. and Schinelli, S. 1998. Coexpression of phospholipase A2 isoforms in rat striatal astrocytes. *Neurosci. Lett.* 247, 83–86.
49. Andersen, J.M., Myhre, O. and Fonnum, F. 2003. Discussion of the role of the extracellular signal-regulated kinase-phospholipase A2 pathway in production of reactive oxygen species in Alzheimer's disease. *Neurochem. Res.* 28, 319–326.
50. Kalmijn, S., Feskens, E.J., Launer, L.J. and Kromhout, D. 1997. Polyunsaturated fatty acids, antioxidants, and cognitive function in very old men. *Am. J. Epidemiol.* 145, 33–41.
51. Mills, S.C., Windsor, A.C. and Knight, S.C. 2005. The potential interactions between polyunsaturated fatty acids and colonic inflammatory processes. *Clin. Exp. Immunol.* 142, 216–228.
52. Kuboyama, T., Tohda, C. and Komatsu, K. 2006. Withanoside, I.V. and its active metabolite, sominone, attenuate Abeta(25–35)-induced neurodegeneration. *Eur. J. Neurosci.* 23, 17–26.

53. Lynch, A.M., Loane, D.J., Minogue, A.M., Clarke, R.M., Kilroy, D., Nally, R.E., Roche, O.J., O'connell, F. and Lynch, M.A. 2006. Eicosapentaenoic acid confers neuroprotection in the amyloid-beta challenged aged hippocampus. *Neurobiol Aging* (Epub ahead of print).
54. Zhang, J.M., Wu, M.N., Qi, J.S. and Qiao, J.T. 2006. Amyloid beta-protein fragment 31–35 suppresses long-term potentiation in hippocampal CA1 region of rats in vivo. *Synapse* 60, 307–313.
55. Yamaguchi, Y., Miyashita, H., Tsunekawa, H., Mouri, A., Kim, H.C., Saito, K., Matsuno, T., Kawashima, S., Nabeshima, T. 2006. Effects of a novel cognitive enhancer, spiro[imidazo-[1,2-a]pyridine-3,2-indan]-2(3H)-one (ZSET1446), on learning impairments induced by amyloid-beta1-40 in the rat. *J. Pharmacol. Exp. Ther.* 317, 1079–1087.
56. Cheng, G., Whitehead, S.N., Hachinski, V. and Cechetto, D.F. 2006. Effects of pyrrolidine dithiocarbamate on beta-amyloid (25–35)-induced inflammatory responses and memory deficits in the rat. *Neurobiol. Dis.* 23, 140–151.
57. White, J.A., Manelli, A.M., Holmberg, K.H., Van Eldik, L.J. and Ladu, M.J. 2005. Differential effects of oligomeric and fibrillar amyloid-beta 1-42 on astrocyte-mediated inflammation. *Neurobiol. Dis.* 18, 459–465.
58. Paris, D., Town, T., Parker, T.A., Tan, J., Humphrey, J., Crawford, F. and Mullan, M. 1999. Inhibition of Alzheimer's beta-amyloid induced vasoactivity and proinflammatory response in microglia by a cGMP-dependent mechanism. *Exp. Neurol.* 157, 211–221.
59. Bergamaschini, L., Donarini, C., Rossi, E., De Luigi, A., Vergani, C. and De Simoni, M.G. 2002. Heparin attenuates cytotoxic and inflammatory activity of Alzheimer amyloid-beta in vitro. *Neurobiol. Aging* 23, 531–536.
60. Harris-White, M.E., Chu, T., Miller, S.A., Simmons, M., Teter, B., Nash, D., Cole, G.M. and Frautschy, S.A. 2001 Estrogen (E2) and glucocorticoid (Gc) effects on microglia and A beta clearance in vitro and in vivo. *Neurochem. Int.* 39, 435–448.
61. Hsiao, K., Chapman, P., Nilsen, S., Eckman, S., Harigaya, Y., Younkin, S., Yang, F. and Cole, G. 1996. Correlative memory deficits, Abeta elevation, and amyloid plaques in transgenic mice, *Science* 274, 99–102.
62. Kawarabayashi, T., Younkin, L.H., Saido, T.C., Shoji, M., Ashe, K.H. and Younkin, S.G. 2001. Age-dependent changes in brain, CSF, and plasma amyloid (beta) protein in the Tg2576 transgenic mouse model of Alzheimer's disease, *J. Neurosci.* 21, 372–381.
63. Yao, Y., Chinnici, C., Tang, H., Trojanowski, J.Q., Lee, V.M. and Pratico, D. 2004. Brain inflammation and oxidative stress in a transgenic mouse model of Alzheimer-like brain amyloidosis. *J. Neuroinflammation* 1, 21.
64. Apelt, J., Bigl, M., Wunderlich, P. and Schliebs, R. 2004. Aging-related increase in oxidative stress correlates with developmental pattern of beta-secretase activity and beta-amyloid plaque formation in transgenic Tg2576 mice with Alzheimer-like pathology. *Int. J. Dev. Neurosci.* 22, 475–484.
65. Abbas, N., Bednar, I., Mix, E., Marie, S., Paterson, D., Ljungberg, A., Morris, C., Winblad, B., Nordberg, A. and Zhu, J. 2002. Up-regulation of the inflammatory cytokines IFN-gamma and IL-12 and down-regulation of IL-4 in cerebral cortex regions of APP(SWE) transgenic mice. *J. Neuroimmunol.* 126, 50–57.
66. Garcia-Alloza, M., Robbins, E.M., Zhang-Nunes, S.X., Purcell, S.M., Betensky, R.A., Raju, S., Prada, C., Greenberg, S.M., Bacskai, B.J. and Frosch, M.P. 2006. Characterization of amyloid deposition in the APPswe/PS1dE9 mouse model of Alzheimer disease. *Neurobiol. Dis.* 24, 516–524.

67. Ozmen, L., Woolley, M., Albientz, A., Miss, M.T., Nelboeck, P., Malherbe, P., Czech, C., Gruninger-Leitch, F., Brockhaus, M., Ballard, T. and Jacobsen, H. 2005. BACE/APPV717F double-transgenic mice develop cerebral amyloidosis and inflammation. *Neurodegener. Dis.* 2, 284–298.
68. Heneka, M.T., Ramanathan, M., Jacobs, A.H., Dumitrescu-Ozimek, L., Bilkei-Gorzo, A., Debeir, T., Sastre, M., Galldiks, N., Zimmer, A., Hoehn, M., Heiss, W.D., Klockgether, T. and Staufenbiel M. 2006. Locus ceruleus degeneration promotes Alzheimer pathogenesis in amyloid precursor protein 23 transgenic mice. *J. Neurosci.* 26, 1343–1354.
69. Richards, J.G., Higgins, G.A., Ouagazzal, A.M., Ozmen, L., Kew, J.N., Bohrmann, B., Malherbe, P., Brockhaus, M., Loetscher, H., Czech, C., Huber, G., Bluethmann, H., Jacobsen, H. and Kemp, J.A. 2003. PS2APP transgenic mice, coexpressing hPS2mut and hAPPswe, show age-related cognitive deficits associated with discrete brain amyloid deposition and inflammation. *J. Neurosci.* 23, 8989–9003.
70. Machova, E., Jakubik, J., Michal, P., Oksman, M., Iivonen, H., Tanila, H. and Dolezal, V. 2006. Impairment of muscarinic transmission in transgenic APPswe/PS1dE9 mice. *Neurobiol. Aging* (Epub ahead of print).
71. Savonenko, A., Xu, G.M., Melnikova, T., Morton, J.L., Gonzales, V., Wong, M.P., Price, D.L., Tang, F., Markowska, A.L. and Borchelt, D.R. 2005. Episodic-like memory deficits in the APPswe/PS1dE9 mouse model of Alzheimer's disease: relationships to beta-amyloid deposition and neurotransmitter abnormalities. *Neurobiol. Dis.* 18, 602–617.
72. Griffin, W.S. and Mrak, R.E. 2002. Interleukin-1 in the genesis and progression of and risk for development of neuronal degeneration in Alzheimer's disease. *J. Leukoc. Biol.* 72, 233–238.
73. Casamenti, F., Prosperi, C., Scali, C., Giovannelli, L., Colivicchi, M.A., Faussone-Pellegrini, M.S. and Pepeu, G. 1999. Interleukin-1beta activates forebrain glial cells and increases nitric oxide production and cortical glutamate and GABA release in vivo: implications for Alzheimer's disease. *Neuroscience* 91, 831–842.
74. Song, C. 2002. The effect of thymectomy and IL-1 on memory: a implication between depression and immunity. *Brain Behav. Immun.* 6, 557–568.
75. Song, C. and Horrobin, D. 2004. Omega-3 fatty acid ethyl-eicosapentaenoate, but not soybean oil, attenuates memory impairment induced by central IL-1beta administration. *J. Lipid. Res.* 45, 1112–1121.
76. Song, C. 2004. The role of IL-1beta in Alzheimer's disease: studies on an animal model and therapeutic effects of n-3 fatty acids. *International Congress of Biological Psychiatry*, Sydney, February 9–13. *World J. Biol. Psychiatr.* 5, (suppl), 21.
77. Hemmer, K., Fransen, L., Vanderstichele, H., Vanmechelen, E. and Heuschling, P. 2001. An in vitro model for the study of microglia-induced neurodegeneration: involvement of nitric oxide and tumor necrosis factor-alpha. *Neurochem. Int.* 38, 557–565.
78. Hashimoto, M., Hossain, S., Agdul, H. and Shido, O. 2005. Docosahexaenoic acid-induced amelioration on impairment of memory learning in amyloid beta-infused rats relates to the decreases of amyloid beta and cholesterol levels in detergent-insoluble membrane fractions. *Biochim. Biophys. Acta.* 1738, 91–98.
79. Lim, G.P., Calon, F., Morihara, T., Yang, F., Teter, B., Ubeda, O., Salem, N. Jr., Frautschy, S.A. and Cole, G.M. 2005. A diet enriched with the omega-3 fatty acid docosahexaenoic acid reduces amyloid burden in an aged Alzheimer mouse model. *J. Neurosci.* 25, 3032–3040.

80. Oksman, M., Iivonen, H., Hogyes, E., Amtul, Z., Penke, B., Leenders, I., Broersen, L., Lutjohann, D., Hartmann, T. and Tanila, H. 2006. Impact of different saturated fatty acid, polyunsaturated fatty acid and cholesterol containing diets on beta-amyloid accumulation in APP/PS1 transgenic mice. *Neurobiol. Dis.* 23, 563–572.
81. Mucke, L. and Pitas, R.E. 2004. Food for thought: essential fatty acid protects against neuronal deficits in transgenic mouse model of AD. *Neuron* 43, 596–599.
82. Cole, G.M., Morihara, T., Lim, G.P., Yang, F., Begum, A. and Frautschy, S.A. 2004. NSAID and antioxidant prevention of Alzheimer's disease: lessons from in vitro and animal models. *Ann. N.Y. Acad. Sci.* 1035, 68–84.
83. Song, C., Li, X., Kang, Z. and Kadotomi, Y. 2007. Omega-3 fatty acid ethyl-eicosapentaenoate attenuates IL-1beta-induced changes in dopamine and metabolites in the shell of the nucleus accumbens: involved with PLA2 activity and corticosterone secretion. *Neuropsychopharmacology* 32, 736–744.
84. Kou, W., Luchtman, D. and Song, C. 2006. Eicosapentaenoic acid (EPA) increases the cell viability of fully differentiated SH-SY5Y cells through up-regulation of tyrosine kinase receptor B expression. *Int. J. Neuropsychopharmacol.* 9 (1Suppl), S125.
85. Florent, S., Malaplate-Armand, C., Youssef, I., Kriem, B., Koziel, V., Escanye, M.C., Fifre, A., Sponne, I., Leininger-Muller, B., Olivier, J.L., Pillot, T. and Oster, T. 2006. Docosahexaenoic acid prevents neuronal apoptosis induced by soluble amyloid-beta oligomers. *J. Neurochem.* 96, 385–395.
86. Greenamyre, J.T., Sherer, T.B., Betarbet, R. and Panov, A.V. 2001. Complex I and Parkinson's disease. *IUBMB Life* 52, 135–141.
87. Bywood, P.T. and Johnson, S.M. 2003. Mitochondrial complex inhibitors preferentially damage substantia nigra dopamine neurons in rat brain slices. *Exp. Neurol.* 179, 47–59.
88. Seniuk, N.A., Tatton, W.G. and Greenwood, C.E. 1990. Dose-dependent destruction of the coeruleus-cortical and nigral-striatal projections by MPTP. *Brain Res.* 527, 7–20. Erratum in: *Brain Res.* 1990. 535, 360. *Brain Res.* 1991. 552, 357.
89. Moratalla, R., Quinn, B., DeLanney, L.E., Irwin, I., Langston, J.W. and Graybiel, A.M. 1992. Differential vulnerability of primate caudate-putamen and striosome-matrix dopamine systems to the neurotoxic effects of 1-methyl-4-phenyl-1,2,3,6-tetrahydropyridine. *Proc. Natl. Acad. Sci. USA* 89, 3859–3863.
90. Okuno, T., Nakatsuji, Y., Kumanogoh, A., Moriya, M., Ichinose, H., Sumi, H., Fujimura, H., Kikutani, H. and Sakoda, S. 2005. Loss of dopaminergic neurons by the induction of inducible nitric oxide synthase and cyclooxygenase-2 via CD 40: relevance to Parkinson's disease. *J. Neurosci. Res.* 81, 874–882.
91. Hunot, S., Vila, M., Teismann, P., Davis, R.J., Hirsch, E.C., Przedborski, S., Rakic, P. and Flavell, R.A. 2004. JNK-mediated induction of cyclooxygenase 2 is required for neurodegeneration in a mouse model of Parkinson's disease. *Proc. Natl. Acad. Sci. USA* 101, 665–670. Epub.
92. Klivenyi, P., Beal, M.F., Ferrante, R.J., Andreassen, O.A., Wermer, M., Chin, M.R. and Bonventre, J.V. 1998. Mice deficient in group IV cytosolic phospholipase A2 are resistant to MPTP neurotoxicity. *J. Neurochem.* 71, 2634–2637.
93. Yoshinaga, N., Yasuda, Y., Murayama, T. and Nomura, Y. 2000. Possible involvement of cytosolic phospholipase A(2) in cell death induced by 1-methyl-4-phenylpyridinium ion, a dopaminergic neurotoxin, in GH3 cells. *Brain Res.* 855, 244–251.
94. Przedborski, S., Jackson-Lewis, V., Naini, A.B., Jakowec, M., Petzinger, G., Miller, R. and Akram, M. 2001. The parkinsonian toxin 1-methyl-4-phenyl-1,2,3,6-tetrahydropyridine (MPTP): a technical review of its utility and safety. *J. Neurochem.* 76, 1265–1274.

95. Depino, A.M., Earl, C., Kaczmarczyk, E., Ferrari, C., Besedovsky, H., del Rey, A., Pitossi, J. and Oertel, W.H. 2003. Microglial activation with atypical proinflammatory cytokine expression in a rat model of Parkinson's disease. *Eur. J. Neurosci.* 18, 2731–2742.
96. Carrasco, E., Casper, D. and Werner, P. 2005. Dopaminergic neurotoxicity by 6-OHDA and MPP^+: differential requirement for neuronal cyclooxygenase activity. *J. Neurosci. Res.* 81, 121–131.
97. Wang, X.J. and Xu, J.X. 2005. Salvianic acid A protects human neuroblastoma SH-SY5Y cells against MPP^+-induced cytotoxicity. *Neurosci. Res.* 51, 129–138.
98. Dennis, J. and Bennett, J.P. Jr. 2003. Interactions among nitric oxide and Bcl-family proteins after MPP^+ exposure of SH-SY5Y neural cells I: MPP^+ increases mitochondrial NO and Bax protein. *J. Neurosci. Res.* 72, 76–88.
99. Gao, H.M., Liu, B., Zhang, W. and Hong, J.S. 2003. Critical role of microglial NADPH oxidase-derived free radicals in the in vitro MPTP model of Parkinson's disease. *FASEB J.* 17, 1954–1956.
100. Luchtman, D. and Song, C. 2007. Protective effects of eicosapentanoic acid (EPA) in MPP^+ and MPTP-induced cell and animal models of Parkinson's disease. *World Congress of Biopsychiatry.* Chile.
101. Arimoto, T., Choi, D.Y., Lu, X., Liu, M., Nguyen, X.V., Zheng, N., Stewart, C.A., Kim, H.C. and Bing, G. 2006. Interleukin-10 protects against inflammation-mediated degeneration of dopaminergic neurons in substantia nigra. *Neurobiol. Aging* (Epub ahead of print).
102. Shavali, S., Combs, C.K. and Ebadi, M. 2006. Reactive macrophages increase oxidative stress and alpha-synuclein nitration during death of dopaminergic neuronal cells in co-culture: relevance to Parkinson's disease. *Neurochem. Res.* 31, 85–94.
103. Iravani, M.M., Leung, C.C., Sadeghian, M., Haddon, C.O., Rose, S. and Jenner, P. 2005. The acute and the long-term effects of nigral lipopolysaccharide administration on dopaminergic dysfunction and glial cell activation. *Eur. J. Neurosci.* 22, 317–330.
104. Boeckler, F., Leng, A., Mura, A., Bettinetti, L., Feldon, J., Gmeiner, P. and Ferger, B. 2003. Attenuation of 1-methyl-4-phenyl-1,2,3,6-tetrahydropyridine (MPTP) neurotoxicity by the novel selective dopamine D3-receptor partial agonist FAUC 329 predominantly in the nucleus accumbens of mice. *Biochem. Pharmacol.* 66, 1025–1032.
105. Haik, K.L., Shear, D.A., Schroeder, U., Sabel, B.A. and Dunbar, G.L. 2000. Quinolinic acid released from polymeric brain implants causes behavioral and neuroanatomical alterations in a rodent model of Huntington's disease. *Exp. Neurol.* 163, 430–439.
106. Ryu, J.K., Choi, H.B. and McLarnon, J.G. 2006. Combined minocycline plus pyruvate treatment enhances effects of each agent to inhibit inflammation, oxidative damage, and neuronal loss in an excitotoxic animal model of Huntington's disease. *Neuroscience* 141, 1835–1848.
107. Mangiarini, L., Sathasivam, K., Seller, M., Cozens, B., Harper, A., Hetherington, C., Lawton, M., Trottier, Y., Lehrach, H., Davies, S.W. and Bates, G.P. 1996. Exon 1 of the HD gene with an expanded CAG repeat is sufficient to cause a progressive neurological phenotype in transgenic mice. *Cell* 87, 493–506.
108. Crocker, S.F., Costain, W.J. and Robertson, H.A. 2006. DNA microarray analysis of striatal gene expression in symptomatic transgenic Huntington's mice (R6/2) reveals neuroinflammation and insulin associations. *Brain Res.* 1088, 176–186.

109. Sipione, S., Rigamonti, D., Valenza, M., Zuccato, C., Conti, L., Pritchard, J., Kooperberg, C., Olson, J.M. and Cattaneo, E. 2002. Early transcriptional profiles in huntingtin-inducible striatal cells by microarray analyses. *Hum. Mol. Genet.* 11, 1953–1965.
110. Clifford, J.J., Drago, J., Natoli, A.L., Wong, J.Y., Kinsella, A., Waddington, J.L. and Vaddadi, K.S. 2002. Essential fatty acids given from conception prevent topographies of motor deficit in a transgenic model of Huntington's disease. *Neuroscience* 109, 81–88.
111. Wu, D.C., Re, D.B., Nagai, M., Ischiropoulos, H. and Przedborski, S. 2006. The inflammatory NADPH oxidase enzyme modulates motor neuron degeneration in amyotrophic lateral sclerosis mice. *Proc. Natl. Acad. Sci. USA* 103, 12132–12137.
112. Sargsyan, S.A., Monk, P.N. and Shaw, P.J. 2005. Microglia as potential contributors to motor neuron injury in amyotrophic lateral sclerosis. *Glia* 51, 241–253.
113. Guillemin, G.J., Meininger, V. and Brew, B.J. 2005. Implications for the kynurenine pathway and quinolinic acid in amyotrophic lateral sclerosis. *Neurodegener. Dis.* 2, 166–176.
114. Azari, M.F., Profyris, C., Le Grande, M.R., Lopes, C., Hirst, J., Petratos, S. and Cheema, S.S. 2005. Effects of intraperitoneal injection of Rofecoxib in a mouse model of ALS. *Eur. J. Neurol.* 12, 357–364.
115. Veldink, J.H., Kalmijn, S. and Groeneveld, G.J. 2006. Intake of polyunsaturated fatty acids and vitamin E reduce the risk of developing ALS. *J. Neurol. Neurosurg. Psychiatr.* (Epub ahead of print).
116. Verdier, Y., Zarandi, M. and Penke, B. 2004. Amyloid beta-peptide interactions with neuronal and glial cell plasma membrane: binding sites and implications for Alzheimer's disease. *J. Pept. Sci* 10, 229–248.
117. Block, M.L. and Hong, J.S. 2005. Microglia and inflammation-mediated neurodegeneration: multiple triggers with a common mechanism. *Prop. Neurobiol*, 76, 77–78.

5 Microencapsulation of Marine Lipids as a Vehicle for Functional Food Delivery

Yulai Jin, Chi Perrie, Wei Zhang, Corrine Van Diepen, Jonathan Curtis, and Colin J. Barrow

CONTENTS

5.1 INTRODUCTION

Eicosapentaenoic acid (EPA) and docosahexaenoic acid (DHA) are long-chain polyunsaturated fatty acids (PUFAs) of the omega-3 class that are normally obtained in the diet from fish. The health benefits attributed to consuming EPA and DHA from fish or fish oils are well established and include protection against coronary heart disease and various inflammatory-mediated diseases [1]. Governments from many countries have encouraged their citizens to consume omega-3 fatty acids and particularly EPA and DHA [2]. For instance, the Canadian government has recommended that each person between the ages of 25 and 49 should consume a total of 1.5 g of omega-3 PUFA daily. Similarly, in May 2003, the U.S. Office of Management and Budget urged the departments of Health and Human Services (HHS) and Agriculture (USDA) to include the benefits of omega-3 fatty acids when revising the Dietary Guidelines for Americans [3]. Organizations such as the American Heart Association (AHA) and the International Society for the Study of Fatty Acids and Lipids (ISSFAL) have specific recommendations for EPA and DHA consumption; the most recent being a recommendation of 500 mg/day from ISSFAL.

Humans can acquire marine lipids by eating oily fish, omega-3 fortified foods, or by taking fish oil supplements. Those who dislike either the taste of fish or the idea of taking a daily supplement often have a low daily intake of EPA and DHA. This is the case for many North Americans with the average intake of EPA and DHA being estimated as less than 200 mg/day [4,5]. Therefore the average North American is deficient in these long-chain fatty acids (LCFAs) and would benefit from increased consumption. This provides an impetus for food companies to produce healthy foods fortified with EPA and DHA to address this deficiency and positively impact consumer health. Omega-3 fortified foods also offer food companies a way of growing relatively stagnant food categories and increasing margins [6]. One challenge facing food and nutraceutical companies is to determine how much EPA and DHA is to be added to have a significant health benefit for consumers. While the recommended dietary allowance (RDA) for omega-3 fatty acids has not been established in the United States, clinical studies do indicate that health benefit positively correlates with amount consumed and blood phospholipid levels of EPA and DHA.

The major technical hurdle for the incorporation of efficacious quantities of EPA and DHA into foods is to prevent lipid oxidation and the related off-fishy smell and flavors associated with lipid degradation [7]. A variety of strategies have been attempted to prevent lipid oxidation in omega-3 fortified foods, with encapsulation (such as dry powder and liquid emulsion products) being the most established and successful approach [8]. To avoid mouth detection and altered food texture, encapsulated oil particles must be smaller than 100 μm [9]. Consequently, microencapsulation is the term generally used for encapsulating marine lipids for foods and the resulting product can be referred to as encapsulated oil, microcapsule, powder, or particle.

Achieving microencapsulation success for a particular food application depends on three factors: materials, technologies, and properties of the microcapsules. Materials include fish oil and shell constituents. Technologies are the microencapsulation "know how" or process. The number of properties to be evaluated in microcapsules can vary depending on the food application. Particle size, sensory, and oxygen barrier at a given temperature and moisture level are some examples of the microcapsule's properties. These three factors will be discussed later in the chapter in detail. The following section will describe the current status of microencapsulation of marine lipids based on the work of various research institutions and companies producing encapsulated marine lipids and marine lipid–fortified foods.

5.2 MICROENCAPSULATION MATERIALS

Microencapsulation is the process by which fine particles, droplets, or bubbles are coated. Microencapsulating marine lipids is advantageous in that the coating creates a free-flowing powder and a barrier to oxidation, thereby masking off-flavor. The shelf life of the active ingredients, in this case marine lipids containing EPA and DHA, is also increased within the food matrix as the coating decreases the susceptibility of these oils to oxidative degradation. The resulting microcapsules are core-shell particles with an equivalent diameter approximately from 0.1 to 100 μm, while those less than 0.1 μm are called nanocapsules according to the International Union of Pure and Applied Chemistry (IUPAC)[10]. In industry, the definition of core-shell particles for microcapsules and nanocapsules is in the range of 1–1000 μm and less than 1 μm, respectively [11].

While the particle size varies, microcapsules also differ in structural design. Depending on the technology used, different core-shell assemblies can be achieved [12]. One core of active material with one shell around it forms the simple core-shell structure (Figure 5.1a). A shell may contain more than one material to meet the requirement of a microencapsulation technology and the desired properties of the microcapsules. Rather than a single core, many smaller cores may form the core, as in the case of multicore having a simple shell (Figure 5.1b). Multiple-layer

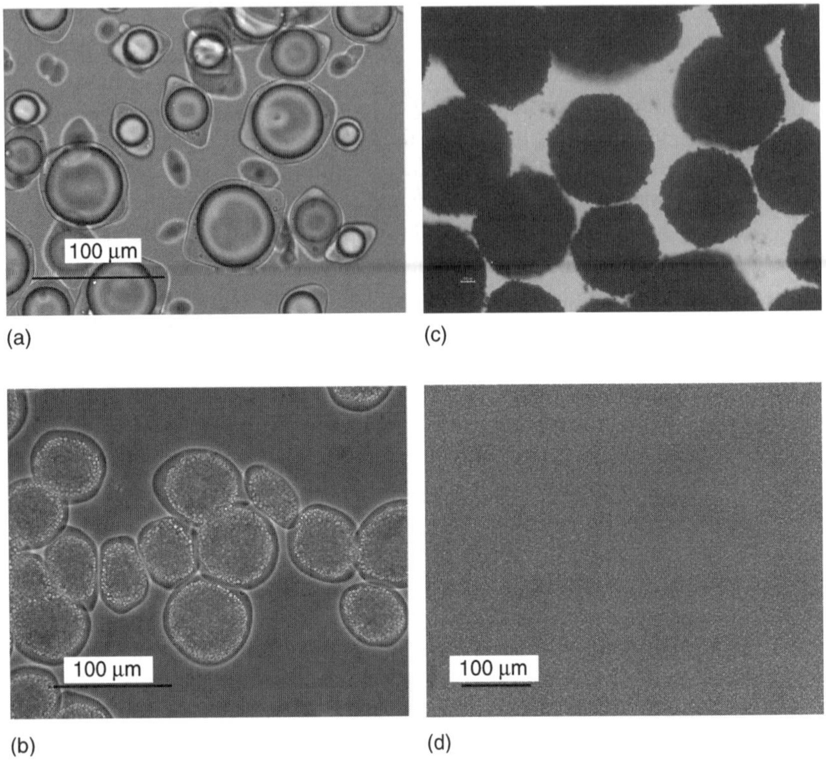

FIGURE 5.1 Examples of the core-shell structures. (a) Simple core-shell structure. (b) Multicore with inner and outer shells. (c) Spray-dried emulsion (dry state) (bar = 100 μm) (d) Spray-dried emulsion (rehydrated) (bar = 100 μm).

shells are also possible to render a specific functionality to the microcapsules. Shell can also be designed as a continuous matrix in which droplets of the active material are dispersed throughout. For the development of a microencapsulated technology targeted at a specific food application multicomponent shell materials can be mixed and matched to meet the required performance criteria. The technology mostly used in the industry and research is spray-drying emulsions base on various food materials. These products readily "dissolve" (indeed disperse) in water (Figures 5.1c and 5.1d) to restore the emulsion form, without true core-shell structures.

Marine lipids must be dispersed (i.e., emulsified) in the shell materials, forming droplets of single-core or multicore structure. Although both synthetics and biopolymers can be used as shell materials, only those that are food-grade biomaterials can be utilized in marine lipid microencapsulation for use in healthy foods. These shell materials include carbohydrates (e.g., starches and modified starches, cellulose and its derivatives, maltodextrins, cyclodextrins, alginate, gum arabic [gum *Acacia*], pectins, and carrageenans), proteins (e.g., gelatins, albumin, soy proteins, and whey proteins), and lipids (e.g., beeswax, carnauba wax, and phospholipids). Since both the oil core and shell materials contribute to the success of microencapsulation process and final powder forms for a particular food application, their most critical properties will be discussed in this section. A few selected categories of shell materials that are currently being used in the microencapsulation of marine lipids will also be discussed.

5.2.1 Quality of Lipids for Microencapsulation

Marine lipids are highly susceptible to oxidation (i.e., autoxidation initiated by metal ions and photoinduced oxidation), which leads to the formation of off-flavor as well as the loss of nutritional value. It is important to attain high-quality oil with no rancidity or fishy off-flavor before microencapsulation because the sensory quality of fish oil normally correlates with sensory quality of the finished microcapsules. The color and contaminant levels of fish oils containing EPA and DHA can be improved through filtering, deodorizing, and winterizing the unrefined oil. To improve the shelf life of the oil antioxidants such as rosemary extract or tocopherols are normally added. A final deodorization step, such as steam deodorization, is required to produce an oil of good sensory quality. This final deodorization step removes oil degradation components such as aldehydes that are responsible for most of the off-flavor associated with partially degraded fish oils. Great care should also be taken during packaging (e.g., flushing with nitrogen) and storing (e.g., using containers that block light and heat) marine lipids prior to microencapsulation to ensure a good sensory quality of the oil after microencapsulation.

The Council for Responsible Nutrition (CRN) in Washington, DC (now the Global Organization for EPA and DHA omega 3, GOED omega 3) suggests that the quality of marine lipids for the supplement market should be based on four standards: free fatty acid (FFA) (acid value or AV), primary oxidation products (peroxide value or PV), secondary oxidation products (anisidine value or AnV), and total oxidation (Totox). FFAs are released from the acylglycerols or ethyl esters due to hydrolysis. The upper limit for FFAs in fresh oil is typically 0.05%. High levels of FFAs in marine lipids reflect a quality loss of the product [13]. Indeed, FFAs exert a prooxidative effect on marine lipids [14]. PV is a direct measurement of the hydroperoxides and is expressed as the milliequivalent (meq) of peroxide-oxygen combined per kilogram of lipid. PV values of $\leq$ 1.0 meq/kg are typical for deodorized fresh oil. AnV is another measurement of the extent of oxidative deterioration. The AnV value is determined by spectrophotometric assay (at a wavelength of 350 nm) of aldehydes and ketones in the lipid by reaction with *p*-anisidine solution. An AnV value approaching 10 indicates that considerable oxidation has occurred and the accumulation of rancid compounds. Therefore, PV represents oxidation at present and AnV represents accumulative oxidation. TV is an assessment of the total oxidation ($TV = 2PV + AnV$). Detailed analytical procedures for lipids are available from sources such as the *Official Methods and Recommended Practices of the American Oil Chemists' Society (AOCS Methods).*

However, even oils that satisfy these values are often unacceptable for microencapsulation purposes as it takes only very minor quantities of specific aldehyde degradation products before they can be detected in microencapsulated oils. A variety of more sensitive analytical methods have been developed to measure low levels of strong smelling or tasting aldehydes present in processed fish oils.

The natural profile of marine lipids extracted from fish is normally given as 18% EPA triacylglycerol and 12% DHA triglyceride (TG 18/12), although in most fish used for producing fish oil supplements the levels are actually lower. These standard processed fish oils can also be concentrated to for example 30% EPA ethyl ester and 20% DHA ethyl ester (EE 30/20), or 40% EPA ethyl ester and 20% DHA ethyl ester (EE 40/20) by conversion to ethyl esters and reaction with acylglycerol to reform triacylglycerol concentrates. Encapsulating the concentrated TG oil, in comparison to TG18/12, is more challenging because the higher concentration of EPA + DHA and increased processing make the oils more prone to oxidative deterioration.

Fish oils contain a mixture of EPA and DHA. Sardine, anchovy, and menhaden oils contain more a ratio of approximately 2:1 EPA to DHA, while tuna oils contain more DHA with a ratio of EPA to DHA of approximately 1:5. Microbial oils used commercially contain mainly DHA with very little EPA. There has been some debate regarding the relative stability and sensory properties of oils with different EPA to DHA ratios. For instance, a study on oxidative stability of PUFAs (99% purity) suggested that stability increased with higher degree of unsaturation so that DHA was more stable than EPA [15]. However, an opposite trend was found in an earlier study with ethyl EPA and ethyl DHA (94.5 and 94.1% purity, respectively) where oxygen uptake of ethyl DHA was 1.6 times faster than that of EPA [16]. A comparative study of fish oil and DHA microbial oil showed no difference in stability if antioxidants are removed [17]. It appears that if there is any difference in stability of oils containing different ratios of EPA and DHA, including algal oils, then this difference is minimal and all oils containing EPA, DHA, or a combination, need to be microencapsulated for stabilization and use in foods.

5.2.2 Requirements for Shell Materials

When selecting shell materials for a specific encapsulation process, some general properties such as sensory quality, emulsification activity, and barrier to oxygen and water vapor should be considered. For use in foods, the selected shell materials must be food grade, low in cost, and consistently available in large quantities.

Meeting food regulations in specific jurisdictions is also required when microencapsulated oil is destined for foods. In European Union (EU) countries, directives under the EU food laws set requirements for safety and the purpose for food additives. In the United States, the Food and Drug Administration (FDA) regulates food additives and ingredients to ensure that they are generally recognized as safe (GRAS). In Canada, an equivalent regulatory organization is the Canadian Food Inspection Agency. Each regulatory group has its own list of acceptable food ingredients and excipients that are often approved for certain purposes and at specific levels. This makes the development of a microencapsulated food ingredient a difficult technology and regulatory challenge for use worldwide.

5.2.2.1 Organoleptic Acceptance

Shell materials usually carry the characteristic odor of their source and the processes, which have been used to produce them. They may also transfer their own sensory properties to the microencapsulated marine lipid powders as well as to the final food products. Therefore, it is desirable to select those materials, particularly biomaterials, which possess bland odor and flavor for a broad spectrum of food applications. Examples include starch and modified starch, cellulose and its derivatives, gum arabic, alginate, carrageenan, pectin, gelatin as well as whey and soy protein isolates. Some biopolymers are good encapsulation materials for specific food applications because of their compatible organoleptic characteristics. For instance, whey protein–based microcapsules are most suitable for dairy products, while pork-gelatin-based microcapsules are compatible with meat products.

5.2.2.2 Emulsification and Emulsion Stability

Emulsification is the process of dispersing one immiscible liquid in another liquid by some mechanical means. The resulting droplets can be water-in-oil (W/O) or oil-in-water (O/W), as in the case of marine lipids. Of the three regions of an O/W emulsion (the dispersed oil phase, the continuous water phase, and the oil–water interface), the oil–water interface is the most critical in dictating emulsification and emulsion stability. Water has high interfacial tension (γ) and therefore, to disperse oil in water, high shear force is required. The smaller the droplet size, the higher the shear force, according to the Laplace pressure $\Delta P = 2\ \gamma/r$, where r is the radius of the droplets [18].

Because pure liquids cannot produce stable emulsions, the presence of an emulsifying agent or emulsifier is necessary for marine lipids to be dispersed into droplets and stabilized in water that contains one or more shell materials. Ideally, the shell materials should lower the interfacial tension of the dispersion system to promote emulsification. An example is the protein class such as gelatin. Their surface tension lowering is attributed to the presence of both hydrophobic and hydrophilic side chains, which extend readily into the hydrophobic and the hydrophilic sides of the interface, respectively. Another advantage of proteins is their tendency to reside at the oil–water interface because the free energy at the interface is lower than that in the bulk aqueous phase [19].

The emulsifying properties of proteins are affected by pH, ionic strength, temperature, the presence of sugars, and low-molecular-weight surfactants [19]. At pH values close to the isoelectric point (IP), proteins are poor emulsifiers but good emulsion stabilizers [19]. The latter might be due to the lack of electrostatic repulsive forces that promotes protein adsorption and formation of a viscoelastic film at the interface. In practice, protein shell materials are usually employed outside the range of IP for improved solubility as well as enhanced emulsification.

The emulsifying properties within the protein class depend on their chemical makeup and protein loading (i.e., concentration at the interface). For instance, caseinate is a better emulsifying agent than whey, soy protein isolate (SPI), and egg white proteins [20,21]. In contrast, upon cross-linking with transglutaminase, SPI exhibits highest emulsion stability, in comparison to other proteins such as caseinate, soluble whey protein, and whey protein isolate (WPI) [22]. Based on the flexibility of the polymeric chain and its unfolding potential [23], proteins adsorb more strongly at the interface, in comparison to polysaccharides. In the absence of proteins, the limited emulsion stability rendered by polysaccharides such as gum arabic, alginate, and modified starch, depend on their high viscosity, electrostatic interactions, as well as steric hindrance [21,24]. Consequently, it may be advantageous to add a polysaccharide to a protein to enhance emulsion stability. A case in point, carboxymethylcellulose (CMC) protects WPI-stabilized O/W emulsion against coalescence but promotes the flocculation of the oil droplets [25].

Besides food-grade biopolymers, low-molecular-weight emulsifiers can also be used, particularly when carbohydrates with low emulsifying capacity such as starch and maltodextrin are used as shell materials alone. Such surface-active agents are categorized as nonionic, anionic, cationic, or amphophilic. These emulsifiers are characterized by hydrophile–lipophile balance (HLB), which varies from 1 to 40 [18]. A rule of thumb is that lipophilic (hydrophobic) surfactants (HLB <7) are suitable for W/O emulsions while hydrophilic surfactants (HLB >7) are usually selected for O/W emulsions [23]. Examples are glycerol monostearate (HLB 3.8), sorbitan monostearate (span 60, HLB 4.7) for the former, and polyoxyethylene sorbitan monooleate (Tween 80, HLB 16) for the latter, respectively [18]. Because soybean lecithin products vary in HLB from 2 to about 10 [26], they can be used for both O/W and W/O systems. After comparison of the microencapsulation efficiency and yield of DHA in corn starch, blends of Tween 85 and Tween 20 have been found to be superior to some other surfactants [27]. A note of caution, low-molecular-weight emulsifiers may exhibit competitive adsorption in protein-stabilized emulsions, thus causing destabilization of the emulsions [28].

5.2.2.3 Glass Transition Temperature

Biopolymers can behave like glass or crystal, thus showing two main transition temperatures, glass transition temperature T_g and melting temperature T_m. In the glasslike case, a biopolymer undergoes a gradual structural change, from glassy solid state to a rubbery state, as it is heating up through the glass transition temperature. In the crystallike case, the orderly structure abruptly disappears at the melting point [29]. Differential scanning calorimetry (DSC), which determines heat capacity of materials, is a common method to measure glass transition temperature. Other properties of materials, such as rheological or dielectric, also change during the glass transition. Carbohydrate shell materials possess clear glass transitions observed by DSC, while proteins exhibit less obvious transitions over a broader temperature range [30].

Since molecular mobility and reaction rates are low at temperatures below T_g, the shell materials should be more stable in the glassy state. For marine lipids, the glass transition theory implies that their stability against oxidation is maintained while the encapsulating shell materials remained in the glassy state. A recent study, however, failed to prove the direct relationship between molecular mobility and lipid oxidation rate in maltodextrin-based microcapsules by spray drying [31].

In addition to heat, glass transition theory attributes moisture content (water activity) to increasing the permeability of a given shell material. Water plasticizes biopolymers and reduces their glass transition temperatures, which is unfavorable for stability. It has been found that a high water activity increases the oxidation rate of microencapsulated lipids [31,32]. One approach for estimating activation energy from the diffusivity of carbohydrates and carbohydrate blends has also been proposed as a means of selecting shell materials with better protection against oxidation [33].

5.2.2.4 Oxygen and Water Vapor Permeabilities

Shells with good oxygen barrier are necessary to protect microencapsulated lipids from oxidation. In other words, the shells must possess low oxygen permeability because permeability is directly proportional to the diffusion rate of small molecules such as oxygen and water across the shells. Since oxygen permeability increases with water activity, the desired shells must be low in both oxygen and moisture permeability, for a given architecture of the microcapsule as well as shell/matrix thickness.

It is common practice to select materials and test optimal formulations by monitoring the lipid oxidation products after encapsulation. In industry, screening of materials is usually carried out in the film form as it is more economical than in the microcapsule form. Films made from lipids possess the lowest water vapor permeability, followed by carbohydrate films; both lower than protein-based films [34]. In contrast, protein-based films generally exhibit lower oxygen permeability than that of carbohydrates and lipids [34,35]. To meet both low oxygen and water vapor permeability, blending materials from two or more categories of biopolymers (e.g., a protein and a carbohydrate) may be necessary.

5.2.3 Proteins

One of the most often used proteins is gelatin, particularly type A gelatin that has higher IP values than type B. Gelatin can be used as one of the primary components for preparing single-core or multicore microcapsules by complex coacervation, which will be described in Section 5.3 [36–41]. Its gelation point is below 35°C. When gelatin is used in the spray drying of marine lipids, it shows better product stability against oxidation than whey protein concentrate (WPC), soy protein hydrolysate, and SPI [42]. Gelatin has also been blended with carbohydrates as encapsulating material during spray chilling [43]. Cross-linking is required to compensate for the thermal reversible nature of gelatin at the gelation point.

Whey and milk proteins alone, or blended with carbohydrates, have been studied for their microencapsulation properties using various process technologies [21,44–51]. Whey and milk proteins are capable of forming and stabilizing fish oil emulsions, thus protecting PUFAs against oxidation [52,53]. Blending with carbohydrates can help improve the drying performance [54].

SPIs have also been used to produce microencapsulated fish oil and other lipids [22,42,55–57]. In addition, an equal amount of maltodextrin with a dextrose equivalent (DE) of 29 was found to improve the encapsulation efficiency and stability of SPI-based microcapsules [55]. Heat-denatured soy and milk proteins are able to undergo irreversible thermal gelation and therefore do not require additional chemical cross-linking. The globular protein β-lactoglobulin as well as soy proteins have shown the capability of binding various flavor compounds and small ligands improving the sensory properties of microcapsules incorporating these proteins [58–60]. While it is unclear if and how these proteins improve the sensory stability of the encapsulated marine lipids, it is possible that amine or thiol amino acid side chains react with and capture the oxidative off-flavor compounds such as aldehydes. Alternatively, the hydrophobic regions in these proteins may adsorb the aldehyde components when present at low levels.

Albumins are the least heat-stable proteins [61]. An ovalbumin has been used to make fish oil emulsion followed by spray or freeze drying [62]. Owing to lower porosity of the matrices, spray-dried fish oil microcapsules in ovalbumin were found to be more stable than those that were freeze dried. Gliadin and glutenin, two protein components present in gluten, can be dissolved in an aqueous ethanol solution and used for fish oil encapsulation by solvent evaporation method [63]. Gluten fractions have certain advantages over milk proteins and gelatins because of their insolubility in cold or hot water and the formation of disulfide linkages that promote molecular networks without the need for cross-linking agents. However, the use of organic solvents adds cost to the process and gluten intolerance is of concern to some people, such as individuals with coeliac disease.

5.2.4 Carbohydrates

Carbohydrates possess moderate barrier properties against diffusion of oxygen and moisture. Most carbohydrate shell materials are poor emulsifiers but exhibit enhanced stability of protein-based emulsions. Carbohydrates studied for the purpose of lipid encapsulation include gum arabic [36–40,64,65], derivatives from starch [21,55,64,66–70], cellulose derivatives [36,39,71–74], carrageenan [36], and alginate [36,41,43,75].

Gum arabic is of special interest since its protein component (2–5% by weight) enables the polymer to act as a good emulsifier and emulsion stabilizer. The unique properties of gum arabic are its high solubility in water and low viscosity of its solutions at high concentration. These features improve productivity by shortening the spray-drying process. Noticeably, the emulsion viscosity is exponentially increased with increasing lipid payload [65]. Besides spray drying of emulsions, gum arabic is also used extensively in complex coacervation as a polyanionic component interacting with polycationic shell materials such as gelatin [36–40]. Because gum arabic, the natural exudation of acacia trees, has a limited supply, other gums or carbohydrates have been studied as replacements.

Although raw starch materials have been used for microencapsulation of flavor compounds, vitamin C, and other materials [76–81], little work has been done on marine lipid encapsulation using unmodified starch. Maltodextrins (hydrolyzed starch segments) with various DE values have proved very useful in marine lipid encapsulation applications, particularly in spray-drying technology [21,55,64,68,70]. The low DE (i.e., high-molecular-weight) maltodextrins exhibit high glass transition temperatures as well as high viscosities. Maltodextrins provide good oxidative stability of encapsulated lipids but possess limited emulsifying ability and emulsion stability [82,83]. Thus, these materials are usually combined with other shell materials such as proteins and gum arabic for optimal encapsulation results [33,55,64]. Modified starch (sodium octenyl succinate starch) has also been used for marine lipid encapsulation, due to its improved emulsifying property.

Another group of starch derivatives, cyclodextrins, have also been explored for their use as encapsulants of PUFAs [66–69,84,85]. Commercial α-, β-, and γ-cyclodextrins are composed of six, seven, and eight glucose units, respectively. Because these compounds can complex with guest molecules such as PUFAs that are attracted to the interior of cyclodextrins by hydrophobic interactions, the process is often referred to as inclusion encapsulation or molecular encapsulation. The optimal inclusion loading is around 25% oil at an oil to cyclodextrin molecular ratio of 2:1 [86]. Interestingly, the cyclodextrin complexes can reduce unpleasant taste and odor [69], presumably because of their ability to absorb flavor compounds such as aldehydes [87].

Cellulosic shell materials that have been studied include carboxymethylcellulose (CMC) and cellulose esters such as ethyl cellulose and hydropropylcellulose (HPC) [36,39,71–74]. CMC can be used as the polyanionic component for complex coacervation while HPC has better barrier properties and can be incorporated into shell materials or matrices.

Pectin and alginate are other carbohydrates that have been experimented with as shell materials for microencapsulation [36,41,43,57,75]. These materials are capable of gelation through ionic interactions with calcium, which is advantageous when the shells or matrices require hardening.

5.2.5 Lipids

Lipids are natural compounds grouped together because of their nonpolar structure and relative insolubility in water. Waxes, the simple lipids formed by the esterification of fatty acids and long-chain alcohols, are commonly used to control the desiccation of fresh fruits and vegetables [88]. Waxes are often combined with hydrocolloids to counteract their brittleness. The trade-off is a reduction in water vapor permeability. Phospholipids are excellent emulsifying agents because they contain a glycerol backbone that is bound to two fatty acids (hydrophobic) and a phosphoric acid (hydrophilic). Lecithin is one of the most common emulsifiers used in food products. Phospholipids have also been used to encapsulate and extend the shelf life of PUFAs [89–93]. Other food lipids such as butter can also be used to blend with marine lipids for protection against oxidation [94]. Overall, lipids have been extensively used as edible films for foods but sparingly used as encapsulating materials for PUFAs.

5.2.6 Other Materials

Biological cells can be potentially used for the encapsulation of marine lipids [95–97]. For example, microbial cells have been commercially used for flavor encapsulation by Micap PLC. Other minor components that are added to the microencapsulating shell for a specific functionality include emulsifying and cross-linking agents, antioxidants, and chelating agents. Cross-linking agents can be used to harden the shells where applicable. Formaldehyde, glutaraldehyde, and transglutaminase are good cross-linkers for protein-based shells, while calcium is effective and economic for gum-based shells and matrices such as alginate and pectin. Noticeably, formaldehyde and glutaraldehyde are not GRAS. Emulsifiers such as lecithin, mono- and diacylglycerols, Tween 20, and sucrose fatty acid ester may be added to promote emulsification and emulsion stabilization, particularly when carbohydrates are used as shell materials alone [27,55,68,98]. Proteins are good emulsifiers and usually provide excellent emulsion capacity and stability. Antioxidants play an important role in minimizing oxidation of the marine lipids. Lipophilic antioxidants such as tocopherols and rosemary extracts are often incorporated into the lipids for improved stability, while water-soluble antioxidants, for example, ascorbate and phenolic acids, can be added in the aqueous phase before marine lipids are emulsified [99–104]. Glucose oxidase, other oxidoreductases, and green tea polyphenols have also been used as oxygen scavengers in shell and oil, respectively, to protect lipids against oxidation [66,105–107].

To prevent the prooxidants such as iron and copper from triggering autoxidation, chelators can be used as well. Ethylenediaminetetraacetic acid (EDTA), citric acid, and sodium tripolyphosphate are often added to the aqueous phase and cause partition of metal ions from oil phase and interface to the solution [63,108–111].

5.3 MICROENCAPSULATION TECHNOLOGIES

The many technologies for microencapsulation can be divided into two categories, one which uses a liquid as a suspending medium, such as in complex coacervation, and one which uses a gas as a suspending medium into which a liquid phase is sprayed, such as fluidized bed coating [112]. The former are made using a chemical process and the latter utilize a mechanical process [113]. Although there are many different processes, all include three main steps: (1) dispersion or emulsion formulation, (2) capsule wall deposition, and (3) capsule isolation. This section will focus on technologies that produce microcapsules with a diameter of 500 μm or less. While more complex products and processes are being developed or are in the product pipeline, microencapsulation of marine lipids remains a relatively new endeavor with much to be learned in applying the technologies to this specific area. As of yet, not all of the technologies explained below have been applied to encapsulating marine lipids.

5.3.1 Mechanical Processes

5.3.1.1 Spray-Drying of Marine Lipid Emulsions

To date, the technology of spray-drying emulsion is most commonly used in marine lipid encapsulation. Developed in the 1930s, spray drying has become one of the oldest encapsulation methods and also the most commonly used method in the food industry. The particle diameter is generally 10–300 μm, although most often in the larger end of this size range [113,114] owing to agglomeration of the individual oil droplets (Figures 5.1c and 5.1d). The structure of spray-dried microcapsules consists of active core material entrapped in a protective polymer or melt matrix. The first step in the process is to emulsify the core material in a concentrated shell solution until 1–3 μm or smaller oil droplets are stabilized. Besides fish oil, the core material is generally a water-immiscible flavor, vitamin, animal fat, or plant oil. The emulsion is then fed as droplets, via an atomizer, into a heated chamber. The rapid evaporation of water from the coating keeps the encapsulated core at far below 100°C, despite the fact that the inlet temperature can be as high as 200–300°C.

Several variables may affect the spray-drying process. The initial consideration is to find a suitable material, which has good emulsifying properties, low viscosity at high solid levels, low hygroscopy (i.e., low moisture adsorption from the environment), and can form a good film for coating. A low-cost product is also desirable, with a bland taste, stable supply, and good protection for the encapsulated material. The in-feed solids level and the inlet and outlet spray-drying temperatures are other important determinants for retention and stability of the marine oil–core material.

Spray-drying offers many advantages. The process is economical, flexible, and uses readily available equipment to produce large amounts of microcapsules. Some disadvantages are that core loading is usually low, only 20–30%, to avoid high surface oil, difficulty in finding suitable shell materials or emulsifiers, and the instability of the reconstituted droplets in adverse food processing such as homogenization and heat treatment. Heinzelmann et al. [115,116] investigated freeze drying as an alternative for the microencapsulation of fish oil by spray drying. The lower drying temperature of freeze drying, in comparison to spray drying, might provide additional protection against oxidation.

5.3.1.2 Spray Chilling

Spray chilling, cooling, and congealing are variations of conventional spray drying [113]. These processes are similar in that they both involve emulsifying the core material into a liquid shell material and spraying it through heated nozzles into a controlled environment. The main differences are the temperature of the air used in the spray dryer and the type of coating [114].

Spray chilling uses air that has been cooled to ambient or refrigerated temperatures below the solidification point of a molten fat or wax coating, to avoid affecting the polymorphism of fat. This process can be used for encapsulating water-soluble core materials in hot-melt fat, such as encapsulation of minerals, water-soluble vitamins, enzymes, acidulants, and some flavors. The microcapsules are insoluble in water because of the lipid coatings. The process is also used by DSM to encapsulate fat-soluble vitamins, by spray chilling of emulsion droplets. Therefore, it is potentially useful in marine lipid encapsulation.

5.3.1.3 Other Mechanical Processes

Fluidized bed coating is a process developed in the 1950s for the pharmaceutical industry. Microencapsulation is accomplished by spraying a liquid onto a solid core or porous particles that are suspended by a moving stream of hot or cold air. Core particle diameter ranges from 50 to 500 μm, and are usually greater than 100 μm to allow complete coating. Thus, this technology is only applicable as a secondary process of encapsulated marine lipids as it is not capable of encapsulating oil droplets directly.

Coextrusion and spinning disk are potentially useful in preparing microencapsulated marine lipids. Coextrusion is a drop-forming method to produce small particles. It begins with extruding a heated aqueous polymer solution through an outer tube and the oil to be encapsulated through an inner tube into a moving stream of a carrier fluid [113]. Coextrusion process varies depending on the nature of solidification. In one type of process, solidification takes place in a liquid phase. In another, solidification takes place in a gas phase. The resulting capsules can be collected on a moving bed of fine-grained starch, cushioning the impact and absorbing coating moisture [114]. One advantage of coextrusion over spray drying of emulsion is that the core and shell may be miscible. Although immiscible materials are more preferable, the rapid flash off in a hot air tower can solidify the shell before it has a chance to coalesce with the core because of their miscibility [117].

The particle size of coextruded microcapsules ranges from 150 to 2000 μm. This is a disadvantage when dealing with marine lipids for food fortification because larger than 100 μm microcapsules can change the mouthfeel of food. Consequently, this method has not yet been used for the microencapsulation of marine lipids for food applications.

Rotating disk, also known as spinning disk, is a process that was developed in the 1980s, with microcapsule size ranging from 30 μm to 2 mm [114,118]. It involves suspending core particles in a shell solution, and then pouring the suspension through a rotating disk apparatus. Centrifugal force generated by the rotating disk forces the emulsion to the outer edge of the disk to make discrete droplets. Excess liquid is atomized and separated from the product and recycled. Chilling and drying solidifies the particles.

If the shell is a hot melt formulation (e.g., molten wax), the microcapsule cools and hardens as it falls through the gas phase. If the shell formulation is an aqueous polymer solution that can be gelled by ions or a combination of ions and cooling, the coated particles fall into a curing bath. The particles can then be dried to a powder. Disk geometry, diameter and speed of the rotation, as well as volume flow rate of liquid across the disk are parameters that affect the size of the microcapsules. This process is best suited for encapsulating solids or solidlike materials. Therefore, like fluidized bed coating, the marine lipid particles must first be preformed, which increases operating costs. However, these two methods are particularly useful when additional protection by coating is needed.

5.3.2 Chemical Processes

5.3.2.1 Complex Coacervation

Also called phase separation, coacervation was developed in the 1950s to produce a two-component ink system for carbonless copy papers [119]. Simple coacervation involves the use of a substance such as gelatin where separation of a gelatin-rich phase and a phase almost completely devoid of gelatin occurs. Marine lipids are dispersed in a gelatin solution and a pH adjustment causes the gelatin to coacervate and adhere to small core droplets. As the mixture is cooled, the gelatin shell is hardened [117].

Complex coacervation is so far the most successful technology of protecting the marine lipid core material from oxidation while delivering high levels of loading. It differs from simple coacervation in that there are two oppositely charged polymers instead of one participating in the phase separation process. The coacervate is rich in polymers and weak in water, and the supernatant is low in polymer concentration [36]. The deposition of the coacervate onto the core droplets to form the microcapsule shell and the microcapsule shape and size depend on several processing factors. They are polymer type, molecular weight, charge density, concentration of polymers and their ratio, pH and temperature of the system, and cooling rate. Recent studies also reveal the important role of surfactants in promoting encapsulation efficiency [120]. Since all these processing parameters are interrelated, optimization becomes a great challenge [113].

Solidification follows shell formation to strengthen the microcapsule shell. It can be achieved by thermal, cross-linking, or desolvation techniques. Finally, microcapsules can be collected by filtration or centrifugation, solvent washed, and dried by techniques such as spray drying or fluidized-bed drying to yield free-flowing, discrete particles.

When complex coacervation was first developed for delivery of marine lipids, the focus was on "single-core" microcapsules (Figure 5.1a). One of the problems with this type of microcapsule is its susceptibility to rupture especially during food processing operations such as homogenization and extrusion. Rupture can be minimized by replacing a single core with many smaller cores that agglomerate into one multicore (Figure 5.1b). In this case, each oil droplet is protected not only by its own shell but also by the shell surrounding the multicore. The process of making "multicore" microcapsules by complex coacervation for the encapsulation of PUFAs is described in a U.S. patent application [41]. The process leads to microencapsulated powders with very low surface oil. This is most likely because the multicore is really an agglomeration of smaller capsules surrounded by an additional shell material. Most of the surface nonencapsulated oil is trapped on the surface of the smaller capsules that are subsequently entrapped by the external shell. Surface oil levels from this process are less than 0.2% of total oil, which is much lower than surface oil observed in spray-dried emulsions or single-core microcapsules. Surface oil is readily oxidized and so low surface oil is very important in microencapsulation of omega-3 oils, where a small amount of degraded oil has a significant impact on powder sensory properties.

Complex coacervation generates small particle sizes, from about 1 to 1000 μm, with relatively easy control of the average particle size within this range. In comparison to other encapsulation process, complex coacervation also generates the highest oil to shell material ratio, or payload, usually 60%. The higher payload, as compared to the formation of emulsions that normally have payloads of 20–30%, is a significant advantage both in terms of cost and impact on food. Less powder is required to deliver a similar omega-3 content when the payload is higher. However, as a batch process, it can be time consuming and this expense partially offsets the savings associated with using less shell material [112].

5.3.2.2 Double Emulsification and Gelation

The process of double emulsification involves making two emulsions in succession. It can be either water–oil–water (W/O/W) or oil–water–oil (O/W/O). For marine lipids, the O/W/O technique is used. In the first step of this process, a stable O/W emulsion is prepared with the aid of a water-soluble emulsifier, and in the second step, the O/W emulsion is dispersed in an oil phase using an oil-soluble emulsifier [121].

Shim and coworkers [122] studied the development of O/W/O double emulsification using an enzymatic gelation method of microencapsulation for fish oil. Fish oil containing DHA and SPI were used as core and shell material, respectively. O/W emulsions were prepared by dispersing fish oil into a 10% isolated soy protein solution. This emulsion was subsequently dispersed into corn oil containing 3% emulsifier to form an O/W/O double emulsion. The gelation of O/W/O emulsion was completed in the presence of microbial transglutaminase for 4-h incubation at 37°C. The most stable O/W emulsion was obtained with the combination of fish oil to wall material in the ratio of 1:2. Furthermore, the stability of O/W/O emulsion was highly affected by the type of emulsifier. The capsules were reported to be 23 μm in mean diameter, spherical in shape, and dent-free in surface. Such a microencapsulation process consisting of double emulsification and subsequent enzymatic gelation might be suitable for preparing protein-based microcapsules containing sensitive ingredients such as marine lipids [122].

5.3.2.3 Liposome Entrapment

By definition, liposomes are spherical vesicles that form with the hydration of surfactants such as phospholipids. When mixed with water under low shear, phospholipids arrange themselves in sheets, the heads of the molecules facing upward and the tails downward. By joining tails and tails, these sheets form a bilayer membrane that contains some water inside a phospholipid vesicle [123]. The main advantage of using liposomes for entrapment is that they can contain both water-soluble and water-insoluble materials at the same time. The water-soluble materials are added to water that is used for phospholipid hydration. As the liposomes are formed, these materials become trapped in the center. The wall of the liposomes, being a phospholipid membrane, can hold water-insoluble materials such as marine lipids. The topologically closed shell(s) of liposomes can provide a barrier [124]. Liposomes have been found to protect DHA from peroxidation [92]. Since liposome entrapment does not require the use of other surfactants or emulsifiers, it enables special encapsulation whereby the presence of these materials is a drawback. Also, as a solvent-free process, it shows good potential in terms of lowering production costs for the microencapsulation of nutraceuticals.

Liposome entrapment is a flexible encapsulation method. It can be custom designed for the encapsulation of a wide array of core materials such as proteins, peptides, polynucleotides, hormones, and lipid-soluble compounds. Process parameters such as solution pH, agitation method and rate, type of cation, and ratio of phospholipid to core material can also be varied.

Besides spherical shape, liposomes can also take a spiral or cochleate form. Nanocochleates (less than 500 nm) are formed as a result of the condensation of small unilamellar, negatively charged liposomes. In this process, the core material such as fish oil is added to a suspension of liposomes containing a negatively charged phospholipid, phosphatidylserine in this case. The addition of metal ions such as calcium induces the collapse and fusion of liposomes into large sheets made up of lipid bilayers. Since these sheets have hydrophobic surfaces, to minimize their interactions with water, they tend to roll-up into cigarlike shapes (cochleates). Each cochleate consists of a series of solid layers so that the core material remains intact and stable, although the outer layers of the cochleate may be exposed to harsh environmental conditions [89].

5.4 PROPERTIES OF MICROCAPSULES

In the literature, microcapsules have been evaluated by various methods [114, 125–128]. Some of the essential characteristics and properties of microencapsulated marine lipids that are necessary to meet specific food application criteria will be discussed in this section. These properties are physical, such as particle size and size distribution, payload, compressibility and burst pressure; chemical, such as lipid content, omega-3 content, lipid oxidative stability and chemical stability of wall materials; and organoleptic, such as off-odor and off-flavor. The numerous methods and instruments associated with the essential characteristics and properties of microcapsules will also be described briefly.

Other microcapsule properties, such as solute permeability, viscoelasticity, and electrophoresis mobility, while important to applications in the areas of pharmaceutics and cosmetics, are not as critical for microencapsulated marine lipids and will not be discussed. Similarly, control of microbiological contaminants and other pathogens that fall within the guidelines of good manufacturing practice will not be discussed here.

5.4.1 Physical Properties

An intact and integrated microcapsule is supported by its physical and mechanical properties. Its functionality depends on these properties. As mentioned in Section 5.1, to avoid altered mouthfeel of food, the encapsulated marine lipids must be small in size, normally less than 100 μm. The particle size distribution should also be narrow to tightly control the properties and performance of the microcapsules. For instance, small deviation of compressive strength from the norm favors small size distribution. Particle size and wall (shell) thickness can be measured using an optical microscope, which can be equipped with a phase contrast device for enhanced images [129,130]. A particle image analyzer equipped with charge-coupled device (CCD) camera has also been used to facilitate size determination [52]. There are other methods and instruments available to determine particle size over a wide range, including laser beam diffraction, light scattering techniques, and sieving with a set of sieves of different sizes [130].

Particle size analysis can be applied early in the microencapsulation process. For example, during the emulsification step it can be used to assess the impact of encapsulating materials and other process parameters (e.g., processing time, stirring rate, or homogenization pressure) on the emulsion droplets and their change with time (Figure 5.2). High-resolution imaging systems, such as scanning and transmission electron microscopes, atomic force microscopes, and other imaging methods can also be used to determine particle size, although these techniques also provide information on the internal and surface structure of microcapsules [52,131,132]. In particular, confocal laser scanning microscope (CLSM) is an effective tool for investigating the various domains inside the microcapsules [133, 134]. With selected fluorescent markers, each component of a microcapsule, particularly the oil phase and the shell can be visualized using CLSM.

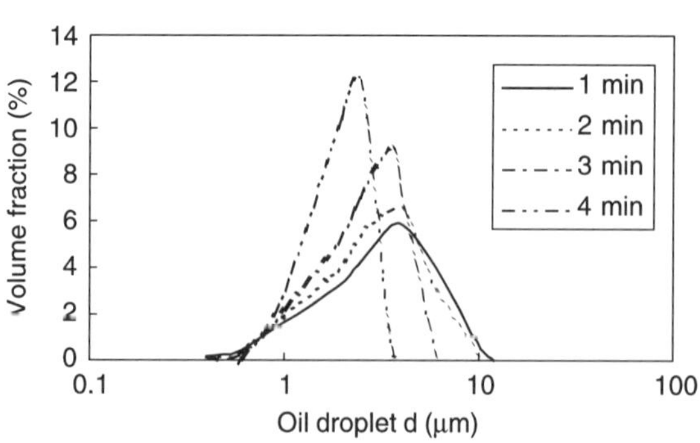

FIGURE 5.2 Fish oil droplet size distributions in emulsions containing 10% w/v gelatin, emulsified at 10,000 rpm for 1–4 min.

A computational image analyzer can be used to calculate the volumes of the different phases. Results from a nondestructive image analysis such as CLSM correlate well with results obtained by extraction and chemical analysis.

Mechanical strength of microcapsules is an important property for success in food processing that involves high shear force and pressure, which can damage microcapsules if they are not mechanically strong. Mechanical strength is generally positively related to particle size and wall thickness [129]. Multicore microcapsules in general have higher mechanical strength than single-core microcapsules because of their increased compressibility and also the strength provided by the outer shells surrounding inner shells of individual oil droplet. Mechanical strength can be measured by a variety of methods. Some methods involve determining percent breakage in a sample containing many microcapsules, rather than determining the breaking point of a single microcapsule. The most common method for measuring mechanical strength uses a texture analyzer to press microcapsules until they burst. The force at burst and two other force-related indicators (the displacement at bursting and the force at a defined displacement) can be extracted from a single measurement. In terms of mechanical stability of single-core microcapsules, an optimal wall thickness was found to be around 18% of the particle radius [135]. This finding was based on water-saturated microcapsules, which can behave differently from dry microcapsules. Micromanipulation is another technique for evaluating mechanical stability, specifically for microcapsules as small as 1 μm. Using this technique a negative pressure can be applied to a single microcapsule via a microcapillary. Burst events and wall deformations of microcapsules are then observed under microscope [136].

Microcapsules with good free-flowing properties are desirable in most food applications since free-flowing powders are easier to blend with other dry ingredients and flow better during food processing. This flow property can be characterized by static angle of repose measured by the fixed funnel and free-standing cone method [137]. A small repose angle indicates good flowing property. Compressibility is another method to assess flow properties of powder. It compares packed (tapped) bulk density to aerated (loose) bulk density [52]. Noticeably, moisture content can also be an important factor for flowability. A high moisture content can affect the powder flowability. Dried microcapsules normally contain some moisture because of the hydrophilic nature of the wall materials and can be hydroscopic so that incorrect storage can compromise flow behavior.

5.4.2 Chemical Properties

As mentioned in Section 5.2.1, oxidative stability of the microcapsules depends, to a large extent, on the quality of the marine lipids. The same methods that measure oil quality in fish oils, including acid value, peroxide value, and anisidine value, can be applied to those contained in microcapsules. However, before these methods can be applied, the oil needs to be extracted from the microcapsule after chemical or enzymatic digestion of the shell material. Processes used for this extraction can significantly impact the oil quality and therefore measurements on extracted oil can be misleading. The extraction of marine lipids from encapsulated powders can be achieved in several ways, depending upon the shell material composition. Sometimes oil can be simply extracted from powder using solvents such as boiling water [138], *n*-hexane extraction [52], or a mixture of organic solvents [21]. However, when the microcapsules are formed using complex coacervation and are cross-linked, the shell material must undergo acid or protease digestion. Normally, these methods can be applied without significant degradation of EPA and DHA so that measurement is accurate.

Payload is an important property as it represents the percentage of marine lipids encapsulated within the microcapsules. Payload dictates how much microencapsulated ingredient should be added to meet the required loading of EPA and DHA per food serving. A higher payload means less shell material is required to deliver a specific dosage of EPA and DHA, so that both cost and impact on food properties are minimized. Payload determinations normally require degradation of the shell material either chemically or enzymatically and subsequent extraction followed by gas chromatographic quantification of the marine lipids.

Closely related to payload is surface oil (free oil), which is the portion of marine lipids that has not been encapsulated into the core of microcapsules, and is absorbed on the surface [52]. Free oil can also be considered the inverse of encapsulation efficiency, which is expressed as a percentage of real payload over theoretical value. Furthermore, free oil contributes to the oxidative off-note of the encapsulated lipids and must be kept to a minimum, typically below 0.1–1% (w/w). The percentage of surface oil present on dried microencapsulated powder can be determined by extraction with organic solvent such as hexane or pentane, and subsequent quantification by weighing the collected lipids or by using Fourier Transform Infrared spectroscopy (FTIR) [21].

Although a variety of oxidative products are generated during the oxidation of marine lipids, one or two can be selected for a comparative study [17]. For instance, based on the propanal detected by gas chromatography in headspace, there was no difference in oxidative stability between an algal oil (42% DHA) and a typical fish oil (12% DHA). Oxidation as a function of time and temperature was slower for the algal oil, only in the presence of an antioxidant. In general, it has been difficult to interpret the oxidative stability results so abundantly available in literature. This is because of inconsistent oxidation conditions, incomparable analytical methods, and irrelevant stability methods.

Elemental analysis of iron, copper, arsenic, lead, cadmium, and mercury can be carried out by standardized methods found elsewhere. The same can be said for the determination of water activity, antioxidant residue, or polychlorinated biphenyls (PCB). New and advanced analytical methodologies and instruments for characterizing various types of microencapsulation processes and products are continually being developed. Electron spin resonance (ESR) spectroscopy, nuclear magnetic resonance (NMR) spectroscopy, ultrasonic absorption, radioactive tracers, and fluorescence quenching are just a few examples. At present, these techniques are mainly used for the characterization of liposomal types of microcapsules [127].

5.4.3 Sensory

Sensory tests, also referred to as organoleptic tests, have gained significant advancements over the last century. Despite modernization of the sensory process, human panelists have been, and still are, the main measuring instruments. Inevitably, variability and bias are a component of their results [139]. Important parameters provided by sensory tests include odor, flavor, appearance, and texture.

While it is important to evaluate the sensory properties of an encapsulating powder, it is the sensory in foods that will ultimately dictate acceptance into specific applications. Extensive research has been conducted on food sensory testing, but a limited number of articles are available concerning the use of microencapsulated marine lipids. One example used spray-dried marine lipids to fortify brown soda bread that was taste tested by an untrained panel. A triangle test and subsequent paired preference tests were unable to determine if alterations of bread flavor were caused by the addition of marine lipids encapsulates [21].

At present, it is unclear whether sophisticated modern instruments such as GC-MS can replace sensory panels for better accuracy and reliability. Various analytical sensory techniques can determine levels of volatile components selected as off-odor markers and there has been some correlation with the results of sensory panels, although these methods appear to be more accurate for straight oil rather than microencapsulated oil [17,140]. The lower correlation of analytical and human sensory for microencapsulated oil is probably related to the added complexity of having shell material that may have its own sensory properties.

5.4.4 Shelf Life

Shelf life or storage stability refers mainly to long-term oxidative stability of the encapsulated oil, either by itself or in foods. Oxidative stability must be constantly monitored to ensure product quality and appropriate measures must be implemented to prevent oxidation reaction. As in the case of fish oils, powders must be sealed and stored in a cold room (e.g., 4°C) and away from light when not in use. With effective barrier layers, powders can possess low oxygen permeation and prolonged shelf life. For instance, microcapsules prepared by spray drying with shell material of highly branched cyclic dextrin showed enhanced oxidative stability when compared with maltodextrins, based on PV over a period of 80 days. Powders are also expected to be more stable when the encapsulating oils have increased shelf life in the presence of antioxidants. Experiments show that antioxidants can increase shelf life of fish oil–fortified mayonnaise to an average of 49 days at room temperature and 89 days under refrigeration [141]. Significant improvement of oxidative stability by microencapsulation of spray-dried emulsion of PUFAs has been reported. Shelf life of infant formula and bread containing microencapsulated PUFAs can be extended to more than 2 years [142].

To enable faster development of an optimized technology, microcapsules are usually subjected to an accelerated oxidation test first. Only those passing this test

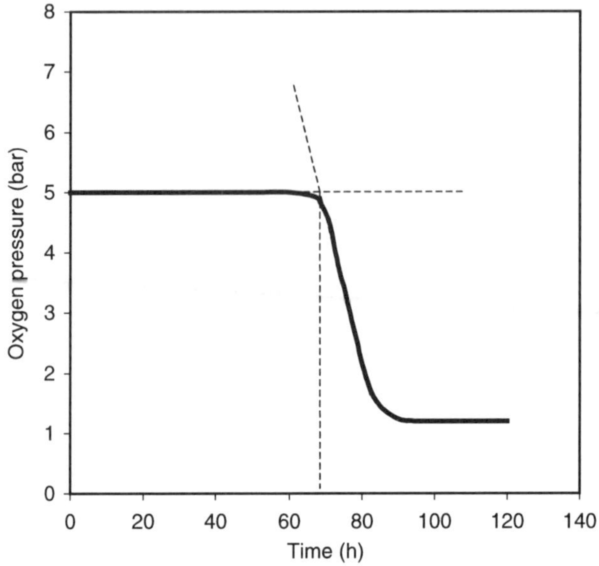

FIGURE 5.3 Illustration of the derivation of the induction period of a microcapsule sample based on the slope of the oxygen pressure drop.

will be considered for further shelf life test. In accelerated tests, a small amount of powder is exposed to pure oxygen at a high pressure and a high temperature, for example, 5 bars (500 kPa) and 65°C. At the onset of oxidation, the pressure drops and the time is recorded as induction period, as illustrated in Figure 5.3. This may last from a few hours to more than a week depending on the protective properties inherent in the microcapsules.

5.4.5 Bioavailability

The bioavailability of microencapsulated marine lipids added to food has been compared to marine lipid–containing soft gel capsules. The fortified foods include liquid-based foods such as milk shakes and soups, as well as baked foods such as bread and biscuits [143,144]. The results indicate no significant difference in blood lipid levels between the two groups of participants for carbohydrate or protein shell materials. Bioavailability was also demonstrated with low-dose omega-3 marine lipids, as little as 60 mg/day, microencapsulated and fortified in bread [145].

5.5 CURRENT MICROENCAPSULATION STATUS

5.5.1 Status Based on Developments from Research Institutes

Although numerous articles can be found in the literature that describe technological development and progress in the area of microencapsulation, only a limited number of them deal with encapsulated marine lipids. Spray drying of emulsions is the dominant technology for microencapsulating marine lipids, presumably because of low processing cost and fewer restrictions placed on the characteristics and properties of shell materials. In one example, marine lipids containing 5.5% EPA and 17% DHA, together with tocopherol and ascorbyl palmitate, were microencapsulated via spray drying with shell material composed of a combination of maltodextrins, a highly branched cyclic dextrin (HBCD), WPI, and sodium caseinate. In another example, the compositions of sodium caseinate with maltodextrin or HBCD as shell materials improved oxidative stability. In this study it was found that a payload at 69% provides better oxidative protection when compared with either 46 or 83% payload. However, it was not clear what mechanism is responsible for this oxidative stability.

Similarly, menhaden and herring oils, with either α-tocopherol or Trolox C added as an antioxidant, were spray dried using sodium caseinate and polysaccharides of various DE values as shell materials. It was found that high DE values of polysaccharides correlated positively with increased oxidative stability within 28 days of storage. An unexpected finding was that the oxidative stability decreased at 4°C. This decrease in stability may be due to increased dissolved oxygen concentrations at the lower temperature [138].

In another recent study, cod-liver oil was spray dried with skim milk powder, sodium caseinate, whey powder, whey protein concentrate, modified starch, and maltodextrin, or combinations of these materials at a wall to core ratio of 4:1 [21]. Spray-dried powders were heated for 30 min at various temperatures from 60 up to 235°C, and the contents of EPA and DHA were measured using GC-MS. Compared to unencapsulated oil, EPA was partially stabilized, particularly with shell materials containing maltodextrin mixed with whey protein concentrate. This was not the case for DHA content, which degraded similarly in the encapsulated product and the unencapsulated starting oil. Among the formulations of shell materials tested, the oil retention of the dried powders varied from 17 to 74%, while encapsulation efficiency varied from 4.7 to 98%.

In a separate study, sand eel oil was microencapsulated by emulsifying caseinate with lactose or skim milk powder, followed by spray drying. It was found that the presence of entrapped air promoted oxidation of the encapsulated oil. Caseinate-based powders contained higher volumes of entrapped air than skim milk–based powders. This entrapped air could be eliminated from commercial preparations of microcapsules, meaning that the problem was process-related and could be avoided by using appropriate encapsulation process conditions [53].

Sand eel oil was also microencapsulated using a freezing and drying process with lactose and maltodextrin as shell materials. Frozen pellets of oil emulsion were extruded and ground to produce a powder. One would expect that the grinding of pellets would cause the release of a certain amount of encapsulated oil, leading to a decreased oxidative stability. However, the experiment showed the opposite trend, which is not readily explainable [115,116].

Liposomes have also proved to confer some stability on marine oils containing EPA and DHA. Fish oil containing about 20% EPA and 33% DHA were subjected to liposome entrapment in a recent study. Liposomes of up to 5 μm were prepared and were found to be stable for one month in storage [93]. An alternative method that has achieved commercial success as a method for stabilizing oils containing EPA and DHA for delivery into functional foods is complex coacervation. While complex coacervation was first developed more than 50 years ago, it has not been used widely for the encapsulation of marine lipids. In recent publications, only a few articles exist which deal with complex coacervation as a means to microencapsulate cod-liver oil for oyster farming purpose [130,146]. The resulting powder, which was used in an artificial diet, had particle size in the range of 5–30 μm. Complex coacervation, if done correctly, has several advantages over other methods for the encapsulation and stabilization of marine oils for delivery into foods. The application of complex coacervation has been proven commercially by Ocean Nutrition Canada as discussed in Section 5.3.

5.5.2 Status Based on Commercialization by Companies

The current status of the use of commercial application of microencapsulation for the stabilization and delivery of marine oils into foods will be addressed in this section. Food ingredient companies have developed several different encapsulation processes to commercial scale and there are several food products fortified with EPA and DHA available in the market in several countries, with the number of available products increasing rapidly.

5.5.2.1 Encapsulated Oils

Microencapsulation of marine lipids by companies centers around two processes, spray-dried emulsions and spray-dried complex coacervates. The formation of spray-dried emulsions is the most commonly practiced technology, in part because of its low cost and short processing time. The shell may contain one or more biopolymers depending upon both the properties required and the regulatory requirements for the targeted food product in the countries it is to be sold. An example of a microencapsulated powder containing a single biopolymer is OmegaDry 1510 that used to be sold by Wacker Biochem Corp., which uses gamma cyclodextrin as the shell material component. When two biopolymers are used, one is typically a protein and the other a carbohydrate. The proteins render a good oxygen barrier, while the carbohydrates provide protection from moisture, not as effectively as lipids, but better than proteins. Furthermore, polymers can be mixed together or superimposed to improve the oxygen or moisture barrier properties of the resulting powders, irrespective of the encapsulation technology. For instance, in DSM infant formula powder, fish oil is dispersed in a matrix of fish gelatin and carbohydrate and coated with cornstarch during spray drying for improved dispersibility [147]. Similarly, Nu-Mega Ingredients (joint venture between Clover and Food Spectrum) uses a combination of whey

TABLE 5.1
Representatives of Fish Oil Powders Commercially Available around the World

Company	Powder Name	Payload (%)	Total Omega-3 (%)	Oil Source
BASF	Dry n-3 18:12, Dry n-3 5:25	25	8.75	Fish
Nu-Mega	Driphorm HiDHA Bake D101	25	7.40	Tuna
	Driphorm 50 D201	50	15.80	
DSM	ROPUFA '10' n-3 INF	N/A	9	Fish
	ROPUFA '10' n-3 Food	N/A	9	
Lipid Nutrition	Marinol omega-3	50	19.5	Fish
Ocean Nutrition	Omega-3 powder	60	22.0	Fish
Canada	DHA powder	60	22.0	Bonito

protein and lactose, and in some cases a small amount of pectin, as encapsulating materials [57].

Currently, only Ocean Nutrition Canada uses complex coacervation to manufacture fish oil powder by its patented technology. Gelatin and gum arabic are the two common polymers used in this type of encapsulation. Higher payload of fish oil by complex coacervation, as compared to spray drying, is a significant advantage of this technology (refer to Table 5.1), since it enables less powder to be added to fortify with a specific level of EPA and DHA. A lower level of powder addition has less impact on food texture and flavor, and also decreases the cost of the ingredient on a per serving basis. Omega-3 powders from Ocean Nutrition have 60% payload whereas most commercial spray-dried emulsions usually have a 25 to 50% payload. The amount of EPA and DHA per gram of powder can be increased by using concentrated oil instead of 1812TG or 0525TG found in natural oil source. As the trend is to pack as much omega-3 as possible into the powder, to minimize its impact on foods and to reduce the cost, many companies are working on improved technologies that enable the stable delivery of omega-3 concentrates in microencapsulated form. Concentrates are the fastest growing segment of the supplement market and hence there is considerable interest in adding these to foods. However, concentrates are less stable than unconcentrated triglyceride oils and so stabilization is more challenging. A number of companies produce omega-3 concentrates at a cost that is appropriate to the food industry, including Ocean Nutrition Canada, DSM, and Pronova.

Encochleation described in Section 5.3 is by far the most expensive method to produce fish oil powder. It is in part due to the high cost of soy-derived phosphatidylserine (PS) as the shell material. At current PS prices and payloads, the use of PS as a microencapsualtion shell material does not provide a reasonable cost per serving to enable its use by food companies. From the perspective of materials, since PS is a phospholipid it should provide the best moisture barrier protection for the encapsulated marine lipids, and allow the latter to be used in wet food applications. According to BioDelivery Sciences International, the experimental cochleates known as Bioral omega-3 powder have a payload of 30% and up to 300 mg of EPA + DHA per serving, which can be added to a variety of beverages without fishy taste or odor [148]. Other advantageous features of cochleates are micron size and low density, which minimizes powder settlement in a liquid food host such as milk or juice. However, the small particle sizes mean that likely stability of this technology is limited and so it is unlikely to be useful for foods with a long shelf life.

5.5.2.2 Marine Lipid–Fortified Foods

In a 2001 survey of products, which incorporate omega-3 fatty acids from plant or fish sources into foods (Table 5.2), dairy ranked third (15%) after pet food (29%) and dietary supplements (24%) [149]. Baby food, baked foods, and beverage attained the lowest ranking, 2% each. This ranking may not hold true today based on the continuous launching of new products since 2001 around the world, particularly in Europe and Australia or New Zealand. In the last couple of years, major new launches have occurred in breads and dairy products and also in infant formula. The products that belong to the seven food categories summarized in Table 5.3 contain only marine lipids as the omega-3 source.

5.5.2.2.1 Infant Formula

Human milk is rich in PUFAs. By comparison, cow's milk is low in PUFAs [150]. Infant formula therefore must be enriched with these fatty acids, particularly DHA, to promote the proper development of the retina and brain tissue of infants

TABLE 5.2
Distribution of Global Food Categories That Contain Omega-3 in 2001

Food Category	Number of Products	Distribution (%)
Pet food	50	29
Healthcare (dietary supplements)	42	24
Dairy	26	15
Processed fish, meat, and egg products	24	14
Snacks	6	3
Meals and meal centers	5	3
Baby food	3	2
Baked foods	3	2
Beverage	3	2
Other	12	7
Total	174	100

Source: Mintel's global new products database, www.gnpd.com.

TABLE 5.3
Representation of Omega-3 Fortified Foods Marketed by 2005

Food Category	Manufacturer	Country	Product Name or Type	Format of Omega-3
Infant formula	Nutricia	Australia, New Zealand	Karicare	Powder
	Mead Johnson	USA	Enfamil	?
Baked goods	Warburton	England	Women's Bread	Powder
	George Weston Foods	Australia, New Zealand	TipTopUp	Powder
Nutritional bars	ZonePerfect Nutrition	USA	ZonePerfect Bars	Powder
	Dr. Barry Sears	USA	OmegaZone Nutrition Bars	Powder
Milk and milk-based products	Parmalat	Italy	Parmalat UHT Milk	Oil
	Brownes	Australia, New Zealand	Heart Plus Milk	Powder?
Nonmilk beverages	Ross	USA	ProSure	Oil
	Symrise	Germany	?	?
Spread	Arla	Denmark	Blue Gaio	?
	Meadow Lea-Hi	Australia, New Zealand	Omega Margarine	Powder
Processed meats	Hans Continental Smallgoods	Australia, New Zealand	Strasburg	Powder
	Birds Eye Smart Choice	Australia, New Zealand	Fish fillet	Powder

during the first 2 years. In fact, the World Health Organization has endorsed this type of omega-3 fortification [151].

Many infant formula manufacturers use an oil mixture from Martek Biosciences Corp., which contain DHA and arachidonic acid, an omega-6 fatty acid. Others use fish oil powders such as BASF Dry n-3 5:25 or DSM ROPUFA "10" n-3 INF or ONC's MEG-3 powder in their formulation. While there have been at least 60 countries selling omega-3 fortified infant formulas since 1996, the first time the FDA approved their use in the United States was in 2002. Enfamil from Mead Johnson, Farley from Heinz, and Karicare from Nutricia, to name just a few, are among the more recognized brands around the world.

5.5.2.2.2 Baked Goods

Bread is among the first staple food that has been successfully fortified with omega-3 oil and powder. The carbon dioxide released during proofing and baking aids in protecting the oil from oxidation. Consequently, the EPA and DHA levels added to bread are typically higher than for most other foods, particularly wet foods. Bread fortified with omega-3 is marketed mainly in Europe, Asia, and

Australia [152], while in 2004, Wagman's in the United States had also launched several new products containing Ocean Nutrition Canada's MEG-3 ingredient. The most successful bread product containing EPA and DHA currently on the market is TipTopUp from George Weston Foods in Australia, which contains 27 mg omega-3 per slice, and has an approximate 15% of the Australian bread market. Both TipTopUp and Warburtons Women's bread in the United Kingdom use fish oil powder as the omega-3 source.

Fazer bakeries, one of Finland's leading bakery company, leveraged Finnish consumer's high level of trust for its long-established name by introducing the "iLove" rye breads and multigrain toasts in 2002 with emphasis on the overall health benefits [153]. Biosan Activa, a type of wafer that contains omega-3 and eight essential vitamins, was launched in Spain by Industrias Rodriguez in 2000 [154]. The bread category is expected to be one of the fastest growing omega-3 fortified foods, partly because it is a dry food that has a relatively short shelf life and so is a reasonably easy application in terms of ingredient stability requirements.

5.5.2.2.3 Nutritional Bars

An example in this food category is the introduction of gelatin-encapsulated fish oil into the ZonePerfect bars from ZonePerfect Nutrition in the United States. It should be noted that only 3 mg of EPA + DHA are added to each of the 50 g bars [155]. Another example is the OmegaZone nutrition bars from Dr. Barry Sears, which is designed to stabilize the blood sugar of people with diabetes. Each 55 g bar contains 160 mg of microencapsulated fish oil or 64 mg EPA and 32 mg DHA [156]. In both of these cases, the added EPA + DHA level is well below the 3 g/day recommended by the American Heart Association [157].

5.5.2.2.4 Milk and Milk-Based Products

Typically, omega-3 oil rather than powder is introduced to the milk formulation, and subjected to high-pressure (5–15 MPa) homogenization to ensure uniform dispersion [158]. It is believed that the high content of protein in milk can help to preserve its organoleptic quality in the presence of fish oil. However, the stability of fish oil in milk is still poor and so has been mainly applied to single-serve packs rather than multiple-serving containers. Consequently, encapsulated fish oil in powder or emulsion form rather than straight fish oil has been added to some milk products to provide a longer shelf life especially in multiple-serving products.

In 1998, Parmalat Finanziaria S.P.A. introduced milk fortified with omega-3 fatty acids [159]. This type of milk is processed at ultrahigh temperature (UHT, 150°C from 4 to 6 s) and subjected to aseptic packaging, and therefore, has a shell life of 3 months. The Parmalat milk is semiskimmed and contains 80 mg of omega-3 per 100 mL serving. UHT milk manufactured by Puleva in Spain also enjoys commercial success selling about 40,000 L of omega-3 enriched milk in 2002. In the pasteurization subcategory, there is Aux Omega milk from Candia in France, which contains 190 mg/250 mL serving. Heart Plus milk from Brownes in Australia is another example of pasteurized milk fortified with omega-3 with

150 mg EPA and DHA per 250 mL serving size. At least one powdered milk containing omega-3 and omega-6 from Nestle USA, Leite Omega Plus, has been marketed in South America (e.g., Brazil, Argentina) and Southeast Asia (e.g., Malaysia, Philippines) [160].

While omega-3 fortified milk is readily available in Europe, few products are currently available in North America when the manuscript was prepared. This slow trend suggests that the North American population, in comparison to European and other continental populations, are slower to recognize and are less inclined to pay a premium for omega-3 as a healthy component in their staple foods. According to one supplier, 65% of Scandinavians consume some form of omega-3 products in comparison to 3% North Americans [7]. Some companies, especially in Canada where regulations used to favor flax seed oil over fish oil, have launched products containing α-linolenic acid rather than EPA and DHA. For example, in late September 2003, Natrel in Quebec launched their partially skimmed milk that is fortified with flax seed oil [161]. It is believed that omega-3 oils will be fortified in more milk products in North America.

Yogurt drink Sadafco contains Nu-Mega's powder as the omega-3 delivery vehicle. It is marketed in Saudi Arabia. In 2001, two Gaio products, a junket and a yogurt drink, from Arla Foods in Sweden were launched [162]. To provide consumers with a choice, these products come either plain or with the addition of omega-3 fatty acids. Similarly, a chilled crème caramel dessert from Dhul in Spain is available in three varieties: Omega-3 (enriched with omega-3 and vitamin E), Fibra (enriched with fiber and vitamins A, D, and E), and Calcio (enriched with calcium and vitamins A, D, and E) [154].

5.5.2.2.5 Nonmilk Beverages

Similar to previously described food categories, nonmilk beverages utilize both fish oil and its powder for omega-3 fortification. ProSure from Ross is a nutritional beverage containing EPA, and is claimed to counteract the weight loss caused by cancer [163]. In this case, a highly refined and deodorized sardine oil supplies the EPA. It is surmised that the high level (1 g) of EPA contained in each 250 mL can of ProSure is possible, in part owing to the organoleptic protection of protein (16 g).

In the category of juice containing omega-3, there are only a few products available commercially. Bertrams Fruchtsafte is one example [161]. Symrise from Germany also produces omega-3 fortified beverages. Tidal Wave drink from Naked Juice in the United States is being reformulated. In a previous formulation, they used fish oil rather than powder to obtain 242 mg of EPA and DHA per 8 oz serving. Major juice companies are working toward the fortification of omega-3 in beverages, and various new launches are expected.

The small number of omega-3 fortified juices suggests that better moisture protection must be developed for fish oil powder. One approach might be to add lipids such as waxes to the edible proteins or carbohydrates that are currently used to form the shell of commercial omega-3 powders. Another approach is to form the shell entirely of lipids as in the case of the experimental phosphatidylserine-based cochleates.

5.5.2.2.6 *Spread*

Avoiding fishy flavor in spreads is probably more difficult than in other foods because of their small serving size, requiring high fortification levels. Companies such as Unilever are developing new products with omega-3 from fish oil. Blue Gaio is a spread fortified with omega-3 fatty acids from Arla, a food manufacturer in Denmark. SmartBalance OmegaPlus is a buttery spread launched in November 2003 in the United States by GFA Brands, Inc. In this case, menhaden fish is the omega-3 oil source, which is claimed to be odorless and tasteless [164]. The menhaden oil is introduced directly to the patented spread formulation developed by Brandeis University in Massachusetts. Another example of a spread, which contains unencapsulated fish oil is Yamega spread from Israel [165]. At 620 mg of omega-3 fatty acids (190 mg EPA and 430 mg DHA) per 1 g serving size, this spread has some negative odor and taste issues and has not gained wide consumer acceptance.

In Australia, Meadow Lea-Hi Omega margarine contains 250 mg omega-3 per teaspoon (5 g serving size). This means that eight teaspoons of this spread provide an equivalence of 2000 mg omega-3 in 100 g fresh salmon [166]. Seachange margarine, also from Australia, contains a smaller dosage of omega-3, 200 mg per teaspoon.

5.5.2.2.7 *Processed Meats*

Most of the omega-3 fortified foods in this category are being marketed in Australia and New Zealand. Nu-Mega's fish oil powder is the main source for this region [167]. According to the manufacturer, Hans Continental Smallgoods, a single serving (66.7 g) of their sliced chicken, Strasburg (similar to salami), or bacon, provides 60 mg or 20% of the suggested daily consumption of EPA and DHA. Although the omega-3 content is insignificant, the taste is fine, making these products a welcome addition to the luncheon meats market. Birds Eye Smart Choice markets breaded fish fillets, which contains 178 mg of DHA and EPA in every two fillets.

5.5.2.3 Status Assessment

Based on the proliferation of marine lipid–fortified products in seven food categories for human consumption, microencapsulation could become a profitable and effective way to deliver marine lipids to the population at large. The limited number of products in certain food categories such as juices or breakfast cereals, however, indicates that the commercialization of microencapsulated omega-3 powder in foods is still at an early stage. Furthermore, the low level of EPA and DHA incorporated into many food categories, in comparison to that proposed by various national and international organizations, suggests that elimination of fishy taste and smell remains a challenge. While a number of microencapsulation technologies and edible polymeric shell materials can be applied to produce encapsulated oils that are suitable for short shelf life foods like bread or UHT milk, the same cannot be said for other staple foods like cereal or salad dressing where longer shelf life or exposure to high moisture is a common occurrence.

There are few technologies currently available that can deliver and stabilize efficacious doses of omega-3 oils in multiserving foods while still conforming to regulatory cost requirements for food products.

5.6 SUMMARY

Given the well-documented health benefits of omega-3 fatty acids, microencapsulation is a good vehicle for delivering marine lipids, particularly EPA and DHA, to the public at large, from infancy to adulthood. There is currently a proliferation of functional foods fortified with marine lipids around the world, particular in Europe and Australia or New Zealand.

Microencapsulated oils work best in dry food applications such as baked goods, nutritional bars, using low dosages of EPA and DHA, or with shorter shelf life such as powder mixes. Achieving good sensory properties in wet food applications such as juices, higher EPA and DHA (>250 mg per serving), or longer shelf life such as cereals requires further development to improve current materials and processes.

Of the three main types of edible biopolymers reserved for shell materials, proteins and carbohydrates are most commonly used by nutraceutical companies and research institutions. Similarly, while there are numerous technologies to encapsulate marine lipids, only spray drying and complex coacervation are currently used commercially to stabilize EPA and DHA for food delivery. Encochleating marine lipids with phosphatidylserine holds good promise because the phospholipid shell provides a good moisture barrier and particles size is submicron, which is important for preventing setting in certain drink products. However, the cost of shell material and the limited stability of these small particles currently limit the commercialization of encochleation technologies.

The current strong demand for fortification of foods with marine oils, especially EPA and DHA, is creating a pull for research organizations and companies to develop technologies to deliver these oils into foods in a manner that does not impact the taste and smell of the food. The shelf life of the microencapsulated ingredient must be equal to or longer than the shelf life of the food. The ingredient must also be acceptable in a regulatory sense and have little impact on the cost of the food. There is no doubt that the demand will result in continued development of new and improved technologies to microencapsulate marine oils for delivery in food products.

ADDENDUM

During the preparation of this chapter, the marine lipids research and new product development have achieved great success, and the market and consumer awareness have grown amazingly. Mintel estimates that the number of omega-3 food products has increased from 120 in the year of 2004 to 250 in 2006. Some excellent review papers have also been published recently, such as Garg et al., 2006, *J. Food Sci.*, 71(5): R66–R71, and Whelan and Rust, 2006, *Annu. Rev. Nutr.*, 26: 75–103.

REFERENCES

1. Chow, C.K., *Fatty Acids in Foods and Their Health Implications*. 2nd Edn., Marcel Dekker, New York, 2000, Chap. 40.
2. Dahm, L., Fish oils provide fatty acids that are critical for health, 1999. Retrieved from http://www.foodprocessing.com/Web_First/fp.nsf/ArticleID/DTOS - 4LNTBE/ on December 11, 2003.
3. Executive Office of the President, To save lives, OMB urges revising dietary guidelines, 2003. Retrieved from http://www.whitehouse.gov/omb/pubpress/2003-13.pdf on January 30, 2004.
4. Raper, N.R., Cronin, F.J. and Exler, J., Omega-3 fatty acid content of the US food supply, *J. Am. Coll. Nutr.*, 11, 304–308, 1992.
5. Kris-Etherton, P.M., Taylor, D.S., Yu-Poth, S., Huth, P., Moriarty, K., Fishell, V., Hargrove, R.L., Zhao, G. and Etherton, T.D., Polyunsaturated fatty acids in the food chain in the United States, *Am. J. Clin. Nutr.*, 71, 179s–188s, 2000.
6. AROQ, The quest for health boosts Europe's functional foods market to £2bn, 2003. Retrieved from http://just-food.com/features_detail.asp?art=806 on December 11, 2003.
7. Schutt, E., A true snake oil: Omega 3's gaining ground as ingredient with many potential benefits and indications, 2002. Retrieved from http://www.nutraceuticalsworld.com/jan022.htm on January 29, 2004.
8. Vasishtha, N., Microencapsulation: delivering a market advantage, prepared foods, 2002. Retrieved from http://www.preparedfoods.com/CDA/ArticleInformation/coverstory/BNPCoverStoryItem/0,1229,114559,00.html on December 23, 2003.
9. Truelstrup-Hansen, L, Allan-Wojtas, P.M., Jin, Y.L. and Paulson, A.T., Survival of free and Ca-alginate microencapsulated *Bifidobacterium* spp. in simulated gastrointestinal conditions, *Food Microbiol.*, 19, 35–45, 2002.
10. Alemán, J.V., Gilbert, R.G., Hess, M., Horie, K., Kubisa, P., Meisel, I., Mormann, W., Penczek, S. and Slomkowski, S., Terminology of polymerization processes and polymers in dispersed systems, Draft 26, August 2003. Retrieved from http://www.kcpc.usyd.edu.au/resources/IUPAC_polym_coll_terminology.pdf on December 10, 2003.
11. Thies, C., A survey of microencapsulation processes, in: *Microencapsulation: Methods and Industrial Applications*. Benita, S., Ed., Marcel Dekker, New York, 1996, Chap. 1.
12. Gibbs, B.F., Kermasha, S., Alli, I. and Mulligan, C.N., Encapsulation in the food industry: a review, *Int. J. Food Sci. Nutr.*, 50, 213–224, 1999.
13. Food and Agricultural Organization (FAO), The Production of Fish Meal and Oil, Fishery Industries Division, Fish Tech Paper 12, Rome, ITA, 1986.
14. Aubourg, S.P., Fluorescence study of the pro-oxidant effect of free fatty acids on marine lipids, *J. Sci. Food Agric.*, 81, 385–390, 2001.
15. Miyashita, K., Nara, E. and Ota, T., Oxidative stability of polyunsaturated fatty acids in an aqueous solution, *Biosci. Biotech. Biochem.*, 57, 1638–1640. 1993.
16. Cho, S.Y., Miyashita, K., Miyazawa, T., Fujimoto, K. and Kaneda, T., Autoxidation of ethyl eicosapentaenoate and docosahexaenoate, *J. Am. Oil Chem. Soc.*, 64, 876–879, 1987.
17. Frankel E. N., Satué-Gracia, T., Meyer, A.S. and German, J.B., Oxidative stability of fish and algae oils containing long-chain polyunsaturated fatty acids in bulk and in oil-in-water emulsions, *J. Agric. Food Chem.*, 50, 2094–2099, 2002.
18. Walstra, P., Dispersed systems: basic considerations, in: *Food Chemistry*. 2nd Edn., Fennema, O.R., Ed., Marcel Dekker, New York, 1996, Chap. 3.

19. Damodaran, S., Amino acids, peptides, and proteins, in: *Food Chemistry*. 2nd Edn., Fennema, O.R., Ed., Marcel Dekker, New York, 1996, Chap. 6.
20. Mangino, M. and Litchfiled, J., Protein functionality. Handouts of *Food Science and Nutrition 605: Advanced Food Chemistry*, 1999. Retrieved from http://class.fst.ohio-state.edu/FST605/ lectures/Lect12.html on February 17, 2001.
21. Ní Néill, M.E. and Younger, K.M., Microencapsulation of marine oils with a view to food fortification, in: *Special Publication—Royal Society of Chemistry: 215 (Functional Foods)*, Sadler, M. and Saltmarsh, M., Eds., Royal Society of Chemistry, Cambridge, GBR, 1998, pp. 149–158.
22. Cho, Y.-H., Shim, H.K. and Park, J., Encapsulation of fish oil by an enzymatic gelation process using transglutaminase cross-linked proteins, *J. Food Sci.*, 68, 2717–2723, 2003.
23. Linden, G. and Lorient, D., Functional properties, in: *New Ingredients in Food Processing: Biochemistry and Agriculture*. Translated by Rosengarten, M., CRC Press, New York, 1999, Chap. 2.
24. Dickinson, E. and Euston S.R., Stability of food emulsions containing both protein and polysaccharide, in: *Food Polymers, Gels, and Colloids*. Dickinson, E., Ed., Royal Society of Chemistry, Cambridge, GBR, 1991, pp. 132–146.
25. Girard, M., Turgeon, S.L. and Paquin, P. Emulsifying properties of whey protein-carboxymethylcellulose complexes, *J. Food Sci.*, 67, 113–119, 2002.
26. American Lecithin Company, Lecithin applications, 2003. Retrieved from http://www.americanlecithin.com/leci_appfood.html on December 12, 2003.
27. Chang, P.S., Microencapsulation and oxidative stability of docosahexaenoic acid, in *ACS Symposium Series 674 (Flavor and Lipid Chemistry of Seafoods)*, American Chemical Society, Washington, DC, 1997, pp. 264–273.
28. Chen, J. and Dickinson, E., Protein/surfactant interfacial interactions Part 3. Competitive adsorption of protein + surfactant in emulsions, *Coll. Surf. A: Physicochem. Eng. Asp.*, 101, 77–85, 1995.
29. Fennema, O.R., Water and ice, in: *Food Chemistry*. 3rd Edn., Fennema, O.R., Ed., Marcel Dekker, New York, 1996, Chap. 2.
30. Roos, Y.H., Glass transition. Course handout of *FC3003: water activity, glass transition, food stability*, University College Cork, Cork, IRL, 2000. Retrieved from http://www.ucc.ie/acad/departments/foodtech/5.PDF on December 2, 2003.
31. Grattard, N., Salaun, F., Champion, D., Roudaut, G. and Le Meste, M., Influence of physical state and molecular mobility of freeze-dried maltodextrin matrices on the oxidation rate of encapsulated lipids, *J. Food Sci.*, 67, 3002–3010, 2002.
32. Beristain, C.I., Azuara, E. and Vernon-Carter, E.J., Effect of water activity on the stability to oxidation of spray-dried encapsulated orange peel oil using mesquite gum (*Prosopis Juliflora*) as wall material, *J. Food Sci.*, 67, 206–211, 2002.
33. Pérez-Alonso, C., Báez-González, J.G., Beristain, C.I., Vernon-Carter, E.J. and Vizcarra-Mendoza, M.G., Estimation of the activation energy of carbohydrate polymers blends as selection criteria for their use as wall material for spray-dried microcapsules, *Carbohydr. Polym.*, 53, 197–203, 2003.
34. McHugh, T.H. and Krochta, J.M., Permeability properties of edible films, in: *Edible Coatings and Films to Improve Food Quality*, Krochta, J.M., Baldwin, E.A. and Nisperos-Carriedo, M.O., Eds., Technomic Publishing, Lancaster, PA, 1994, Chap. 7.
35. Miller, K.S. and Krochta, J.M., Oxygen and aroma barrier properties of edible films: a review, *Trends Food Sci. Technol.*, 8, 228–237, 1997.
36. David, J., Lefrancois, C. and Ridoux, C., Microcapsules based on gelatin and polysaccharides and process for obtaining same, US Pat. 5051304, 1991.

37. Fabri, D. and Marrs, M., Microencapsulation of sensitive oils. *Leatherhead Food International Research Report. Number RR827*, 2003.
38. Lamprecht, A., Schafer, U. and Lehr, C.-M., Influences of process parameters on preparation of microparticle used as a carrier system for ω-3 unsaturated fatty acid ethyl esters used in supplementary nutrition, *J. Microencapsulation,* 18, 347–357, 2001.
39. Soper, J.C. and Thomas, M.T., Enzymatically protein encapsulating oil particles by complex coacervation, US Pat. 6325951, 2001.
40. Tang, H., Zhao, Y., Zhang, X., Zou, J. and Liao, N., Study on preparation of capsules of microencapsulated fish oil, *J. Chinese Hospital Pharm. (Zhongguo Yiyuan Yaoxue Zazhi),* 18(1), 25–26, 1998.
41. Yan, N., Encapsulated agglomeration of microcapsules and method for the preparation thereof, US Pat. Appl. 2003/0193102, 2003.
42. Yin, X. and Xu, S., Microencapsulation of EPA and DHA: wall-material selection, *Food Fermentation Ind. (Shiping yu Fajao Gongyi),* 26, 33–36, 2000.
43. Zheng, B. and Yi, R., Preparation of fish oil microcapsule, Chinese Publication of Innovations and Patents (*Faming Zhuanli Shenqing Gongkai Shuomingshu*), CN1120933. Application number CN94-117364, 1996.
44. Dairy Management Inc., Whey protein-based microencapsulating agents, 2002. Retrieved from http://www.doitwithdairy.com/infolib/ingspecsheet/factmicro.htm on August 15, 2002.
45. Heinzelmann, K., Franke, K., Jensen, B. and Haahr, A.-M., Protection of fish oil from oxidation by microencapsulation using freeze-drying techniques, *Eur. J. Lipid Sci. Technol.*, 102, 114–121, 2000.
46. Hogan, S.A., McNamee, B.F., O'Riordan, E.D and O'Sullivan, M., Microencapsulating properties of whey protein concentrate 75, *J. Food Sci.*, 66, 675–680, 2001.
47. Lee, S.J. and Rosenberg, M., Whey protein-based microcapsules prepared by double emulsification and heat gelation, *Lebensm.-Wiss. u.-Technol.*, 33, 80–88, 2000.
48. Weinbreck, F.C.J., de Kruif, C.G. and Schrooyen, P., Complex coacervates containing whey proteins, WO 03/106014 A1, 2003.
49. Satpathy, G. and Rosenberg, M., Encapsulation of chlorothiazide in whey proteins: effects of wall-to-core ratio and cross-linking conditions on microcapsule properties and drug release, *J. Microencapsulation,* 20, 227–245, 2003.
50. Velasco, J., Dobarganes, M.C. and Marquez-Ruiz, G., Oxidation of free and encapsulated oil fractions in dried microencapsulated fish oils, *Grasas y Aceites (Sevilla),* 51, 439–446, 2000.
51. Young, S.L., Sarda, S. and Rosenberg, M., Microencapsulating properties of whey proteins. 2. Combination of whey proteins with carbohydrates, *J. Dairy Sci.*, 76, 2878–2885, 1993.
52. Kagami, Y., Sugimura, S., Fujishima, N., Matsuda, K., Kometani, T. and Matsumura, Y., Oxidative stability, structure, and physical characteristic of microcapsules formed by spray drying of fish oil with protein and dextrin wall materials, *J. Food Sci.*, 68, 2248–2255, 2003.
53. Keogh, M.K., O'Kennedy, B.T., Kelly, J., Auty, M.A., Kelly, P.M., Fureby, A. and Haahr, A.-M., Stability to oxidation of spray-dried fish oil powder microencapsulated using milk ingredients, *J. Food Sci.*, 66, 217–224, 2001.
54. Sheu, T.Y. and Rosenberg, M., Microencapsulation by spray drying ethyl caprylate in whey protein and carbohydrate wall systems, *J. Food Sci.*, 60, 98–103, 1995.

55. Zhu, X. and Xu, S., Study on microencapsulation of fish oil with soybean protein by spray-drying method (I)—selection of microcapsule's wall compositions, *China Lipids (Zhongguo Youzhi)*, 23(5), 31–34, 1998.
56. Zhu, X. and Xu, S., Microencapsulation of fish oil with soybean protein based wall by spray-drying process (2)—study on process for microencapsulation of fish oil, *China Lipids (Zhongguo Youzhi)*, 23(6), 35–38, 1998.
57. Augustin, M.A. and Sanguansri, L., Encapsulation of food ingredients, US Pat. Appl. 2003/0185960, 2003.
58. Fischer, N. and Widder, S., How proteins influence food flavor, *Food Technol.*, 51, 68–70, 1997.
59. Sawyer, L., Brownlow, S., Polikarpov, I. and Wu, S.Y., β-Lactoglobulin: structure studies, biological clues, *Int. Dairy J.*, 8, 65–72, 1998.
60. Yang, J., Powers, J.R., Clark, S., Dunker, A.K. and Swanson, B.G., Ligand and flavor binding functional properties of β-lactoglobulin in the molten globule state induced by high pressure, *J. Food Sci.*, 68, 444–452, 2003.
61. Johnson, T.M. and Zabik, M.E., Gelation properties of albumen proteins, singly and in combination, *Poultry Sci.*, 60, 2071–2083, 1981.
62. Taguchi, K., Iwami, K., Ibuki, F., and Kawabata, M., Oxidative stability of sardine oil embedded in spray-dried egg white powder and its use for n-3 unsaturated fatty acid fortification of cookies, *Biosci. Biotechnol. Biochem.*, 56, 560–563, 1992.
63. Yajima, M., Stabilized oil and fat powder, Eur. Pat. Appl. EP372669, 1990.
64. Kong, B., Zheng, D., Lin, S., Zhou, S. and Liu, W., Microencapsulation of fish oil, *Food Ind. Sci. Technol. (Shipin Gongye Keji)*, No. 5, 8–10, 1999.
65. McNamee, B.F., O'Riordan, E.D. and O'Sullivan, M., Emulsification and microencapsulation properties of gum arabic, *J. Agric. Food Chem.*, 46, 4551–4555, 1998.
66. Cao, D., and Song, W., Stability of fish oil emulsion, *J. Univ. Wuxi Light Ind. (Wuxi Qinggong Daxue Xuebao)*, 19(1), 65–68, 2000.
67. Deng, Y., Microencapsulation of ω-3 polyunsaturated fatty acid with compound wall material, *Food Fermentation Ind. (Shipin Yu Fajiao Gongye)*, 27(6), 30–34, 2001.
68. Qi, H., Remmert, M., Shieh, W. and Hedges, A., Dry, edible oil and starch composition, US Pat. 6638557, 2003.
69. Reuscher, H., Stabilized PUFA triglycerides for nutraceuticals and functional foods using β-cyclodextrin, in: *Cyclodextrin: From Basic Research to Market. The 10th Inernational Cyclodextrin Symposium*, Szejtli, J., Ed., Wacker Biochem Corp., Adrian, MI, 2000, pp. 609–617
70. Saleeb, F.Z. and Arora, V.K., Method of preparing glass stabilized material, US Pat. 5972395, 1999.
71. Bakker, M.A.E., Galema, S.A. and Visser, A., Microcapsules of gelatin and carboxycellulose. Eur. Pat. Appl. EP0937496, 1999.
72. Kantor, M.L., Steiner, S.S. and Pack, H.M., Microencapsulation of fish oil, Eur. Pat. Appl. EP0336662, 1989.
73. Koh, G.-L. and Tucker, I.G., Characterization of sodium carboxymethylcellulose-gelatin complex coacervation by viscosity, turbidity and coacervate wet weight and volume measurements, *J. Pharm. Pharmcol.*, 40, 233–236, 1988.
74. Porzio, M.A. and Madsen, M.G., Double encapsulation process and flavorant compositions prepared thereby, PCT Int. Appl. WO97/13416, 1997.
75. Chan, L.W., Lim, L.T. and Heng, P.W.S., Microencapsulation of oils using sodium alginate, *J. Microencapsulation*, 17, 757–766, 2000.
76. Fanta, G., Knutson, C., Eskins, K. and Felker, F., Starch microcapsules for delivery of active agents, US Pat. Appl. 2001/0006698, 2001.

77. Jeon, Y.-J., Casanthan, T., Temelli, F., Song, B.-K., The suitability of barley and corn starches in their native and chemically modified forms for volatile meat flavor encapsulation, *Food Res. Int.*, 36, 349–355, 2003.
78. Korus, J., Tomasik, P. and Lii, C.Y., Microcapsules from starch granules, *J. Microencapsulation,* 20, 47–56, 2003.
79. Lii, C.Y., Tomasik, P., Hung, W.L., Yen, M.T. and Lai, M.-F., Granular starches as dietary fibre and natural microcapsules, *Int. J. Food Sci. Technol.*, 38, 677–685, 2003.
80. Qi, Z.H. and Xu, A., Starch-based ingredients for flavor encapsulation, *Cereal Foods World,* 44, 460–465, 1999.
81. Trindade, M.A. and Grosso, C.R.F., The stability of ascorbic acid microencapsulated in granules of rice starch and in gum arabic. *J. Microencapsulation*, 17, 169–176, 2000.
82. Kenyon, M.M., Modified starch, maltodextrin, and corn syrup as wall materials for food encapsulation, in: *Encapsulation and Controlled Release of Food Ingredients, ACS Symposium Series 590,* Risch, S.J. and Reineccius, G.A., Eds., American Chemical Society, Washington, DC, 1995. pp. 42–50.
83. King, A.H., Encapsulation of food ingredients, in: *Encapsulation and Controlled Release of Food Ingredients. ACS Symposium Series 590.* Risch, S.J. and Reineccius, G.A., Eds., American Chemical Society, Washington, DC, 1995, pp. 26–30.
84. Masaichiro, M., Compositions containing unsaturated fatty acid compounds and method of stabilizing unsaturated fatty acid compounds, US Pat. 4564475, 1986.
85. Wagu, M., Hayashi, S. and Kodama, K., Inclusion compound of eicosapentaenoic acid or docosahexaenoic acid with cyclodextrin, US Pat. 4438106, 1984.
86. O'Donnell, C.D., New encapsulating molecule improves taste, *Prepared Foods*, July 2001. Retrieved from http://www.preparedfoods.com/archives/2001/2001_7/0701r&d.htm on June 2, 2002.
87. Suloff, E.C., Marcy, J.E., Blakistone, B.A., Duncan, S.E., Long, T.E. and O'Keefe, S.F., Sorption behavior of selected aldehydes-scavenging agents in poly(ethylene terephathalate) blends, *J. Food Sci.*, 68, 2028–2033, 2003.
88. Morillon, V., Debeaufort, F., Blond, G., Capelle, M., and Voilley A., Factors affecting the moisture permeability of lipid-based edible films: a review, *Crit. Rev. Food Sci. Nutr.,* 42, 67–89, 2002.
89. BioDelivery Sciences International, Micronutrient encapsulation for the processed food industry, 2003. Retrieved from http://www.biodeliverysciences.com/Applications/ Processed_Food/processed_food.html on November 12, 2003.
90. Haynes, L.C., Levine, H. and Finley, J.W., Liposome composition for the stabilization of oxidizable substances, US Pat., 5015483, 1991.
91. Hayward, J.A., Levine, D.M. and Simon, S.R., Microcapsules containing liposomes, PCT Int. Appl. WO87/01587, 1987.
92. Kubo, K., Sekine, S. and Saito, M., Docosahexaenoic acid-containing phosphatidylethanolamine in the external layer of liposomes protects docosahexaenoic acid from 2,2′-azobis(2-aminopropane)dihydrochloride-mediated lipid peroxidation. *Arch. Biochem. Biophys.*, 410, 141–148, 2003.
93. Moussaoui, N., Cansell, M. and Denizot, A., Marinosomes, marine lipid based liposomes: physical characterization and potential application in cosmetics, *Int. J. Pharm.,* 242, 361–365, 2002.
94. Shiota, M., Konishi, H. and Tatsumi, K., Oxidative stability of fish oil blended with butter, *J. Dairy Sci.*, 82, 1877–1881, 1999.

95. Benoit, J.-P., Rolland, H., Thies, C. and Van De Velde, V., Encapsulation of active particles by coating of a central core, Eur. Pat. Appl. EP0706821, 1996.
96. Bishop, J.R.P., Nelson, G. and Lamb, J., Microencapsulation in yeast cells, *J. Microencapsulation,* 15,761–773, 1998.
97. Ishiguro, M., Shimura, Y. and Ishiwaki, N., Process for treating yeast with β-1, 3-glucanase to produce microcapsules for enclosing hydrophobic liquids, US Pat. 5521089, 1996.
98. Fomuso, L.B., Corredig, M. and Akoh, C.C., Effect of emulsifier on oxidation properties of fish oil-based structured lipid emulsions, *J. Agric. Food Chem.*, 50, 2957–2961, 2002.
99. De Koning, A.J. and Milkovitch, S., The storage behaviour of a number of fish oil health capsules at ambient temperature, *S. Afr. J. Food Sci. Nutr.*, 1, 7–8, 1989.
100. Han, D., Yi, O.S. and Shin, H.K., Antioxidative effect of ascorbic acid solubilized in oils via reversed micelles, *J. Food Sci.*, 55, 247–249, 1990.
101. Jakobsson, M. and Sivik, B., Oxidative stability of fish oil included in a microemulsion, *J. Dispersion Sci. Technol.*, 15, 611–619, 1994.
102. Kulås, E., Olsen, E. and Ackman, R.G., Effect of α-, γ-, and δ-tocopherol on the distribution of volatile secondary oxidation products in fish oil, *Eur. J. Lipid Sci. Technol.*, 104, 520–529, 2002.
103. Watanabe, Y., Fang, X., Minemoto, Y., Adachi, S. and Matsuno, R., Suppressive effect of saturated acyl L-ascorbate on the oxidation of linoleic acid encapsulated with maltodextrin or gum arabic by spray-drying, *J. Agric. Food Chem.*, 50, 3984–3987, 2002.
104. Yoshioka, K., Yamada, A. and Wada, S., Influence of rosemary extract on the oxidative stability of tuna orbital oil and on the effect in vivo of the oxidized oil on rat liver, *J. Oleo Sci.*, 51, 73–81, 2002.
105. Antrim, R.L., Lioyd, N.E. and Taylor, J.B., Stabilized highly unsaturated fatty acids and derivatives of such acids, Eur. Pat. Appl. EP0338499, 1989.
106. Koketsu, M. and Satoh, Y.-I., Antioxidative activity of green tea polyphenols in edible oils, *J. Food Lipids.*, 4, 1–9, 1997.
107. Wanasundara, U.N. and Shahidi, F., Stabilization of seal blubber and menhaden oils with green tea catechins, *J. Am. Oil Chem. Soc.*, 73, 1183–1190, 1996.
108. Antrim, R.L. and Taylor, J.B., Stabilized emulsions containing highly unsaturated oils, US Pat. 4963385, 1990.
109. Cho, Y.-J., Alamed, J., McClements, D.J. and Decker, E.A., Ability of chelators to alter the physical location and prooxidant activity of iron in oil-in-water emulsions, *J. Food Sci.*, 68, 1952–1957, 2003.
110. Let, M.B., Jacobsen, C., Frankel, E.N. and Meyer, A.S., Oxidative flavour deterioration of fish oil enriched milk, *Eur. J. Lipid Sci. Technol.*, 105, 518–528, 2003.
111. Mancuso, J.R., McClements, D.J. and Decker, E.A., The effects of surfactant type, pH, and chelators on the oxidation of salmon oil-in-water emulsions, *J. Agric. Food Chem.*, 47, 4112–4116, 1999.
112. Schrooyen, P.M.M., van der Meer, R. and De Kruif, C.G., Microencapsulation: its application in nutrition, *Proc. Nutr. Soc.*, 60, 475–479, 2001.
113. Thies, C., Microencapsulation: what it is and purpose, in: *Microencapsulation of Food Ingredients.* Vilstrup, P., Ed., Leatherhead Food International, Leatherhead, GBR, 2001, Chap. 1.
114. Shahidi, F. and Han, X.-Q., Encapsulation of food ingredients, *Crit. Rev. Food Sci. Nutr.,* 33, 501–547, 1993.

115. Heinzelmann, K. and Franke, K., Using freezing and drying techniques of emulsions for the microencapsulation of fish oil to improve oxidation stability, *Coll. Surf. B: Biointerfaces,* 12, 223–229, 1999.
116. Heinzelmann, K., Franke, K., Velasco, J. and Márquez-Ruiz, G., Microencapsulation of fish oil by freeze-drying techniques and influence of process parameters on oxidative stability during storage, *Eur. Food Res. Technol.*, 211, 234–239, 2000.
117. Hegenbart, S., Encapsulated ingredients keep problems covered, *Food Product Design*, April 1993. Retrieved from http://www.foodproductdesign.com/archive/1993/0493CS.html on December 10, 2003.
118. Sparks, R.E., Jacobs, I.C., and Mason, N.S., Polymeric Delivery Systems, *ACS Symposium Series 520*, El-Nokaly, M.A., Piatt, D.M., and Charpentier, B.C., Eds., American Chemical Society, Washington, DC, 1993, pp. 143–153.
119. Bungenberg de Jong, H.G., Complex colloid systems, in: *Colloid Science*, Vol II, Kruyt, H.R., Ed., Elsevier, New York, 1949, Chaps. 8 and 10.
120. Mayya, K.S., Bhattacharyya, A. and Argillier, J.-F., Micro-encapsulation by complex coacervation: influence of surfactant, *Polym. Int.,* 52, 644–647, 2003.
121. Kondo, T., Microcapsules: their science and technology. Part I. Various preparation methods, *J. Oleo Sci.*, 50, 1–11, 2001.
122. Shim, H.K., Kim, S.Y. and Park, J., Development of encapsulation process for fish oil double emulsification and enzymatic gelation. Presented at the 2002 Annual Meeting of Institute of Food Technologists, June 19, Anaheim, CA. Retrieved from http://ift.confex.com/ift/2002/techprogram/paper_13053.htm on December 10, 2003.
123. Collaborative Laboratories, Liposomes controlled-delivery systems, 2004. Retrieved from http://www.collabo.com/liposome.htm on February 23, 2004.
124. Martin, F.J. and Huang, T., Stealth liposomal technology: current therapies & future directions, *Drug Delivery Technology*, 3(5), 2003. Retrieved from http://www.drugdeliverytech.com/cgi-bin/articles.cgi?idArticle=155 on February 23, 2004.
125. Thies, C., Characterization of microcapsules, in: *Proceedings of International Symposium on Controlled Release of Bioactive Material,* Roseman, T.J., Peppas, N.A., and Gabelnick, H.L., Eds., 20, 157–158, 1993.
126. Thies, C., Microcapsule characterization, in: *Microencapsulation of Food Ingredients*, Vilstrup, P., Ed., Leatherhead Food International, Leatherhead, GBR, 2001, Chap. 2.
127. Kondo T., Microcapsules: their science and technology. Part II, properties, *J. Oleo. Sci.*, 50, 81–95, 2001.
128. Hurst, W.J., *Methods of Analysis for Functional Foods and Nutraceuticals*, CRC Press LLC, New York, 2002.
129. Rehor A., Canaple, L., Zhang, Z. and Hunkeler, D., The compressive deformation of multicomponent microcapsules: influence of size, membrane thickness, and compression speed, *J. Biomater. Sci. Polym. Edn.*, 12, 157–170, 2001.
130. Jouzel, B., Pennarun, A.-L., Prost, C., Renard, D., Poncelet, D. and Demaimay, M., Encapsulation of a lipid precursor, the eicosapentaenoic acid, to study the development of the crassostrea gigas oyster flavours, *J. Microencapsulation,* 20, 35–46, 2003.
131. Xu, K., Hercules, D.M., Lacik, I. and Wang, T.G., Atomic force microscopy used for the surface characterization of microcapsules immunoisolation devices, *J. Biomaterials Res.*, 41, 461–467, 1998.
132. Uddin, M.S., Hawlander, M.N.A. and Zhu, H.J., Microencapsulation of ascorbic acid: effect of process variables on product characteristics, *J. Microencapsulation,* 18, 199–209, 2001.

133. Lamprecht. A., Schafer, U.F. and Lehr, C.-M., Characterization of microcapsules by confocal laser scanning microscopy: structure, capsule wall composition and encapsulation rate, *Eur. J. Pharm. Biopharm.,* 49, 1–9, 2000.
134. Schmitt C., Sanchez, C., Lamprecht, A., Renard, D., Lehr, C., de Kruif, C.G. and Hardy, J., Study of beta-lactoglobulin/acacia gum complex coacervation by diffusing-wave spectroscopy and confocal scanning laser microscopy, *Coll. Surf. B: Biointerfaces,* 20, 267–280, 2001.
135. Rosinski, S., Grigorescu, G., Lewinska, D., Ritzen, L.G., Viernstein, H., Teunou, E., Poncelet, D., Zhang, Z., Fan, X., Serp, D., Marison, I. and Hunkeler, D., Characterization of microcapsules: recommended methods based on round-robin testing, *J. Microencapsulation,* 19, 641–659, 2002.
136. Zhang, Z., Saunders, R. and Thomas, C.R., Mechanical strength of single microcapsules determined by a novel micromanipulation technique, *J. Microencapsulation,* 16, 117–124, 1999.
137. Hegde, R.P., Rheingold, J.L., Welch, S. and Rhodes, C.T., Studies of powder flow using a recording powder flowmeter and measurement of the dynamic angle of repose, *J. Pharm. Sci.,* 74, 11–15, 1985.
138. Hogan S.A., O'Riordan, E.D. and O'Riordan, M., Microencapsulation and oxidative stability of spray-dried fish oil emulsions, *J. Microencapsulation,* 20, 675–688, 2003.
139. Meilgaard, M., Civille, G.V. and Carr, B.T., *Sensory Evaluation Techniques,* 3rd Edn., CRC Press, Boca Raton, FL, 1999.
140. Hsieh, T.C.Y., Williams, S.S., Vejaphan, W. and Meyers, S.P., Characterization of volatile components of menhaden fish (*Brevoortia tyrannus*) Oil, *J. AOCS,* 66, 114–117, 1989.
141. Jafar, S.S., Hultin, H.O., Bimbo, A.P., Crowther, J.B. and Barlow, S.M. Stability by antioxidants of mayonnaise made from fish oil, *J. Food Lipids,* 1, 295–311, 1994.
142. Andersen, S., Microencapsulated omega-3 fatty acids from marine sources, *Lipid Technol.*, 7, 81–85, 1995.
143. Higgins, S., Carroll, Y.L., O'Brien, N.M. and Morrissey, P.A., Use of microencapsulated fish oil as a means of increasing n-3 polyunsaturated fatty acid intake, *J. Hum. Nutr. Diet.*, 12, 265–271, 1999.
144. Wallace, J.M.W., McCabe, A.J., Robson, P.J., Keogh, M.K., Murray, C.A., Kelly, P.M., Marquez-Ruiz, G., McGlynn, H., Gilmore, W.S. and Strain, J.J., Bioavailability of n-3 polyunsaturated fatty acids (PUFA) in foods enriched with microencapsulated fish oil, *Ann. Nutr. Metab.*, 44, 157–162, 2000.
145. Yep Y.L., Li, D., Mann, N.J., Bode, O. and Sinclair, A.J., Bread enriched with microencapsulated tuna oil increases plasma docosahexaenoic acid and total omega-3 fatty acids in humans, *Asia Pacific J. Clin. Nutr.,* 11, 285–291, 2002.
146. Numaguchi, K., Effect of an artificial diet on early spat growth of the Japanese pearl oyster *Pinctada fucata martensii*, *Fish. Sci.*, 68, 694–696, 2002.
147. DSM Nutritional Products, ROPUFA '10' n-3 INF powder, 2001. Retrieved from http://www.dsmnutrafacts.com/food/prod/Ropufa_10n-3_powder.jsp on November 12, 2003.
148. Bonitz, S., BioDelivery Sciences International Inc., Personal communication, 2003.
149. O'Donnell, C., Popular applications—ingredients in use: omega-3 fatty acids, *prepared foods*, June 2002. Retrieved from http://www.preparedfoods.com/CDA/ArticleInformation/features/BNP__Features__Item/0,1231,114450,00.html on December 11, 2003.
150. Candia, Lipids in milk, 2003. Retrieved from http://www.candia.fr/candiaus/nutrition/lipid.pdf on December 5, 2003.

151. Food and Agriculture Organization of the United Nations/World Health Organization, Lipids in early development, in *Fats and Oils in Human Nutrition, Report 57*, 49, 1994.
152. Drummond, H. and Palmer, M., Microencapsulation of tuna oil: opening the door to consumer health benefits, *Food Technol.*, 14, 540–541, 2003.
153. Press Information Worldwide, Fazer bakeries introduces new functional products at Anuga, 2001. Retrieved from http://www.pressi.com/ru/release/38106 on January 30, 2004.
154. Jago, D., Continental cuisine: functional, organic foods in Europe in 2000, 2001. Retrieved from http://www.findarticles.com/cf_dls/m3289/4_170/74829297/p1/article.jhtml on Febrauary 2, 2004.
155. ZonePerfect, Frequently asked questions, 2003. Retrieved from http://store.zoneperfect.com/site/content/bars_faq.asp on December 12, 2003.
156. OmegaZone Nutrition Bars, OmegaZone bars ingredients, 2003. Retrieved from http://www.healthypetnet.com/products/people/omega_bars_ingredients.asp?realname= on December 15, 2003.
157. Harris, W.S. and Appel, L.J., New guidelines focus on fish, fish oil, omega-3 fatty acids, 2002. Retrieved from http://www.americanheart.org/presenter.jhtml?identifier=3006581 on February 2, 2004.
158. Berry, D., Fortification update: beyond [Ca.sup.$^{2+}$], A & D. (use of omega-3 fatty acids and soy protein), 2002. Retrieved from http://www.findarticles.com/cf_dls/m3301/12_101/69279212/p1/article.jhtml on December 11, 2003.
159. Mellentin, J. and Heasman, M., Functional dairy transforms EU market, 2002. Retrieved from http://www.ffnmag.com/ASP/47/Display-Article on January 30, 2004.
160. O'Donnell, C., Ingredients in use: omega-3 fatty acids, 2003. Retrieved from http://www.preparedfoods.com/CDA/ArticleInformation/features/BNP_Features_Item/0,1231,114450,00.html on February 2, 2004.
161. Foulds, J., Functional drinks, Natrel milks omega-3 benefits, *Zenith Int. Newslett.*, Issue 20, p. 5, 2003.
162. New Nutrition Business, New Gaio products in Sweden, 2001. Retrieved from http://www.new-nutrition.com/newspage/240801c.htm on February 3, 2004.
163. ProSure, Compare products, 2003. Retrieved from http://www.prosure.com/compareProducts.asp on January 30, 2004.
164. Omega Protein Corporation, New buttery spread contains healthy omega-3s: GFA Brands, Inc. Announces Smart Balance® Omega Plus with omega-3 from the sea™ and natural plant sterols, 2003. Retrieved from http://biz.yahoo.com/prnews/031117/clm005_1.html on December 3, 2003.
165. Ivam Innovative Products, Dr. Milli's natural health products, 2003. Retrieved from http://www.ivamgroup.com/naturalproducts.html on January 29, 2004.
166. Perth Diet Clinic, Hans Strassburg 97% fat free with essential omega-3, 2003. Retrieved from http://www.perthdietclinic.com.au/article.asp?GroupID=41&ArticleID=303 on January 30, 2004.
167. Sovereign Publications Ltd., Numega: award winning omega-3 fish oil microencapsulation technology, 2003. Retrieved from http://www.sovereign-publications.com/nu-mega.htm on December 2, 2003.

6 Chitin and Chitosan

Rosalee S. Rasmussen and Michael T. Morrissey

CONTENTS

6.1 INTRODUCTION

Chitin and chitosan are natural, nontoxic, biodegradable compounds with a broad range of commercial applications. These biopolymers possess properties that can be exploited by a variety of fields, such as agriculture, chemistry, medicine, biotechnology, the pulp and paper industry, cosmetics, water treatment, and foods [1,2]. However, this review will be focused solely on the use of chitin and chitosan as marine-derived nutraceuticals in the food and food supplement industries. As nutraceuticals, chitin and chitosan possess numerous potentially valuable properties. Many studies have reported immuno-modulatory effects in the form of antibacterial, anti-inflammatory, antioxidant, anticarcinogenic, or antiulcer activity [3–5]. Recent studies have also indicated a possible use for chitin and chitosan in renal disease recovery [6]. However, the major nutraceutical application of chitin and chitosan thus far has been as a dietary fiber supplement. Although there has been some inconsistency with regard to the ability of these biopolymers to contribute to weight loss, a number of studies have shown reductions in cholesterol levels and a subsequent reduced risk for cardiovascular disease [7]. This overview will begin by introducing the general structures and properties of chitin and chitosan followed by a discussion of their various nutraceutical applications.

6.2 CHITIN

Chitin is an aminopolysaccharide made up of *N*-acetyl-D-glucosamine residues joined by β-(1–4) linkages (Figure 6.1). It is second only to cellulose as the most abundant biopolymer in nature and differs structurally from cellulose solely in the substitution of acetamide groups at the C-2 position in place of hydroxyl groups [4,5]. Chitin is an important structural component in the hard outer coatings of both terrestrial and aquatic organisms. It forms a part of the exoskeletons of crustaceans, mollusks, and arthropods, and is found in the cell walls of many fungi and algae [8]. The estimated worldwide annual production of chitin in nature is around 10^{11} tons and industrial use has been estimated at 10,000 tons annually [2].

Currently, the most exploited sources of chitin are the processing discards of shellfish and marine crustaceans, especially crab, lobster, shrimp, krill, oysters, and squid (Table 6.1) [1–4]. Large quantities of shellfish are harvested each year, with total U.S. commercial landings for 2004 amounting to 1910 million pounds [9]. Processing wastes resulting from shellfish and crustacean fisheries can be excessive, with shell discards from Dungeness crab (*Cancer magister*) and Pacific shrimp (*Pandalus borealis*) amounting to around 78 million pounds annually in the United States alone [4]. On average, marine crustacean shells contain around

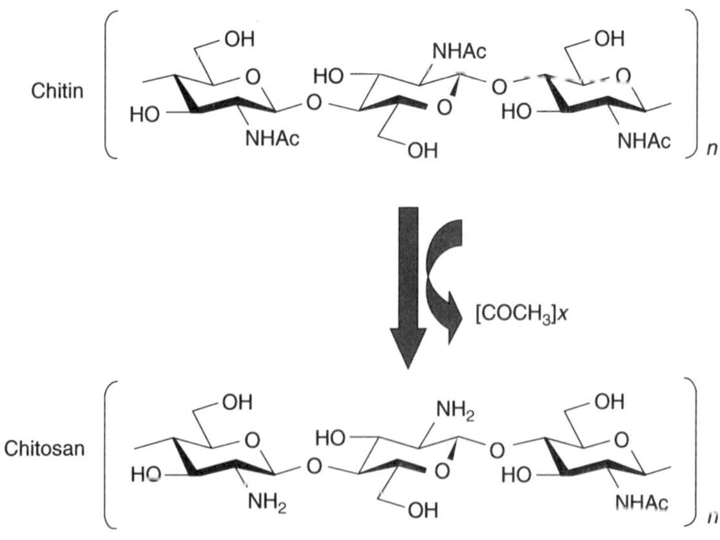

FIGURE 6.1 Chemical structure of chitin and its conversion to chitosan. Chitosan is a partially deacetylated form of chitin and generally contains *N*-acetyl groups on 5–30% of its glucosamine units. (Adapted from Kurita, K., *Mar. Biotechnol.*, 8(3), 203–226, 2006.)

TABLE 6.1
Marine Sources of Chitin and Percentage (Dry Weight Basis) Found in Shell Discards

Chitin Source	Chitin in Shell Wastes (%)
Clam/oyster	3–6
Crab:	
Collinectes sapidus	13.5
Chinoecetes opilio	26.6
Shrimp:	
Pandalus borealis	17.0
Crangon crangon	17.8
Penaeus monodon	40.4
Crawfish:	
Procamborus clarkia	29.8
Prawn	33.0
Squid pen	20–40
Krill:	
Euphausia superba	41.0

Source: Adapted from Synowiecki, J. and Al-Khateeb, N.A., *Crit. Rev. Food Sci. Nutr.*, 43(2), 145–171, 2003; Kurita, K., *Mar. Biotechnol.*, 8(3), 203–226, 2006.

15–40% chitin (dry weight), with the other major components being proteins and calcium carbonate [2]. Antarctic krill and squid skeleton are typically rich in chitin (~40%), while the dry shells of crab, lobster, crayfish, shrimp, and prawns generally contain around 14–35% chitin [3,10,11].

A major setback to the utilization of chitin for commercial applications has been the need for more cost-effective isolation procedures. Thus far, extraction of chitin from natural sources has proven to be a relatively laborious process that can create additional waste issues [2,12]. In nature, chitin is typically associated in a complex matrix with other polysaccharides, such as cellulose, mannan, polygalactosamine, and glucan, all of which must be removed during the extraction procedure [3]. Chitin is not soluble in ordinary solvents but is relatively stable under mild acidic or basic conditions. Therefore, the other compounds in the exoskeleton matrix must first be degraded and removed prior to isolation of the chitin residue. In a commercial setting, shell discards are initially treated with a dilute acid to remove salts and minerals (especially calcium carbonate), and then the sample undergoes several treatments with a hot alkaline solution to remove any remaining organic compounds. The residue left behind is then dried and chitin is obtained in a flake or powder form that is almost colorless or off-white [2,4]. Chitin can be partially deacetylated to form a variety of derivatives, such as chitosan, oligosaccharides, and glucosamine. Owing to their increased water solubility, these chitin derivatives may be more advantageous as nutraceutical agents than the fully acetylated, insoluble form of chitin.

6.3 CHITOSAN

Chitosan is actually the collective name for a specific group of partially deacetylated chitin molecules [13]. Chitosans are linear polysaccharides containing a mixture of both D-glucosamine and *N*-acetyl-D-glucosamine monomers that are joined by β-(1–4) linkages (Figure 6.1). These large, cationic copolymers tend to have degrees of *N*-acetylation ranging from 5 to 30% [2,7]. Commercial chitosan ranges in molecular weight (MW) from around 100,000 to 1,000,000 Da and is generally available in one of the three morphological forms (i.e., amorphous, semicrystalline, and crystalline). The pure microcrystalline form of chitosan (MW: 10,000–300,000 Da) has an especially large adsorption area and is used as a gellike dispersion or as a fine powder for the creation of chitosan molecules with a specific structure and size. The amino groups of chitosan have a p*K*a of ~6.5 (typical range is 6.2–6.8), giving the molecule an overall positive charge and solubility in acidic conditions. Although chitosans are generally insoluble in water, they do interact with negatively charged molecules such as fatty acids, bile acids, phospholipids, proteins, and anionic polysaccharides (e.g., carrageenan, pectin, and alginate) [3]. Chitosan salts can be water soluble depending on factors such as pH and degree of acetylation [14]. Owing to its inherent properties and its ability to interact with a range of biomolecules, chitosan has a number of commercial applications in areas such as bioadhesives (i.e., wound dressings), waste treatment, food preservation (i.e., edible films), food additives, pharmaceuticals, and nutraceuticals [5,8,15].

Chitosan is approved for use as a food additive or dietary supplement in countries such as Japan, England, the United States, Italy, Portugal, and Finland [1,6,14,16]. It has been sold as a nonprescription weight-loss supplement for over 20 years in Europe and Japan and has entered the U.S. market over the past 10 years [6]. In 1992, chitosan was accepted as a functional food ingredient by Japan's health department. They determined that it met the required attributes for functional foods, including illness prevention, promotion of longevity, immune enhancement, and biorhythm control [16]. Although the U.S. Food and Drug Administration (FDA) has not approved chitosan for use as a drug, it is regulated as a dietary supplement [6]. The main producers of chitosan are Japan and the United States, with smaller operations in India, Italy, and Poland. The market price of chitosan has been reported to range widely, from around US$ 5/kg for use as a low-grade product in agriculture to approximately US$ 200/kg for a high-quality, ultra-pure grade product to be used in the health-care industry [16]. A major application of chitosan is in the nutraceuticals industry, which utilized about one-half of the total chitosan produced in the year 2000 [2]. Chitosan has been shown to provide a number of health benefits because of its activity as a dietary fiber, an antimicrobial agent, and an immune system booster [4,5]. Many of the chitosan products currently available in the food industry are marketed as fat reducers and cholesterol-lowering agents, although there is a good deal of controversy surrounding the effectiveness of chitosan at blocking fat absorption.

6.4 NUTRACEUTICAL APPLICATIONS

Chitin and chitosan have a variety of nutraceutical applications, including immune-enhancement, disease recovery, and use as dietary fiber (Table 6.2). Their ability to be useful in these areas is largely dependent on factors such as MW and degree of *N*-acetylation, which can greatly influence solubility along with interactions with other biomolecules [4]. This section will discuss the nutraceutical properties of chitin and chitosan with regard to their antibacterial, anti-inflammatory, antioxidant, anticarcinogenic, and antiulcer activities, along with their applications in renal disease recovery and as a dietary fiber.

6.4.1 Antimicrobial Activity

Chitin and chitosan have been reported to exhibit antimicrobial activity against a range of fungi, yeast, and bacteria [5,17]. Chitosan is believed to have greater antimicrobial activity than chitin because of its positive charge at low pH, allowing it to disrupt the negatively charged cell membranes of bacteria [5]. Chitosan is also known for its ability to chelate trace metals, which helps to inhibit microbial growth and production of toxins. The antimicrobial properties of chitosan have primarily been exploited by the food industry for use in agriculture and as edible films in food preservation; however, they may prove to be useful in nutraceuticals for the ability to stimulate host defenses against a variety of bacteria, fungi, and yeast [17].

6.4.2 Anti-inflammatory Activity

Inflammation is the basic immune response of the body to injury, infection, or irritation, and it is characterized by symptoms such as redness, heat, swelling, pain, and organ dysfunction. Disorders can arise when an injurious agent persists or when the immune system attacks its own tissue, as is the case in autoimmune diseases. Chronic or prolonged inflammation can lead to complications and diseases such as rheumatoid or osteoarthritis, ulcerative colitis, appendicitis, psoriasis, gastritis, and meningitis. Anti-inflammatory nutraceuticals, such as omega-3 polyunsaturated fatty acids (PUFA), soy isoflavones, plants and spices, and possibly chitosan, can help to prevent or alleviate these types of complications by inhibiting activity of proinflammatory agents [18–21].

Although very few studies have investigated the anti-inflammatory activity of chitosan, a few promising papers have recently been published. One study reported that chitosan was able to prevent pulmonary inflammation by inhibiting type 2 helper T cells (Th2) and reducing levels of interleukin-4 (IL-4) and IL-5, all of which are important components of the immune response to allergens [22]. In the first of a series of two studies, a water-soluble form of chitosan was reported to possess anti-inflammatory activity by inhibiting the production of tumor necrosis factor-α (TNF-α) and IL-6, two proinflammatory cytokines, in human astrocytoma cells [23]. Both cytokines are normally synthesized and

TABLE 6.2
Nutraceutical Properties of Chitin and Chitosan

Compound	Nutraceutical Property	Mode of Action	Potential Health Benefits	Research Stage	Reference(s)
Chitosan (most effective) and chitin	Antimicrobial activity	Variety of proposed mechanisms: disruption of negatively charged cell membranes, chelate trace metals	Stimulate host defenses against a range of bacteria, fungi, and yeast	Widely used as an antibacterial agent and edible film for food preservation	[5,17]
Chitosan	Anti-inflammatory activity	Inhibits activity of Th2 and mast cells; reduces levels of proinflammatory cytokines (IL-4, IL-5, IL-6, and TNF-α)	Prevention or alleviation of inflammatory disorders (e.g., rheumatoid or osteoarthritis, Crohn disease, appendicitis, allergies)	Lab studies *in vivo* (rats) and *in vitro* (human astrocytoma cells and mast cells)	[22,23,24]
Chitosan (most effective) and chitin	Antioxidant activity	Chelates metal ions that catalyze oxidation reactions	Prevention or alleviation of age-related health disorders (e.g., arthritis, cancer, stroke, atherosclerosis, retinal damage, heart attack)	Lab studies *in vitro*	[30,32]
Chitin and chitosan	Anticarcinogenic activity	Adsorption of mutagens	Protective effect against potential DNA mutagens	Lab studies *in vitro*	[39]
Low MW chitosan (most effective), chitin, and high MW chitosan	Antiulcer activity	Prevents changes in gastric acidity and peptic activity; may coat the ulcerated area and neutralize protons and pepsin in the gastric juices, may help to maintain activity of antioxidative enzymes	Prevention of gastric mucosal injury and acceleration of gastric ulcer healing	Lab studies *in vivo* (rodents)	[40,43]

Chitosan	Renal disease prevention and recovery	May improve renal function by binding and removing uremic toxins and nitrogenous wastes; lowers uric acid levels in blood and urine; may reduce occurrence of risk factors (i.e., hypertension and type II diabetes)	Reductions in total serum cholesterol, urea, and creatinine levels; increases in serum hemoglobin levels and overall health (e.g., physical strength and appetite)	Unblinded clinical trials; lab studies *in vivo* (rats) and *in vitro* (porcine renal cells)	[3,38,45,47,51, 52,53,54]
Chitosan	Prevention of type II diabetes	Owing to its fiberlike properties, it may slow carbohydrate metabolism	Promotes control of glycemic index	Lab studies *in vitro*	[3,52]
Chitosan	Weight reduction	May bind fatty acids and neutral lipids, thus preventing their gastrointestinal absorption; may promote satiety and satiation after a meal	Reduced risk of obesity and obesity-related disorders (e.g., type II diabetes, cardiovascular disease, some cancers)	Clinical and animal trials; effect on weight loss is debatable; currently being sold as weight-loss supplements	[5,6,7,58,63]
Chitosan (relatively low MW most effective) and chitin (less effective)	Reduction in total and LDL cholesterol levels	May bind bile acids, thus preventing their enterohepatic recycling and, ultimately, reducing serum cholesterol levels; may promote satiety and satiation after a meal	Reduced risk of hypercholesterolemia and cholesterol-related disorders (e.g., cardiovascular disease)	Clinical and animal trials; some conflicting results; currently being sold in supplement form and in nutraceutical foods	[1,4,5,7,55]

released by mast cells and play important roles in the inflammatory response, with TNF-α being especially involved in allergic inflammation. In a follow-up study, chitosan was reported to partially inhibit the secretion of both IL-6 and TNF-α from mast cells, and it was suggested that water-soluble chitosan has the potential to reduce the allergic inflammatory response by down-modulating the Ca^{2+}-induced activation of mast cells [24]. Since mast cells are necessary for allergic reactions and have been implicated in a number of neuroinflammatory diseases, chitosan nutraceuticals may help to prevent or alleviate some of these complications [25].

Unregulated overproduction of IL-6 is associated with a number of complications, including rheumatoid arthritis, Crohn's disease, Castleman's disease, and tumorigenesis [18,26]. Therefore, partial inhibition of IL-6 may help to prevent or alleviate some of these inflammatory diseases. TNF-α may also have a significant role, as biological agents that inhibit it have recently been shown to successfully treat some of the diseases listed above [26,27]. The ability of chitosan to reduce and delay an inflammatory response by partial inhibition of TH2, IL-6, and TNF-α could be valuable in the nutraceuticals industry for those interested in provided anti-inflammatory foods for the prevention or alleviation of inflammatory diseases.

6.4.3 Antioxidant Activity

Reactive oxygen species (ROS) and free radicals are naturally generated in the body during aerobic metabolism and can cause oxidation of lipids, proteins, sugars, sterols, and nucleic acids [28]. During the aging process, the body's natural antioxidant defense systems weaken, resulting in the accumulation of ROS and free radicals. ROS activity has been strongly associated with a number of age-related health disorders, including arthritis, cancer, stroke, atherosclerosis, retinal damage, and heart attack. Antioxidative nutraceuticals, such as tocopherols, ascorbic acid, carotenoids, polyphenols, and possibly chitosan, can help to minimize oxidative damage and reduce the risk for age-related disorders by preventing the accumulation of ROS and free radicals. These nutraceuticals can serve as antioxidative enzymes, singlet oxygen quenchers, metal chelators, or hydrogen donors [28].

Although the antioxidant activity of chitosan has been demonstrated in a number of studies, most research has been related to extending the shelf life of food products rather than providing health benefits to consumers. For example, chitosan has been reported to reduce lipid oxidation in seafoods, giving it a potential use in the food industry for improving the shelf life and stability of products such as chilled fish fillets [1,29]. However, chitosan and its derivatives may also prove to be valuable antioxidants in the nutraceuticals industry. As described by Agullo et al. [3], antioxidants can be divided into two major categories: primary antioxidants, which involve the use of a phenol group during the initial part of the oxidative reactions, and secondary antioxidants, which can chelate metal ions that catalyze the oxidative reactions. The secondary antioxidant properties of chitosan have been demonstrated in a number of studies [3,30,31]. Chitin and chitosan are known to adsorb metal cations, with chitosan exhibiting an increased affinity as a result of its free amino groups [2]. The ability to chelate metal ions involved in oxidation reactions can be highly useful in food preparation and storage, health care, water treatment, and pharmaceutical products [1]. A protective effect of chitosan against whole-body x-ray irradiation was recently reported for mice [32]. Mice fed a diet containing 5% insoluble chitosan showed ~20% higher survival rates, with significant improvements in hematopoietic activity and leukocytogenesis, compared to mice on the control diet. It was suggested that chitosan acted by scavenging free radicals and chelating the Fe(II) or Cu(I) ions involved in oxidative reactions. Low MW chitosan was recently reported to be more effective at scavenging superoxide and hydroxyl radicals compared to high MW chitosan, with sulfated chitosans showing the greatest effect [31].

Manipulation of the chitosan molecule may prove to further increase its antioxidant properties. Agullo et al. [3] recently reported the possibility of inducing primary antioxidant effects by the addition of a phenol group to the chitosan molecule. The resulting compound would possess both primary and secondary antioxidant properties, making it a potentially valuable antioxidative nutraceutical for the prevention of oxidative stress and age-related health disorders.

6.4.4 Anticarcinogenic Activity

Next to cardiovascular disease, cancer is the second leading cause of death in most industrialized countries, including the United States. Incorporation of anticarcinogenic and antioxidative nutraceuticals, such as phytochemicals (i.e., flavonoids, polyphenols, retinoids, etc.), into the diet can help reduce the occurrence of some cancers by a variety of mechanisms, including prevention of DNA mutagenesis and induction of apoptosis [33,34]. In addition to the ability to decrease oxidative stress and subsequent DNA damage, some nutraceuticals can adsorb mutagens and thereby inhibit their carcinogenic activity. For example, some dietary fibers have been reported to have a protective effect on human cancers through adsorption of a variety of direct- and indirect-acting mutagens [35].

Most of the anticarcinogenic studies involving chitin and chitosan have been focused on their oligomers, which have been shown to inhibit heavy metal–induced genotoxicity and possess growth inhibitory and antimetastatic effects against a variety of cancerous tumors [4,5,36–38]. However, one study reported that chitin and chitosan themselves possess antigenotoxic activities against four different mutagens [39]. Since chitin and chitosan are considered to be dietary fibers, the authors investigated their ability to adsorb the mutagens 4-nitroquinoline-*N*-oxide, dinitropyrene, mitomycin C, and adriamycin. The mutagens were successfully adsorbed in both ionic and nonionic solutions, indicating that chitin and chitosan act not only as ion exchange units, but also have some nonionic activity. The protective effect of chitin and chitosan against environmental mutagens could be quite useful in the nutraceuticals field. For example, a dietary supplement of chitin/chitosan could potentially adsorb potential carcinogens and transport them out of the digestive tract, thereby providing protection against a variety of cancers.

6.4.5 Antiulcer Activity

Ulcers are open, nonhealing sores that can occur in a number of locations throughout the body. Peptic ulcers generally occur as sores or holes in the lining of the stomach or the duodenum, and are characterized by symptoms such as abdominal pain, nausea, bloating, and loss of appetite [40]. These ulcers are a result of imbalances in pepsin, an aggressive stomach acid, and the protective mucosal coating, which is weakened by the bacteria *Helicobacter pylori*. Chitin and chitosan are widely used as wound dressings to promote rapid healing of external cuts or burns [41]. Since similar agents used for healing skin ulcers have also been reported to be effective at preventing gastric mucosal injury and gastric ulcers, chitin and chitosan were also expected to exhibit some protective effects [42].

One study investigated the effects of chitin, high MW chitosan, and low MW chitosan on gastric mucosal injury and healing from gastric ulcers in mice [43]. Low MW chitosan (MW: 25,000–50,000 Da) showed the greatest antiulcer effect, with a dose-dependent prevention of ethanol-induced mucosal injury and acceleration of gastric ulcer healing. The effects of chitin (MW: >1,000,000 Da) and high MW chitosan (MW: 500,000–1,000,000 Da) were much less potent, possibly due to differences in acid solubility or in the strength/duration of adhesion to mucosa or to the ulcerated site. Low MW chitosan may have been more effective because of its ability to dissolve more easily in acid, whereas the less-soluble chitin and high MW chitosan may have passed to the small intestine more quickly. It was proposed that solubilized chitosan may have exerted its protective effect by coating the ulcerated area and neutralizing H^+ and pepsin in the gastric juices.

In a more recent study, the effects of ulcer induction by HCl-ethanol in rats were studied in the presence and absence of chitin and chitosan [40]. In the absence of chitin and chitosan, significant increases were observed in the volume and acidity of gastric fluid and the level of lipid peroxidation. Also, there were significant decreases in glutathione levels, activity of antioxidant enzymes, and peptic activity in the gastric mucosa. However, rats that were pretreated with chitin or chitosan (2% of feed) did not experience significant ulcerogenic effects of ethanol-HCl treatment and they maintained near-normal levels of gastric acidity and peptic activity. Interestingly, in this study chitin showed a stronger protective effect than chitosan. The authors suggested that chitin and chitosan prevented ulcer formation as a result of their antioxidant capacity, ability to neutralize gastric juices, and ability to help maintain activity of antiperoxidative enzymes. Although there is a need for more research in this area, including clinical trials, chitosan may prove to be valuable in the nutraceuticals industry for those interested in producing supplements that might help to either prevent or promote healing of gastric ulcers.

6.4.6 Renal Disease Recovery

Kidney disease is a major human health concern that can lead to further complications such as decreased kidney function, kidney failure, and cardiovascular disease [44]. A number of studies have indicated a possible use of chitosan in recovery from renal disease. In one unblinded study, 40 hemodialysis patients with chronic renal failure were treated with 10 chitosan tablets (45 mg/tablet) three times daily for 12 weeks [45]. The results showed reductions in total serum cholesterol, urea and creatinine levels and increases in serum hemoglobin levels along with feelings of physical strength and appetite. Additionally, no clinically significant adverse effects were reported. It was suggested that chitosan may improve renal function by binding and removing uremic toxins (e.g., uric acid, urea, and creatinine) and nitrogenous wastes in the gastrointestinal tract. Under normal conditions, these compounds are excreted by the kidneys; however, during renal failure they can accumulate and negatively affect renal functioning [46].

Studies in rats have reported that supplementation with chitosan results in significantly lower serum levels of uric acid and urea nitrogen as compared to a control, and chitosan is already being used as a food supplement for the prevention and treatment of hyperuricemia [47–49]. The addition of copper to chitosan has been reported to result in an increased adsorption capacity for urea, and greater effects were observed with increased degrees of acetylation and decreased MW [50]. Additionally, Yoon et al. [51] reported that chitosan was able to increase the release of renal dipeptidase from porcine renal proximal tubule cells, thereby promoting recovery from renal injury. Although the above research indicates a potential use of chitosan as a nutraceutical in the protection and recovery from renal disease, placebo-controlled clinical trials must be carried out to discern the long-term safety and pharmacokinetics of chitosan for renal disease recovery [6].

Interestingly, chitosan may also show some protective effects against the development of renal disease. Owing to its fiberlike properties, chitosan has been shown to reduce the occurrence of a number of precursors to both hypertension and type II diabetes, two of the major risk factors for kidney disease [3,52–54]. These properties will be discussed in more detail in the following section.

6.4.7 Dietary Fiber

Dietary fibers are nondigestible carbohydrates that are derived from the edible parts of plants or animals [3,55]. They absorb water and facilitate movement of food through the digestive system. Dietary fibers have been reported to provide beneficial health effects in their ability to lower serum cholesterol and reduce the risk of coronary heart disease [55]. Although chitosan is not generally thought of as a classical dietary fiber, it is considered to have fiber characteristics because of its limited digestibility by humans [3,7].

Dietary fibers can be classified into two major categories in regard to water solubility (i.e., soluble and insoluble), both of which have high water–holding capacity and distinct physiological roles [3]. Insoluble fiber does not readily ferment in the colon, allowing for the formation of softer stools that are able to pass through the gastrointestinal system more quickly. Soluble fiber, which also has a high water-holding capacity, undergoes some extent of fermentation in the large intestine and is able to increase the viscosity of the intestinal environment.

Chitosan has been reported to have similar characteristics as vegetable-derived soluble fibers and to have a high water-holding capacity (around 20 times its initial weight) *in vitro* [52].

Diets high in fiber have been shown to promote glycemic control and reduce the risk for type II diabetes [56,57]. Fibers help to slow carbohydrate metabolism, thereby reducing the rate of glucose absorption in the small intestine. Chitosan was recently reported to slow glucose dialysis *in vitro*, giving it a potential use in nutraceuticals for controlling glycemic index and reducing the risk for type II diabetes [3,52].

6.4.7.1 Weight and Cholesterol Reduction

Both obesity and high cholesterol are major threats to human health and can lead to a myriad of complications, including cardiovascular disease, type II diabetes, and some cancers. Currently, approximately two-thirds of the U.S. population is considered to be either overweight or obese, and methods for reducing weight and cholesterol levels are in high demand [58–60]. Chitosan may offer some benefit in this area as a nutraceutical food or supplement with the potential to help reduce total cholesterol levels and body weight.

6.4.7.1.1 Theory behind Weight and Cholesterol Reduction

Chitosan is currently being sold as a nondigestible weight-loss supplement with the claim that it can bind and entrap many times its weight in fats, thereby inhibiting fat uptake by the digestive tract and reducing occurrence of obesity, high cholesterol, and hypertension. The mechanistic theory behind these claims is that, upon entering the human stomach, the $-NH_2$ groups of chitosan take on protons to become $-NH_3^+$. In the aqueous, acidic environment, the chitosan molecule swells to form a positively charged gel that interacts with the negatively charged carboxylic groups of fatty acids (e.g., oleic, linoleic, palmitic, stearic, and linolenic acids) and bile acids (e.g., cholic, deoxycholic, and lithocholic acids). Also, chitosan could interfere with digestion of neutral lipids (i.e., cholesterol and other sterols) through hydrophobic interactions. As a result of these electrostatic and hydrophobic interactions, large polymer compounds are formed that are poorly digested by the human gastrointestinal system [7]. Although chitosan is soluble in the acidic conditions of the stomach, it begins to precipitate once it enters the small intestine at pH 6.3–6.8 [3,7]. As the polysaccharide chains of chitosan aggregate, they entrap the acids, cholesterols, or lipids they are bound to, forming a nondigestible complex that is excreted in the feces. This proposed mechanism for chitosan acting within the gastrointestinal system may promote weight loss by reducing metabolism of dietary fats and cholesterols.

In a similar manner, chitosan may help to reduce total serum cholesterol levels by entrapping bile acids. Normally, cholic acid and other bile acids are produced by the oxidation of cholesterol in the liver, then stored in the gallbladder and secreted into the intestine. Production of bile acids from cholesterol is a major route for cholesterol consumption, with about half of the cholesterol synthesized by the body being used for bile acids. Typically, around 90% of the bile acids used in the intestine are reabsorbed and recycled through enterohepatic circulation. However, when bile acids become entrapped by chitosan, they are subsequently removed from this recycling system and the liver must manufacture additional bile acids from cholesterol [7,61]. This results in a decrease in the cholesterol content of the liver, which activates expression of low-density lipoprotein (LDL) receptors and subsequently increases uptake of LDL from the bloodstream to the liver. The overall effect is a reduction in blood levels of total and LDL cholesterols. A similar mechanism has been reported for other soluble fibers, such as oat bran and bean fiber and the fiber supplement psyllium mucilloid, which reduces total and LDL cholesterol levels through increased excretion of bile acids and decreased production of cholesterol and fatty acids in the liver [56,57,62].

Dietary fibers also play an important role in controlling obesity and energy intake by initiating early signals of satiation and prolonging signals of satiety. In addition to the proposed role of chitosan in blocking lipid absorption, chitosan may help to reduce overeating by promoting feelings of satisfaction and fullness following a meal [58].

6.4.7.1.2 Weight Studies

Studies into the effectiveness of chitosan at significantly reducing body weight have reported conflicting results [5,63]. This section will highlight some of the major studies involving the effects of chitosan on reducing body weight.

6.4.7.1.2.1 Studies Reporting an Effect

Early studies provided evidence that chitosan may exhibit a hypolipidemic effect through its ability to reduce fat absorption in the gastrointestinal tract [64–67]. A study involving rats subjected to chitosan supplementation showed increases in fecal fat excretion compared to the control and glucomannan feeding, providing evidence for the theory that chitosan binds and sequesters fats in the gastrointestinal system [68]. The ability of chitosan to bind fats may be a contributing factor to the accelerated weight loss observed in humans on a hypocaloric diet supplemented with chitosan. A number of Italian studies have reported an effect of chitosan combined with dietary changes for reducing excess weight gain, hyperlipidemia, hypercholesterolemia, and blood pressure in overweight or obese subjects [69–74]. In general, the subjects maintained hypocaloric diets consisting of around 1000–1100 kCal/day that were supplemented with 2 g of chitosan or a placebo per day (1 g prior each of their two main meals) [6]. Although both the chitosan and control groups showed reductions in the parameters analyzed, those taking chitosan tended to have significantly greater reductions.

More recently, a double-blind, placebo-controlled human trial reported an effect of chitosan (LipoSan Ultra™) on weight and body mass index (BMI) [59]. The study took place in the United States and involved 59 overweight and mildly obese females (21–55 years in age) who took chitosan or placebo supplements (1.5 g/dose) just prior to the two biggest meals of the day. Participants maintained their normal, high-fat diets during the 8-week-long trial. Although there were no reported differences in dietary intake of calories and fat, subjects taking chitosan supplements lost an average of 1.0 kg with a 1% reduction in BMI, while those in the placebo group gained an average of 1.5 kg. The major side effect of the chitosan supplement was gastrointestinal discomfort, including flatulence, stool bulkiness, bloating, nausea, and heartburn.

In another human trial, 50 obese women were given a daily supplement of chitosan or a placebo in combination with advice for a low-calorie diet, physical activity, and behavior modification [53]. After 6 months, the women taking the chitosan supplements showed significantly greater reductions in body weight as compared to women taking the placebo. Those taking chitosan also experienced decreases in systolic and diastolic blood pressure, with no significant differences in LDL or total cholesterol levels. No adverse effects were reported as a result of chitosan supplementation.

Since most trials showing an effect of chitosan supplementation on weight loss were carried out in conjunction with dietary changes, chitosan supplement manufacturers often promote the effectiveness of their product in combination with a low-calorie diet [75]. However, chitosan may also prevent weight gain and, subsequently, obesity in individuals consuming high-fat diets. Recently, a low MW form of water-soluble chitosan (46 kDa) was reported to prevent increases in body weight, white adipose tissue weights, and hepatic levels of cholesterol and triacylglycerols in mice fed a high-fat diet for 20 weeks [76]. The 46 kDa chitosan was also reported to increase fecal excretion of fat and bile acids compared to the control group, and exhibit an inhibitory effect on pancreatic lipase *in vitro*. Therefore, it was suggested that weight gain was prevented due to decreased absorption of dietary lipids resulting from reduced activity of pancreatic lipase.

6.4.7.1.2.2 Studies Reporting No Effect

Although the studies discussed above report promising weight-loss results, a large number of studies have reported no effect of chitosan supplementation on body weight. Therefore the true effectiveness of chitosan as a weight-loss supplement remains in question. In contrast to most of the studies reporting weight loss because of chitosan, many of the studies documenting no effect of chitosan involved subjects without dietary restrictions. For example, Pittler et al. [77] reported no effect of chitosan on body weight in 30 overweight males and females (18–60 years in age) from the United Kingdom. This study was a double-blind, randomized trial in which subjects did not undergo dietary changes. The major side effect of chitosan supplementation was constipation in several of the participants. A similar study out of Singapore reported no effect of a chitosan salt (Absorbitol™) on obesity or plasma lipids in 68 hypercholesterolemic obese males and females; however, some gastrointestinal side effects were reported [78]. Participants in the double-blind, 12-week study were given no dietary restrictions. A study out of Japan involving 90 females (34–70 years in age) with mild to moderate hypercholesterolemia reported no effect of chitosan supplementation on body weight or body mass index [79]. This study also did not involve dietary alterations.

A recent randomized, double-blind study out of New Zealand did report some effect of chitosan in combination with dietary and lifestyle advice compared to a placebo with advice; however, the effect was marginal and not deemed to be clinically significant [80]. The authors also reported no increases in fecal fat excretion as a result of chitosan supplementation. The trial involved 250 overweight or obese males and females (≥18 years in age) and was carried out over the course of 24 weeks.

In a review of supplemental products used for weight loss, Lenz and Hamilton [81] indicated the need for more long-term, well-designed studies investigating the effects of chitosan on body weight. Owing to the lack of consistent information in this area, the authors stated that they were not able to make a positive recommendation for the use of chitosan as a weight-loss supplement. Likewise, in a series of meta-analyses involving double-blind, randomized controlled trials with overweight or obese subjects, firm evidence for an effect of chitosan on weight loss could not be established [60,82,83]. Although in the initial meta-analysis a 3.3-fold effect was found for chitosan in reducing weight, the five trials that were included were said to contain serious methodological errors and the effectiveness of chitosan on body weight could not be established beyond reasonable doubt [82]. In a literature search conducted up to January 2004, four additional studies were found and added to the original meta-analysis [60]. Once again, the authors did not find strong evidence for chitosan supplementation, concluding that considerable doubt exists as to whether chitosan has the ability to reduce body weight in humans.

Recently, Mhurchu et al. [63] conducted a systematic review of 14 randomized, controlled trials that have examined the use of chitosan as a weight-loss supplement for overweight, obese, and hypercholesterolemic individuals. All trials were at least 4 weeks in duration and were graded for quality in terms of adequate concealment of randomization assignments. When all 14 trials were analyzed together, the results showed that chitosan supplementation results in a small, but statistically significant ($p < 0.00001$), weight reduction of 1.7 kg compared to the placebo. However, when only the four studies deemed to be of high quality were included in the analysis, the average weight reduction was much smaller (0.6 kg) and not statistically significant ($p < 0.11$). These four studies were by Pittler et al. [77], Schiller et al. [59], Bokura and Kobayashi [79], and Mhurchu et al. [80]. The authors concluded that their analysis of these studies demonstrates that the effect of chitosan on body weight is minor and not likely to be considered significant in a clinical sense [63].

6.4.7.1.3 *Cholesterol Studies*

One of the main beneficial effects of chitosan supplements thus far appears to be a mild to moderate reduction in total serum cholesterol levels. The ability of chitosan to lower cholesterol levels could prove to be important in the prevention and treatment of cardiovascular disease, which includes hypercholesterolemia as a major risk factor [13,79].

6.4.7.1.3.1 *Animal Trials*

Early studies in rats indicated that chitosan exhibits hypocholesterolemic and hypolipidemic activity through its ability to lower cholesterol and triacylglycerol levels in the blood plasma and in the liver [64,84–89]. During these trials and others, it was found that the MW and degree of acetylation of dietary chitosan are important factors in determining cholesterol- and lipid-absorption abilities [7]. Although chitosans with relatively low MW (5–120 kDa) and a moderate degree of acetylation have shown the greatest cholesterol-lowering potential, short oligomers with less than six glucosamine residues have been deemed to be ineffective [7,88]. Also, chitosan was reported to be much more effective than chitin, which has a relatively large MW. In regard to degree of acetylation, 79% deacetylated chitosan was reported to be more effective at lowering cholesterol than 92% deacetylated chitosan [90].

When chitosan was incorporated into the diet of hamsters, significant drops in food intake and cholesterol levels were observed [90]. Additionally, formation of cholesterol gallstones was inhibited by chitosan at the highest dose (8% of diet). The reductions in cholesterol could be attributed to the ability of chitosan to entrap and increase the fecal excretion of neutral sterols and bile acids [86,91,92]. However, a similar study reporting reduced serum cholesterol levels in rats fed a chitosan-enriched diet for 3 weeks found no effect of chitosan on fecal excretion of neutral sterols [93].

Chitosan has been reported to lead to a 46% reduction in blood cholesterol levels when incorporated at 5% into the diet of hypercholesterolemic mice [94]. Also, atherogenesis, which begins with entrapment of LDL in blood vessel walls, was significantly inhibited by 42 and 50% in the whole aorta and the aortic arch, respectively, in mice on the chitosan-supplemented diet. In another study, both normal and diabetic mice receiving chitosan (5% of diet) supplementation for 4 weeks experienced significant reductions in total cholesterol and blood glucose levels [95]. More recently, chitosan supplementation in rats was reported to decrease cholesterol absorption and lead to a greater excretion of bile acids as compared to a cellulose-containing diet [68]. Also, chitosan supplementation resulted in decreased cholesterol levels in the liver, an effect that was heightened with the addition of glucomannan, another dietary fiber.

Although the cholesterol-lowering effects of chitosan have largely been attributed to the idea that it entraps compounds such as neutral lipids and bile acids, a recent study suggested that satiation and satiety may play a more important role. Chitosan, cellulose, and cholestyramine (a positive pharmaceutical control) were all reported to reduce increases in serum cholesterol level in rats fed diets high in fat and cholesterol for 3 weeks [55]. The control group accumulated liver cholesterol at a rate of 4 mg/mouse/day, whereas diets supplemented with chitosan and cellulose showed accumulation rates of 2 and 3 mg/mouse/day, respectively. The authors reported that supplementation with chitosan or cellulose reduced food (cholesterol) intake, but did not alter intestinal absorption of cholesterol or fecal excretion of sterols. Dietary fibers are known for their ability to contribute to feelings of fullness after meals and slow sugar absorption, thereby balancing blood sugar and insulin levels [5,96]. Therefore, van Bennekum et al. [55] suggested that chitosan and cellulose supplementation resulted in decreased food consumption in mice by inducing satiety and satiation.

6.4.7.1.3.2 Clinical Trials

Among human trials that have found an effect of chitosan supplementation on cholesterol, total serum cholesterol and LDL levels have generally been reported to be reduced by 5.8–42.6 and 15.1–35.1%, respectively [7]. In one of the first trials reporting a hypercholesterolemic effect of chitosan in humans, eight healthy males consuming chitosan-enriched biscuits (3–6 g chitosan/day) experienced significant decreases in total serum cholesterol and increases in serum high-density lipoprotein (HDL) cholesterol and excretion of bile acids (cholic and chenodeoxycholic acid) in the feces [97]. The chitosan ingestion period lasted two weeks, with subjects consuming three biscuits per day during the first week and six per day in the second week. In 1999, a double-blind, randomized study reported no significant differences in serum cholesterol, triacylglycerols, or BMI after a 4 week period of chitosan supplementation [77]. However, in a slightly longer study, supplementation with microcrystalline chitosan was reported to result in significant reductions in serum LDL cholesterol after 8 weeks [98]. A comparison of these studies indicates that (a) chitosan may require more than 4 weeks to show a pronounced effect on cholesterol levels and (b) microcrystalline chitosan may be more effective at reducing cholesterol levels due to its especially large adsorption capability discussed earlier.

Recently, chitosan supplementation was reported to result in significant but mild reductions in total cholesterol after 8 weeks, with tests at 4 weeks showing no significant differences [79]. The study analyzed the effects of chitosan on hypercholesterolemic women aged 34–70 years. Chitosan supplementation also resulted in reductions in LDL cholesterol in a subgroup of women over 60 years of age. In the study, 41 women were given 1.2 g/day of chitosan and 43 women took a placebo. The participants maintained their normal diet and consumption levels and were tested at weeks 4 and 8 for serum lipid levels, body weight, and adverse effects. The authors reported that there were no serious events and few adverse effects. As reported in a previous study [98], there were no significant changes in serum triacylglycerol levels. Bokura and Kobayashi [79] concluded that their results indicate chitosan is a safe and mildly effective supplement for reducing cholesterol levels in women.

Despite previous results showing an effect of microcrystalline chitosan on LDL cholesterol levels [98], a more recent double-blind, randomized study did not find an effect of microcrystalline chitosan on plasma levels of cholesterol, triacylglycerols, or glucose [99]. The participants were middle-aged men and women with moderately increased plasma total cholesterol levels and no dietary restrictions. The study was more long-term (10 months) than previous ones and involved healthy adults (N = 130) rather than obese or overweight individuals. These results indicate that chitosan may not be useful for cholesterol reduction in subjects with a healthy weight, as most studies that have reported an effect of chitosan on cholesterol involved obese and overweight subjects. Also, a reexamination of the data showed no influence of the apolipoprotein E genotype, which has previously been shown to alter the effects of hypocholesterolemic drugs and dietary supplements [100].

As discussed previously, Gallaher et al. [68] reported the ability of chitosan to reduce cholesterol levels in rats, with a more pronounced effect when taken in conjunction with the dietary fiber glucomannan. To examine the mechanism by which glucomannan and chitosan exert this hypocholesterolemic effect, Gallaher et al. [101] conducted a study involving 21 overweight human subjects. The participants added a chitosan and glucomannan supplement to their diets, and maintained their normal intake of calories, fat, and dietary fiber over the course of 4 weeks. The results showed reductions in total serum, HDL, and LDL cholesterol levels. Although there were no changes in fecal fat excretion, there was a trend toward greater excretion of neutral sterols and bile acids, indicating an ability of the dietary fiber supplement to bind and sequester cholesterol and its derivatives found within the gastrointestinal system.

6.4.8 Chitin and Chitosan Nutraceutical Products on the Market

Chitosan as a cholesterol-lowering nutraceutical can either be taken in supplement form or added to certain foods (Table 6.3). For example, Japan produces a variety of chitosan-enriched foods marketed for their hypocholesterolemic effects, including potato chips, noodles, soybean sauce and paste, dietary cookies, and vinegar products [1,5,15]. As discussed previously, a study out of Japan reported significant reductions in total serum cholesterol in adult males consuming chitosan-enriched biscuits (3–6 g/day chitosan) for 4 weeks [97]. However, it has been suggested that chitosan-enriched foods should not undergo thermal treatment such as cooking or baking because this would induce the Maillard reaction, resulting in the destruction of the primary amino groups of chitosan and thereby reducing its functionality as a dietary fiber [1,102]. Chitosan preparations that can be taken orally are now available and are targeted at hypercholesterolemic overweight individuals. The most effective preparations contain chitosan molecules with a relatively low MW (~8000 Da) [4].

Despite uncertainty regarding its weight-reducing capabilities, chitosan is frequently sold in tablet form in health food stores and on the Internet as a weight-loss supplement with claims that it can eliminate fat from the digestive system and carry it out of the body (Table 6.3) [16,75]. Other substances, including appetite suppressants, amino acids, stimulants, or carnitine are often added to these chitosan products [6]. The effectiveness of chitosan supplements may be influenced by the timing of ingestion prior to meals, as it is important to allow the tablets to dissolve and become dispersed in the stomach. According to Subasinghe [16], the weight-loss supplements are advised to be taken 5–10 min prior to consumption of a meal, with an adequate amount of water to disperse the chitosan for maximum adsorption of fats, bile acids, and neutral sterols. However, it has been suggested that chitosan intake too close to mealtime will not give the compound time to solubilize and individuals should therefore ingest the tablets 30–60 min before mealtimes [59,103]. Since different forms of chitosan show varying solubility, the optimal ingestion time will also vary depending on the properties of specific chitosan supplements. Individuals on a low to medium fat diet are generally advised

TABLE 6.3
Some Examples of Chitosan Nutraceutical Products on the Market

Product Name	Manufacturer/Distributor	Health Benefits Claimed	Recommended Dosage	Amount of Chitosan
Chitosan Gold Tea	Kitto Life	Decreased cholesterol, improved liver function, cancer prevention, antibacterial activity	1–2 teabags, 3–4 × per day	4.4 g per tea bag
KL Chitosan Oligosaccharide 100	Kitto Life	Immune enhancement, anticancer and antibacterial activity, cholesterol and blood sugar control, acceleration of calcium absorption	3 capsules 3 × per day	250 mg per capsule
ChitoClear®	Primex	Dietary fiber, antioxidant, antimicrobial agent, edible film	Marketed as a GRAS-approved food ingredient	N/A
Chitosan HD™ (high density)/LipoSan Ultra	Whole Health/Primex	Decreased cholesterol, binds/absorbs fat and lipids at three to five times the rate of regular chitosan	4 capsules per day	500 mg per capsule
Chitosan Plus	Doctor's Trust	Weight management/loss, cholesterol reduction	2 capsules 30 min before lunch and dinner	250 mg per capsule
Chitosan Plus	Universal Nutrition	Fat blocker, reduction in LDL and VLDL cholesterol levels	1–2 capsules before each meal	500 mg per capsule
Diet Chitosan	Source Naturals	Reduced absorption of dietary fats, maintains healthy serum cholesterol levels	4 capsules 30 min before lunch and dinner	250 mg per capsule
FBlock	Absolute Nutrition	Fat trimming diet formula	2 capsules 15 min before meals	500 mg per capsule
Chitosan Diet Formula	Optimum Nutrition	Weight-loss aid	2 capsules 30 min before meals	500 mg per capsule
Chit-O-Power	Chitopower	Immune-enhancer, minimizes effects of aging, prevents illness, controls biorhythm	N/A	100 mg, 250 mg, or 500 mg capsules

to take 2–4 capsules (250–500 mg) at a time, while those on a high fat diet are advised to take 6–8 capsules (750–1000 mg) [16].

6.4.9 Challenges to the Industry: Side Effects and Production Costs

In general, short-term supplementation with chitosan has proven to be safe with no major side effects being reported in clinical trials [6,7]. However, safety trials in mice have indicated that chitosan may induce a number of biological and physiological changes, including alterations in the normal morphology of macrophages and in the normal flora of the intestine, decreased body weights, and physical inactivity [104]. In humans, chitosan ingestion over the course of two weeks has been shown to alter the normal fecal microbiota and decrease production of secondary metabolites by intestinal flora [97,105]. This ability of chitosan to interfere with microflora activity is most likely a result of the antimicrobial properties of chitosan. Based on these observations, it was suggested that long-term clinical use of chitosan and chitin should be carried out with special care. Also, since the major source of chitin and chitosan products is shellfish, these nutraceuticals cannot be recommended for people who are allergic to crustaceans [7].

Another side effect of chitin and chitosan supplementation may be interference with the normal absorption of valuable dietary components. It has been proposed that, in addition to entrapping lipids and sterols, the chitosan networks in the gastrointestinal system also entrap important fat-soluble vitamins and minerals, such as vitamins A, D, E, and K, along with carotenoids such as beta-carotene, lutein, and zeaxanthin [106]. It has also been suggested that dietary chitosan supplements reduce calcium availability by speeding its metabolism and urinary excretion. According to Koide [106], reported side effects of chitosan supplementation include reductions in bone mineral content, plasma vitamin E levels, and growth. In one study, massive and continuous intake of chitosan in rats led to decreases in the levels of serum vitamin E along with reduced absorption of minerals, such as calcium, iron, and magnesium [107]. The rats were administered chitosan as 5% of their diet (wt./wt.), which is approximately eight times the recommended dosage in humans. It was also reported that supplementation with sodium ascorbate prevented the observed reductions in iron and magnesium and improved calcium absorption.

Owing to possibly harmful deficiencies in humans, it was recommended that individuals taking chitosan supplements on a regular basis also take calcium and vitamin D and E supplements, and children should be periodically monitored for healthy growth and vitamin or mineral deficiencies [106]. Also, because of possible decreases in vitamin D and calcium absorption, consumption of chitin or chitosan by pregnant women was deemed to be risky.

More recently, intake of chitosan was reported to not to interfere with absorption of the vitamins A, D, and E, and beta-carotene, and vitamin K levels were actually found to increase significantly following 4 weeks of chitosan supplementation [6,77,95]. Intestinal absorption of mineral salts and trace metals such as iron and zinc was also not affected by administration of chitosan in a number of studies lasting from 4 to 8 weeks [69–71,73,77,95]. These studies were conducted in humans at much lower chitosan dosages than studies reporting vitamin and mineral deficiencies in rats. To verify the safety of chitosan as a long-term dietary supplement, there is a need for extended clinical studies that monitor vitamin and mineral levels during chitosan treatment.

The most common side effect of chitosan supplementation is gastrointestinal discomfort, most likely due to the fermentation of fat in the large intestine. Adverse events include flatulence, bloating, constipation, nausea, and diarrhea [99,106]. According to Ylitalo et al. [7], around 2.6–5.4% of subjects in human chitosan trials have reported mild or transitory nausea and constipation. In an effort to counter some of these uncomfortable side effects while at the same time increasing the lipid-lowering ability of chitosan, Geremias et al. [108] studied the addition of either *Aloe vera* L. or *Brassica olearaceae* L. to chitosan supplements. *A. vera* L. was chosen because of its laxative and hypolipidemic properties, while *B. olearaceae* was chosen for its hypocholesterolemic activity and association with reduced risk for coronary disease. Rats were fed a diet supplemented with either hydrosoluble chitosan (HC), HC + *A. vera*, or HC + *B. olearaceae* for 35 days. Although none of the supplements resulted in changes in food or water consumption or weight gain, they all reduced blood lipid levels, LDL cholesterol levels (with reductions ranging from 78 to 95%), and blood glucose levels. Also, rats fed the diet with HC + *A. vera* showed a significant increase in HDL cholesterol (by 42%) and rats fed the diet with HC + *B. olearaceae* showed significant reductions in levels of very low-density lipoprotein (VLDL) and triglyceride in the blood serum. Based on the results, the authors concluded that the supplement containing HC + *B. olearaceae* was the most effective of the three treatments.

In addition to overcoming some of the minor but unpleasant side effects of chitin and chitosan supplementation, the industry is looking toward developing more economical methods for extraction of these compounds from marine crustaceans. Currently, the costs of production are relatively high, extraction yields are often low (~3–5% of raw material), and seasonal variations in crustacean harvesting give way to a limited and inconsistent supply. Additionally, the alkali deproteinization step in the isolation of chitin creates excess waste streams [4,109]. In an attempt to overcome some of these processing drawbacks, a new technology involving fermentation of shrimp waste with the lactic acid bacteria *Lactobacillus plantarum* has been developed [110]. This method allows for the isolation of solid chitin along with a liquid mixture of shrimp amino acids, minerals, and pigments. The use of lactic acid bacteria offers an exciting alternative to the more harsh chemical methods normally used in the extraction of chitin from crustacean discards.

6.5 CONCLUSIONS

Chitin and chitosan are primarily extracted from the exoskeletons of marine crustaceans and shellfish, including shrimp, oysters, and squid. Owing to its high MW and degree of acetylation, chitin is largely insoluble and is generally deacetylated to form a variety of derivatives that have greater applications in nutraceuticals. Chitosan is a deacetylated form of chitin that interacts with a number of biomolecules, including fatty acids, bile acids, neutral lipids, proteins, and anionic polysaccharides. Chitosan is marketed globally as a nutraceutical and is often sold as a weight-loss or cholesterol-reducing supplement. Chitin and chitosan have been reported to act as dietary fibers and to contribute to reductions in weight and cholesterol by blocking absorption of compounds such as fats and bile acids. However, these claims have been met with criticism, and a number of meta-analyses have determined that the effect is not clinically significant, especially in regard to weight loss. In recent years, chitin and chitosan have been found to possess a variety of properties that give them the potential to be highly effective nutraceuticals. Reported health benefits include accelerated renal disease recovery and antibacterial, anti-inflammatory, antioxidant, anticarcinogenic, and antiulcer activities. Although chitin and chitosan have yet to be fully exploited for their nutraceutical properties, there is much potential for growth and development in this field.

REFERENCES

1. Alasalvar, C., Shahidi, F. and Quantick, P. 2002. Food and health applications of marine nutraceuticals: a review. In: Alasalvar C, Taylor T. Seafoods: Quality, Technology, and Nutraceutical Applications. Berlin: Springer-Verlag. pp. 175–204.
2. Kurita, K. 2006. Chitin and chitosan: functional biopolymers from marine crustaceans. *Mar. Biotechnol.* 8(3), 203–226.
3. Agullo, E., Rodriguez, M.S., Ramos, V. and Albertengo, L. 2003. Present and future role of chitin and chitosan in food. *Macromol. Biosci.* 3, 521–530.
4. Synowiecki, J. and Al-Khateeb, N.A. 2003. Production, properties, and some new applications of chitin and its derivatives. *Crit. Rev. Food Sci. Nutr.* 43(2), 145–171.
5. Shahidi, F. and Abuzaytoun, R. 2005. Chitin, chitosan, and co-products: chemistry, production, applications, and health effects. *Adv. Food Nutr. Res.* 49, 93–135.
6. Novak, K., Cupp, M.J. and Tracy, T.S. 2003. Chitosan. In: Cupp MJ, Tracy TS. Dietary Supplements: Toxicology and Clinical Pharmacology. Totowa, NJ: Humana Press. pp. 33–40.
7. Ylitalo, R., Lehtinen, S., Wuolijoki, E., Ylitalo, P. and Lehtimaki, T. 2002. Cholesterol-lowering properties and safety of chitosan. *Arzneimittel-Forschung* 52(1), 1–7.
8. Rasmussen, R.S. and Morrissey, M.T. 2007. Marine biotechnology for production of food ingredients. *Adv. Food Nutri. Res.* 52 (May), Chapter 5, 237–292.
9. Johnson, H.M. 2005. 2005 Annual Report on the United States Seafood Industry. 13th ed. Jacksonville, OR: H.M. Johnson & Associates. p. 103.
10. Haard, N.F., Simpson, B.K. and Sikorski, Z.E. 1994. Biotechnological applications of seafood proteins and other nitrogenous compounds. In: Sikorski ZE, Pan BS, Shahidi F. Seafood Proteins. New York: Chapman and Hall. pp. 194–216.
11. No, H.K. and Meyers, S.O. 1997. Chitin composition. In: Muzzarelli RAA, Peter MG. Chitin Handbook. Grottammare: European Chitin Society. pp. 475–489.
12. Khor, E. 2001. The sources and production of chitin. In: Khor E. Chitin: Fulfilling a Biomaterials Promise. 1st ed. Oxford: Elsevier Science. pp. 63–72.
13. Tharanathan, R.N. and Kittur, F.S. 2003. Chitin—the undisputed biomolecule of great potential. *Crit. Rev. Food Sci. Nutr.* 43(1), 61–87.
14. Illum, L. 1998. Chitosan and its use as a pharmaceutical excipient. *Pharm. Res.* 15, 1326–1331.

15. Shahidi, F., Kamil, J.Y.V.A. and Jeon, Y.J. 1999. Food applications of chitin and chitosans. *Trends Food Sci. Technol.* 10, 37–51.
16. Subasinghe, S. 1999. Chitin from shellfish waste: health benefits overshadow industrial uses! *INFOFISH Int.* 3, 58–65.
17. Lim, S.H. and Hudson, S.M. 2003. Review of chitosan and its derivatives as antimicrobial agents and their uses as textile chemicals. *J. Macromol. Sci.* C43, 223–269.
18. Dijsselbloem, N., Vanden Berghe, W., De Naeyer, A. and Haegeman, G. 2004. Soy isoflavone phyto-pharmaceuticals in interleukin-6 affections. Multi-purpose nutraceuticals at the crossroad of hormone replacement, anti-cancer and anti-inflammatory therapy. *Biochem. Pharmacol.* 68, 1171–1185.
19. Goggs, R., Vaughan-Thomas, A., Clegg, P.D., Carter, S.D., Innes, J.F., Mobasheri, A., Shakibaei, M., Schwab, W. and Bondy, C.A. 2005. Nutraceutical therapies for degenerative joint diseases: a critical review. *Crit. Rev. Food Sci. Nutr.* 45, 145–164.
20. Srinivasan, K. 2005. Role of spices beyond food flavoring: nutraceuticals with multiple health effects. *Food Rev. Int.* 21(2), 167–188.
21. Cheeke, P., Piacente, S. and Oleszek, W. 2006. Anti-inflammatory and anti-arthritic effects of yucca schidigera: a review. *J. Inflamm.* 3, 6.
22. Hall, G., Lund, L., Lamb, J.R. and Jarman, E.R. 2002. Kinetics and mode of peptide delivery via the respiratory mucosa determine the outcome of activation versus TH2 immunity in allergic inflammation of the airways. *J. Allerg. Clin. Immunol.* 110, 883–890.
23. Kim, M.S., Sung, M.J., Seo, S.B., Yoo, S.J., Lim, W.K. and Kim, H.M. 2002. Water-soluble chitosan inhibits the production of pro-inflammatory cytokine in human astrocytoma cells activated by amyloid beta peptide and interleukin-1beta. *Neurosci. Lett.* 321, 105–109.
24. Kim, M.S., You, H.J., You, M.K., Kim, N.S., Shim, B.S. and Kim, H.M. 2004. Inhibitory effect of water-soluble chitosan on TNF-alpha and IL-8 secretion from HMC-1 cells. *Immunopharmacol. Immunotoxicol.* 26, 401–409.
25. Theoharides, T.C. and Cochrane, D.E. 2004. Critical role of mast cells in inflammatory diseases and the effect of acute stress. *J. Neuroimmunol.* 146(1–2), 1–12.
26. Nishimoto, N. and Kishimoto, T. 2004. Inhibition of IL-6 for the treatment of inflammatory diseases. *Curr. Opin. Pharmacol.* 4, 386–391.
27. Feldmann, M. and Maini, R.N. 2003. Lasker clinical medical research award. TNF defined as a therapeutic target for rheumatoid arthritis and other autoimmune diseases. *Nat. Med.* 9, 1245–1250.
28. Lee, J., Koo, N. and Min, D.B. 2004. Reactive oxygen species, aging, and antioxidative nutraceuticals. *Compr. Rev. Food Sci. Food Saf.* 3, 21–33.
29. Kamil, J.Y.V.A., Jeon, Y.J. and Shahidi, F. 2000. Control of oxidation of cooked herring by chitosan. Paper presented at the Annual Meeting of the Institute of Food Technologists, Dallas, TX, book of abstracts, p. 95.
30. Mitani, T., Moriyama, A. and Ishii, H. 1992. Heavy metal uptake by swollen chitosan beads. *Biosci. Biotechnol. Biochem.* 56, 985.
31. Xing, R., Liu, S., Guo, Z., Yu, H., Wang, P., Li, C., Li, Z. and Li, P. 2005. Relevance of molecular weight of chitosan and its derivatives and their antioxidant activities in vitro. *Bioorganic Med. Chem.* 13(5), 1573–1577.
32. Nishimura, Y., Kim, H.S., Ikota, N., Arima, H., Bom, H.S., Kim, Y.H., Watanabe, Y., Yukawa, M. and Ozawa, T. 2003. Radioprotective effect of chitosan in sub-lethally X-ray irradiated mice. *J. Radiat. Res.* 44, 53–58.
33. Liu, R.H. 2003. Health benefits of fruit and vegetables are from additive and synergistic combinations of phytochemicals. *Am. J. Clin. Nutr.* 78, 517S–520S.

34. Ferrari, C.K. 2004. Functional foods, herbs and nutraceuticals: towards biochemical mechanisms of healthy aging. *Biogerontology* 5, 275–289.
35. Karakaya, S. and Kavas, A. 1999. Adsorption of direct-acting and indirect-acting mutagens by various dietary fibers. *Int. J. Food Sci. Nutr.* 50, 319–323.
36. Tsukada, K., Matsumoto, T., Aizawa, K., Tokoro, A., Naruse, R., Suzuki, S. and Suzuki, M. 1990. Antimetastatic and growth-inhibitory effects of *N*-acetylchitohexaose in mice bearing Lewis lung carcinoma. *Jpn. J. Cancer Res.* 81, 259–265.
37. Pae, H.O., Seo, W.G., Kim, N.Y., Oh, G.S., Kim, G.E., Kim, Y.H., Kwak, H.J., Yun, Y.G., Jun, C.D. and Chung, H.T. 2001. Induction of granulocytic differentiation in acute promyelocytic leukemia cells (HL-60) by water-soluble chitosan oligomer. *Leuk. Res.* 25(4), 339–346.
38. Yoon, H.J., Park, H.S., Bom, H.S., Roh, Y.B., Kim, J.S. and Kim, Y.H. 2005. Chitosan oligosaccharide inhibits $^{203}HgCl_2$-induced genotoxicity in mice: micronuclei occurrence and chromosomal aberration. *Arch. Pharm. Res.* 28, 1079–1085.
39. Ohe, T. 1996. Antigenotoxic activities of chitin and chitosan as assayed by sister chromatid exchange. *Sci. Total Environ.* 181, 1–5.
40. Anandan, R., Nair, P.G. and Mathew, S. 2004. Anti-ulcerogenic effect of chitin and chitosan on mucosal antioxidant defence system in HCl-ethanol-induced ulcer in rats. *J. Pharm. Pharmacol.* 56(2), 265–269.
41. Khor, E. 2001. Chitin biomedical applications. In: Khor E. Chitin: Fulfilling a Biomaterials Promise. 1st ed. Oxford: Elsevier Science. pp. 9–44.
42. Yanagawa, A., Hukumura, T., Shimada, J., Mizushima, Y., Kumagaya, M. and Nakajima, K. 1994. The new application of gastric protective agents to wound healing. *Jpn. J. Inflamm.* 14, 43–51.
43. Ito, M., Ban, A. and Ishihara, M. 2000. Anti-ulcer effects of chitin and chitosan, healthy foods, in rats. *Jpn. J. Pharmacol.* 82(3), 218–225.
44. Levey, A.S., Coresh, J., Balk, E., Kausz, A.T., Levin, A., Steffes, M.W., Hogg, R.J., Perrone, R.D., Lau, J. and Eknoyan, G. 2003. National Kidney Foundation practice guidelines for chronic kidney disease: evaluation, classification, and stratification. *Ann. Intern. Med.* 139, 137–147.
45. Jing, S.B., Li, L., Ji, D., Takiguchi, Y. and Yamaguchi, T. 1997. Effect of chitosan on renal function in patients with chronic renal failure. *J. Pharm. Pharmacol.* 49(7), 721–723.
46. Vanholder, R., De Smet, R., Glorieux, G., Argiles, A., Baurmeister, U., Brunet, P., Clark, W., Cohen, G., De Deyn, P.P., Deppisch, R., Descamps-Latscha, B., Henle, T., Jorres, A., Lemke, H.D., Massy, Z.A., Passlick-Deetjen, J., Rodriguez, M., Stegmayr, B., Stenvinkel, P., Tetta, C., Wanner, C. and Zidek, W. 2003. Review on uremic toxins: classification, concentration, and interindividual variability. *Kidney Int.* 63, 1934–1943.
47. Muzzarelli, R.A.A. and De Vincenzi, M. 1997. Chitosans as dietary food additives. In: Goosen MFA. Applications of Chitin and Chitosan. Lancaster, PA: Technomic Publishing Company. pp. 115–127.
48. Koguchi, T., Nakajima, H., Wada, M., Yamamoto, Y., Innami, S., Maekawa, A. and Tadokor, T. 2002. Dietary fiber suppresses elevations of uric acid and allantoin in serum and urine induced by dietary RNA and increases its excretion to feces in rats. *J. Nutr. Sci. Vitaminol.* 48, 184–193.
49. Koguchi, T., Koguchi, H., Nakajima, H., Takano, S., Yamamoto, Y., Innami, S., Maekawa, A. and Tadokoro, T. 2004. Dietary fiber suppresses elevation of uric acid and urea nitrogen concentrations in serum of rats with renal dysfunction induced by dietary adenine. *Int. J. Vitam. Nutr. Res.* 74, 253–263.

50. Zhou, Y.G., Yang, Y.D., Guo, X.M. and Chen, G.R. 2003. Effect of molecular weight and degree of deacetylation of chitosan on urea adsorption properties of copper chitosan. *J. Appl. Polym. Sci.* 89, 1520–1523.
51. Yoon, H.J., Kim, Y.H., Park, S.W., Lee, H.B. and Park, H.S. 2003. Chitosan increases the release of renal dipeptidase from porcine renal proximal tubule cells. *Korean J. Biol. Sci.* 7, 309–315.
52. Albertengo, L. 2002. Quitina y Quitosano el el Tracto Digestivo Humano. Interaccion con los Macronutrientes. Tesis Doctoral, Universidad nacional del Sur., Argentina.
53. Zahorska-Markiewicz, B., Krotkiewski, M., Olszanecka-Glinianowicz, M. and Zurakowski, A. 2002. Effect of chitosan in complex management of obesity. *Polish Med. J.* 13(74), 129–132.
54. Jafar, T.H., Stark, P.C., Schmid, C.H., Landa, M., Maschio, G., de Jong, P.E., de Zeeuw, D., Shahinfar, S., Toto, R. and Levey, A.S. 2003. Progression of chronic kidney disease: the role of blood pressure control, proteinuria, and angiotensin-converting enzyme inhibition: a patient-level meta-analysis. *Ann. Intern. Med.* 139, 244–252.
55. van Bennekum, A.M., Nguyen, D.V., Schulthess, G., Hauser, H. and Phillips, M.C. 2005. Mechanisms of cholesterol-lowering effects of dietary insoluble fibres: relationships with intestinal and hepatic cholesterol parameters. *Brit. J. Nutr.* 94, 331–337.
56. Anderson, J.W., Smith, B.M. and Gustafson, N.J. 1994. Health benefits and practical aspects of high-fiber diets. *Am. J. Clin. Nutr.* 59, 1242S–1247S.
57. Hu, F.B., van Dam, R.M. and Liu, S. 2001. Diet and risk of type II diabetes: the role of types of fat and carbohydrate. *Diabetologia* 44, 805–817.
58. Burton-Freeman, B. 2000. Dietary fiber and energy regulation. *J. Nutr.* 130, 272S–275S.
59. Schiller, R.N., Barrager, E., Schauss, A.G. and Nichols, E.J. 2001. A randomized, double-blind, placebo-controlled study examining the effects of a rapidly soluble chitosan dietary supplement on weight loss and body composition in overweight and mildly obese individuals. *J. Am. Nutraceut. Assoc.* 4, 42–49.
60. Pittler, M.H. and Ernst, E. 2005. Complementary therapies for reducing body weight: a systematic review. *Int. J. Obesity* 29, 1030–1038.
61. Lee, J.K., Kim, S.U. and Kim, J.H. 1999. Modification of chitosan to improve its hypocholesterolemic capacity. *Biosci. Biotechnol. Biochem.* 63, 833–839.
62. Anderson, J.W., Allgood, L.D., Lawrence, A., Altringer, L.A., Jerdack, G.R., Hengehold, D.A. and Morel, J.G. 2000. Cholesterol-lowering effects of psyllium intake adjunctive to diet therapy in men and women with hypercholesterolemia: meta-analysis of 8 controlled trials. *Am. J. Clin. Nutr.* 71, 472–479.
63. Mhurchu, C.N., Dunshea-Mooij, C., Bennett, D. and Rodgers, A. 2005. Effect of chitosan on weight loss in overweight and obese individuals: a systematic review of randomized controlled trials. *Obesity Rev.* 6, 35–42.
64. Nagyvary, J.J., Falk, J.D., Hill, M.L., Schmidt, M.L., Wilkins, A.K. and Bradbury, E.L. 1979. The hypolipidemic activity of chitosan and other polysaccharides in rats. *Nutr. Rep. Int.* 20, 677–684.
65. Nauss, J.L., Thompson, J.L. and Nagyvary, J. 1983. The binding of micellar lipids to chitosan. *Lipids* 18, 714–719.
66. Vahouny, G.V., Satchithanandam, S., Cassidy, M.M., Lightfoot, F.B. and Furda, I. 1983. Comparative effects of chitosan and cholestyramine on lymphatic absorption of lipids in the rat. *Am. J. Clin. Nutr.* 38, 278–284.

67. Kanauchi, O., Deuchi, K., Imasato, Y., Shizukuishi, M. and Kobayashi, E. 1995. Mechanism for the inhibition of fat digestion by chitosan and for the synergistic effect of ascorbate. *Biosci. Biotechnol. Biochem.* 59, 786–790.
68. Gallaher, C.M., Munion, J., Hesslink, R., Jr., Wise, J. and Gallaher, D.D. 2000. Cholesterol reduction by glucomannan and chitosan is mediated by changes in cholesterol absorption and bile acid and fat excretion in rats. *J. Nutr.* 130, 2753–2759.
69. Giustina, A. and Ventura, P. 1995. Weight-reducing regimens in obese subjects: effects of a new dietary fiber integrator. *Acta. Toxicol. Ther.* 16, 199–214.
70. Sciutto, A.M. and Colombo, P. 1995. Lipid-lowering effect of chitosan dietary integrator and hypocaloric diet in obese subjects. *Acta. Toxicol. Ther.* 16, 215–230.
71. Colombo, P. and Sciutto, A.M. 1996. Nutritional aspects of chitosan employment in hypocaloric diet. *Acta. Toxicol. Ther.* 17, 278–302.
72. Macchi, G. 1996. A new approach to treatment of obesity: chitosan's effects on body weight reduction and plasma cholesterol's levels. *Acta. Toxicol. Ther.* 17, 303–320.
73. Veneroni, G., Veneroni, F., Contos, S., Tripodi, S., De Bernardi, M., Guarino, C. and Marletta, M. 1996. Effects of a new chitosan on hyperlipidemia and overweight in obese patients. In: Muzzarelli RAA. Chitin Enzymology. Ancona, Italy: Atec Edizioni. pp. 63–67.
74. Ventura, P. 1996. Lipid lowering activity of chitosan, a new dietary integrator. In: Muzzarelli RAA. Chitin Enzymology. Ancona, Italy: Atec Edizioni. pp. 55–62.
75. Ohr, L.M. 2005. Riding the nutraceuticals wave. *Food Technol.* 59, 95–96.
76. Sumiyoshi, M. and Kimura, Y. 2006. Low molecular weight chitosan inhibits obesity induced by feeding a high-fat diet long-term in mice. *J. Pharm. Pharmacol.* 58(2), 201–207.
77. Pittler, M.H., Abbot, N.C., Harkness, E.F. and Ernst, E. 1999. Randomized, double-blind trial of chitosan for body weight reduction. *Eur. J. Clin. Nutr.* 53(5), 379–381.
78. Ho, S.C., Tai, E.S., Eng, P.H., Tan, C.E. and Fok, A.C. 2001. In the absence of dietary surveillance, chitosan does not reduce plasma lipids or obesity in hypercholesterolaemic obese Asian subjects. *Singapore Med. J.* 42, 6–10.
79. Bokura, H. and Kobayashi, S. 2003. Chitosan decreases total cholesterol in women: a randomized, double-blind, placebo-controlled trial. *Eur. J. Clin. Nutr.* 57, 721–725
80. Mhurchu, C.N., Poppitt, S.D., McGill, A.T., Leahy, F.E., Bennett, D.A., Lin, R.B., Ormrod, D., Ward, L., Strik, C. and Rodgers, A. 2004. The effect of the dietary supplement, chitosan, on body weight: a randomised controlled trial in 250 overweight and obese adults. *Int. J. Obesity* 28, 1149–1156.
81. Lenz, T.L. and Hamilton, W.R. 2004. Supplemental products used for weight loss. *J. Am. Pharm. Assoc.* 44, 59–68.
82. Ernst, E. and Pittler, M.H. 1998. Chitosan as a treatment for body weight reduction? A meta-analysis. *Perfusion* 11, 461–465.
83. Pittler, M.H. and Ernst, E. 2004. Dietary supplements for body-weight reduction: a systematic review. *Am. J. Clin. Nutr.* 79, 529–536.
84. Sugano, M., Fujikawa, T., Hiratsuji, Y. and Hasegawa, Y. 1978. Hypocholesterolemic effects of chitosan in cholesterol-fed rats. *Nutr. Rep. Int.* 18, 531–537.
85. Kobayashi, T., Otsuka, S. and Vugari, Y. 1979. Effect of chitosan on serum and liver cholesterol levels in cholesterol-fed rats. *Nutr. Rep. Int.* 19, 327–334.
86. Sugano, M., Fujiwara, T., Hiratsuji, Y., Nakashima, K., Fukuda, N. and Hasegawa, Y. 1980. A novel use of chitosan as a hypocholesterolemic agent in rats. *Am. J. Clin. Nutr.* 33, 787–793.
87. Jennings, C.D., Boleyn, K., Bridges, S.R., Wood, P.J. and Anderson, J.W. 1988. A comparison of the lipid-lowering and intestinal morphological effects of cholestyramine, chitosan, and oat gum in rats. *Prod. Soc. Exp. Biol. Med.* 189, 13–20.

88. Sugano, M., Watanabe, S., Kishi, A., Izumi, M. and Ohtakara, A. 1988. Hypocholesterolemication of chitosans with different viscosity in rats. *Lipids* 23, 187–191.
89. Zacour, A.C., Silva, M.E., Cecon, P.R., Bambirra, E.A. and Vieira, E.C. 1992. Effect of dietary chitosan on cholesterol absorption and metabolism in rats. *J. Nutr. Sci. Vitaminol.* 38, 609–661.
90. Trautwein, E.A., Jurgensen, U. and Erbersdobler, H.F. 1997. Cholesterol-lowering and gallstone-preventing action of chitosans with different degrees of deacetylation in hamsters fed cholesterol-rich diets. *Nutr. Res.* 17, 1053–1065.
91. Ebihara, K. and Schneeman, B.O. 1989. Interaction of bile acids, phospholipids, cholesterol and triglyceride with dietary fibers in the small intestine of rats. *J. Nutr.* 119, 1100–1106.
92. Hirano, S., Akiyama, Y., Ogura, M. and Ayaki, Y. 1992. The regulation of serum cholesterol level by oral administration of chitosan in rabbits. In: Tokura S, Azuma I. Chitin Derivatives in Life Science. Sapporo, Japan: Japanese Chitin Society. pp. 115–120.
93. Fukada, Y., Kimura, K. and Ayaki, Y. 1993. Effect of chitosan feeding on intestinal bile acid metabolism in rats. *Lipids* 26, 395–399.
94. Ormrod, D.J., Holmes, C.C. and Miller, T.E. 1998. Dietary chitosan inhibits hypercholesterolaemia and atherogenesis in the apolipoprotein E-deficient mouse model of atherosclerosis. *Atherosclerosis* 138, 329–334.
95. Muzzarelli, R.A. 1999. Clinical and biochemical evaluation of chitosan for hypercholesterolemia and overweight control. *EXS* 87, 293–304.
96. Hughes, K. 2002. Chitosan and dietary fibers. *Prepared Foods* July, NS11–NS14.
97. Maezaki, Y., Tsuji, K. and Nakagawa, Y. 1993. Hypocholesterolemic effect of chitosan in adult males. *Biosci. Biotechnol. Biochem.* 579, 1439–1444.
98. Wuolijoki, E., Hirvela, T. and Ylitalo, P. 1999. Decrease in serum LDL cholesterol with microcrystalline chitosan. *Meth. Find Exp. Clin. Pharmac.* 21(5), 357–361.
99. Metso, S., Ylitalo, R., Nikkila, M., Wuolijoki, E., Ylitalo, P. and Lehtimaki, T. 2003. The effect of long-term microcrystalline chitosan therapy on plasma lipids and glucose concentrations in subjects with increased plasma total cholesterol: a randomised placebo-controlled double-blind crossover trial in healthy men and women. *Eur. J. Clin. Pharmacol.* 59, 741–746.
100. Lehtimaki, T., Metso, S., Ylitalo, R., Rontu, R., Nikkila, M., Wuolijoki, E. and Ylitalo, P. 2005. Microcrystalline chitosan is ineffective to decrease plasma lipids in both apolipoprotein E epsilon 4 carriers and non-carriers: a long-term placebo-controlled trial in hypercholesterolaemic volunteers. *Basic Clin. Pharmacol. Toxicol.* 97, 98–103.
101. Gallaher, D.D., Gallaher, C.M., Mahrt, G.J., Carr, T.P., Hollingshead, C.H., Hesslink, R., Jr. and Wise, J. 2002. A glucomannan and chitosan fiber supplement decreases plasma cholesterol and increases cholesterol excretion in overweight normocholesterolemic humans. *J. Am. Coll. Nutr.* 21, 428–433.
102. Tanaka, M., Huang, J.R., Chiu, W.K., Ishizaki, S. and Taguchi, T. 1993. Effect of the Maillard reaction on functional properties of chitosan. *Nippon Suisan Gakkaishi* 59, 1915–1921.
103. Kanauchi, O., Deuchi, K., Imasato, Y. and Kobayashi, E. 1994. Increasing effect of a chitosan and ascorbic acid mixture on fecal dietary fat excretion. *Biosci. Biotechnol. Biochem.* 58, 1617–1620.
104. Tanaka, Y., Tanioka, S., Tanaka, M., Tanigawa, T., Kitamura, Y., Minami, S., Okamoto, Y., Miyashita, M. and Nanno, M. 1997. Effects of chitin and chitosan particles on BALB/c mice by oral and parenteral administration. *Biomaterials* 18(8), 591–595.

105. Terada, A., Hara, H., Sato, D., Higashi, T., Nakayama, S., Tsuji, K., Sakamoto, K., Ishioka, E., Maezaki, Y., Tsugita, T., Takekawa, T. and Mitsuoka, T. 1995. Effect of dietary chitosan on faecal microbiota and faecal metabolites of humans. *Microb. Ecol. Health Dis.* 8(1), 15–21.
106. Koide, S.S. 1998. Chitin-chitosan: properties, benefits and risks. *Nutr. Res.* 18, 1091–1101.
107. Deuchi, K., Kanauchi, O., Shizukuishi, M. and Kobayashi, E. 1995. Continuous and massive intake of chitosan affects mineral and fat-soluble vitamin status in rats fed on a high-fat diet. *Biosci. Biotechnol. Biochem.* 59, 1211–1216.
108. Geremias, R., Pedrosa, R.C., Locatelli, C., de Favere, V.T., Coury-Pedrosa, R. and Laranjeira, M.C.M. 2006. Lipid lowering activity of hydrosoluble chitosan and association with *Aloe vera* L. and *Brassica olearaceae* L. *Phytother. Res.* 20, 288–293.
109. Ludlow, M.E. 2001. World applications of chitin and chitosan. Presented at the Annual Meeting of the Institute of Food Technologists (IFT). New Orleans, LA. Session 82, Food byproducts and waste utilization.
110. Rao, M.S. and Stevens, W.F. 2005. Chitin production by *Lactobacillus* fermentation of shrimp biowaste in a drum reactor and its chemical conversion to chitosan. *J. Chem. Technol. Biotechnol.* 80, 1080–1087.

7 Production of Bioactive Chitosan Oligosaccharides and Their Potential Use as Nutraceuticals

Se-Kwon Kim, Niranjan Rajapakse, and Fereidoon Shahidi

CONTENTS

7.1 INTRODUCTION

Chitosan and its derivatives have many interesting properties that make them attractive for a wide variety of applications in many fields such as food [1], cosmetics [2], biomedicine [3], agriculture, and wastewater management. Their antibacterial, antifungal, and antiviral properties make them particularly useful for biomedical applications, such as wound dressings, surgical sutures, and as aids in cataract surgery and periodontal disease treatment. Moreover, researches have shown that chitosan derivatives are nontoxic and nonallergenic; so the body is not likely to reject them as foreign invaders. Chitosan is commercially obtained by deacetylation of chitin, the second most abundant natural biopolymer

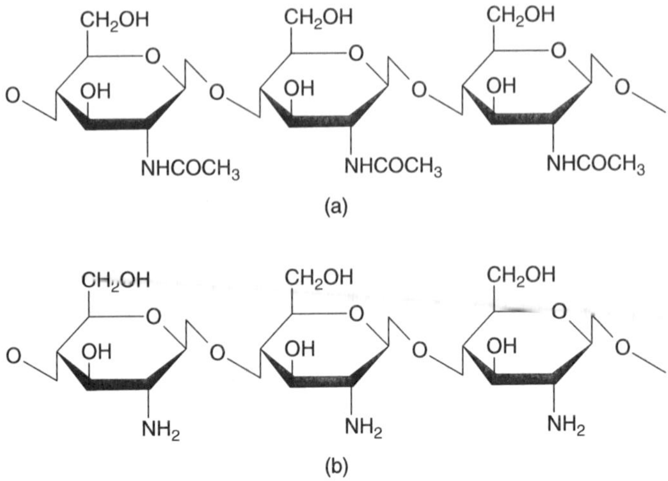

FIGURE 7.1 Chemical structures of (a) chitin and (b) chitosan.

on earth after cellulose. Chitin, a major structural component in the exoskeleton of crustaceans, insects, and cell walls of fungi is identified as a polymer of *N*-acetylglucosamine (β-1,4 linked 2-acetamido-D-glucosamine). Therefore, chitosan, the *N*-deacetylated form of chitin, can be named as a polymer of β-1,4 linked D-glucosamine units (Figure 7.1).

Normally chitin is water-insoluble, but chitosan can be solubilized in water at slightly acidic pH. Unlike chitin and chitosan, chitosan oligosaccharides (COSs), the hydrolyzed products of chitosan are readily soluble in water because of their shorter chain length and free amino groups of D-glucosamine units [4]. Therefore, over the past decade, researchers in Asia, Europe, and North America have tested COSs in biomedical applications. Chitosan-based researches have also been focused on the food and nutrition areas, including development of edible films and coatings to preserve the quality and texture of foods. In addition, numerous studies have also been undertaken to find out unraveled new properties of COSs and sequential developments were achieved in continuous production of COSs in large scale. This chapter provides an overview of the development of new continuous production methods of COSs by enzymatic means and some biological activities of COSs those may contribute to improve their value as nutraceuticals and functional food ingredients.

7.2 CHITOSAN OLIGOSACCHARIDES: A POTENTIAL SOURCE FOR NUTRACEUTICALS

Although chitin and chitosan are known to have important functional properties in many areas, their application as nutraceuticals is restricted because of poor absorption through the human intestine. This is mainly because chitinase- or chitosanase-like enzymes that are required to break down chitin and chitosan into smaller chain molecules for absorption are absent in the human digestive track. In addition, the insoluble nature and high viscosity of chitinous material at neutral pH act as major barriers for their absorption into the body. However, results of early studies on chitin and chitosan conducted *in vitro* and *in vivo* demonstrated that they are capable of binding dietary fats and thereby prevent fat absorption from the gut. Despite their beneficial effects to act as antihypertensive [5], hypocholesterolemic [6] and weight-loss materials, chelation of some metal ions from the dietary sources becomes unfavorable for human nutrition. However, there is a renewed interest to identify biological properties of COSs because they can be readily absorbed into the bloodstream. Observations have proven that COSs exert many favorable biological properties, including increase of body resistance to diseases [7,8] that makes them attractive for a wide variety of health applications. Owing to their health benefits, COSs have the potential to be used as physiological functional foods. Several production methods of COSs have already been developed and some methods are capable of producing safe and nontoxic COSs suitable for human consumption. Especially, the higher-grade COSs are highly purified and their safety has been verified scientifically through many tests for medical and food applications. Korea, China, and Japan have a long history for using chitosan and COSs for their health benefits. Currently, COSs are available in the market as different commercial products highlighting their beneficial effects. However, the high price and limited knowledge about the beneficial health effects of COSs may limit their widespread use among consumers.

The proven bioactivities and possibility of large-scale and safe production of COSs as well as availability of raw materials have stimulated the potential use of COSs as nutraceuticals. In large-scale production of COSs, exoskeletons of crab and shrimp are utilized as starting material. Every year a large volume of crustacean exoskeletons is discarded as processing waste. According to the fisheries statistics of Food and Agriculture Organization, the global crustacean harvest is increasing annually, and more than 40% of the catch is utilized for processing. Hence, a considerable amount of crustacean shells is discarded as by-products. Therefore, the raw material for production of COSs is readily available at low cost.

7.3 PRODUCTION OF BIOACTIVE CHITOSAN OLIGOSACCHARIDES

The degree of polymerization (DP) is crucial for the bioactivities of COSs and COSs with relatively high DP (five to seven D-glucosamine units) are favored in this regard [9,10]. Therefore, bioactivities of COSs relate to their structural features. For the production of COSs, both chemical and enzymatic hydrolysis can be employed. In the chemical method, COSs are produced by partial hydrolysis of chitosan with concentrated HCl. However, experimental results have shown that chemical hydrolysis produces low yields of COSs and a larger amount of monomeric D-glucosamine units. In addition, the COSs prepared by acid hydrolysis may not be suitable for human consumption because of possibility of production of toxic compounds during hydrolysis [11]. Therefore, enzymatic hydrolysis of chitosan for production of COSs has become the preferred method during the past few decades. Chitosanases obtained from microbes produce a relatively higher proportion of COSs from chitosan. Initially, enzymatic hydrolysis was carried out in batch reactors, where chitosanase was mixed with its substrate, and allowed to break down glycosidic bonds of chitosan under optimum conditions [12]. This batch method had some disadvantages such as low yields and higher cost associated with the use of large quantities of expensive chitosanase enzyme. Later, it was found that a number of different enzymes can also be used for the hydrolysis of chitosan. For instance, Lysozyme and chitinase can act on partially *N*-acetylated chitosan by recognizing *N*-acetylglucosamine residues in the chitosan sequence [13,14]. However, due to inefficient production of desirable chain lengths of COSs, these methods were not applicable for large-scale production of bioactive COSs, thus limiting the potential use of COSs as nutraceuticals.

In 1998, a new method for producing COSs with higher degree of polymerization by means of a column reactor packed with an immobilized enzyme was introduced [15]. However, in this method the yield of COSs was lower, because the immobilized enzyme showed a lower affinity and lower reaction rate than the free enzyme. Therefore, an enzyme reactor system together with an ultrafiltration (UF) membrane reactor was tested to produce COSs with higher degrees of polymerization and yield [16]. The optimum conditions for the production of relatively high proportions of COSs were determined by changing reaction temperature, incubation time, permeation rate, and amount of reducing sugar, which indicates the degree of hydrolysis. This reactor system could hydrolyze at least 11 batches of substrate for the same amount of enzyme used in the batch reactor and was effective for the production of relatively longer-chain oligosaccharides that have shown interesting biological activities in other studies. The most important factor in the UF reactor system was the control of permeation rate that determines the components of the resultant oligosaccharides. However, this method did not allow continuous production of COSs and resulted in increased transmembrane pressure, possibly because of the fouling of membrane by highly viscous chitosan and accumulation of substrate. Therefore, to improve this method to a more effective continuous production system, efforts were made to reduce chitosan viscosity prior to treatment in the UF membrane system.

The continuous production of COSs was found to be feasible with the combination of the above two methods in which a column reactor packed with immobilized enzyme was coupled to a UF membrane reactor and the new system was named as dual reactor system [17]. In this system, production of COSs may be performed in two steps (Figure 7.2). In the first step, chitosan is partially hydrolyzed by the immobilized enzyme packed in the column reactor and it is supplied to the enzyme reactor in UF membrane reactor system for the production of COSs. As expected, the low viscosity of partially hydrolyzed chitosan did not create fouling problems under the controlled conditions, and continuous production

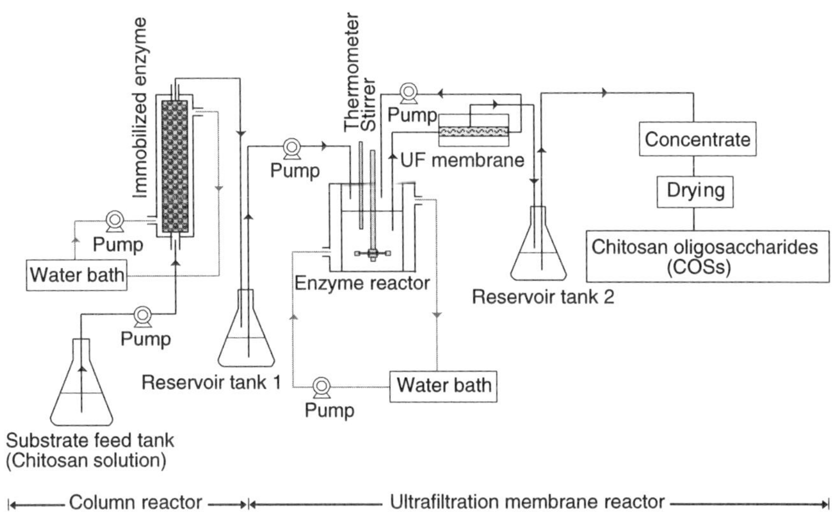

FIGURE 7.2 Schematic diagram of the dual reactor system developed for continuous production of chitosan oligosaccharides. (From Jeon, Y.J. and Kim, S.K., *Process Biochem.*, 35, 623–632, 2000. With permission.)

of COSs was achieved. A greater productivity per unit enzyme, ability to control molecular weight distribution, and more efficient continuous production process were obtained by utilizing the dual reactor compared to conventional methods. Therefore, this method is commonly used to produce different molecular size COSs to study their bioactivities.

In recent years, many researchers have attempted to utilize enzymatic methods to produce bioactive COSs. Currently, few methods are used to produce COSs with a particular molecular size distribution. In these methods complex enzyme systems have been used to produce COSs with a preferred degree of polymerization range [18]. Later, Kuroiwa et al. [19] reported optimum conditions for production of pentamers and hexamers of COSs using a packed-bed enzyme reactor. In addition to enzymatic methods, several other approaches such as chemoenzymatic synthesis of chitosan oligomers have also been identified to produce bioactive COSs [20]. Although COSs can be produced using different methods, enzymatic hydrolysis of chitosan is the most reliable and effective method to obtain bioactive COSs with higher purity, used for human consumption.

7.4 BIOLOGICAL ACTIVITIES OF CHITOSAN OLIGOSACCHARIDES

Low-molecular-weight water-soluble chitosan or COSs are known to have many desirable biological activities such as antifungal activity [21], antibacterial activity [17,22,23], antitumor activity [24,25], immuno-enhancing effects [26], and protective effects against infection [4]. The molecular weight or chain length, which is generally referred to as the degree of polymerization (DP) and the degree of acetylation (DA) are considered as principal characteristics of COSs related to their biological activities. Therefore, by altering substrate and enzyme conditions, desired properties of COSs that are necessary to exhibit different biological activities can be obtained. As described earlier [15–17], many researchers have demonstrated that enzymatic hydrolysis is useful for preparation of COSs as it gives greater yields of oligomers with higher DP than those in acid hydrolysis [11]. In this chapter, several biological activities of COSs prepared by enzymatic hydrolysis of differentially acetylated chitosans are discussed. These beneficial biological activities of COSs may encourage their utilization as functional foods as well as in improving food quality.

7.4.1 Antibacterial Activity

Chitosans are capable of inhibiting the growth of some microorganisms including several bacterial strains. Positively charged amino groups on chitosan are presumed to be the reason for this bioactivity. The cationic amino groups may form polyelectrolyte complexes with negatively charged carboxylic anion groups present on cell walls of bacteria, thus inhibiting their growth and functions. In addition, antibacterial activity of COSs has been shown to be greatly dependent on their DP. Therefore, to identify the antibacterial activities of COSs, three fractions of COSs with different molecular weights were prepared by employing the dual reactor system and tested against Gram-negative, Gram-positive, and lactic acid bacterial strains [15,23]. COSs with molecular weights higher than 1 kDa seemed to be suitable for antibacterial activity; inhibitory effects were slightly varied depending on the type of bacteria (Table 7.1). In addition, COSs exhibited a more effective suppression against Gram-positive bacteria compared to Gram-negative bacteria. Interestingly, COSs were more effective against pathogenic bacteria associated with human diseases in comparison to nonpathogenic bacteria. For examples, *Streptococcus mutans*, which induces tooth decay, was completely inhibited by 0.1% of COSs and *Staphylococcus aureus*, which is responsible for pimples on human skin, was inhibited 93–100% at the same concentration of COSs. In the case of Gram-negative bacteria, all fractions of COSs could effectively inhibit *Salmonella typhi* that causes typhoid fever.

In addition to COSs, hetero-COSs (oligosaccharides derived from partially deacetylated chitosan) also have proven their ability to serve as antibacterial compounds. Antibacterial activity of hetero-COSs is dependent on the degree of deacetylation as well as the molecular size. To identify the effect of these two parameters on antibacterial activity of hetero-COSs, differentially deacetylated chitosans and their oligosaccharides were tested against a number of bacterial strains [27,28]. Interestingly, hetero-COSs followed the same pattern of inhibition as that exhibited by COSs in other studies. It was also found that treatment with all hetero-COSs could reduce the growth of bacteria significantly. However, high-molecular-weight hetero-COSs were more effective in inhibiting bacterial growth. In addition, 75% deacetylated hetero-COSs rendered a better antibacterial activity

TABLE 7.1
Antibacterial Activities of COS Fractions with Different Molecular Weight Ranges

Bacteria		Antibacterial Activity (%)[a]		
		HMWCOSs[b]	MMWCOSs[c]	LMWCOSs[d]
Gram-negative bacteria	*Escherichia coli*	98 ± 0	62 ± 6	51 ± 7
	Escherichia coli O-157	71 ± 3	56 ± 4	60 ± 2
	Salmonella typhi	91 ± 2	88 ± 0	89 ± 0
	Pseudomonas aeruginosa	47 ± 5	35 ± 5	22 ± 8
Gram-positive bacteria	*Streptococcus mutans*	100 ± 0	99 ± 0	99 ± 0
	Staphylococcus aureus	97 ± 3	95 ± 0	93 ± 9
	Staphylococcus epidermidis	82 ± 0	57 ± 3	23 ± 1
	Bacillus subtilis	63 ± 5	60 ± 5	63 ± 7
	Micrococcus luteus	70 ± 0	67 ± 3	63 ± 7

[a] Following the incubation of bacterial culture with 0.1% different COSs fractions, the number of colonies formed on the medium was calculated as a percentage compared to the control.
[b] High-molecular-weight COSs (molecular weight range 10–5 kDa).
[c] Medium-molecular-weight COSs (molecular weight range 5–1kDa).
[d] Low-molecular-weight COSs (molecular weight less than 1 kDa).
Source: Jeon, Y.J., Park, P.J. and Kim, S.K., *Carbohyd. Polym.*, 44, 71–76, 2001.

than that of 50 and 90% deacetylated oligomers. Especially, the growth of *Vibrio parahaemolyticus*, a food-borne Gram-negative bacterial strain, was inhibited by the 75% deacetylated hetero-COSs at a high rate [24]. Therefore, antibacterial property of COSs and deacetylated forms of them better reflects their functional properties, which could stimulate their use as potential nutraceuticals.

7.4.2 Antitumor Activity

Early studies demonstrated that chitosan and COSs could inhibit the growth of tumor cells by exerting immuno-enhancing effects. Some studies have suggested that the observed antitumor activity is not because of the direct killing of tumor cells but possibly because of the increased production of lymphokines [29]. *In vivo* studies carried with hexamer COSs ingested mice have shown significant antimetastatic effects on lung carcinoma [26]. However, the antitumor activities observed in COSs also depend on their structural characteristics such as degree of deacetylation and molecular weight. Thus, a study was carried out to identify the antitumor activities of different molecular weight COSs prepared using UF membrane reactor system. The mean molecular weight COSs ranging from 1.5 to 5.5 kDa could effectively inhibit the growth of Sarcoma 180 solid (S180) or Uterine cervix carcinoma No. 14 (U14) tumor cell–bearing mice [25]. In addition, the optimum dose of this COS (89% deacetylated) for inhibiting these tumors was approximately 20 mg/kg/day, which resulted in 66.6 and 73.6% tumor inhibition rates against S180 and U14-bearing mice, respectively (Table 7.2). Many reports suggest that these antitumor compounds exert effects on immune system to stimulate leucocytes, cytotoxic T cells, and natural killer cells. The observed increase in thymus weight of S180 and U14 tumor cell–bearing animals after the COSs treatment implied an improvement in the immune system particularly by activation of T lymphocytes. Furthermore, studies on antitumor activity of chitosans and their derivatives revealed that partially deacetylated chitin and carboxymethylchitin with appropriate degrees of substitution were effective toward controlling various tumor cells [30]. Unlike many other biological molecules, COSs could exert their biological activities following oral administration and effects were more or less similar to those of intraperitoneal injection. Qin et al. [31] have demonstrated that water-soluble COSs prepared with a mixture of tetramer and pentamer could inhibit the growth of S180 tumor cells in mice after oral and intraperitoneal administration. Therefore, COSs and their *N*-acetylated analogs that are soluble in basic physiologic environments may serve as good candidates for developing nutraceuticals. Further, very few reports have hypothesized that free amine groups of COSs play an important role for antitumor activity in tumor-bearing animals.

TABLE 7.2
Effect of COSs on Thymus Growth and Tumor Growth Inhibition in BALB/c Mice

		Sarcoma 180 (S180)		Uterine Cervix Carcinoma (U14)	
Sample	**Dose (mg/kg/day)**	**Thymus (mg/10g)**	**Tumor Inhibition (%)**	**Thymus (mg/10g)**	**Tumor Inhibition (%)**
Control		29.2 ± 9.5		25.4 ± 11.5	
HMWCOSs[a]	50	6.5 ± 4.5	–	14.5 ± 4.8	–
	20	31.4 ± 15.9	12.7	21.6 ± 11.9	11.9
	10	31.6 ± 8.5	61.7	25.9 ± 9.0	15.3
MMWCOSs[b]	50	41.4 ± 5.5	66.6	36.7 ± 14.0	73.6
	20	37.1 ± 9.7	35.5	33.8 ± 8.4	61.4
	10	34.4 ± 16.7	28.4	28.0 ± 10.0	26.6
LMWCOSs[c]	50	31.9 ± 6.7	12.4	27.8 ± 6.7	27.1
	20	34.1 ± 6.6	15.3	31.6 ± 17.5	7.8
	10	33.2 ± 10.5	5.7	30.9 ± 7.5	3.5

[a] High-molecular-weight COSs (molecular weight range 12.0–6.5 kDa).
[b] Medium-molecular-weight COSs (molecular weight range 5.5–1.5 kDa).
[c] Low-molecular-weight COSs (molecular weight range 1.4–0.5 kDa).
Source: Jeon, Y.J. and Kim, S.K., *J. Microbiol. Biotechnol.*, 12, 503–507, 2002.

This was attributed to the improvement in the antitumor activity with the increase of deacetylation of COSs.

7.4.3 Radical Scavenging Activity

Aerobic organisms must deal with free radicals that are generated from sequential reduction of oxygen during the normal course of aerobic metabolism. Uncontrolled formation of these free radicals is toxic as they cause cellular damage leading to a number of pathological conditions including atherosclerosis, arthritis, diabetes, and carcinogenesis [32]. The body has developed natural antioxidant systems to fight against these free radicals; however, the capacity of such systems gradually decreases with ageing, resulting in imbalances in the redox status. Therefore, the body must be nourished with a diet that includes adequate antioxidants. Scavengers of free radicals are preventive antioxidants, and presence of radical scavenging compounds break the oxidative sequence at different levels. Therefore, there has been a growing interest to identify natural antioxidant compounds from many sources to overcome the radical-mediated deleterious effects in biological systems. Many biological compounds including carbohydrates, peptides, and some phenolic compounds have been identified as potent radical scavengers. In addition, COSs and hetero-COSs have shown radical scavenging properties depending on their DA and molecular weight [33,34]. Low-molecular-weight COSs (1–3 kDa) scavenged different radicals as evidenced by electron spin trapping technique using ESR spectroscopy. In addition, highly deacetylated (90%) COSs are more effective for scavenging DPPH, hydroxyl, superoxide, and carbon-centered radicals (Figure 7.3). Especially, the hydroxyl radical, which is one of the most reactive free radicals involved in the oxidation of biomolecules such as lipids and proteins can be effectively scavenged by these COSs. However, scavenging effects on these harmful free radical species *in vivo* should be studied to identify the precise application of COSs or their heteroderivatives as radical scavengers in the human body.

7.4.4 Other Biological Activities

In addition to the bioactivities discussed above, COSs and hetero-COSs have been shown to possess some other important properties such as anticoagulant activity [35,36], angiotensin I converting enzyme (ACE)-inhibitory activity [37], calcium absorption acceleration activity [38], and antifungal activity [21]. Similar to other observed biological activities of COSs, these properties are also dependent on the DA. For antifungal activity, chitosan was more effective than its oligomers. However, low-molecular-weight COSs had better inhibition effects on some fungal species than low- and medium-molecular-weight COSs (Figure 7.4).

In two separate studies, COSs were found to exert anticoagulant and ACE-inhibitory activities regardless of their molecular weights. ACE, a dipeptidylcarboxypeptidase present in mammals plays a major role in high blood pressure. Owing to the undesirable side effects of synthetic ACE inhibitors, increased

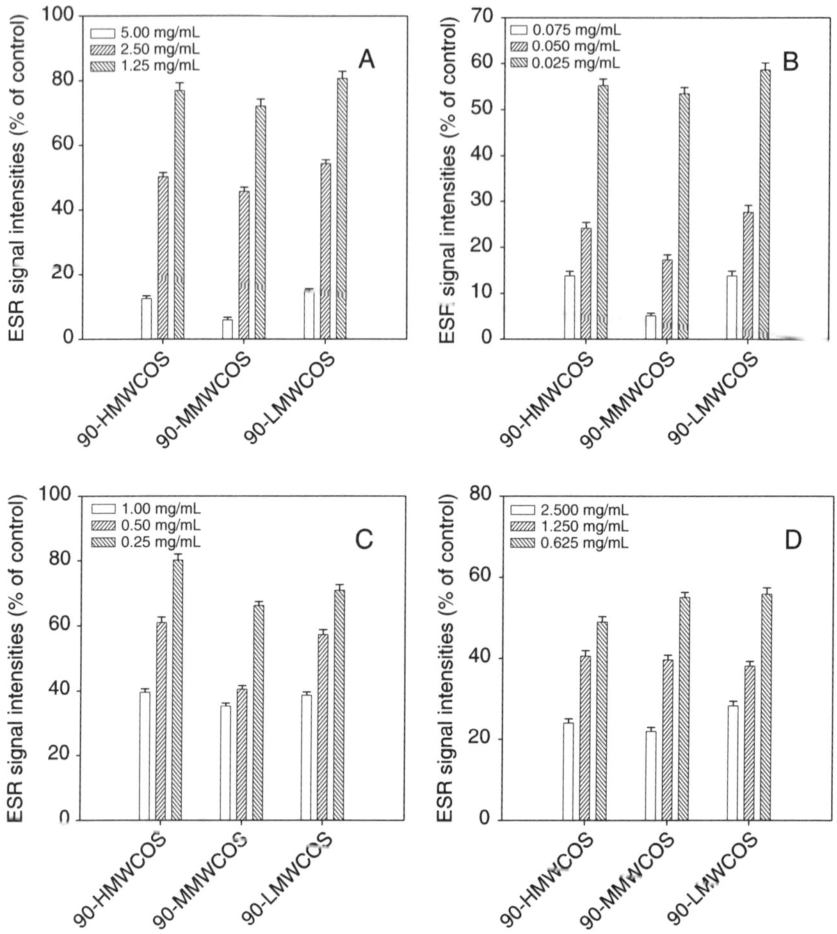

FIGURE 7.3 DPPH (A), hydroxyl (B), superoxide, and (C) carbon-centered radical (D) scavenging activities of COSs tested with their different molecular weight fractions. Values represent means ±S.E ($n = 3$); 90-HMWCOS, 90% deacetylated high-molecular-weight COS (MW 10,000–5,000 kDa); 90-MMWCOS, 90% deacetylated medium-molecular-weight COS (MW 5000–1000 kDa); 90-LMWCOS, 90% deacetylated low-molecular-weight COS (MW <1000 kDa). (From Je, J.Y., Park, P.J. and Kim, S.K., *Food Chem. Toxicol.*, 42, 381–387, 2004.)

attention has been paid toward natural and safe ACE inhibitors such as peptides, carbohydrates, and phenolic compounds. In addition to their inhibitory activity, these natural compounds may offer other beneficial effects to be used them as physiological functional foods or nutraceuticals. COSs with a relatively low degree of deacetylation (50%) are better ACE inhibitors than other deacetylated forms.

Heparin, a widely used anticoagulant polysaccharide in clinical therapy, has been shown to exert a number of complications in patients, and for a long time

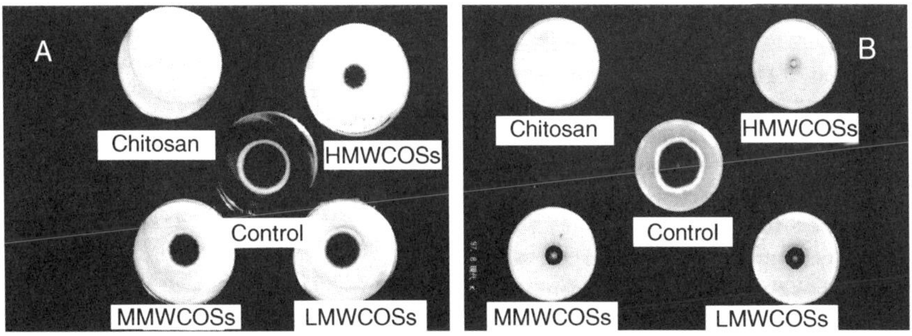

FIGURE 7.4 Antifungal activity of chitosan and COSs. Different molecular weights of COSs affect the growth of *Aspergillus niger* (A) and *Alternaria mali* (B). (From Kim, S.K., *Food Ind. Nut.*, 8, 1–8, 2003. With permission.)

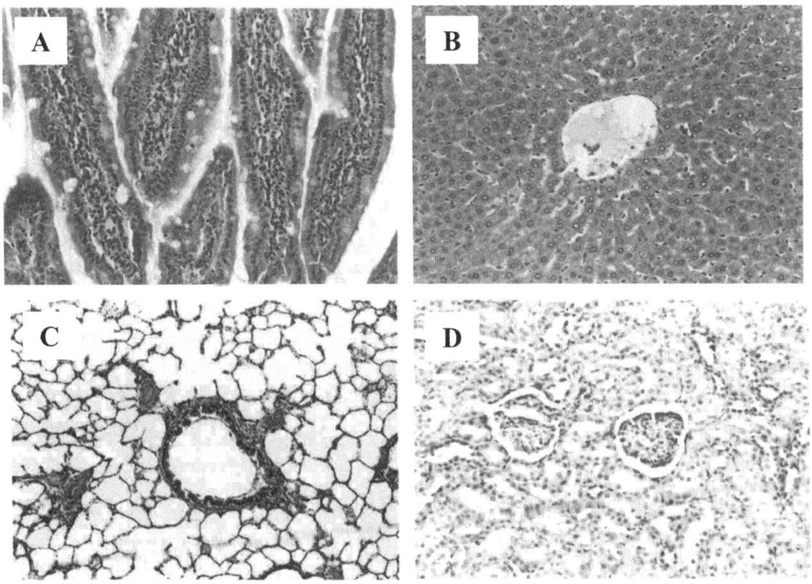

FIGURE 7.5 Microscopic views of small intestine (A), liver (B), lung (C), and kidney (D) sections obtained from Sprague–Dawley rats after 4 weeks of COSs (5000 mg/kg) ingestion. (From Jeon, Y.J. and Kim, S.K., *J. Chitin Chitosan,* 4, 115–120, 1999. With permission.)

researchers have been searching for better natural alternate sources. In this regard, sulfated forms of COSs have displayed promising results for acting as anticoagulants [35]. However, their low activity compared to heparin does not negate their use as natural anticoagulants, because they do not impart any unfavorable side effects. In addition, COSs with high degrees of deacetylation had a better performance as anticoagulants than their less deacetylated counterparts.

7.5 SAFETY OF COSs

Even though COSs have very strong biological properties *in vitro*, little information is available on their cytotoxicity and bioavailability in the human body. Interestingly, experiments on subacute toxicity of COSs in Sprague–Dawley (SD) rats revealed that COSs do not induce any mortalities, or change in blood chemistry, urinalysis, and body weights of rats [39]. Therefore, it can be presumed that COSs may not have any influence on acute toxicity and side effects in humans. Furthermore, histopathological findings revealed that COSs do not cause any lesions in tissues of rats under the tested dose range of COSs (Figure 7.5) [40]. These findings further confirm that intake of COSs will not cause any undesirable side effects, at least in animals.

COSs have become popular during the past few decades because of their biological effects related to their structural properties. In addition, availability of raw materials at low cost and recently developed methods ensure continuous production of COSs. Therefore, it is expected that these biomolecules play an important role in fulfilling the current demand for nutraceuticals.

REFERENCES

1. Zikakis, J.P., Saylor, W.W. and Austin, P.R. (Eds.) 1984. Chitin and chitosan, *Jpn. Soc. Chitin Chitosan* 57–67.
2. Majeti, N.V. and Kumar, R. 2000. A review of chitin and chitosan. *React. Funct. Polym.* 46, 1–27.
3. Felt, O., Buri, P. and Gurny, R. 1998. Chitosan: A unique polysaccharide for drug delivery. *Drug. Dev. Ind. Pharm.* 24, 979–993.
4. Jeon, Y.J., Shahidi, F. and Kim, S.K. 2000. Preparation of chitin and chitosan oligomers and their applications in physiological functional foods. *Food Rev. Int.* 61, 159–176.
5. Huang, R., Mendis, E. and Kim, S.K. 2005. Improvement of ACE inhibitory activity of chitooligosaccharides (COS) by carboxyl modification. *Bioorg. Med. Chem.* 13, 3649–3655.
6. Maezaki, Y., Tsuji, K., Nakagawa, N., Kawai, Y., Akimoto, M., Tsugita, T., Takekawa, W., Terada, A., Hara, H. and Mitsoka, T. 1993. Hypocholestromic effect of chitosan in adult males. *Biosci. Biotech. Biochem.* 57, 1439–1446.
7. Okamoto, Y., Inoue, A., Miyatake, K., Ogihara, K., Shigemasa, Y. and Minami, S. 2003. Effects of chitin/chitosan and their oligomers/monomers on migrations of macrophages. *Macromol. Biosci.* 3, 587–590.
8. Tokoro, A., Kobayashi, M., Tatekawa, N., Suzuki, S. and Suzuki, M. 1989. Protective effect of *N*-acetyl chitohexaose on *Listeria monocytogenes* infection in mice. *Microbiol. Immunol.* 33, 357–367.
9. Suzuki, K., Mikami, T., Okawa, Y., Tokoro, A., Suzuki, S. and Suzuki, M. 1986. Antitumor effect of hexa-*N*-acetylchitohexaose and chitohexaose. *Carbohydr. Res.* 151, 403–408.
10. Kendra, D.F. and Hadwiger, L.A. 1984. Characterization of the smallest chitosan oligomer that is maximally antifungal to *Fusarium solani* and elicits pisatin formation in *Pisum sativum*. *Exp. Mycol.* 8, 276–281.
11. Uchida, Y., Izume, M. and Ohtakara, A. 1989. Preparation of chitosan oligomers with purified chitosanase and its application. In: Braek, G.; Anthonsen, T.; Sandford, P., (Eds.): Chitosan: Barking, UK: Elsevier Applied Science, 372–382.

12. Izume, M. and Ohtakara, A. 1987. Preparation of D-glucosamine oligosaccharides by the enzymatic hydrolysis of chitosan. *Agric. Biol. Chem.* 51, 1189–1191.
13. Aiba, S. 1993. Studies on: 6. Relationship between *N*-acetyle group distribution pattern and chitinase digestability of partially *N*-acetylated chitosans. *Int. J. Biol. Macromol.* 15, 241–245.
14. Aiba, S. 1994. Preparation of *N*-acetyl chitooligosaccharides from lysozymic hydrolysate of partially *N*-acetylated chitosans. *Carbohydr. Res.* 261, 297–306.
15. Jeon, Y.J., Park, P.J., Byun, H.G., Song, B.K. and Kim, S.K. 1998. Production of chitosan oligosaccharides using chitin-immobilized enzyme. *Korean J. Biotechnol. Bioeng.* 13, 147–154.
16. Jeon, Y.J. and Kim, S.K. 2000. Production of chitooligosaccharides using ultrafiltration membrane reactor and their antibacterial activity. *Carbohyd. Polym.* 41, 133–141.
17. Jeon, Y.J. and Kim, S.K. 2000. Continuous production of chitooligosaccharides using a dual reactor system. *Process Biochem.* 35, 623–632.
18. Zhang, H., Du, Y., Yu, X., Mitsutomi, M. and Aiba, S. 1999. Preparation of chitooligosaccharides from chitosan by a complex enzyme. *Carbohydr. Res.* 320, 257–260.
19. Kuroiwa, T., Ichikawa, S., Sato, S. and Mukataka, S. 2003. Improvement of the yield of physiologically active oligosaccharides in continuous hydrolysis of chitosan using immobilized chitosanases. *Biotechnol. Bioeng.* 84, 121–127.
20. Akiyama, K., Kawazu, K. and Kobayashi, A. 1995. A novel method for chemoenzymatic synthesis of elicitor-active chitosan oligomers and partially *N*-deacetylated chitin oligomers using *N*-acylated chitotrioses as substrates in a lysozyme-catalyzed transglycosylation reaction system. *Carbohydr. Res.* 279, 151–160.
21. Hirano, S. and Nagao, N. 1989. Effects of chitosan, pectic acid, lysozyme and chitinase on the growth of several phytopathogens. *Agric. Biol. Chem.* 53, 3065–3066.
22. Jeon, Y.J. and Kim, S.K. 2001. Effect of antimicrobial activity by chitosan oligosaccharides *N*-conjugated with asparagines. *J. Microbiol. Biotechnol.* 11, 281–286.
23. Jeon, Y.J., Park, P.J. and Kim, S.K. 2001. Antimicrobial effect of chitooligosaccharides produced by bioreactor. *Carbohydr. Polym.* 44, 71–76.
24. Nam, M.Y., Shon, Y.H., Kim, S.K., Kim, C.H. and Nam, K.S. 1999. Inhibitory effect of chitosan oligosaccharides on the growth of tumor cells. *J. Chitin Chitosan* 4, 184–188.
25. Jeon, Y.J. and Kim, S.K. 2002. Antitumor activity of chisan oligosaccharides produced in an ultra filtration membrane reactor system. *J. Microbiol. Biotechnol.* 12, 503–507.
26. Tsukada, K., Matsumoto, T., Aizawa, K., Tokoro, A., Naruse, R., Suzuki, S. and Suzuki, M. 1990. Antimetastatic and growth inhibitory effects of N-acetyl chitohexaose in mice bearing Lewis lung carcinoma. *Jpn. J. Cancer Res.* 81, 259–265.
27. Park, P.J., Je, J.Y., Byun, H.G., Moon, S.H. and Kim, S.K. 2004. Antimicrobial activity of hetero-chitosans and their oligosaccharides with different molecular weights. *J. Microbiol. Biotechnol.* 14, 317–323.
28. Park, P.J., Lee, H.K. and Kim, S.K. 2004. Preparation of hetero-chitooligosaccharides and their antimicrobial activity on *Vibrio parahaemolyticus*. *J. Microbiol. Biotechnol.* 14, 41–47.
29. Tokoro, A., Tatewaki, N., Suzuki, K., Mikami, T., Suzuki, S. and Suzuki, M. 1988. Growth-inhibitory effect of hexa-*N*-acetylchitohexaose and chitohexaose and Meth-A solid tumor. *Chem. Pharm. Bull.* 36, 784–790.
30. Nishimura, S., Nishi, N., Tokura, S., Nishimura, K. and Azuma, I. 1986. Bioactive chitin derivatives. Activation of mouse-peritoneal macrophages by *O*-(carboxymethyl)chitins. *Carbohydr. Res.* 146, 251–258.

31. Qin, C., Du, Y., Xiao, L., Li, Z. and Gao, X. 2002. Enzymic preparation of water-soluble chitosan and their antitumor activity. *Int. J. Biol. Macromol.* 31, 111–117.
32. Halliwell, B. 1994. Free radicals, antioxidants, and human disease: Curiosity, cause, or consequences? *Lancet* 344, 721–724.
33. Park, P.J., Je, J.Y. and Kim, S.K. 2003. Free radical scavenging activity of chito-oligosaccharides by electron spin resonance spectrometry. *J. Agric. Food Chem.* 51, 4624–4627.
34. Je, J.Y., Park, P.J. and Kim, S.K. 2004. Free radical scavenging properties of hetero-chitooligosaccharides using an ESR spectroscopy. *Food Chem. Toxicol.* 42, 381–387.
35. Park, P.J., Je, J.Y., Jung, W.K., Ahn, C.B. and Kim, S.K. 2004. Anticoagulant activity of hetero-chitosan and their oligosaccharide sulfates. *Eur. Food Res. Tech.* 219, 529–533.
36. Kim, S.K. 2003. Development of novel bioactive substances from fishery byproducts. *Food Ind. Nut.* 8, 1–8.
37. Park, P.J., Je, J.Y. and Kim, S.K. 2003. Angiotensin I converting enzyme (ACE) inhibitory activity of hetero-chitooligosaccharides prepared from partially different deacetylated chitosans. *J. Agric. Food Chem.* 51, 4930–4934.
38. Jeon, Y.J., Kim, G.H., Park, P.J. and Kim, S.K. 1999. Calcium absorption accelerating effect of chitosan oligosaccharides prepared by ultrafiltration membrane enzymatic reactor. *J. Korean Fish. Soc.* 32, 247–251.
39. Kim, S.K., Park, P.J., Yang, H.P. and Han, S.S. 2001. Subacute toxicity of chitosan oligosaccharide in Sprague-Dawley rats. *Arzneim-Forsch Drug Res.* 51, 769–774.
40. Jeon, Y.J. and Kim, S.K. 1999. Effects of chitooligosaccharides on acute oral toxicity. *J. Chitin Chitosan* 4, 115–120.

8 Glucosamine Production and Health Benefits

Jaroslav A. Kralovec and Colin J. Barrow

CONTENTS

8.1 INTRODUCTION

8.1.1 BACKGROUND

Increases in chronic diseases like cardiovascular disease, type 2 diabetes, cancer, psychiatric disorders, and arthritis are contributing to rapidly mounting health care costs in affluent societies. For instance, the total health care expenditure in North America is currently close to 1 trillion dollars. Drug expenses are the second largest contributor to this total, behind hospital costs, contributing in excess of 100 billion dollars. The current health care system focuses on treatment rather than prevention. There is growing evidence that prevention through correct diet and exercise can significantly impact health and decrease health care costs. As we learn more about the interrelationships among genetics, food, health, and disease, prevention will become a more important part of the health care system. There is an increasing perception that nutraceuticals, including supplements and fortified foods, will play a very important role in this evolutionary change [1].

Osteoarthritis is the most common form of arthritis and a major source of pain and disability, particularly in the elderly. Although not life threatening, osteoarthritis is one of the medical conditions that put a great financial burden on society, and therefore preventative measures and early intervention with nutraceuticals may result in significant benefits to both society and the individual. Osteoarthritis, also called degenerative joint disease, is a chronic degenerative disease where the cartilage protecting the ends of the bones in synovial joints deteriorates. The smooth cartilage becomes rough resulting in friction and inflammation. The damage is in the form of structural changes and erosion, resulting in the loss of mechanical properties, leading to pain and stiffness. The physiological changes of bone and cartilage loss increase significantly with age and the pathogenesis is not adequately understood. Because cartilage is a vascular, anervic, and alymphatic tissue, the supply of nutrients is more sensitive to any interference, particularly acute traumatic joint injury, injuries resulting from chronic overuse, problems resulting from mechanical misalignment, atherosclerosis of blood vessels contributing to cartilage nurturing and drug use. Abnormal loads lead to ischemia, which results in the generation of radicals. The radicals degrade synovial fluid hyaluronan and cartilage macromolecules are digested by synoviocytes that mediate cytokine response. Cytokines not only stimulate the function of chondrocytes to catabolize cartilage but can also stimulate glycosaminoglycan (GAG) synthesis [2]. Unfortunately, all current treatments are only palliative and benefit varies by individual genetics and disease severity. The disease is typically treated with nonsteroidal anti-inflammatory drugs (NSAID), since they are very effective in relieving stiffness and pain [3,4], however their use is limited because of their toxicity. Data suggest that NSAID are also toxic to the articular cartilage and actually contribute to the production of catabolic cytokines and the prevention of anabolic cytokines. COX-2 inhibitors, another category of drugs employed for treatment of osteoarthritis [5,6], are also known to have many side effects. However, unlike traditional NSAID, they do not block the action of COX-1,

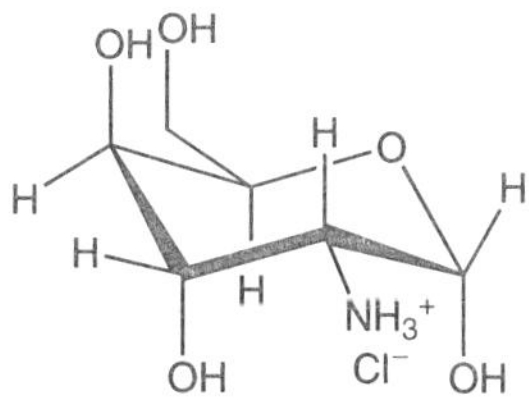

FIGURE 8.1 The structure of D-glucosamine hydrochloride's prevalent β-anomer.

an enzyme known to protect the stomach lining and thus they are viewed as being more selective. Recently, some COX-2 inhibitors have been found to increase the risk of cardiovascular disease, a side effect that is particularly problematic for osteoarthritis sufferers who are obese and inactive, partly due to their arthritis pain. In fact, in late 2004, Merck withdrew its arthritis drug Vioxx® from the market worldwide owing to a clinical trial showing an increased relative risk of cardiovascular events such as heart attack and stroke in subjects taking the drug. Clearly, an alternative to NSAID and COX-2 inhibitors that is side effect free and has at least some efficacy for osteoarthritis sufferers is very desirable. There is a variety of clinical evidence that glucosamine (Figure 8.1) is this alternative, being safe and seemingly efficacious for the prevention and treatment of OA.

8.1.2 Glucosamine as Drug or Supplement

Glucosamine is one of the most thoroughly studied supplements with regard to its health benefits, although it is not approved as a drug in any form in Canada or the United States. Glucosamine in either the chloride or sulfate form, is marketed as a dietary supplement in Canada and the United States, but is not approved for use in disease prevention or treatment and does not carry a health claim, as it does in some European countries. Up to 6000 metric tons of glucosamine are consumed annually and the majority of sales are generated in the United States (about $900 million estimated in 2004) [7]. There is still skepticism surrounding the efficacy and therapeutic spectrum of glucosamine, especially within the medical community. This has been fueled by exaggerating significance of the data from clinical trials and unsubstantiated claims associated with promotional enthusiasm [8–10]. For instance, the International League Against Rheumatism labeled glucosamine as a symptomatic, "slow-acting drug" for osteoarthritis and decided to remove the attribute chondroprotective owing to unproven disease-modifying activity in humans [11].

The advantages of glucosamine over many nutritional supplements are its defined chemical structure and low molecular weight, two typical characteristics of conventional effective therapeutics. Also, glucosamine is normally sold as an ingredient with greater than 99% purity. These features distinguish glucosamine from a vast majority of nutraceuticals. Many nutraceuticals, especially herbals, are often poorly defined mixtures and are difficult to standardize, especially in cases where the chemical structures of the bioactive compounds have not been determined. Another advantage of glucosamine over many supplement products is that it is a natural component of the human body, so that the body has developed mechanisms for processing and incorporating or disposing of this compound. The natural presence of glucosamine in the body is a major reason why glucosamine supplementation is associated with minimal or no side effects.

There are no conventional food sources of glucosamine. Glucosamine is normally produced biosynthetically in the body; however, in instances of OA there appears to be a deficiency for which glucosamine supplementation is beneficial. Most glucosamine used in supplements is obtained from shrimp or crab shell by depolymerization of chitin. Glucosamine is available commercially in several forms, including in combination with herbs, vitamins, creatine, chondroitin sulfate, ascorbic acid, manganese, or dimethylsulfone. Some of the most popular combinations contain chondroitin sulfate [12]. Chondroitin sulfate is a source of several building blocks for cartilage including glucosamine. Chondroitin sulfate stimulates cartilage regeneration *in vitro* but clinical evidence that it improves the symptomology of OA is limited [13]. Osteo Bi-Flex® (Rexall Sundown, Boca Raton, FL), a glucosamine and chondroitin sulfate–based formulation, was the number one selling dietary supplement for joints before the acquisition of the company by NBTY resulted in a combined product called Flex-A-Min. In 2003, Flex-A-Min controlled 35% of glucosamine supplement retail market. Some food manufacturers started fortifying fruit juices with glucosamine. For example, SoBe Sport System® line from Pepsi is a glucosamine-containing drink that became the official beverage of the U.S. track and field national team in 2000. Coca-Cola and Procter & Gamble also launched a glucosamine containing drink Elations®. Other glucosamine-containing drinks are available to customers, including Super Glucosamine and Glucosamine Drink Mix from Action-Labs, Logic Juice4Joints from The Health Company in the United Kingdom and JointJuice in the United States. Glucosamine is not generally recognized as safe (GRAS) ingredient in the United States and therefore cannot be used in foods or normal beverages. However, beverages containing glucosamine are normally sold as supplements although the FDA has become concerned about defining these drinks as supplement and has been targeting these products. The major reasons that glucosamine is not an approved GRAS ingredient is that there have been no clinical studies to show that its consumption is safe in young children, pregnant women, and diabetics.

8.2 CHEMISTRY OF GLUCOSAMINE

Production starting from shrimp and crab shell normally uses hydrochloric acid and results in the formation of glucosamine hydrochloride. Glucosamine hydrochloride can be converted into glucosamine sulfate (see Section 8.2.3). The following section will deal exclusively with glucosamine hydrochloride because glucosamine sulfate is inherently unstable, and currently known glucosamine sulfate mixed salts are not rigorously defined chemical substances. In fact, many commercially available glucosamine sulfate products are actually blends of glucosamine hydrochloride and potassium sulfate, rather than an actual glucosamine sulfate salt.

8.2.1 Chemical and Physical Properties of D-Glucosamine Hydrochloride

Glucosamine hydrochloride has a CAS Registry Number of (66-84-2). Its chemical abstract service name is glucosamine hydrochloride (9C), although synonyms and trade names include 2-amino-2-deoxy-D-glucopyranose, chitosamine hydrochloride, D-glucosamine hydrochloride, and D-(+)-glucosamine hydrochloride. Structurally, glucosamine is an amino sugar with molecular formula $C_6H_{13}NO_5HCl$ and molecular mass 215.63 Da. In its pure form, it is a white crystalline powder with a melting point of 190–194°C. D-Glucosamine hydrochloride tends to decompose rather than melt. It is highly soluble in water, with a solubility of 100 mg/mL at 20°C. Similar to glucose, it is optically active with an optical rotation of $[\alpha]_D^{20}$ + 100° initial, +72.5° final, in the α-form and optical rotation of $[\beta]_D^{20}$ +25° initial, +72.6° final in the β-form [14]. D-Glucosamine hydrochloride is a faintly sweet, slightly astringent tasting compound and has minimal impact on the taste of a beverage when used at a 1.5 g dosage per serving. Because of its good solubility and minimal taste D-glucosamine hydrochloride is an excellent ingredient for functional beverages.

8.2.2 Production of Glucosamine

8.2.2.1 Biological Processes for Producing Glucosamine

Typically, glucosamine is produced from chitin that is isolated from shells of marine crustaceans, and the vast majority of the processes are based on chemical processing of shellfish. The following section is dedicated to this robust technology. Glucosamine can also be successfully produced from shellfish using enzymatic processes, and recently more attention has been paid to this technology. For example, Biopolymer Engineering (Eagan, MN) has established a glucosamine manufacturing facility that is based on enzymatic hydrolysis of chitin that has been extracted from langostino lobsters [15]. Glucosamine has also been successfully produced from chitin that was obtained from biomass by fermentation. In the 1970s, a Japanese company patented a process based on hydrochloric acid hydrolysis of deproteinated culture filtrate of *Azobacter* [16]. Fermentation strategies have experienced resurgence lately. Cargill Inc. (Minneapolis, MN) and the biotech company Bio-Technical Resources (Manitowoc, WI, a division of Arkion Life Sciences LLC, Wilmington, DE) have led the way [17–21]. Cargill patented a process in which glucosamine was produced by hydrolysis of chitin that had been isolated from fungal biomass derived from *Aspergillus* sp., *Penicillium* sp., *Mucor* sp., and the combination of these [18]. The patent claims that 15% hydrochloric acid hydrolysis of the biomass results in glucosamine of 97% purity with a 15% yield in 8 h. This is surprising since data from our laboratories [22] and the data of others [23] indicate that at least 30% hydrochloride acid is required to completely hydrolyze chitin oligomers. However, the rate of hydrolysis depends on the type of chitin, the degree and randomization of deacetylation, the molecular size and the crystallinity type. The physicochemical properties of fungal chitin are usually different from that obtained from shellfish and so hydrolysis may require milder conditions. Currently, Cargill makes glucosamine commercially from *Aspergillus niger* (Regensure®), which is kosher and approved in Japan as a food ingredient under FOSHU. The company promotes the product as having a "shellfish-free status." Czechs patented the production of glucosamine based on hydrolysis of polysaccharide obtained from wet mycelium (*A. niger*) from the production of citric acid [24]. Other companies such as Arkion Life Sciences have applied for patents related to methods and materials for producing glucosamine by fermentation of genetically modified microorganisms [25,26]. Solheim [20] reported a process of making glucosamine from glucose sources such as starch, glycogen, fructose, and a source of amine.

8.2.2.2 Chemical Processes for Producing Glucosamine

Hydrochloric acid is the standard chemical used for the hydrolysis of chitin to glucosamine, though there have been attempts to carry out the hydrolysis with other acids such as sulfuric acid and less commonly hydrogen fluoride, phosphoric acid, or nitric acid. After several hours at 20°C, aqueous hydrogen fluoride typically leads to the production of a mixture of oligomers. After 24 h, ^{13}C-NMR spectrum shows the presence of glucofuranosyl oxazolinium ion, that is a precursor to *N*-acetylglucosamine (NAG) [27]. Recently, a Japanese group filed a patent application describing a method for the industrial production of glucosamine and its oligomers using concentrated hydrogen fluoride [28]. However, hydrogen fluoride is a costly and extremely strong acid that cannot be used with glass or stainless-steel vessels, and so the commercial use of this acid is not feasible. A Chinese group has recently patented a process for the direct formation of glucosamine based on hydrolysis of chitin with 20–80% sulfuric acid [29]. Chitin hydrolysis with sulfuric acid is of interest to glucosamine manufacturers since the successful direct preparation of glucosamine using sulfuric acid would not require the glass-lined reactors required for glucosamine produced using hydrochloric acid. Also, it may be possible to produce glucosamine sulfate directly, rather than by subsequent reaction or blending with potassium sulfate.

The original reference to the first preparation of glucosamine goes back to Lederhose [30] who obtained it by hydrolysis of lobster shells. German chemists continued to be interested in the production and chemistry of glucosamine as witnessed by papers from the early twentieth century [31–33]. Hudson and Dale [34] used a crude process for obtaining glucosamine from a mixture of crab and lobster shells by employing hydrochloric acid, but not involving the deproteination step. Komori [35] prepared glucosamine from cicad larvae and ovomucid using potassium hydroxide for deproteination and subsequent hydrochloric acid chitin hydrolysis. van Alphen [36] isolated glucosamine from lobster shells after a simple demineralization step with concentrated hydrochloric acid and researchers from Atlantic Canada [37] later used similar methodology. Stacey and Webber [38] described the simple production of glucosamine from crab shells of low-protein content, involving only the demineralization step using 2 M hydrochloric acid, followed by treatment of the dried matter with concentrated hydrochloric acid. In a similar manner, Purchase and Braun [39] isolated glucosamine from crab shells and Ingle et al. [40] from fish canning waste. Oeriu et al. [41] used diluted hydrochloric acid to demineralize and 10% sodium hydroxide to deproteinize the shells to get chitin before hydrolyzing it with 10 M hydrochloric acid. Both the demineralization and deproteination steps implemented before the actual hydrolysis of chitin were also included in a procedure for the isolation of glucosamine from prawn shell waste [42]. A Russian group later discovered that glucosamine increases the bioavailability of tetracycline antibiotics resulting in higher concentration of tetracycline in the serum, isolated glucosamine from shrimp (*Penaeus semisulcatus*) shells using a similar methodology [43]. Novikov [44–46] further developed methods for the preparation of glucosamine from shell and summarized these in papers that are widely cited. The Chinese manufacturers are well known for their efforts in glucosamine production and they are currently the world's largest suppliers of glucosamine. Unfortunately, their work has mostly been published in Chinese literature making it not easily accessible to Western researchers and producers [47–55]. One of the Chinese papers describes preparation of glucosamine sulfate from glucosamine chloride, using anion exchange chromatography [56].

8.2.2.3 Standard Industrial Process for Producing Glucosamine

Currently, the most common industrial processes for manufacturing glucosamine start from crustacean shells and are based on the process shown in Figure 8.2. In one variation of the process, known as the HCl–NaOH–HCl variant or acid–base–acid variant, the exoskeletons are shredded to 0.5–5 mm in size and demineralized using dilute hydrochloric acid. Typically, 2 M hydrochloric acid is used, but contrary to the preparation of chitosan, a higher acid strength up to 3 M can be used. The demineralized material is separated from the liquid by filtration using a filter press and then deproteinated using hot diluted sodium hydroxide, typically 1.5 M at 90°C. Chitin is then filtered off and hydrolyzed, typically with 10 M hydrochloric aid. The other variant of this process is theoretically simpler since it involves the base–acid–acid (NaOH–HCl–HCl) sequence and thus only one base–acid change, whereas the first variant involved one acid–base and one base–acid change. Deproteination is more effective on demineralized material and thus the choice is very much dependent on the type of shells being used and particularly on their calcium content.

It is not advisable to use low hydrochloric acid concentrations to hydrolyze chitin since the rate of hydrolysis slows down when the concentration of acid is below 9 M. Our results using a variety of acid strengths from 4 to 10 M hydrochloric acid demonstrated that lower concentrations of acid led to incomplete hydrolysis and the production of chitosan oligomers, which can be readily identified by HPLC.

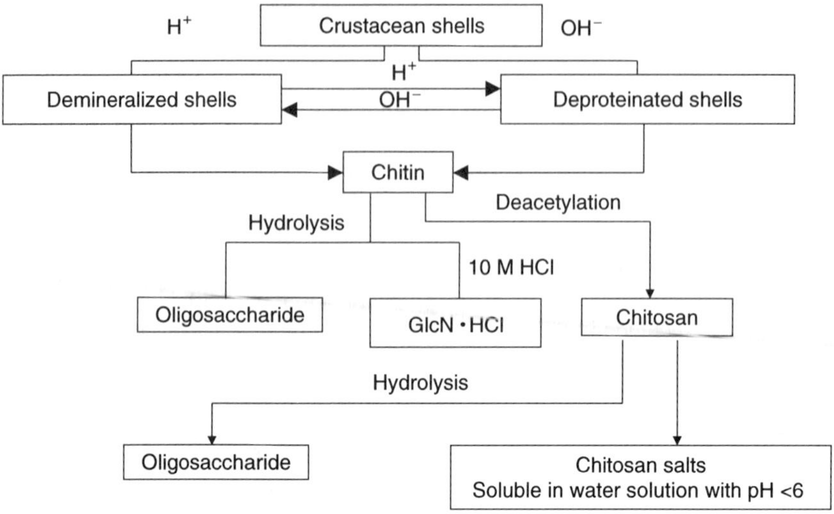

FIGURE 8.2 Summary of the steps normally involved in the chemical manufacturing of D-glucosamine hydrochloride.

Incomplete hydrolysis can also be the result of insufficient reaction time even in concentrated solutions of 10 M hydrochloric acid. Since this hydrolysis method involves deproteination steps, it leads to the production of relatively clean glucosamine that could be easily crystallized from water or dilute hydrochloric acid. The effective removal of hydrochloric acid during drying is important and can be challenging because of the constant boiling temperature of the binary acid and water mixture (109°C/760 torr for 6 M hydrochloric acid). A number of manufacturers recrystallize from ethanol rather than water or the acid. D-glucosamine hydrochloride is sparingly soluble in ethanol and so acid can be completely removed by ethanol recrystallization and subsequent ethanol washing. Ethanol washing results in less significant yield loss than does washing with water.

Several important criteria must be considered when analyzing glucosamine in solutions and in the final dried product. Since glucosamine has a strong buffering capacity, analysis for the hydrochloric acid content is not straightforward. An accurate determination of free acid in a glucosamine aqueous solution can be achieved by the determination of the total acid content by acid–base titration combined with the acid content in the residue upon exhaustive drying. It is also important to monitor for the presence of oligomers to establish that complete hydrolysis has occurred. Oligomer analysis can be performed using an amino acid analyzer or HPLC equipped with a cation exchange column, after product derivatization with an amino reactive reagent such as ninhydrin (unpublished results). One can also use a combination of HPLC with UV detection after derivatization with 6-aminoquinolyl-*N*-hydroxysuccinimidyl carbamate (AQCO pre-column derivatization) and FTIR [57,58].

8.2.3 Production of Glucosamine Sulfate Mixed Salts

Production of various sulfate salts of glucosamine is a specialty of Rotta researchers. Rovati [59] patented the production of glucosamine sulfate from a glucosamine base by sulfation using fuming sulfuric acid fortified with 20% sulfur trioxide. The sulfate with a melting point of 116°C was prepared in an 86% yield. It had to be dried in a vacuum rotary drier at 50°C, ground and stored with silica gel to avoid moisture contact because of the hygroscopic nature of this salt. More than two decades later, Chinese workers prepared the same glucosamine sulfate salt in almost a 94% yield [60]. Rotta researchers were the first to successfully deal with the easily oxidizable and hygroscopic nature of glucosamine sulfate. They succeeded in stabilizing glucosamine sulfate by making a mixed salt with sodium chloride in a 1:2 molar ratio [61]. Finnish researchers extended this to other salts and they executed this in a more straightforward manner by producing glucosamine sulfate *in situ* and mixing it with sodium, potassium, or magnesium chlorides or bromides [62]. Schleck et al. [63] prepared mixed glucosamine sulfate salts by mixing glucosamine hydrochloride and potassium sulfate. He distinguished a mixed salt compound from simple mixtures of salt by specific gravity. The mixed salt compounds should be within 0.9–1.00 g/cc, while the mixtures have density greater than 1.1 g/cc [63]. Rotta researchers basically used the same methodology for making mixed glucosamine salts from glucosamine hydrochloride and sulfates of calcium, potassium, sodium, and magnesium [64]. Recently, they made mixed glucosamine salts from glucosamine hydrochloride and a stoichiometric amount of sulfate [65].

8.2.4 Current Manufacturing of Glucosamine in North America and Japan

Until recently, Pfanstiehl Laboratories (Waukegan, IL), a company specializing in carbohydrate chemistry, was the only North American manufacturer of D-glucosamine hydrochloride. Phanstiehl is a company with more than 40 years experience in the production of this amino sugar and its glucosamine was selected for use in the recently executed NIH clinical trial. Chinese producers put the company under pressure since their glucosamine flooded the European and North American markets and drove the price down to as low as $4 per kg at the end of 2003 [66]. In early 2005, the price recovered considerably and has been over $20 per kg, partly because of a shortage of shell starting material in China. Biopolymer Engineering (Eagan, MN) established a new glucosamine plant housing class 10,000 and class 1,000 clean rooms (U.S. Federal Standard 209E Class of Cleanliness) for production complying with food and pharmaceutical standards. Ocean Nutrition Canada Ltd. also built a state-of-the-art manufacturing plant for glucosamine production, which produced D-glucosamine hydrochloride under both food and drug GMP [67], but this facility was decommissioned in late 2003 because of the low price of glucosamine at that time. Cargill Inc. hope that their "shell-free" glucosamine produced by corn fermentation will give them an advantage over the shell-originated material, although there remain questions as to whether or not this source can compete in price with glucosamine from shell. Japanese Koyo Chemicals holds the largest share of the glucosamine market in Japan; it has made an agreement with Fuso Chemical to establish a 130 ton per year facility in Sakai, Japan. The product is planned to be an ultrahigh purity glucosamine produced by a process involving nanotechnology, originally designed for the production of ultra-high purity colloidal silica for electronic application. The company will expand its output to 1500 ton a year and target the high-value Japanese market [68].

8.3 BIOLOGY OF GLUCOSAMINE

8.3.1 BIOCHEMISTRY AND PHARMACOKINETICS OF GLUCOSAMINE

Since this chapter contributes to a book on marine nutraceuticals, only references to oral administration of glucosamine are considered. D-glucosamine is produced *in vivo* via the hexosamine biosynthetic pathway. Glucose enters the cell with the help of a glucose transporter and is metabolized to fructose-6-phosphate by hexokinase and then converted by glutamine fructose-6-phosphate amidotransferase (GFAT), where glutamine serves as a donor for the amino group. If glucosamine is added to cells, it is taken up by a glucose transporter and phosphorylated to directly produce glucosamine-6-phosphate, bypassing the rate-limiting GFAT [69]. Glucosamine is a good substrate for kinase and studies confirm that glucosamine is incorporated into GAG [70]. Glucosamine is then used for the production of hyaluronic acid and other GAG after its conversion into galactosamine using epimerases [71]. Moreover, there seems to be a linkage between GAG synthesis and collagen production as indicated in the simultaneous increase in synthesis of collagen [72]. Thus the cells can either synthesize glucosamine from glucose or receive it from the circulation and therefore exogenous glucosamine can be helpful. Zupanets et al. [73] studied the effects of exogenous glucosamine on metabolic and repair processes. He investigated this phenomenon in a model of posttraumatic osteoarthritis in the articular cartilage and on a model of posttraumatic keratitis in the cornea. The drug stimulated repair and inhibited dystrophic posttraumatic processes in the connective tissue structures.

A detailed pharmacokinetic study revealed that after oral administration glucosamine was absorbed in 15 min. The peaks in red blood cell and plasma concentrations were achieved in 4 h and then progressively declined at a constant rate between 8 and 48 h. The drug is selectively taken up by liver, bone, and cartilage, and more than 30% of it is concentrated in skeletal tissue. It is metabolized by the liver, has no active metabolism, and more than 82% of it is converted into CO_2 and then exhaled. Approximately 6 and 5% are excreted in the urine and feces, respectively [74].

A relatively recent article on administration, distribution, metabolism, and elimination (ADME) of glucosamine [75] reported that after oral administration of 7.5 g of "cold" glucosamine in a single dose, the plasma levels were below 3 μg/mL. When a single dose of glucosamine (314 mg) was spiked with ^{14}C-labeled glucosamine, the radioactivity was detected in plasma globulin after

1.5 h, and reached a peak after 9 h of administration. Half of the radioactivity was eliminated in 58 h. The absolute oral bioavailability was 44%. More than 88% of the administered dose was absorbed through the gastrointestinal tract. The total excretion within 24 h in the urine was about 1.2%, mostly happening in the first 8 h after administration. When the dose was reduced to 25% and administered daily for 7 days, the daily urinary excretion in the last 24 h reached 2.2%. The study also showed that a single dose of 1884 mg or the same amount split in three equivalent tid doses was equivalent. The study also demonstrated that there was no evidence of interference of glucosamine with ADME of glucose. It is important to realize that radioactivity *per se* is not necessarily a marker for an intact glucosamine molecule. Most of the studies were carried out with a ^{14}C-labeled glucosamine, labeled at a single carbon normally in position C-1, thus it could only be concluded that glucosamine or glucosamine metabolite still containing the original C-1 site was detected. Glucosamine is likely biosynthetically oxidized to glucosamine *N*-oxide, then converted into glucorolactone before it undergoes further structural modification.

8.3.2 Effects on Cartilage Rebuilding

8.3.2.1 Preclinical Studies

The first associations between glucosamine supplementation and cartilage growth can be traced to the 1950s at the Karolinska Institute in Stockholm, Sweden. Their experiments demonstrated that the supplementation of cultured cartilage cells with D-glucosamine hydrochloride increased the production of chondroitin sulfate by about three times compared to the control. Galactosamine, a structural isomer of glucosamine, stimulated the production only two times. Stimulation of ^{35}S uptake in chondroitin sulfate was completely abolished by glucose under physiological conditions [76]. The effect of glucosamine on osteoarthritis was first reported in Germany in 1969 [77]. These experiments involved glucosamine in injectable form. Later Karzel and Domenjoz [78] published a paper showing the effect of glucosamine salts on the secretion of GAG by cultured murine embryonic fibroblasts. D-glucosamine hydrochloride, D-glucosamine sulfate, and D-glucosamine iodide each had a pronounced effect on the synthesis of GAG. NAG and *N*-acetylgalactosamine were considerably less effective, while glucuronic acid was ineffective. Galactosamine primarily stimulated the formation of hyaluronic acid rather than the production of chondroitins. No sulfate was added to the cultures [70]. Nonsteroidal anti-inflammatory drugs inhibited the synthesis of GAG and protein and this inhibition was partially rescued by glucosamine. Glucosamine stimulated the incorporation of sulfate and proline into articular cartilage cultures [79]. Shortly after this study, the first human clinical studies were performed using 500 mg glucosamine sulfate capsules produced by Rottapharm Laboratorium (Monza, Italy) [80–82].

8.3.2.2 Clinical Studies

At least 20 randomized clinical trials were published between 1980 and 2004, with most of them being performed in Italy, Czech Republic, and Germany.

The others were conducted in France, Russia, Thailand, Portugal, Philippines, China, Japan, Canada, and the United States. All of the studies, with the exception of the first Canadian study, used glucosamine sulfate and not glucosamine hydrochloride. Many of them had some sort of involvement from Rotta Research Laboratorium, the early manufacturer of glucosamine sulfate.

T. McAlindon from Boston University School of Medicine (Boston, MA) and T.E. Towheed from the Department of Medicine at Queen's University (Kingston, ON), are two established experts who have been critically evaluating the rigorosity of glucosamine clinical trials for some time. In 1998, McAlindon [83] reviewed the performance of glucosamine in six studies and concluded that it was superior over a placebo. Insufficient information about study design and conduct prevented a conclusive evaluation. In the same year Towheed [84] reviewed nine randomized clinical trials with glucosamine and found that seven of them were superior to placebo. In two trials comparing glucosamine with ibuprofen, glucosamine was equal to ibuprofen in one trial and in the other it was superior to ibuprofen. Later, Towheed et al. [85] completed a systematic review of 12 randomized clinical trials of parallel group design, 11 of them were published between 1980 and 1999, 8 compared glucosamine sulfate and 1 compared glucosamine hydrochloride with placebo, 1 to placebo and NSAID and 1 to NSAID only [85]. Mean trial duration was 6.25 weeks involving 1481 patients with a mean age of 61 years. The method of administration varied as well as the dosage. In eight trials, oral glucosamine was used as a 500 mg tid; in others a 400 mg i.m., i.v., or intraarticular (i.a.) was administered either daily or twice a week. Rotta Research Laboratory sponsored nine of these clinical trials. Nine of the trials studied the knee, one studied multiple sites, and two did not indicate the site being studied. The safety profile was excellent but long-term safety was not evaluated.

The first controlled studies assessing the effect of glucosamine sulfate on osteoarthritis were small [80–82,86]. In one of the randomized double-blind placebo-controlled trials, 24 patients with osteoarthritis of the knee receiving glucosamine sulfate at 500 mg tid were observed for 6–8 weeks. A significantly higher proportion of the patients experienced subjective improvement with regard to articular pain and joint tenderness, but with no progress in movement restriction [80]. A similar study for a shorter duration (30 days) was extended to 80 patients, and although there was a larger score improvement in glucosamine-treated patients, the results were not tested for statistical significance and placebo effects were too high. The results were supported by data from electron microscopy [81]. The trial was included as a well-designed study in a recent meta-analysis [69]. Crolle compared two groups of seniors with 15 subjects in each group. One group received 400 mg glucosamine sulfate i.m. daily for a week followed with an oral administration of 500 mg of glucosamine sulfate tid for 2 weeks; the other group received i.m. injection of an antirheumatic, antiarthritic medication daily for a week followed with a daily oral administration of placebo for 2 weeks. The assessment of pain during the passive and active movements concluded that patients receiving glucosamine did 15% better compared to the placebo group [82]. Similar trends were observed in a similar trial reported by D'Ambrosio et al. [86]. The limited size and duration of both trials preclude a definitive conclusion as with the trial executed by Pujalte et al. [80], so none of these trials passed the evaluation by Kayne et al. [69].

A large, multicenter, open-labeled trial was conducted in Portugal with 1208 ambulatory patients. A dose of 500 mg glucosamine sulfate tid was administered for a mean period of 50 days. Sixteen percent of the patients also took an inflammatory medication or acetaminophen during the trial. About 84% of the patients had arthritis of the spine, hip, or knee. Eighty-eight percent of patients reported no side effects while the rest experienced very mild discomfort such as heartburn, diarrhea, etc. [87]. Although the patients showed improvements in their symptoms of pain at rest, while standing and during limited active movements, it is possible that the open-labeled design may have led to a biased interpretation of the results.

A randomized placebo-controlled, double-blind trial involving 252 patients older than 18 years with osteoarthritis of the knee was performed in Berlin, FRG. The patients had suffered from the disease for at least 6 months prior to the study and the Lequesne (LQ) index had to be at least 4 points. The patients were not allowed to take NSAID 2 weeks prior to starting the trial and then received 500 mg glucosamine sulfate tid orally. After 4 weeks of treatment, the LQ index decreased by 3.2 and 2.2 points in the glucosamine sulfate and placebo group, respectively. The improvement in pain and limitation of movement was improved by 55% in the treatment group and 38% in placebo group, respectively [88].

The first North American randomized clinical trial evaluating glucosamine was conducted by Houpt with colleagues from Mt. Sinai Hospital and the University of Toronto (Toronto, ON). They carried out an 8 week randomized double-blind placebo-controlled parallel trial in which they used 500 mg of glucosamine hydrochloride on 118 outpatients with primary gonarthritis [89]. The primary endpoint was the degree of change based on Western Ontario-McMaster University total scores (WOMAC) and the relevant subscores. Changes in the primary endpoint of the WOMAC pain subscore favored glucosamine hydrochloride over placebo, but the differences were not statistically significant. The secondary endpoints of pain reduction recorded in diaries and during knee examinations were statistically favorable for glucosamine hydrochloride.

Glucosamine did not perform better than placebo in an American randomized, double-blind parallel group single-center trial, in which 98 patients receiving 500 mg of glucosamine sulfate tid or placebo were followed for 2 months at Veterans Affairs Medical Center, Prescott, AZ. The researchers speculated that the trial might have been negative because subjects were older, heavier, and with longer duration and radiographic severity of OA when compared to subjects in other clinical trials. These results may indicate that patients with a more progressed illness do respond but to a significantly lesser degree as compared to those who are affected less severely [90].

In another 3-year randomized, double-blind placebo-controlled study involving 212 subjects with gonarthritis, a treated group received 1500 mg of glucosamine sulfate daily. The 106 placebo patients had a progressive joint-space narrowing, with a mean joint-space loss of 0.31 mm. There was no significant joint-space loss in the glucosamine sulfate–treated patients. WOMAC scores revealed slight symptom worsening in the placebo patients, as compared with improvement observed in the glucosamine sulfate treatment group [91]. The University of Liege and the WHO Collaborating Centre carried out the study for Public Health and Aspect of Osteoarticular Disorders, Liege, Belgium and was supported by Rotta.

In a clinical trial carried out by the Institute of Rheumatology and the Department of Medicine and Rheumatology, Charles University, Prague, sponsored by the Rotta Research and the Rottapharm Group, 202 patients (45–70 years old) with OA of the knee (using American College of Rheumatology Criteria) were randomized and received either 1500 mg of glucosamine sulfate or placebo, orally once a day. They had mild to moderate osteoarthritis at enrollment; with average joint-space widths of slightly less than 4 mm and a LQ index score less than 9 points. After 3 years, progressive joint-space narrowing was 0.19 mm in the placebo group with no average change in the glucosamine sulfate–treatment group. In the placebo group, improvement of symptoms was modest whereas the treated group improved by 25%. The study demonstrated that long-term treatment with glucosamine sulfate inhibited the progression of gonarthritis [92].

In another randomized, placebo-controlled, double-blind trial of the efficacy of glucosamine sulfate in managing pain in gonarthritis, 80 patients were recruited from a rheumatology outpatient clinic. They received either 1500 mg of glucosamine sulfate or placebo daily for 6 weeks, but there was no evidence that glucosamine sulfate was more effective than the placebo [9].

Recently, a study designed to investigate the effect of glucosamine sulfate on long-term symptoms and the progression of gonarthritis of postmenopausal women was published. The paper reports the results on a combination of two 3-year, randomized, placebo-controlled, prospective independent studies and their *post hoc* analysis. Differences between joint space at baseline and after 3 years were assessed and the symptoms were scored by the algofunctional WOMAC index. In all, 414 women, 319 of them postmenopausal, were involved in the study. The glucosamine sulfate group did not show narrowing joint space (joint space change 0.003 mm) whereas the placebo group did (join space change –0.33 mm). Based on the WOMAC index, 14.1% showed an improvement, whereas in the placebo group, 5.4% of patients reported worsening. These are long-term studies on a large cohort of patients demonstrating a disease-modifying effect after intervention with glucosamine [93–95]. These studies were supported by Rotta and were collaborated between the above-mentioned Belgian group, Charles University, and the Institute of Rheumatology in Prague.

Results of a recent four-center, 6-month, randomized, double-blind, placebo-controlled glucosamine discontinuation trial conducted in Canada were reported [96]. The study was conducted with 137 current users of glucosamine with gonarthritis who had experienced at least moderate improvement in knee pain after starting glucosamine. The patients received a maximum dose of 1500 mg/day. Follow-up continued for 6 months or until disease flare, whichever occurred first. The study demonstrated no evidence of symptomatic benefit from continued use of glucosamine sulfate. A very recent, smaller and shorter Japanese study, involving the same dose, did not show antirheumatic effect evaluated by conventional tests; however, there were noticeable improvements in symptoms [97].

The most prominent clinical trial to date was a 24-week clinical trial funded by the National Center for Complementary and Alternative Medicine and the National Institute of Arthritis and Musculoskeletal and Skin Disease (NIAMS/NCCAM). It was a multicenter five-arm placebo-controlled study called the Glucosamine Arthritis Intervention Trial (GAIT). This study involved 1583 patients with symptomatic knee osteoarthritis. They received 1500 mg of glucosamine hydrochloride, 1200 mg chondroitin sulfate, both glucosamine hydrochloride and chondroitin sulfate, 200 mg of celecoxib, or placebo daily. Glucosamine or chondroitin sulfate did not effectively reduce the pain in the overall group of patients but the combination was effective in the subgroup of patients with moderate- to severe pain [98]. It is important to note that several months earlier, Russians published results of a study on 90 women suffering from knee osteoarthritis who were taking in combination with 50 mg of diclofenac (NSAID), 500 mg of glucosamine hydrochloride and 500 mg of chondroitin sulfate twice a day for 1 month, and then once a day for the following 5 months. Both WOMAC and visual analogue scale (VAS) scores were improved after 4 months of therapy. After 1 month of therapy 4.5% patients gave up taking diclofenac, after 4 months the proportion of the patients who discontinued NSAID was raised to 20% and after 6 months to 40% [99].

Many of the clinical trials carried out to date are not long enough in duration to demonstrate long-term benefit. The validity of the most frequently administered oral dose used in the clinical trials (1500 mg daily or 500 mg tid) is unclear. Further dosing studies are required to determine the optimum dosage of glucosamine for OA. Further work is also required to determine whether glucosamine could be used in combination with other drugs such as NSAID that are used for the treatment of OA. Also, little is known about any potential negative interactions between glucosamine and drugs, including drugs for OA such as NSAID and COX-2 inhibitors.

8.3.3 Glucosamine versus Ibuprofen

Several clinical trials have been designed to compare glucosamine sulfate with ibuprofen. In one double-blind trial, 40 patients suffering from gonarthritis received either 1500 mg of glucosamine sulfate or 1200 mg of ibuprofen orally for 8 weeks. No difference in the extent of inflammation between the treatments was found, but there was difference in the level of articular pain. During the first and the second week, ibuprofen was more effective; however, by the fourth week the treatments were equal, and glucosamine sulfate was better than ibuprofen by the eighth week [100].

A larger clinical trial comparing ibuprofen with glucosamine sulfate was conducted on 200 patients who suffered from arthritis for at least 5 years. These patients had suffered with pain for at least 3 months and had an LQ index of at least 7 points. They received either glucosamine sulfate 500 tid or ibuprofen 400 tid and after the first week of treatment, 41% of ibuprofen-treated patients showed a clinical improvement compared to 28% of the glucosamine group. The proportion changed to 39 and 48% after 2 weeks, respectively, and after 3 weeks onward the response was practically equal; 48% for glucosamine compared with 52% for ibuprofen. There were only 6 adverse events in the glucosamine group compared to 35 in the ibuprofen group. Although glucosamine sulfate had a delayed response consistent with the International League Against Rheumatism (ILAR) classification, after 3 weeks both drugs displayed the same efficacy [101].

Canadian researchers (University of Alberta, Edmonton, AB) also compared glucosamine sulfate to ibuprofen for the treatment of temporomandibular joint osteoarthritis in a randomized double-blind controlled 3-month clinical trial. Forty-five patients were treated with either glucosamine sulfate 500 mg tid or ibuprofen (400 mg tid) for 90 days. Both glucosamine sulfate and ibuprofen reduced the pain levels in patients with temporomandibular degenerative joint disease [102].

Viatril-S® (a form of glucosamine sulfate) reduced pain in a Chinese, Rotta-sponsored, double-blind clinical trial on 178 patients. The patients, randomized into two groups, suffered from gonarthritis. They received 500 mg of glucosamine sulfate orally tid or 1200 mg of ibuprofen. After 2 weeks, patients in both groups experienced symptom improvement but glucosamine sulfate was better tolerated and was more efficacious [103].

8.3.4 Glucosamine Formulations

The effects of orally administered D-glucosamine hydrochloride, chondroitin sulfate, and manganese ascorbate (GCMA formulation) were evaluated on the articular cartilage in dogs with a cranial cruciate ligament (CCL). A cross-sectional analysis at 3 months revealed that synovial fluid from the CCL transected knees of GCMA formulation–treated dogs had significantly increased concentrations of synovial fluid chondroitin sulfate 3B3 epitope, and 7D4 epitope in comparison to controls. Furthermore, 7D4 concentrations remained significantly elevated in CCL transected knees of GCMA formulation–treated dogs over the 5-month period. However, the treatment effect of the GCMA formulation was no longer significant when epitope concentrations were expressed as a ratio of CCL-transected to a contralateral nonoperated knee. Reconstruction of the CCL had no major effect on the synovial fluid epitope. The data suggest that 3B3 and TD4 epitopes may act to modulate articular cartilage matrix metabolism *in vivo* [104].

Ocean Nutrition Canada patented a formulation containing ferrous ions, ascorbic acid, and glucosamine, which has a synergistic effect on cartilage development

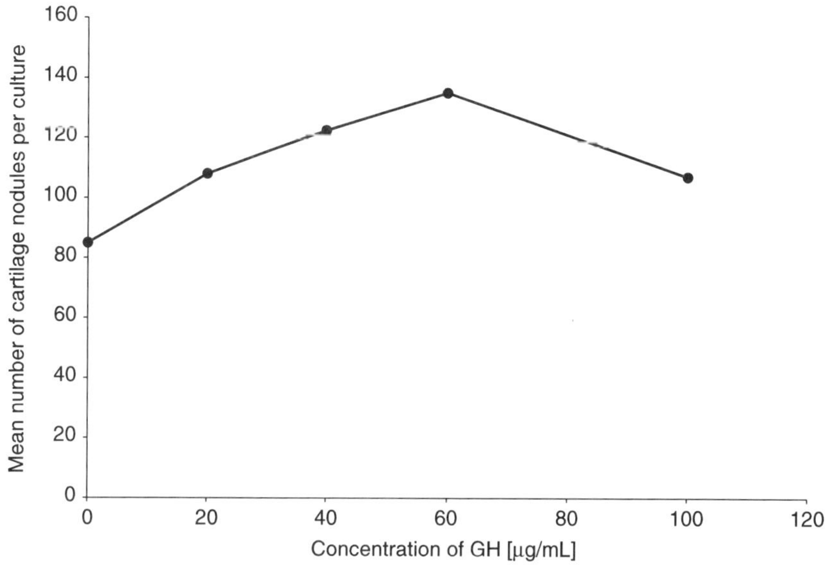

FIGURE 8.3 The effect of $FeSO_4$, ascorbic acid and glucosamine hydrochloride on cartilage development. Cell cultures were prepared as described previously [158]. The medium was fortified with additional AA (75 mg/mL) to give the total concentration of 150 mg/mL, the concentration of $FeSO_4$ was 2 mg/mL, and the concentration of D-glucosamine hydrochloride (GH) varied. The amount of generated cartilage was measured as the mean number of cartilage nodules per culture.

in vitro [105]. Although Levenson [106] reported on the effect of ascorbic acid on chondrocytes, Ocean Nutrition Canada was the first to identify a synergistic effect between ascorbic acid and ferrous ions. Ocean Nutrition Canada's research revealed that a combination of ferrous sulfate (2 μg/mL) and ascorbic acid (150 μg/mL) gave the best response in terms of production of cartilage nodules *in vitro*. When combined with various concentrations of glucosamine hydrochloride, a dose response was observed showing that 60 mg/mL of glucosamine gave optimum cell growth (unpublished results; Figure 8.3).

When stimulation of cartilage growth *in vitro*, as measured by extracted proteoglycan matrix and cell density, was compared with Ocean Nutrition Canada's formulation (ONC-114) and two leading brands (Osteo Bi-Flex and Shiff™), the ONC-114 formulation performed significantly better. Figure 8.4a shows an electromicrogram of the cartilage cells after staining with Alcian blue and this result is also illustrated in the form of a bar graph in Figure 8.4b.

The ascorbic acid and ferrous combination probably promote cartilage development by boosting collagen synthesis, while glucosamine is involved in rebuilding proteoglycan matrix. Cartilage is a composite in which a polyelectrolyte matrix is reinforced by an embedded network of hybrid collagen fibers, containing 65–80% water, 15–25% collagen, and 3–10% proteoglycan [107].

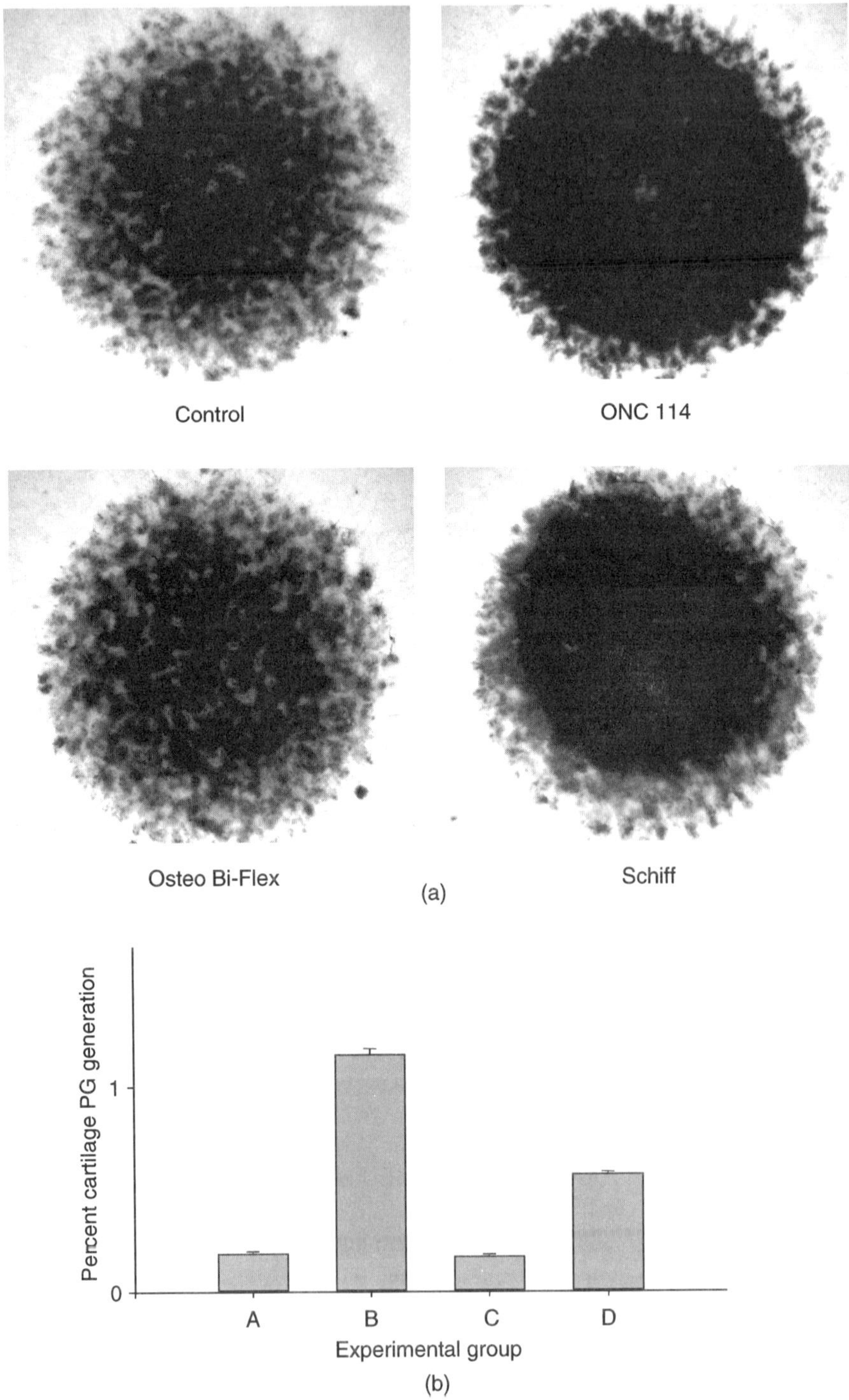
Control
ONC 114
Osteo Bi-Flex
Schiff
(a)
Percent cartilage PG generation
1
0
A
B
C
D
Experimental group
(b)

8.3.5 Glucosamine and Insulin Resistance

There is some concern in the medical community about possible side effects of glucosamine. One possible side effect recently identified is the stimulation of atherosclerosis progress [108]. A potential side effect of glucosamine that requires further investigation is the possible impact of glucosamine on glycemic control. Animal studies indicated that glucosamine might negatively impact glycemic control at very high dose, although this has not yet been seen in humans [109,110]. The potential effect on glycemic control may be mediated via glucosamine interaction with the enzyme that regulates blood sugar levels and may be related to the structural similarity of glucosamine to glucose [111].

A recent study was designed to evaluate possible effects of glucosamine supplementation on glycemic control in a selected population with type 2 diabetes mellitus. This type of diabetes is common in the aging population occurring in 10–15% of people older than 50 years. This is also the population that is to a large extent suffering from osteoarthritis. The study, a 90-day placebo-controlled, double-blind randomized clinical trial, involved 39 patients and demonstrated that oral supplementation of glucosamine sulfate (500 mg, and chondroitin sulfate 400 mg) tid did not result in a clinically significant shift in glucose metabolism in patients with type 2 diabetes mellitus [112]. Two 3-year placebo-controlled trials showed either no effect or a slight lowering effect on serum glucose levels [91,113]. Monauni et al. [114] suggested that glucosamine might increase insulin resistance and suppress insulin secretion. It has also been suggested that the long-term use of glucosamine may affect glucose homeostasis [115]. Until this potential side effect is clarified by further research, it is recommended that people with diabetes taking glucosamine have their blood sugar monitored regularly [116–118].

8.3.6 Glucosamine Hydrochloride versus Glucosamine Sulfate

Almost all clinical trials have been performed with glucosamine sulfate and there is controversy as to whether there are differences in bioactivity between

FIGURE 8.4 A quantitative comparison of proteoglycan produced by ONC-114 as compared to two leading glucosamine brands. The cartilage was produced in the presence of ONC-114 (100 μg/μL glucosamine hydrochloride, 150 μg/mL ascorbic acid and 2 μg/mL of $FeSO_4$) Schiff products contained 100 μg/mL of glucosamine hydrochloride, 33 μg/mL GS, and 33 μg/mL of NAG. Osteo Bi-Flex contained 100 μg/mL glucosamine hydrochloride, 100 μg/ml galactosamine, and 100 μg/mL glucoronic acid. On day four the cartilage was treated with 1 mM retinoic acid (an experimental inducer of cartilage degradation), and after 18 additional hours the cartilage was stained with Alcian blue and the amount of proteoglycan was measured spectrometrically at 600 nm. The amount of proteoglycan produced by ONC-114 was scaled to 100% to enable direct comparison with the other formulations. A is the control that was not treated, B was treated with retinoic acid only, C was treated with retinoic acid and ONC-114, D was treated with RA and Osteo Bi-Flex™, and E was treated with RA and Schiff. The cultures were stained with Alcian blue to visualize cartilage tissue. (b) Quantitation of the cultures in (a) determined by extraction and quantitation of the proteoglycan matrix.

glucosamine hydrochloride and glucosamine sulfate. Most preparations commercially available and sold as glucosamine sulfate are mixtures of glucosamine hydrochloride and an inorganic sulfate such as potassium sulfate, rather than a true glucosamine sulfate. Only some are in the form of a mixed salt, the main representative of which is a product of formula $C_6H_{13}Cl_2K_2SO_4$. Pure glucosamine sulfate is extremely hygroscopic and unstable, and is not sold commercially. The original reference for glucosamine sulfate dates back to 1898, but there are no details on the preparation or physicochemical properties of the compound [119]. Rottapharm researchers are the original inventors of stabilizing freshly prepared glucosamine sulfate with NaCl (e.g., Viatril-S) and other metal halides, and the formation of apparently stable mixed salts.

Any salt of glucosamine is ionized in the acidic stomach and converted into glucosamine hydrochloride. In the small intestine, the pH is elevated to about 6.8 and a significant portion of glucosamine hydrochloride (p*K*a 6.91 at 37°C) [120] is probably converted into the free base. Humans consume sulfur in the form of the amino acids methionine and cysteine that are present in most food proteins and therefore exogenous sulfate supplementation should have little beneficial effect *in vivo*, although a study demonstrated that the depletion of inorganic sulfate impaired synthesis of GAG *in vitro* and exogenous sulfate offsets the depletion [121]. One study indicated that patients with arthritis may be deficient in sulfur, and restoring sulfur levels can result in significant improvements of osteoarthritis conditions in deficient patients [122,123]. In a recent paper Hoffer [124] suggests that sulfate could mediate the therapeutic effect of glucosamine . Thus, it seems that in some cases sulfate supplementation may be useful, but there is no scientific rationale for selecting glucosamine sulfate over glucosamine hydrochloride. Some formulations contain glucosamine hydrochloride and sulfur in the form of dimethylsulfone, commonly known as MSM. MSM not only provides sulfur but also appears to have additional anti-inflammatory properties.

In addition to this glucosamine sulfate versus glucosamine hydrochloride controversy, some also suggest that NAG (*N*-acetylated form of glucosamine) is just as effective as any other form of glucosamine. This appears to be false since evidence indicates that NAG does not have active intestinal transport, but is digested by intestinal bacteria, then binds to lectins in the gut and is excreted in the feces as a lectin–NAG complex [125].

8.4 SUMMARY

Since osteoarthritis is a particularly serious problem in the elderly and the number of sufferers will increase owing to population aging, an effective but less toxic alternative to NSAID is highly desirable. Glucosamine has been studied for a number of years and there is considerable evidence indicating that this compound deserves more attention, especially within the medical community, which tends to be resistant to the use of supplements or other non-FDA approved materials or processes.

Glucosamine sulfate and glucosamine hydrochloride have no major known side effects, while approved NSAID and COX-2 inhibitors drugs have well-established contraindications and adverse effects. For instance, NSAID may erode the digestive tract and cause internal bleeding and or liver failure when taken over a long period of time [126]. Glucosamine, contrary to NSAID, does not irritate the gastrointestinal tract, in fact it may even stimulate the production of protective proteoglycans [127,128]. According to a review article published in the *New England Journal of Medicine*, anti-inflammatory drugs (prescriptions and over-the-counter drugs) alone cause over 16,500 deaths and over 103,000 hospitalizations per year in the United States [129]. NSAID may also actually inhibit cartilage repair and intensify cartilage destruction, while only masking osteoarthritis symptoms [130–132]. Therefore, although NSAID effectively reduce osteoarthritis symptoms such as pain, they appear to exacerbate the condition by inhibiting cartilage formation and speeding up cartilage destruction. Clinical studies have shown that the use of NSAID is associated with worsening osteoarthritis and increased joint destruction [133–135]. Long-term care using NSAID in essence masks pain but does not treat or prevent the underlying condition from worsening. Many of the newer COX-2 medications such as Vioxx and Celebrex® are more selective than the NSAID, but they come with an extensive list of toxicity and side effects. The announcement of a voluntary worldwide withdrawal of Vioxx by Merck (Whitehouse Station, NJ) made on September 30, 2004 caused significant uncertainty around the use of COX-2 inhibitors and left arthritis patients with limited treatment options. The withdrawal decision was based on the data from a 3-year prospective, randomized, placebo-controlled clinical trial study, that indicated an increased relative risk for confirmed cardiovascular events, such as heart attack and stroke, beginning after 18 months of treatment in the patients taking Vioxx compared to those taking placebo [136]. The main concern about COX-2 inhibitors is their potential for blood clotting that is associated with strokes and heart attacks. The withdrawal of Vioxx also brought into question the safety of other COX-2 inhibitors such as Pfizer's drug, Celebrex.

The quality of glucosamine clinical trials varies greatly [137–139], but there are several convincing multicenter placebo-controlled trials published. Applying a quality rating scale to randomized controlled trials used to evaluate the efficacy of NSAID for osteoarthritis of hips and knees indicates that glucosamine performs at least as well as NSAID [140,141]. In addition, glucosamine appears to have an ability to treat the underlying cause of osteoarthritis by aiding in cartilage synthesis and inhibiting cartilage breakdown.

It is not clear whether the combination of glucosamine with chondroitin sulfate is better than glucosamine alone. So far most of the studies demonstrating clinical benefit of chondroitin sulfate have involved direct injection into the site affected with osteoarthritis rather than oral administration. Because chondoitin sulfate is a relatively large polymeric molecule, its bioavailability is uncertain. It is not clear how much orally ingested chondroitin sulfate is essentially destroyed by digestive enzymes or excreted from the body undigested. There is uncertain scientific rationale for improving glucosamine formulations with chondroitin sulfate. The results from the Glucosamine/Chondroitin Arthritis Intervention Trial sponsored by NIH did not provide clear guidance as to the efficacy of the combination versus each ingredient alone, since the combination was applied in the dose that was the sum of glucosamine dose and chondroitin sulfate dose [98].

Glucosamine is typically administered in the dose of 500 mg tid. So far, studies indicate that glucosamine is safe and nontoxic but may in some instances cause minor side effects such as bloating, diarrhea, heartburn, gas, indigestion, and stomach upset. Some of the glucosamine sulfate supplements may contain large amounts of potassium or sodium, which can be a problem for individuals needing to control their intake of these salts, particularly sodium. Glucosamine comes from chitin and is derived from shellfish and thus glucosamine may be problematic for people with allergies to shellfish. Glucosamine hydrochloride originating from shellfish is almost exclusively produced by 10 M HCl (9–12 M), and under these conditions residual protein in chitin is destroyed leaving residues that are well below concentrations causing allergic reactions. With inconsistency in the quality of glucosamine products entering the supplement market, individuals with shellfish allergies should be cautious. Improperly hydrolyzed glucosamine could contain some residue protein and therefore produce an allergic reaction in sensitive individuals. So far only two cases of a possible link of glucosamine to hypersensitivity were reported even though glucosamine is a widely used supplement ingredient [142,143].

Although most of the glucosamine clinical trials conducted to date have in some way been associated with methodical flaws or bias, meta-analyses show a clear trend demonstrating glucosamine efficacy in the symptomatic management of osteoarthritis. There is no definitive clinical evidence, however, for its mechanism of action, including its chondrostimulating effect [144]. *In vitro* studies show that a dose-dependent increase in proteoglycan synthesis occurs after human chondrocytes are challenged with glucosamine indicating at least one mechanism of action [145]. Although both glucosamine sulfate and chondroitin sulfate stimulate chondrocyte growth *in vitro* and in animal models, there is no direct evidence suggesting regeneration of cartilage affected with osteoarthritis [73,146,147]. Glucosamine was shown to impact chondrocyte gene expression, inducing a twofold increase in mRNA coding for proteoglycans such as perlecan and aggrecan [148]. In addition, interleukin-1 aggrecanase–triggered activity can be induced by glutamine or glucosamine [149]. Very recently, glucosamine was found to upregulate TGF-beta-1 mRNA levels. This ability explains the mechanism of glucosamine–chondrocyte interaction, and it could hopefully be exploited in the design of cartilage repair strategies [150].

The current status of glucosamine therapy on osteoarthritis was recently reviewed by Towheed [141,151], and one conclusion from this review is that recent independent studies of glucosamine have had less positive results than earlier industry-sponsored trials for knee osteoarthritis. The clinical trials using Rotta preparations revealed that glucosamine was superior to placebo in the treatment of pain and functional impairment resulting from osteoarthritis, but no difference was found between non-Rotta preparations and placebo [151]. It cannot be concluded that this difference is because of interference of the parties interested in the success of glucosamine in industry-sponsored studies [152]. Further clinical studies are still required to show conclusively the ability of glucosamine to prevent or treat osteoarthritis.

Potential glucosamine antiinflammatory activity also warrants further study. When compared with the anti-inflammatory agent indomethacin, glucosamine had an anti-inflammatory effect on mice models that was 50–300 lower than that of indomethacin, but its toxicity was 1000–4000 times lower, and therefore glucosamine had the higher therapeutic index [127]. Also, glucosamine is not an inhibitor of cyclooxygenase, as many NSAID are, and so does not have the associated side effects identified for this class of drug. Anti-inflammatory effect of glucosamine may be due to synthesis of proteoglycan and an ability to stabilize cell membranes. Glucosamine also inhibits superoxide radicals and lysosomal enzymes [153]. Further evidence for the anti-inflammatory effects of glucosamine is a study that found that adding glucosamine to certain NSAID significantly reduced the required dose of NSAID. For example, combining glucosamine with the drug indomethacin reduced the required amount of the drug by 2–2.7 times [154].

In summary, the preponderance of clinical trial data supports the efficacy and safety of glucosamine for the prevention and treatment of osteoarthritis. However, there are a number of trials that show no therapeutic benefit and several trials that are not adequately controlled. In addition, a number of clinical trials showing efficacy with glucosamine sulfate were supported by the product manufacturers. Before assuming bias in these studies, it should be remembered that drugs available on the market are there as a result of clinical trials funded by the pharmaceutical industry [155]. A number of important questions still remain regarding the optimum dose and the best route of administration. Most clinical trials employed 500 mg tid or 1500 mg daily and so most supplements recommended an oral dose of 1500 mg per day. Although there are no indications of safety issues during long-term use, there is limited long-term prospective safety data for glucosamine. New clinical trials conducted in Europe and the United States may result in some positive information regarding the optimal formulation and doses to be used [156]. Furthermore, additional clinical data is required to confirm the ability of glucosamine to promote cartilage growth or decrease cartilage degradation *in vivo* in humans. In 2004, NIH started recruiting patients for another U.S. government–sponsored clinical trail to evaluate the effects of oral glucosamine on insulin information regarding glucosamine safety [157].

REFERENCES

1. Holub, B. 2004. A preventive model for health care using nutraceuticals and functional foods. *Inform* 15, 90–91.
2. Bucci, L.R. 1994. Chondroprotective agents. Glucosamine salts and chondroitin sulfates. *Townsend Lett. Doct.*, 1, 52–54.
3. Hochberg, M.C., Altman, R.D., Brandt, K.D., Clark, B.M., Dieppe, P.A., Griffin, M.A., Moskowitz, R.W. and Schnitzer, T.J. 1995. Guidelines for the medical management of osteoarthritis. Part I. Osteoarthritis of the hip. American College of Rheumatology. *Arthritis Rheum.* 38, 1535–1540.

4. Hochberg, M.C., Altman, R.D., Brandt, K.D., Clark, B.M., Dieppe, P.A., Griffin, M.R., Moskowitz, R.W. and Schnitzer, T.J. 1995. Guidelines for the medical management of osteoarthritis. Part II. Osteoarthritis of the knee. American College of Rheumatology. *Arthritis Rheum.* 38, 1541–1546.
5. Hinz, B. and Brune, K. 2004. Pain and osteoarthritis: new drugs and mechanisms. *Curr. Opin. Rheumatol.* 16, 628–633.
6. White, W.B., Strand, V., Roberts, R. and Whelton, A. 2004. Effects of the cyclooxygenase-2 specific inhibitor valdecoxib versus nonsteroidal anti-inflammatory agents and placebo on cardiovascular thrombotic events in patients with arthritis. *Am. J. Ther.* 11, 244–250.
7. Challener, C. 2003. Specialty supplements are the bright spot in US dietary supplement market—Focus 2003: food additives/nutraceuticals/vitamins—industry overview. *Chem. Market Rep.* July 14.
8. Towheed, T.E. and Anastassiades, T.P. 2000. Glucosamine and chondroitin for treating symptoms of osteoarthritis. Evidence is widely touted but incomplete. *JAMA* 283, 1483–1484.
9. Hughes, R. and Carr, A. 2002. A randomized, double blind, placebo controlled trial of glucosamine sulphate as an analgesic in osteoarthritis of the knee. *Rheumatology* 41, 279–284.
10. Anonymous 2000. Alternative medicine. Study finds glucosamine doesn't work. *Harv. Health Lett.* 25, 1052–1057.
11. Lequesne, M. 1994. Symptomatic slow acting drugs in osteoarthritis: a novel therapeutic concept? *Rev. Rheum.* (*Engl. Ed.*) 61, 69–73.
12. Henderson, R.W. 1996. Amino sugar and glycosaminoglycan composition for the treatment and repair of connective tissue. US 5,587,363.
13. Pipitone, V.R. 1991. Chondroprotection with chondroitin sulfate. *Drugs Exp. Clin. Res.* 17, 3–7.
14. Levene, P.A. 1925. The configurational relationships of the sugars, hydroxy acids, amino acids and halogen acids. *Chem. Rev.* 2, 179–216.
15. Anonymous 2002. Biopolymer Engineering opens a new plant. *Household Pers. Prod. Ind.* 39(March 25), 132–133.
16. Omata, S., Sakai, T. and Ichihara, Y. 1973. Fermentative production of glucosamine-containing polysaccharide. JP 48010291.
17. Anonymous 2003. Glucosamine gambit. *Chem. Eng. News* 81(February 17), 27–28.
18. Fan, W., Bohlmann, J.A., Trinkle, J.R., Steinke, J.D., Hwang, K.-O. and Henning, J.P. 2002. Glucosamine and method of making glucosamine from microbial biomass. WO 02/066667 A1; US Pat. Appl. 326549/10.
19. Hwang, K-O., Steinke., J.D., Henning, J.P., Bohlman, J.A., Trinkle, J.R. and Fan, W. 2003. Glucosamine and method of making from microbial biomass. US Pat. Appl. 382251/10.
20. Solheim, L. 2004. Cell free production of glucosamine. WO 2004031396. (113)
21. Fosdick, L.E., Bohlman, J.A., Trinkle, J.R. and Ray, B.L. (2003) Glucosamine and method of making glucosamine from microbial biomass. US Pat. Appl. 685125/10.
22. Kralovec, J., Guan, Y., and Boucher, G. 1997. D-Glucosamine hydrochloride. ONC Internal Document.
23. Novikov, V.Y. 1999. Kinetics of formation of D-(+)-glucosamine in acid hydrolysis of chitin. *Zh. Prikl. Khim.* 72, 147–152.
24. Kocourek, J. and Ticha, M. 1983. D-Glucosamine hydrochloride. CS 209258.
25. Berry, A., Burlingame, R.P. and Millis, J.R. 2002. Process and materials for production of glucosamine. US 6,372,457.

26. Deng, M-D., Angerer, J.D., Cyron, D., Grund, A.D., Jerrell, T.A. Jr., Leanna, C., Mathre, O., Rosson, R., Running, J., Severson, D., Song, L. and Wassink, S. 2004. Process and materials for production of glucosamine and *N*-acetylglucosamine. WO 2004003175.
27. Bosso, C., Defaye, J., Domard, A. and Gadelle, A. 1986. The behavior of chitin towards anhydrous hydrogen fluoride. Preparation of β-(1–4)–linked 2-acetamido 2-deoxy α-glucopyranosyl oligosaccharides. *Carbohydr. Res.* 156, 57–68.
28. Fuhimura, H., Kitano, H., Tanimoto, F., Kikuyama, H., Waki, M. and Hashiguchi, S. 2003. Method for producing glucosamine or (and) chitosan oligomer composition. JP 03183296.
29. Xao, L. and Zhang, J. 2002. Preparation of glucosamine sulfate. CN 1350000.
30. Lederhose, G. 1876. Ueber salzsaures Glycosamin. *Berichte* 9, 1200–1201.
31. Oswald, A. 1911. A simple method for preparing glucosamine hydrochloride from ovomucoid. *Z. Physiol. Chem.* 68, 173–180.
32. Neuberg, C. 1913. Minor communications on various subjects. 2. Preparation of D-glucosamine. *Biochemische Zeitschrift* 43, 500–507.
33. Oswald, A. 1915. Isolation of glucosamine hydrochloride from the mucoid of ascites fluid. *Z. Physiol. Chem.* 95, 100–101.
34. Hudson, C.S. and Dale, J.K. 1916. The isomeric penta-acetates of glucosamine and of chondrosamine. *J. Am. Chem. Soc.* 38, 1431–1436.
35. Komori, Y. 1926. Glucosamine compounds. *J. Biochem.* (Japan) 6, 1–20.
36. van Alphen, J. 1929. Preparation of glucosamine hydrochloride. *Chemisch. Weekblad.* 26, 602.
37. Hougland, P.L. and Hiltz R. St-C. 1953. Glucosamine from lobster shells. *Can. Progr. Rep. Atlantic Coast Stat.* 57, 6–8.
38. Stacey, M. and Webber, J.M. 1962. 2-Amino-2deoxy-a-D-glucose (α-D-Glucosamine). In: *Methods in Carbohydrate Chemistry.* Vol. 1. Whistler RL, Wolfrom ML, BeMiller JN, Shafizadech F, eds., Academic Press, New York, pp. 228–230.
39. Purchase, E.R. and Braun, C.E. 1946. D-Glucosamine hydrochloride. *Org. Syntheses* 26, 36-37.
40. Ingle, T.R., Vaidya, S.H. and Pai, M.U. 1973. Production of D-glucosamine hydrochloride (GAH) from fish canning waste. *Res. Ind.* 18, 54–55.
41. Oeriu, S., Lupu, E.R., Dimitriu, M.A. and Craesco, I. 1962. Glucosamine obtained from crustacean shells, and its biological and therapeutic importance. *Ann. Pharm. Franc.* 20, 66–72.
42. Kamasastri, P.V. and Prabhu, P.V. 1961. Preparation of chitin & glucosamine from prawn shell waste. *J. Sci. Ind. Res.* 20D, 466.
43. Nikolaeva, N.E., Sobolev, V.R. and Molochaeva, I.S. 1967. Preparation of glucosamine from shrimp shells, and its use in medicine. *Trudy VNIRO* 165–169.
44. Novikov, V.Y. and Ivanov, A.L. 1995. Method of preparation of D-(+)-glucosamine hydrochloride. RU 2042685.
45. Novikov, V.Y. and Ivanov, A.L. 1997. Method for the preparation of D(+)-glucosamine hydrochloride. *Zh. Prikl. Khim.* 70, 1543–1547.
46. Novikov, V.Y., Chilingarjan, G.G., Kompantsev, V.A. and Samokish, I.I. 1999. Method of preparing antiarthritic glucosamine hydrochloride by hydrolysis of chitin. RU 2141964.
47. Huang, J., Yang, Q. and Liang, S. 1993. Preparation of D-glucosamine hydrochloride from *Cryptotympana pustulata*. *Tianran Chanwu Yanjiu Yu Kaifa* 5, 28–29.
48. Li, N. and Li, J. 1997. Preparation of D-glucosamine hydrochloride. *Zhongguo Yaoke Daxue Xuebao* 28, 56–58.

49. Xia, W., Liu, X. and Zhang, F. 1997. Stability and preparation of D-glucosamine hydrochloride. *Wuxi Qinggong Daxue Xuebao* 16, 29–33.
50. Cao, G. 1998. Preparation of glucosamine hydrochloride from chitin. *Huaxue Shijie* 39, 250–253.
51. Zhou, P., You, Y., Qi, X., Zhang, Z., Geng, Z. and Dai, H. 2000. Preparation and properties of glucosamine hydrochloride. *Shuichan Xuebao (J. Fish China)* 24, 76–80.
52. Chen, X., Yuan, Y. and Lai, X. 2000. Preparation of D-glucosamine hydrochloride. *Shipin Kexue* 21, 34–36.
53. Wang, C. and Li, J. 2002. Preparation of amino glucose hydrochloride from chitin. CN 1335322.
54. Wang, F. 2002. Preparation of D-glucosamine hydrochloride from lobster shell. *Zhonguo Shenghua Yaowu Zazhi* 23, 129–131.
55. Huang, J., Yang, Q. and Liang, S. 1993. Study on preparation of D-glucosamine hydrochloride from *Periostracum cicadae*. *Nat. Prod. Res. Develop.* 5, 28–29.
56. Zhou P, You, Y., Ni, Y., Qi, X. and Tu, S. 2002. Characterization of glucosamine sulfate. *Shanghai Shuichan Daxue Xuebao* 11, 145–148.
57. Cohen, S.A. and Michaud, D.P. 1993. Synthesis of a fluorescent derivatizing reagent 6-aminoquinoyl-*N*-hydroxysuccinimidyl carbamate, and its application for the analysis of hydrolysate amino acids via high performance liquid chromatography. *Anal. Biochem.* 211, 279–287.
58. Cohen, S.A., De Antonis, K. and Michaud, D.P. 1993. Compositional protein analysis using 6-aminoquinoyl-N-hydroxysuccinimidyl carbamate, a novel derivatization reagent. In: *Techniques in Protein Chemistry.* IV, Angeletti RH, ed., Academic Press, San Diego, CA, p. 289.
59. Rovati, L. 1972. Pharmaceutically active glucosamine salts useful in the treatment of osteoarthritis and rheumatoid arthritis. US 3,683,076.
60. Li, J., Xu, J. and Zhu, Y. 1997. Study of D-glucosamine sulfate. *Pharm. Biotechnol.* 4, 102–104.
61. Senin, P., Makovec, F. and Rovati, L. 1987. Stable compounds of glucosamine sulfate. US 4,642,340.
62. Noujua, A.P., Rinne, K.Y. and Tulisalo, J.M. 1987. Mixed salts of glucosamine sulphate and a process for the preparation of the same. Eur. Pat. Appl. 0214642.
63. Schleck, J.R., Burger, C.M. and Chopdekar, V.M. 1998. Glucosamine sulfate potassium chloride and process of preparation thereof. US 5,843,923.
64. De Wan, M. and Volpi, G. 1998. Method of preparing mixed glucosamine salts. US 5,847,107.
65. Pinza, M., Segnalini, F., Marchetti, M., Iacoangeli, T. and De Vita, F. 2003. Method for preparing a glucosamine compound, and compound thus obtained. WO 20033055897.
66. McCoy, M. 2000. Stiff competition in arthritis salve. *Chem Eng. News* April 3, 20–1; http://www.nutraingredients.com/news.
67. Kralovec, J., Wright, J.H.D., McIsaac, S., Quach, H., Curtis, J. and Barrow, C. 2002. Isolation of D-glucosamine from shrimp shells; analysis, assessment and implementation of modern technology. In: *Advance in Chitin Science Vol. V,* Suchiva K, Chandrkrachang S, Methacanon P, Peter MG, eds., National Metal and Materials Technology Center (MTEC) Bangkok, pp. 45–48.
68. Anonymous 2003. Fuso builds nanotech-route glucosamine plant. *Jpn. Chem. Week* 44, August 19, 5.
69. Kayne, S.B., Wadeson, K. and MacAdam, A. 2000. Is glucosamine an effective treatment for osteoarthritis? A meta-analysis. *Pharm. J.* 265, 759–763.

70. Kim, J.J. and Conrad, H. 1974. Effect of D-glucosamine concentration on the kinetics of mucopolysaccharide biosynthesis in cultured chick embryo vertebral cartilage. *J. Biol. Chem.* 249, 3091–3097.
71. Mischiu, L., Marin, A. and Bostinaru, A. 1980. Activity of epimerases in *in vitro* cultured fibroblasts. *Morphol. Embryol.* 26, 231–235.
72. Matalon, R. and Dorfman, A. 1968. The structure of acid mucopolysaccharides produced by Hurler fibroblasts in tissue culture. *Proc. Nat. Acad. Sci.* 60, 179–185.
73. Zupanets, I.A., Bezdetko, N.V., Dedukh, N.V. and Otrishko, I.A. 2002. Experimental study of the effect of glucosamine hydrochloride on metabolic and repair processes in connective tissue structures. *Eksp. Klin. Farmakol.* 65, 67–69.
74. Setnikar, I., Giachetti, C. and Zanolo, G. 1994. Absorption, distribution and excretion of radioactivity after a single intravenous or oral administration of [^{14}C] glucosamine to the rat. *Pharmacotherapeutica* 3, 538–550.
75. Setnikar, I. and Rovati, L.C. 2001. Absorption, distribution, metabolism and excretion of glucosamine sulfate. *Arzneim-Forsch/Drug Res.* 51, 699–725.
76. Roden, L. 1956. Effect of hexosamines on the synthesis of chondroitin sulphuric acid *in vitro. Arkiv. Kemi.* 10, 345–352.
77. Bohne, W. 1969. Glukosamine in der konservativen Anthrosebelhandlung. *Med. Welt* 30, 1668–1671.
78. Karzel, K. and Domenjoz, R. 1971. Effects of hexosamine derivatives and uronic acid derivatives on glycosaminoglycane metabolism of fibroblast cultures. *Pharmacology* 5, 337–345.
79. Vidal y Plana, R.R., Bizzarri, D. and Rovati, A.I. 1978. Articular cartilage pharmacology: I. *In vitro* studies on glucosamine and non-steroidal, anti-inflammatory drugs. *Pharm. Res. Commun.* 10, 557–569.
80. Pujalte, J.M., Llavone, E.P. and Yiescupidez, F.R. 1980. Double-blind clinical evaluation of oral glucosamine sulfate in the basic treatment of osteoarthritis. *Curr. Med. Res. Opinion* 7, 110–114.
81. Drovanti, A., Bignamini, A.A. and Rovati, A.L. 1980. Therapeutic activity of oral glucosamine sulfate in osteoarthrosis: a placebo-controlled double-blind investigation. *Clin. Ther.* 3, 260–272.
82. Crolle, G. and D'Este, E. 1980. Glucosamine sulfate for the management of anthrosis: a controlled clinical investigation. *Curr. Med. Res. Opin.* 7, 104–109.
83. McAlindon, T.E., Gulin, M.P. and Felson, D.T. 1998. Glucosamine and chondroitin treatment for osteoarthritis of the knee or hip: meta analysis and quality assessment of clinical trials. ACR Conference, San Diego, CA, abs. #994.
84. Towheed, T.E. 1998. Glucosamine sulfate in osteoarthritis: a systematic review of clinical trials. *Arthritis Rheum.* 41 Suppl, S198.
85. Towheed, T.E., Anastasiades, T.P., Houp, J., Hochberg, M.C., Wells, G. and Shea, B. 1999. Glucosamine sulfate in osteoarthritis (protocol for a Cochrane review). In: The Cochrane Library Issue 2, Oxford: Update software.
86. D'Ambrosio, E., Casa, B., Bompani, R., Scali, G. and Scali, M. 1981. Glucosamine sulphate: a controlled clinical investigation in arthrosis. *Pharmatherapeutica* 2, 504–508.
87. Tapadinhas, M.J., Rivera, I.C. and Bignamini, A.A. 1982. Oral glucosamine sulphate in the management of arthrosis: report on a multi-centre open investigation in Portugal. *Pharmacotherapeutica* 3, 157–168.
88. Noack, W., Fisher, M., Forster, K., Rovati, L. and Setnikar, I. 1994. Glucosamine sulfate in osteoarthritis of the knee. *Osteoarthritis Cartilage.* 2, 51–59.

89. Houpt, J.B., McMillan, R., Wein, C. and Paget-Dellio, S.D. 1999. Effect of glucosamine hydrochloride in the treatment of pain of osteoarthritis of the knee. *J. Rheumatol.* 2423–2430.
90. Rindone, J.P., Hiller, D., Collacott, E., Nordhaugen, N. and Arriola, G. 2000. Randomized controlled trial of glucosamine for treating osteoarthritis of the knee. *West J. Med* .172, 91–94.
91. Reginster, J.Y., Deroisy, R., Rovati, L.C., Lee, R.L., Lejeune, E., Bruyere, O., Giacovelli, G., Henrotin, Y., Dacre, J.E. and Gossett, C. 2001. Long-term effects of glucosamine sulphate on osteoarthritis progression: a randomized, placebo-controlled clinical trial. *Lancet* 357, 251–256.
92. Pavelka, K., Gatterova, J., Olejarova, M., Machacek, S., Giacovelli, G. and Rovati, L.C. 2002. Glucosamine sulfate use and delay of progression of knee osteoarthritis. A 3-year, randomized, placebo controlled, double blind study. *Arch. Int. Med.* 162, 2113–2123.
93. Bruyere, O., Pavelka, K., Rovati, L.C., Deroisy, R., Olejarova, M., Gatterova, J., Giancovelli, G. and Reginster, J-Y. 2004. Glucosamine sulfate reduces osteoarthritis: evidence from two 3-year studies. *Menopause* 11, 138–143.
94. Anonymous 2003. Postmenopausal women: glucosamine sulfate in arthritis in knee joints. *Deutsche Apotheker-Zeitung* 143, 43–44.
95. Anonymous 2003. Recent studies on glucosamine treatment of osteoarthritis *Aust. J. Pharm.* 84, 325–326.
96. Cibere, J., Kopec, J.A., Thorne, A., Singer, J., Canvin, J., Robinson, D.B., Pope, J., Hong, P., Grant, E. and Esdaile, J.M. 2004. Randomized, double-blind, placebo-controlled glucosamine discontinuation trial in knee osteoarthritis. *Arthritis Rheum.* 51, 738–745.
97. Nakamura, H., Masuko, K., Yudoh, K., Kato, T., Kamada, T. and Kawahara, T. 2007. Effects of glucosamine administration on patients with rheumatoid arthritis. *Rheumatol. Int.* 207, 213–218.
98. Clegg, D.O., Reda, D.J., Harris, C.L., Klein, M.A., O'Dell, J.R., Hooper, M.M., Bradley, J.D., Bingham, C.O. III, Weisman, M.H., Jackson, C.G., Lane, N.E., Cush, J.J., Moreland, L.W., Schumacher, H.R. Jr., Oddis, C.V., Wolfe, F., Molitor, J.A., Yocum, D.E., Schnitzer, T.J., Furst, D.E., Sawitzke, A.D., Shi, H., Brandt, K.D., Moskowitz, R.W. and Williams, H.J. 2006. Glucosamine, chondroitin sulfate, and the two in combination for painful knee osteoarthritis. *N. Engl. J. Med.* 354, 795–808.
99. Alekseeva, L.J., Chichasova, N.V., Benevolenskaia, L.I., Nasonov, E.L. and Mendel, O.I. 2005. Combined medication ARTRA in the treatment of osteaoarthrosis. *Ter. Arkh.* 77, 69–75.
100. Vaz, A.L. 1982. Double blind clinical evaluation of the relative efficiency of ibuprofen and glucosamine sulfate in the management of osteoarthritis of the knee in outpatients. *Curr. Med. Res. Opin.* 8, 145–149.
101. Müller-Fa*f*bender, Bach, G.L., Haase, W., Rovati, L.C. and Setnikar, I. 1994. Glucosamine sulfate compared to ibuprofen in osteoarthritis of the knee. *Osteoarthritis and Cartilage* 2, 61–69.
102. Thie, N.M., Prasad, N.G. and Major, P.W. 2001. Evaluation of glucosamine sulfate compared to ibuprofen for the treatment of temporomandibular joint osteoarthritis: a randomized double blind controlled 3 month trial. *J. Rheumatol.* 28, 1347–1355.
103. Qiu, G.X., Gao, S.N., Giacovelli, G., Rovati, L. and Setnikar, I. 1998. Efficacy and safety of glucosamine sulfate versus ibuprofen in patients with knee osteoarthritis. *Arzneim Forsch* 48, 469–474.

104. Johnson, K.A., Hulse, D.A., Hart, R.C., Kochevar, D. and Chu, Q. 2001. Effects of an orally administered mixture of chondroitin sulfate, glucosamine hydrochloride and manganese ascorbate on synovial fluid chondroitin sulfate 3B3 and 7D4 epitope in a canine cruciate ligament transection model of osteoarthritis. *Osteoarthritis Cartilage* 9, 14–21.
105. Ekanayake, S. 2003. Compositions useful in the treatment of diseases of connective tissues. US 6,632,804.
106. Levenson, G.E. 1969. The effect of ascorbic acid on monolayer cultures of three types of chondrocytes. *Exp. Cell Res.* 55, 225–238.
107. Comper, W. 1991. Physiochemical aspects of cartilage extra cellular matrix. In: *Cartilage: Molecular Aspects*, Hall B and Newman S, ed., CRC press, Boca Raton, FL, pp. 59–96.
108. Goldstain, M.R. 2001. Glucosamine sulphate and osteoarthritis. *Lancet* 357, 1617.
109. Virkamaki, A., Daniels, M.C., Hamalainen, S., Utriainen, T., McClain, D. and Yki-Jarvinen, H. (1997) Activation of the hexosamine pathway by glucosamine *in vivo* induces insulin resistance in multiple insulin sensitive tissues. *Endocrinology* 138, 2501–2507.
110. Rossetti, L., Hawkins, M., Chen, W., Gindi, J. and Barzilai, N. 1995. *In vivo* glucosamine infusion induces insulin resistance in normoglycemic but not in hyperglycemic conscious rats. *J. Clin. Invest.* 96, 132–140.
111. Barzilai, N., Hawkins, M., Angelov, I., Hu, M. and Rossetti, L. 1996. Glucosamine-induced inhibition of liver glucokinase impairs the ability of hyperglycemia to suppress endogenous glucose production. *Diabetes* 45, 1329–1335.
112. Scroggie, D.A., Albright, A. and Harris, M.D. 2003. The effect of glucosamine-chondroitin supplementation on glycosylated hemoglobin levels in patients with type 2 diabetes mellitus—A placebo-controlled, randomized clinical trial. *Arch. Int. Med.* 163, 1587–1590.
113. Rovati, L.C., Annefield, M., Glascovelli, G., Schmid, K. and Setnikar, I. 1999. Glucosamine in osteoarthritis. *Lancet.* 354, 1640–1642.
114. Monauni, T., Zenti, M.G., Cretti, A., Daniels, M.C., Targher, G., Caruso, B., Caputo, M., McClain, D., Del Prato, S., Giaccari, A., Muggeo, M., Bonora, E. and Bonadonna, R.C. 2000. Effects of glucosamine infusion on insulin secretion and insulin action in humans. *Diabetes* 49, 926–935.
115. Chan, N.N., Baldeweg, S., Tan, T.M. and Hurel, S. 2001. Glucosamine sulfate and osteoarthritis. *Lancet* 357, 1617.
116. Adams, M.E. 1999. Hype about glucosamine. *Lancet* 354, 353–354.
117. Tannis, A.J., Barban, J. and Conquer, J.A. 2004. Effect of glucosamine supplementation on fasting and non-fasting plasma glucose and serum insulin concentrations in healthy individuals. *Osteoarthritis Cartilage.* 12, 506–511.
118. Russel, A.I. and MacCarty, M.F. 1999. Glucosamine in osteoarthritis. *Lancet* 354, 1641–1642.
119. Breuer, R. 1898. Ueber das freie Chitosamin. *Chem. Berichte.* 31, 2193–2200.
120. Setnikar, I., Giacchetti, C. and Zanolo, G. 1986. Pharmacokinetics of glucosamine in the dog and in man. *Arzneim-Forsch /Drug Res.* 36, 729–735.
121. van der Kraan, P.M., de Vries, B.J., Vitters, E.L., van den Berg, W.B. and van de Putte, L.B.A. 1988. Inhibition glycosaminoglycan synthesis in anatomically intact rat patellar cartilage by paracetamol-induced sulfate depletion. *Biochem. Pharmacol.* 37, 3683–3690.
122. Sullivan, M.X. and Hess, W.C. 1935. Cystine content of finger nails in arthritis. *J. Bone Joint Surg.* 16, 185–188.

123. Senturia, B.D. 1934. Results of treatment of chronic arthritis and rheumatoid conditions with colloidal sulphur. *J. Bone Joint Surg.* 16, 119–125.
124. Hoffer, L.J., Kaplan, L.N., Hamadeh, M.J., Grigoriu, A.C. and Baron, M. 2001. Sulfate could mediate the therapeutic effect of glucosamine sulfate. *Metabolism.* 50, 767–770.
125. Tesoriere, G., Dones, F., Magistro, D. and Castagnetta, L. 1972. Intestinal absorption of glucosamine and *N*-acetylglucosamine. *Experientia* 28, 770–771.
126. Griffin, M.R., Ray, W.A. and Schaffner, W. 1988. Non-steroidal anti-inflammatory drug use and death from peptic ulcer in elderly persons. *Ann. Int. Med.* 109, 359–363.
127. Setnikar, I., Pacini, M.A. and Revel, L. 1991. Antiarthritic effect of glucosamine sulfate studied in animal models. *Arzneim-Forsch/Drug Res.* 41, 542–545.
128. Moriga, M., Aono, M., Murakami, M. and Uchino, H. 1980. The activity of *N*-acetylglucosamine kinase in rat gastric mucosa. *Gastroenterol. Japonica* 15, 7–13.
129. Wolfe, M.M., Lichtenstein, D.R. and Singh, G. 1999. Gastrointestinal toxicity of nonsteroidal antiinflammatory drugs. *New Engl. J. Med.* 340, 1888–1899.
130. Brandt, K.D. 1987. Effects of nonsteroidal anti-inflammatory drugs on chondrocyte metabolism *in vitro* and *in vivo*. *Am. J. Med.* 83 (Suppl. 5A), 29–34.
131. Shield, M.J. 1993. Anti-inflammatory drugs and their effects on cartilage synthesis and renal function. *Eur. J. Rheumatol. Inflam.* 13, 7–16.
132. Brooks, P.M., Potter, S.R. and Buchanan, W.W. 1982. NSAID and osteoarthritis—help or hindrance. *J. Rheumatol.* 9, 3–5.
133. Newman, N.M. and Ling, R.S. 1985. Acetabular bone destruction related to non-steroidal anti-inflammatory drugs. *Lancet* 2(8445), 11–14.
134. Solomon, L. 1973. Drug induced arthropathy and necrosis of the femoral head. *J. Bone Joint Surg.* 55B, 246–251.
135. Ronnigen, H. and Langeland, N. 1979. Indomethacin treatment in osteoarthritis of the hip joint. *Acta. Orthopaed. Scand.* 50, 169–174.
136. Couzin, J. 2004. Drug safety. Withdrawal of Vioxx casts a shadow over COX-2 inhibitors. *Science* 306(5695), 384–385.
137. Barclay, T.S., Tsourounis, C. and McCart, G.M. 1998. Glucosamine. *Ann. Pharmacother.* 32, 574–579.
138. da Camara, C.C. and Dowless, G.V. 1998. Glucosamine sulfate for osteoarthritis. *Ann. Pharmacother.* 32, 580–587.
139. Heyneman, C.A. and Rhodes, R.S. 1998. Glucosamine for osteoarthritis: cure or conundrum? *Ann. Pharmacother.* 32, 602–603.
140. Gotzsche, P. 1989. Methodology and overt and hidden bias in reports of 196 double-blind trials of non-steroidal anti-inflammatory drugs in rheumatoid arthritis. *Control Clin. Trials* 10, 31–56.
141. Towheed, T.E. 2003. Current status of glucosamine therapy in osteoarthritis. *Arthritis Rheumat.* 49, 601–604.
142. Matheu, V., Gracia Bara, M.T., Pelta, R., Vivas, E. and Rubio, M. 1999. Immediate-hypersensitivity reaction to glucosamine sulfate. *Allergy* 54, 643–650.
143. Tallia, A.F. and Cardone, D.A. 2002. Asthma exacerbation associated with glucosamine-chondroitin supplement. *J. Am. Board Fam. Pract.* 15, 481–484.
144. Zerkak, D. and Dougados, M. 2004. The use of glucosamine therapy in osteoarthritis. *Curr. Rheumatol. Rep.* 6, 41–45.
145. Bassleer, C., Rovati, L. and Franchimont, P. 1998. Stimulation of proteoglycan production by glucosamine sulfate in chondrocytes isolated from human osteoarthritic articular cartilage in vitro. *Osteoarthritis Cartilage* 6, 427–434.

146. Priebe, D., McDiarmid, T., Mackler, L. and Tudiver, F. 2003. Do glucosamine or chondroitin cause regeneration of cartilage in osteoarthritis? *J. Farm Pract.* 52, 237–239.
147. Tudiver, F. 2003. Do glucosamine or chondroitin cause regeneration of cartilage in osteoarthritis? *J. Farm Pract.* 52, 237–239.
148. Jimenez, S.A. and Dodge, G.R. 1997. The effects of glucosamine sulfate on human chondrocyte gene expression. Abstract. ILAR Congress, Singapore.
149. Sandy, J.D., Boyer, H. and Hymer, S.S. 1998. *Control of Chondrocyte Aggrecanase by Glutamine Supply. Transactions.* 44th Annual Meeting, Orthopaedic Research Society, New Orleans, LN, p. 853.
150. Varghese, S., Theprungsirikul, P., Sahani, S., Hwang, N., Yarema, K.J. and Elisseeff, J.H. 2007. Glucosamine modulates chondrocyte proliferation, matrix synthesis, and gene expression. *Osteoarthritis Cartilage.* 15, 59–68.
151. Towheed, T.E., Maxwell, L., Anastassiades, T.P., Shea, B., Houpt, J., Robinson, V., Hochberg, M.C. and Wells, G. 2005. Glucosamine therapy for treating osteoarthritis. Cochrane Database of Systematic Review, CD002946.
152. McAlindon, T. 2003. Why are clinical trials of glucosamine no longer uniformly positive? *Rheum. Dis. Clin. N. Amer.* 29, 789–801.
153. Verbruggen, G., Goemaere, S. and Veys, E.M. 1998. Chondroitin sulfate S/DMOAD (structure/disease modifying osteoarthritis (OA) drug) in the treatment of OA of the finger joints. *Osteoarthritis Cartilage* Suppl A, 37–38.
154. Zupanets, IA., Drogonoz, S.M., Bezdetko, N.U., Rechliman, I.E. and Semenov, A.N. 1991. The influence of glucosamine on the antiexudative effect nonsteroidal anti-inflammatory agents. *Farmakol. Toksikol.* 54, 61–63.
155. Manson, J.J. and Rahman, A. 2004. This house believes that we should advise our patients with osteoarthritis of the knee to take glucosamine. *Rheumatology* 43, 100–101.
156. Reginster, J.Y., Richy, F. and Bruyere, O. 2006. Glucosamine as a pain-modifying drug in osteoarthritis. What is new in 2006? *Rev. Med. Liege* 61, 169–172.
157. http://www.clinicaltrials.gov/ct/show/NCT00065377.

9 Functional and Bioactive Peptides from Hydrolyzed Aquatic Food Proteins

Hordur G. Kristinsson

CONTENTS

9.1 PROCESSING HYDROLYSATES FROM AQUATIC FOODS AND BY-PRODUCTS

Proper utilization of aquatic resources has been an important topic for many decades, and recently the scope of the problem has become even clearer. It is evident that as we enter the twenty-first century, our common food fish stocks are greatly exploited and there is a strong need to better utilize the by-products that remain after processing. Furthermore, there are a number of underutilized species that could be utilized for human and animal consumption, if proper economical technologies were available. The amount of protein from aquatic sources that goes underutilized in this world is staggering. If one merely looks at processing by-products from filleting, these raw materials may contain about 10–20% of the protein in the fish, and in many cases are not utilized for human or animal consumption. A positive step in this direction is fish meal production for animal feed. However, given the proper methods, much higher quality and more nutritional end products could be produced from this leftover protein. A number of methods to better utilize raw materials of aquatic origin have been proposed, one of them being enzymatic hydrolysis to produce what is collectively called fish protein hydrolysates (FPH). In addition to fish, these hydrolysates can be produced from a number of aquatic sources. A large number of research publications on a number of different species and raw materials have been published in connection to FPH, but only relatively few processes have been commercialized. The main obstacle for commercial development of FPH into human food products has been process economics, nonuniformity of final product, functional problems, and off-odors/flavors that can commonly develop during this process [1]. Production of FPH for animal food and feed has been more successful, where final product characteristics are not as stringent. The growing world demand for animal protein–based feed has helped this development.

Evaluating the literature reveals a vast array of different processes that are used to hydrolyze and recover proteins/peptides from aquatic food products. Although the processes may use different enzymes, reaction conditions, etc., they share for the most part the same general process flow. First of all, the nature of the raw material is very important. Generally, raw materials used for enzymatic hydrolysis are inexpensive (hence not always kept under optimal conditions) and very complex. Materials rich in fat or blood and dark muscle are extremely prone to lipid oxidation during and after processing, and also may produce highly undesirable yellow to brown colors [2,3]. Even rather lean raw materials may oxidize substantially if the material has a high level of blood and dark muscle, since both are sources of very potent prooxidants (e.g., hemoglobin and myoglobin). A slightly alkaline prewash (to remove heme proteins and reduce their autoxidation) and addition of water and lipid-soluble antioxidants (e.g., vitamin C and tocopherol, respectively), will reduce oxidation and color problems, but yields will be reduced. Some raw materials also have a high microbial load, and antimicrobial agents or process conditions where microbial growth is reduced (e.g., low pH and low temperature) would be recommended. Even though enzyme inactivation is usually effective in reducing microbial counts, some microbes have the ability to produce toxins that are not destroyed by heat (e.g., histamine), so special process precautions have to be followed for certain species (e.g., scombroid species). Ideally one should work with a clean starting raw material, but until recently this has not been economically feasible. The recently developed pH-shift process makes this a feasible option. In this process, the raw material (e.g., frames or even whole headed and gutted fish) is finely homogenized in water and proteins solubilized using low or high pH. Proteins are then separated from undesirable insoluble materials, such as fat, connective tissue, skin, bones, and cellular membranes (phospholipids), and some toxins (e.g., mercury) using high-speed centrifugation [4]. The proteins are then precipitated by adjusting the pH to 5.5, following another centrifugation. These recovered proteins are a relatively pure protein substrate that can be then further hydrolyzed to make a highly functional and bioactive FPH [5].

Enzyme choice is very important if certain functional or bioactive properties are desired (i.e., certain peptide makeup). A large variety of enzymes are available commercially, both as single enzymes and as preparations of different enzymes. Each enzyme and enzyme preparation has its unique activity and will produce different peptides, which in turn have different functions and activities. There are also yield differences reported between different enzymes [6]. Therefore it is important to select the proper enzyme to achieve the desired final end product and to maximize yield. Enzymes with exopeptidase activities are specific and cleave inside the protein molecule, and produce relatively few but large peptides, while preparations with endopeptidase activity cleave amino acids from the ends of the proteins and result in a higher degree of hydrolysis and a combination of small and large peptides. Usually a combination of the two forms of activities is used. Since different enzyme activities can lead to FPHs with significantly different functions, its important to work with a well-characterized and consistent enzyme mixture. In the past, endogenous enzymes (e.g., from guts or pancreas) were used (and still are) to produce FPH. However, the type of enzymes and their activities may vary greatly from one batch to another, thus making it very difficult to get a consistent hydrolysis reaction and a consistent final product. The advantage, however, of using these preparations is they are normally inexpensive, as they can be prepared as crude extracts on-site. Another advantage is that they often are more active at lower temperatures (especially if they are extracted from cold water fish), thus allowing for an effective reaction while minimizing microbial growth and oxidation reactions. In one study on salmon muscle proteins [7], it was found that an endogenous extract of serine proteases from salmon pyloric caeca was more efficient hydrolyzing the proteins at about 20°C than four other commercial preparations tested. In addition to different mechanisms of hydrolysis, enzymes have a limited range of pH and temperature where they are active, and this information has to be available before designing a hydrolysis process. For example, if low pH is desired to control microbial growth, then an acidic protease such as pepsin or a commercial preparation such as Newlase (Amano Enzymes, Japan) would be used. These days, most enzyme reactions are run at neutral to slightly alkaline pH values and at moderately high temperatures (40–60°C) when producing FPH. Although these pH values are favorable to spoilage microorganisms, they reportedly lead to final products of higher functionality and quality [8]. Reactions at low pH values are known to promote oxidation reactions, which are detrimental to the final product.

It is important that the raw material is very finely homogenized and well dispersed before enzymes are added to allow for proper enzyme access. Homogenization is not recommended after enzyme addition as it may denature the enzymes. The level of water in the reaction mixture is important to allow for proper mixing of enzyme and substrate. Slizyte et al. [9] found that the amount of water used in the reaction mixture was more important than type of enzyme used when hydrolyzing cod by-products. Lower levels of water addition led to lower recoveries. Too little water will produce a system of high viscosity that is difficult to mix and thus enzyme access to substrate will be limited. Given enough time, the enzymes will eventually break down enough protein to reduce viscosity significantly, but this may unnecessarily prolong the reaction. Initially, when enzymes are added, the reaction proceeds rapidly and then it levels off as fewer peptide bonds are available for the enzymes and also as enzymes deactivate during the reaction. The degree of hydrolysis (%DH) during the reaction can be monitored by a variety of different techniques, as summarized by Kristinsson and Rasco [1]. One of the most commonly used techniques is based on maintaining a fixed pH value and from the addition of base (if reaction is above neutrality) or acid (if reaction is below neutrality), one can calculate the %DH, if acid/base normality, protein content, total number of peptide bonds per unit weight protein, and degree of dissociation of amino or carboxyl groups is known [10]. To maintain enzyme activity and get an effective reaction it is important to know the optimal pH, ionic strength, and temperature range of the enzyme(s) used. It is generally not recommended to operate right at the optimal temperature, as this is often close to the denaturation temperature of the enzyme, and thus even though reaction may be quick at first, deactivation may soon follow. Speed and extent of reaction can also be controlled by enzyme to substrate ratio. Normally, the enzyme is the cost-limiting factor, so it is desirable to use as little enzyme as possible. Higher amounts of enzyme do lead to a more rapid hydrolysis and to a higher degree of hydrolysis. Also, the higher the %DH the more the recovery of proteins and peptides from the raw material as more proteins are solubilized [11]. Therefore, enzyme cost versus speed and extent of hydrolysis has to be balanced when designing a hydrolysis process. Adding and comparing enzymes based on weight or volume is a common problem seen in the literature. A study where Atlantic salmon muscle proteins were hydrolyzed with different enzymes adjusted to the same activity units (Azocoll units) demonstrated that different enzymes can result in very different degrees of hydrolysis and have different reaction rates even after their activity has been standardized (Table 9.1) [12]. For example, using Corolase 7089 (Rohm and Haas, Germany) at 29,296 Azocoll units gave 10.64%DH after a 1 h reaction, while at the same Azocoll units, Alcalase only reached 2.91%DH.

Once a desired end %DH is achieved the enzyme has to be deactivated. The most common technique would be to heat and inactivate the enzyme using a time and temperature combination sufficient to go well beyond what is required to irreversibly deactivate the enzyme. Subjecting a fish protein homogenate to 80–90°C for at least 10 min is acceptable for most commercial enzyme preparations used. Higher protein (and fat) contents and reaction solution conditions favorable to the enzyme (e.g., pH and ionic strength) may increase the stability of the enzyme, so process variations may warrant modifications in the deactivation scheme used.

TABLE 9.1
Degree of Hydrolysis Reached after 1 h with Different Enzymes at the Same Enzyme Activity (Azocoll Units)

Enzyme	25000 Azocoll Units (%DH)	100000 Azocoll Units (%DH)
Alcalase 2.4L	2.46	6.42
Flavorzyme 1000L	5.04	9.24
Corolase PN-L	3.61	7.32
Corolase 7089	10.24	13.74
Salmon pyloric caeca enzyme extract	8.63	16.86

Most enzymes sold have specification sheets from the manufacturer reporting their optimal pH, temperature, as well as the thermal scheme needed for inactivation. It is important to keep in mind that these are normally done in dilute systems on very simple substrates that are very different from a homogenized slurry of fish proteins and associated components. Therefore, the reaction optima and inactivation scheme should only be taken as guidelines and not absolute. It is important to establish these optima for each different raw material substrate being worked on. Furthermore, after the reaction mixture has been heat-treated it is important to check the mixture for enzyme activity. Sometimes it may take hours to days for the enzymes to renature after a heat treatment, so it is advisable to check for activity immediately after the reaction mixture has cooled and at some time point later. Using a combination of pH and temperature may be effective. For example, if a neutral protease such as Alcalase or Protamex is used, the system can be acidified (e.g., pH 3) and subjected to a heat treatment after hydrolysis. Since the enzyme is more susceptible well outside its optimal range a milder heat treatment may be used.

After heat inactivation, hydrolyzed proteins are typically separated from insoluble undesirable compounds such as nonhydrolyzed proteins, connective tissue, fat, bones, scales, and skin. This material can be further processed into various products and utilized. The separation or recovery of FPH from this material can be done via various means. A common technique is to subject the system to high-speed centrifugation, where the insoluble material would be forced to the bottom of the centrifuge and the soluble FPH would be decanted. Filters can also be used to separate the small soluble peptides making up the FPH fraction from the insoluble material. The FPH fraction is then often concentrated (by heat) or dried (e.g., with spray driers). To stabilize the FPH, it may be necessary to incorporate antimicrobial compounds or antioxidants. Often the FPH is acidified to minimize microbial growth, but this greatly increases the risk of lipid oxidation, so proper antioxidant strategies are critical for these products if product functionality and palatability is to be maintained.

9.2 FOOD FUNCTIONALITY AND APPLICATIONS OF AQUATIC PROTEIN HYDROLYSATES

9.2.1 INTERACTIONS WITH WATER

As proteins become hydrolyzed during production of FPH they become increasingly more soluble, not only because they are being released from structures that attach them to the myofibrillar matrix but also because they become progressively smaller and have newly exposed amino and carboxyl groups, which readily interact with water molecules. Most studies report high solubilities for FPH. Work with Alcalase on sardines [13] and salmon [11] has demonstrated well over 90% solubility under conditions where nonhydrolyzed fish proteins would have very little solubility. Furthermore, the study with salmon FPH demonstrated that the high solubility (96–100%) is not sensitive to pH, contrary to nonhydrolyzed fish proteins, thus greatly extending the pH range where functional fish proteins could be used in food products. Owing to the good solubility of FPH they have found use and have a potential use in a variety of products. FPH have reportedly been used successfully as milk substitutes in weanling animals [14,15]. High solubility of FPH suggests they could be successfully used in various drink formulas. Good solubility and thus good interactions with water are believed to be behind the ability of FPH to help retain water when added to various food products. Several studies have demonstrated that FPH can improve water holding in raw and cooked food products. Vareltzis and coworkers [16] incorporated FPH into hamburgers and found that it reduced water loss on cooking. Shahidi and coworkers [17] reported a similar finding when incorporating FPH from capelin into minced pork. Work done on incorporating salmon FPH into patties made from minced salmon fillets showed that FPH was more effective than egg albumin or soy proteins in reducing water loss on freezing and thawing [11]. This study also found no clear relationship between %DH and effectiveness of the FPH, but that enzyme type used to make the FPH was a more important factor (Figure 9.1). There are several mechanisms whereby FPH can aid in water binding in a food product. First of all, the increased amount of terminal COOH and NH_3 groups on hydrolysis can bind and hold on to water in the food product. Also, the small peptides can be well distributed in the water phase of the product and may increase the osmotic pressure of the system, which could work against water loss from the product. The peptides may also directly interact with the proteins in the food products, increasing their ability to hold onto more water. Research has shown that FPH may protect proteins from denaturation, which could in part explain their ability to help food systems where proteins are the main contributor of water holding. Fish protein hydrolysates prepared from five different species were found to provide stability to lizard fish myofibrillar proteins on drying, as indicated by measuring ATPase activity [18]. Khan and coworkers [18] also found that surimi containing FPH (5% w/w) had better gelling ability than surimi with no FPH and also had significantly higher calcium-dependent ATPase activity, suggesting a cryoprotective effect of FPH. It is well known that good gel-forming ability of muscle proteins

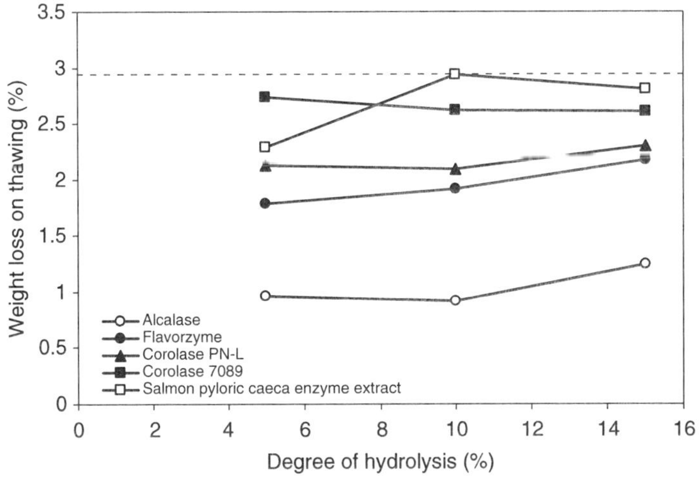

FIGURE 9.1 Water-holding capacity of salmon mince patties containing 1.5% (w/w) salmon fish protein hydrolysates at different %DH produced using different enzyme preparations. The dotted line represents water-holding capacity of salmon mince patties with no added FPH. (Adapted from Kristinsson, H.G. and Rasco, B.A., *J. Agric. Food Chem.* 48, 657–666, 2000.)

is important for proper water holding [19,20]. If FPH can aid in protein gelation, this may also in part explain their positive effect on water holding.

9.2.2 Interactions with Fats and Oils

9.2.2.1 Interfacial and Surface Properties

When proteins are hydrolyzed, they may at low %DH expose more hydrophobic groups than the parent proteins and thus may have better ability to interact with fats and oils. A handful of studies have investigated the oil-binding abilities of FPH. When salmon FPH was compared to egg albumin and soy protein concentrate, it was found to have significantly better oil binding when mixed with vegetable oil [11]. Not surprisingly, lower %DH gave better oil binding and also the type of enzyme used did not have a significant effect on the results. Other workers have reported oil-binding ability of FPH from capelin, shark, and herring by-products [17,21,22].

More research has been conducted on the interfacial activities of FPH, that is, their ability to emulsify oil in water. Good emulsifiers should have the ability to rapidly absorb to an interface of oil and water, and at the interface rapidly rearrange themselves to interact optimally with the two phases, and form a film, which is resistant to forces (mechanical, gravitational, etc.) [23]. Limited hydrolysis is normally desirable as it gives a good balance of hydrophobic and hydrophilic amino acid patches on the polypeptide resulting from hydrolysis [8]. Extensive hydrolysis, which leads to small peptides may give peptides that can very rapidly absorb to the surface, but due to size they cannot stabilize the interface. Lee and coworkers [24] suggested that peptides should not be smaller than 20 amino acid residues if good emulsifying properties are desired. Therefore to obtain good emulsifying properties a limited %DH is suggested, and it also helps to select an enzyme preparation, which has a relatively narrow specificity, thus giving larger peptides. Work by Kristinsson and Rasco [11] are in agreement with the previous statement. They found that as %DH increases, emulsifying properties of salmon FPH decreases. Quaglia and Orban [25] reported a similar connection between %DH and sardine FPH. Recently, Jeon and coworkers [26] performed a study on cod FPH, which was fractionated with ultrafiltration and found that peptide size is of great importance for proper emulsification. In the salmon FPH study [11], it was also found that type of enzyme used to reach a certain %DH played an important role. The enzyme preparation that gave the best emulsifying properties was an endogenous extract from salmon pyloric ceaca. SDS-PAGE studies revealed that this extract led to larger peptides than the other enzymes used, suggesting it had a narrower specificity. An earlier study by Spinelli and coworkers [27,28] found that better emulsifying properties were generated from rockfish FPH when Alcalase was used compared to bromelain. Previous studies on rockfish FPH [27,28] and herring FPH [29] showed that FPH have better emulsifying properties than their parent proteins. Other studies have however demonstrated rather poor emulsifying properties [17,21,22].

Fish protein hydrolysates have also been found to have good surface-active properties, that is, foaming properties. During FPH processing, foaming would not be desired as this would cause a loss in recovery, and normally foaming can be minimized by reducing air incorporation and being careful that mixing and stirring is slow. Good foaming abilities can however be a benefit for the final product if it is to be used in foam based products. Herring FPH was found to have good foam-forming abilities but did have poor foam stability [29]. Shark and capelin FPH were found to have foaming ability, but it was substantially less than that of whey and egg white proteins [17,21]. Surface-active properties of FPH are very sensitive to peptide size, just as interfacial properties are. In a study where cod FPH was fractionated into the following fractions: <30, 10–30, 5–10, 3–5, and <3 kDa, foamability was found to decrease in the same sequence [26].

9.2.2.2 Antioxidative Properties

In recent years, natural antioxidants have seen a surge of interest and increased research efforts. Many studies have shown that proteins and peptides have the ability to delay lipid oxidation reactions in food and living systems. As early as 1990, Hatate and coworkers [30] demonstrated that sardine FPH had antioxidative properties. Later studies by Shahidi and coworkers clearly demonstrated that FPH can have a function as antioxidants in food systems. Capelin FPH added to minced pork muscle reduced the formation of secondary oxidation products (TBARS) in the product by 17.7–60.4% [17]. When capelin FPH was fractionated, both pro- and antioxidative fractions were found [31]. Fractions with higher molecular weight were found to have more antioxidative activity. Work by Wu and coworkers [32] showed that intermediate peptide sizes (1–1.5 kDa) had the highest antioxidative activity. Jeon and coworkers reported that cod FPH fractions below 5 kDa most effectively reduced lipid oxidation in a linoleic acid emulsion system. Studies have furthermore shown that protease type can have an impact on the antioxidative properties of hydrolysates [33]. Kim and coworkers [34] compared the use of three enzymes (Alcalase, Pronase E, and collagenase) on Alaska pollack skin and found that certain fractions from Pronase E hydrolysis had peptides with strong antioxidative activity when tested in a linoleic acid emulsion system. Kristinsson and coworkers have investigated the antioxidative activities of FPH made from catfish protein isolates (made using the alkaline pH-shift process described before). Catfish FPH were found to have high radical scavenging activity, higher as %DH increased and thus peptide sizes were reduced [35]. Higher %DH also led to increased metal-chelating abilities. Heme proteins are known to be the most potent prooxidants in fish systems, so the catfish FPH were tested in a model system made of tilapia Hb and washed tilapia white muscle. The FPH were able to significantly delay the onset of oxidation [36]. Hydrolysates at 15 and 30%DH were significantly more effective than FPH at 5%DH in suppressing lipid oxidation (Figure 9.2). The mechanisms behind the antioxidative function of FPH are not well known, but are very likely because of peptide composition and size. Studies have been performed on manufactured peptides with known amino acid compositions and sequences and have demonstrated that both play a great role in the antioxidative properties of peptides [37].

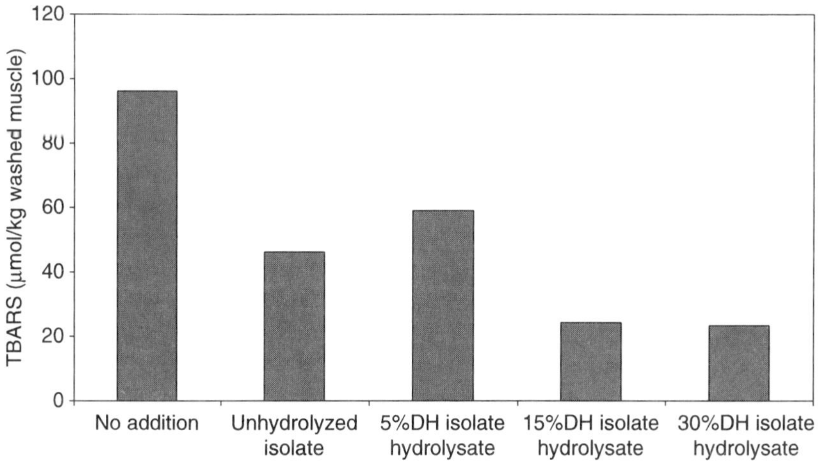

FIGURE 9.2 Effect of hydrolysates made from catfish protein isolate on lipid oxidation (thiobarbituric reactive substances, TBARS) in a model system containing 6 μmol tilapia hemoglobin in a washed tilapia muscle system. The data represents lipid oxidation after 45 h at 4°C.

9.2.3 Effect on Flavor

Hydrolyzed proteins are used in many industries as flavorings. The same holds true for FPH, but its penetration into the market as a flavoring agent has been slow, mainly owing to bitterness and "fishiness" problems that are associated with certain peptides in FPH as well as oxidation- and microbial-related products that form during and after the process [1]. The relationship among the flavor/odor, peptide size, and composition is a very complex one. It has been reported that limited hydrolysis leads to an increase in bitterness while extensive hydrolysis produces less bitterness, and may in some cases produce a flavor-enhancement effect similar to monosodium glutamate and related nucleotides [38–40]. Basic peptides with lysine and asparagine as the C-terminal and second residue, respectively, and with leucine and glycine as the N-terminal residue, have been linked to FPH bitterness [41]. Noguchi and coworkers [42] reported that five acidic tripeptides and three acidic dipeptides were responsible for the MSG-like flavor-enhancement effect of FPH. Red hake hydrolyzed to give small peptides was reported to give a flavor-enhancement effect [43]. Extensive hydrolysis of lobster by-products also reportedly led to FPH with flavor-enhancement properties [44], which could also be related to the high levels of nucleotides naturally present in lobster. A similar MSG-like effect was also reported when proteins in tuna cooking water were hydrolyzed [45]. A handful of companies have capitalized on these properties of FPH and are preparing seafood "extracts" to add to a variety of products (soups, bisques, frozen seafoods, fillings, snacks, imitation seafood products, etc.). Many of these commercial products are hydrolyzed proteins in one form or another.

The problem of bitterness and off-flavor in general is a real one with FPH and can greatly limit its use for both human and animal consumption. There are steps that can be taken to control bitterness during and after processing. Proper %DH can in part control bitterness, but choice of enzyme may in some cases be more important as some studies suggest [39,46,47]. It is recommended to use an enzyme preparation with a balance of endo- and exopeptidase activities. There are certain commercial enzyme preparations available that reportedly produce hydrolysates with less bitterness (e.g., Flavorzyme from Novo Nordisk). There are also certain postprocess steps that can be taken to reduce bitterness and off-flavors. Shahidi and coworkers [17] treated FPH with activated carbon, which removed bitter peptides. Bitter FPH may also be treated with certain enzymes (rich in exopeptidase activities) after processing to reduce bitterness [39,46]. Solvents can also be used to remove bitter compounds from FPH, but these may adversely affect functionality [46,48]. It is also possible to mask bitter flavors by adding certain ingredients to FPH or products containing FPH, such as salt and phosphate [43] and cyclodextrin [49]. Some studies have shown that bitterness and off-odors may not be just because of peptides in FPH but lipid oxidation products. A study on mackerel hydrolysis demonstrated that bitterness development was well correlated with increased lipid oxidation [50]. As discussed before, this can be minimized with proper antioxidative strategies and process modifications (e.g., reducing temperature and operating at neutral to alkaline pH values).

9.3 PHYSIOLOGICAL AND BIOACTIVE PROPERTIES OF AQUATIC PROTEIN HYDROLYSATES

9.3.1 Influence on Animal Growth and Health

Bioactive properties of FPH have received much attention lately. It has been well known for many years that proteins from aquatic sources (e.g., in the form of fish meal) have an excellent effect on animal growth and well-being [1]. This is because of their good balance of amino acids, high digestibility, and good utilization by the animal. For example, Sugiyama and coworkers [39] reported that FPH from sardines had an amino acid score of 100, protein efficiency ratio (PER) of 3.2, net protein ratio (NPR) of 5.2, and biological value (BV) of 86 when fed to rats. This FPH was also found to have a net protein utilization of 85% and digestibility of 99%. A number of FPH from other species have closely mirrored these findings [51–53]. Feeding trials with rats demonstrate that FPH lead to more rapid weight gain and higher weight compared to casein [54,55]. Feeding trials with fish have shown positive effects as well [56,57]. There are, however, a handful of studies that have reported a negative effect on growth [58–60]. Cordova-Murueta and Garcia-Carreno [60] suggested that FPH with too many low-molecular-weight peptides could lead to an imbalance in amino acid absorption and lack of uptake of essential amino acids.

Fish protein hydrolysates are finding some innovative uses in the aquaculture industry. Gildberg and coworkers [61–63] in Norway have found that FPH give fish more disease resistance and stimulates their immune system and could be used in fish vaccines to improve the health status of aquacultured fish. Using FPH early on in the lifecycle of fish species may significantly increase their survival rate [57,63]. Studies have also shown that more FPH addition to fishmeal leads to more fish meal consumption, suggesting FPH may act as a feeding stimulant [64].

9.3.2 Influence on Plant Growth and Propagation

Fish protein hydrolysates have a proven track record as fertilizers for plants. Humans have for ages used fish and silage (endogenously hydrolyzed fish material) as fertilizer with excellent results. Several commercial liquid FPH products are available as plant fertilizers. Their effectiveness as fertilizers has been linked to their favorable amino acid composition and their rapid uptake through roots and leaves when applied to plants (George Pigott, personal communication, 1997). Fish protein hydrolysates have been found to greatly stimulate the production of certain valuable bioactive compounds in certain plant varieties. Mackerel FPH increased the production of rosemaric acid and phenolics in oregano when it was included in the growth media [65]. The same FPH increased the production of an important phenolic compound in anise root cultures, which could help regulate production of important phytochemicals in plants [65]. Vattem and Shetty [66] reported an increased amount of water extractable phenolic compounds in cranberry pomace when FPH was added to the growth media. The use and function of FPH in the plant industry therefore can go beyond just providing basic nutrition, but could be used to stimulate the production of valuable nutraceutical compounds.

9.3.3 Influence on Hypertension

The ability of natural compounds/peptides to influence blood pressure (hypertension) has received great attention. Quite a few studies have demonstrated that FPH will either reduce blood pressure or may have a potential to reduce it. The angiotensin I converting enzyme (ACE) converts angiotensin I to angiotensin II, which plays an instrumental role in elevation of blood pressure [67]. Many workers have found that FPH from various sources inhibit this enzyme to a significant extent [35,68–72]. Shrimp FPH inhibited 57% of ACE activity, while cod and sardine FPH provided close to 30% inhibition. Theodore and Kristinsson [35] reported a ~70–90% inhibition of ACE for catfish FPH (Figure 9.3). There was no clear relationship between %DH or peptide size range and the ability of catfish FPH to inhibit ACE. Other studies, however, suggest that peptide size is important in ACE inhibition. Cod FPH was fractionated and found to reduce ACE activity in the following order: 3 kDa >5 kDa > 10 kDa > 30 kDa, suggesting that low-molecular-weight peptides are more inactivating [26]. Studies with yellowfin sole FPH showed that <5 kDa fractions of were more effective than high-molecular-weight fractions in inhibiting ACE [73]. Je and coworkers [74]

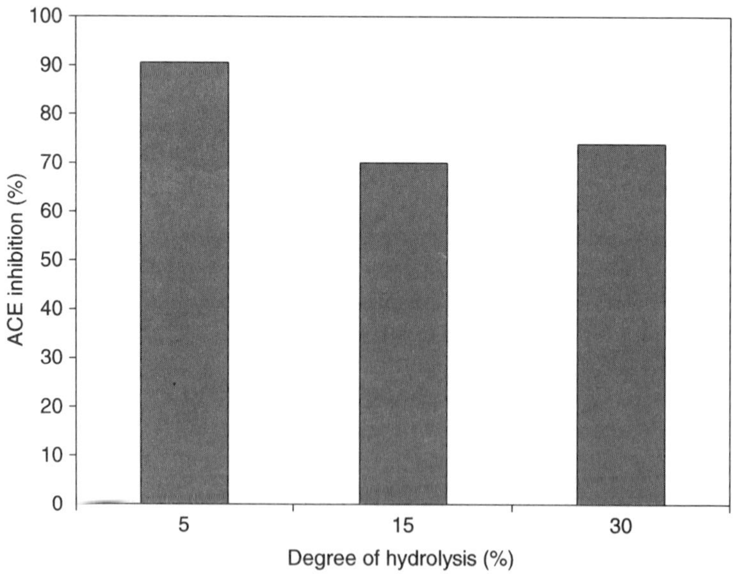

FIGURE 9.3 Ability of hydrolysates at different %DH from catfish protein isolate to inhibit the angiotensin I converting enzyme (ACE) *in vitro*. (Adapted from Theodore, A.E. and Kristinsson, H.G., *J. Sci. Food Agric.*, In press.)

hydrolyzed and fractionated Alaska pollack frame muscle proteins (<1, 1–3, 3–5, 5–10, and 10–30 kDa) and tested the different fractions for ACE-inhibitory activity. The <1 kDa fraction had the highest ACE-inhibitory activity, and a peptide with the following sequence was found to be responsible for the inhibition: Phe-Gly-Ala-Ser-Thr-Arg-Gly-Ala.

There is also evidence that fish-derived peptides may play a role in regulating blood pressure in living systems. Fahmi and coworkers [75] fed 300 mg of purified peptides from sea bream scale hydrolysates to hypertensive rats, which significantly dropped their blood pressure. The peptides were found to be more effective than the commercial hypertension drug enalapril maleate. According to Cheung and coworkers [76], the C-terminal amino acid of peptides make the most important contribution to substrate binding at the ACE active site. The most favorable C-terminal amino acids were tryptophan, tyrosine, phenylalanine, and proline, with the latter two being well-known inhibitors of ACE when found in peptides. In addition to a positive effect on hypertension, FPH from salmon has been found to reduce total cholesterol and increased high-density lipoprotein cholesterol ("good cholesterol") when fed to rats [77].

9.3.4 Other Physiological Influences

A few studies have shown that FPH have antioxidative activity *in vitro* [5,35,78], but very few have tested this *in vivo*. When hypertensive rats were fed FPH, their antioxidant status increased by 35% [79]. The same increase was not seen when the rats were fed casein. Fish protein hydrolysates have also been shown to reduce anxiety and improve learning and memory in humans [80–82]. A stress reduction was also reported for rats fed a commercial FPH (PC60) [83]. Fish peptides have also been found to stimulate the proliferation of white blood cells in humans, and thus could have a positive effect on the immune system [84]. Pacific whiting FPH was found to have a positive effect on the digestive tract and was shown to have a reparative ability as cell growth and injury was significantly reduced [85]. A study by Chuang and coworkers [78] demonstrated, *in vitro*, that FPH inhibits lipoxygenase, which is known to lead to low-density lipoprotein oxidation. The same study also demonstrated that FPH has an ability to reduce viscosity of blood (i.e., thin blood and increase blood flow). Rajapakse and coworkers [86] found a protein or a protein fraction in yellowfin sole FPH which had antiplatelet and anticoagulant properties.

REFERENCES

1. Kristinsson, H.G. and Rasco, B.A. 2000. Fish protein hydrolysates: production, biochemical and functional properties. *Crit. Rev. Food Sci. Nutr.* 32, 1–39.
2. Hultin, H.O. 1994. Oxidation of lipids in seafood. In: *Seafoods: Chemistry, Processing Technology and Quality*, F. Shahidi and J.R. Botta, eds., Blackie Academic and Professional, Glasgow, pp. 49–74.
3. Hultin, H.O. and Kelleher, S.D. 2000. Surimi processing from dark muscle fish. In: *Surimi and Surimi Seafood*, J.W. Park, ed., Marcel Dekker, New York, pp. 59–77.

4. Hultin, H.O., Kristinsson, H.G., Lanier, T.C. and Park, J.W. 2005. Process for recovery of functional proteins by pH shifts. In: *Surimi and Surimi Seafood*, 2nd ed., J.W. Park, ed., Marcel Dekker, New York, pp. 107–139.
5. Theodore, A.E. 2005. Bioactive and functional properties of catfish protein hydrolysates and catfish protein isolates. MS thesis. University of Florida, Gainesville, FL.
6. Kristinsson, H.G. and Rasco, B.A. 2002. Fish protein hydrolysates and their potential use in the food industry. In: *Recent Advances in Marine Biotechnology,* Vol 7, M. Fingerman and R. Nagabhushanam, eds., Science Publishers, Enfield, NH, pp. 157–181.
7. Kristinsson, H.G. and Rasco, B.A. 2000. Kinetics of the hydrolysis of Atlantic salmon (*Salmo salar*) muscle proteins by alkaline proteases and a visceral serine protease mixture. *Proc. Biochem.* 36, 131–139.
8. Kristinsson, H.G. 2004. The production, properties and utilization of fish protein hydrolysates. In: *Food Biochemistry*, 2nd ed., K. Shetty, ed., Marcel Dekker, New York, pp. 1111–1133.
9. Slizyte, R., Dauksas, E., Falch, E. and Rustad, T. 2003. Functional properties of different fractions generated from hydrolysed cod (*Gadus morhua*) by-products. In: *Proceedings of the First Joint Trans-Atlantic Fisheries Technology Conference—TAFT 2003,* June 10–14, The Icelandic Fisheries Laboratories, Reykjavik, Iceland, pp. 301–303.
10. Adler-Nissen, J. 1986. *Enzymic Hydrolysis of Food Proteins*, Elsevier Applied Science Publishers, Barking, UK.
11. Kristinsson, H.G. and Rasco, B.A. 2000. Biochemical and functional properties of Atlantic salmon (*Salmo salar*) muscle proteins hydrolyzed with various alkaline proteases. *J. Agric. Food Chem.* 48, 657–666.
12. Kristinsson, H.G. and Rasco, B.A. 2000c. Hydrolysis of salmon muscle proteins by an enzyme mixture extracted from Atlantic salmon (*Salmo salar*) pyloric caeca. *J. Food Biochem.* 24, 177–187.
13. Quaglia, G.B. and Orban, E. 1987. Influence of the degree of hydrolysis on the solubility of the protein hydrolysates from sardine (*Sardina pilchardus*). *J. Sci. Food Agric.* 38, 271–276.
14. Yanez, E., Ballester, D. and Monckeberg, F. 1976. Enzymatic fish protein hydrolyzate: chemical composition, nutritive value and use as a supplement to cereal protein. *J. Food Sci.* 41, 1289–1292.
15. Rebeca, B.D., Pena-Vera, M.T. and Diaz-Castaneda, M. 1991. Production of fish protein hydrolysates with bacterial proteases; yield and nutritional value. *J. Food Sci.* 56, 309–314.
16. Vareltzis, K., Soultos, N., Zetou, F. and Tsiaras, F. 1990. Proximate composition and quality of a hamburger type product made from minced beef and fish protein concentrate. *Lebensm. Wiss. u. Technol.* 23(2), 112–115.
17. Shahidi, F., Han, X-Q. and Synowiecki, J. 1995. Production and characteristics of protein hydrolysates from capelin (*Mallotus villosus*). *Food Chem.* 53, 285–293.
18. Khan, M.A.A., Hossain, M.A., Hara, K., Osatomi, K., Ishihara, T. and Nozaki, Y. 2003. Effect of enzymatic fish protein hydrolysate from fish scrap on the state of water and denaturation of lizard fish (*Saurida wanieso*) myofibrils during dehydration. *Food Sci. Technol. Res.* 9, 257–263.
19. Feng, Y. and Hultin, H.O. 2001. Effect of pH on the rheological and structural properties of gels of water-washed chicken breast muscle at physiological ionic strength. *J. Agric. Food Chem.* 49, 3927–3935.
20. Kristinsson, H.G. and Hultin, H.O. 2003. Role of pH and ionic strength on water relationships in washed minced chicken-breast muscle gels. *J. Food Sci.* 68, 917–922.

21. Onodenalore, A.C. and Shahidi, F. 1996. Protein dispersions and hydrolysates from shark (*Isurus oxyrinchus*). *J. Aquat. Food Prod. Technol.* 5, 43–59.
22. Sathivel, S., Bechtel, P.J., Babbitt, J., Smiley, S., Crapo, C., Reppond, K.W. and Prinyawiwatkul, W. 2003. Biochemical and functional properties of herring (*Clupea harengus*) byproduct hydrolysates. *J. Food Sci.* 68, 2196–2200.
23. Damodaran, S. 1996. Amino acids, peptides, and proteins. In: *Food Chemistry*, 3rd ed., O.R. Fennema, ed., Marcel Dekker, New York, pp. 321–429.
24. Lee, S.W.M., Shimizu, S. and Kaminogawa, K. 1987. Yamaguchi. Emulsifying properties of a mixture of peptides derived from the enzymatic hydrolysates of β-casein. *Agric. Biol. Chem.* 51, 161–165.
25. Quaglia, G.B. and Orban, E. 1990. Influence of enzymatic hydrolysis on structure and emulsifying properties of sardine (*Sardina pilchardus*) protein hydrolysates. *J. Food Sci.* 55(6), 1571–1573.
26. Jeon, Y-J., Byun, H-G. and Kim, S-E. 1999. Improvement of functional properties of cod frame protein hydrolysate using ultrafiltration membranes. *Proc. Biochem.* 35, 471–478.
27. Spinelli, J., Koury, B. and Miller, R. 1972. Approaches to the utilization of fish for the preparation of protein isolates; isolation and properties of myofibrillar and sarcoplasmic fish protein. *J. Food Sci.* 37, 599–603.
28. Spinelli, J., Koury, B. and Miller, R. 1972. Approaches to the utilization of fish for the preparation of protein isolates; enzymic modifications of myofibrillar fish proteins. *J. Food Sci.* 37, 604–608.
29. Liceaga-Gesualdo, A.M. and Li-Chan, E.C.Y. 1999. Functional properties of fish protein hydrolysate from herring (*Clupea harengus*). *J. Food Sci.* 64(6), 1000–1004.
30. Hatate, H., Numata, Y. and Kochi, M. 1990. Synergistic effect of sardine myofibril protein hydrolyzates with antioxidant. *Nippon Suisan Gakkaishi* 56(6), 1011.
31. Amarowicz, R. and Shahidi, F. 1997. Antioxidant activity of peptide fractions of capelin protein hydrolysates. *Food Chem.* 58, 355–359.
32. Wu, Y.C., Kellems, R.O., Holmes, Z.A. and Nakaue, H.S. 1984. The effect of feeding four fish hydrolyzate meals on broiler performance and carcass sensory characteristics. *Poultry Sci.* 63, 2414–2418.
33. Saiga, A., Tanabe, S., and Nishimura, T. 2003. Antioxidant activity of peptides obtained from porcine myofibrillar proteins by protease treatment. *J. Agric. Food Chem.* 51, 3661–3667.
34. Kim, S-K., Kim, Y-T., Byun, H-G., Nam, K-S., Joo, D-S. and Shahidi, F. 2001. Isolation and characterization of antioxidative peptides from gelatin hydrolysate of Alaska pollack skin. *J. Agric. Food Chem.* 49, 1984–1989.
35. Theodore, A.E. and Kristinsson, H.G. 2005. Bioactive properties of fish protein hydrolysates at varying degrees of hydrolysis made from catfish protein isolates. Annual IFT Meeting Book of Abstracts. July 16–20, New Orleans, LA. Abstract 50-5.
36. Petty H.T., Bello-Malabu, J.I., Theodore, A.E. and Kristinsson, H.G. 2006. Antioxidative potential of soluble and insoluble fractions of protein hydrolysates made from catfish protein isolates. Annual IFT Meeting Book of Abstracts. June 24–28, Orlando, FL, Abstract 020A-13.
37. Saito, K., Jin, D-H., Ogawa, T., Muramoto, K., Hatakeyama, E., Yasuhara, T. and Noki, K. 2003. Antioxidant properties of tripeptide libraries prepared by the combinatorial chemistry. *J. Agric. Food Chem.* 51, 3668–3674.
38. Fujimaki, M., Arai, S., Yamashita, M., Kato, H. and Nogushi, M. 1973. Taste peptide fractionation from a fish protein hydrolysate. *Agric. Biol. Chem.* 37, 2891–2895.

39. Sugiyama, K., Egawa, M., Onzuka, H. and Oba, K. 1991. Characteristics of sardine muscle hydrolysates prepared by various enzymic treatments. *Nippon Suisan Gakkaishi*. 57(3), 475–479.
40. Yu, S.Y. and Fazidah, S. 1994. Enzymic hydrolysis of proteins from *Aristichthys noblis* by protease PAmano3. *Trop. Sci.* 34, 381–391.
41. Hevia, P. and Olcott, H.S. 1977. Flavour of enzyme-solubilized fish protein concentrate fractions. *J. Agric. Food Chem.* 25(4), 772–775.
42. Noguchi, M., Arai, S., Yamashita, M., Kato, H. and Fujimaki, M. 1975. Isolation and identification of acidic oligopeptides in a flavor potentiating fraction from a fish protein hydrolysate. *J. Agric. Food Chem.* 23(1), 49–53.
43. Imm, J.Y. and Lee, C.M. 1999. Production of seafood flavor from red hake (*Urophycis chuss*) by enzymatic hydrolysis. *J. Agric. Food Chem.* 47, 2360–2366.
44. Viera, G.H.F., Martin, A.M., Saker-Sampaiao, S., Omar, S. and Goncalves, R.C.F. 1995. Studies on the enzymatic hydrolysis of Brazilian lobster (*Panulirus* spp.) processing wastes. *J. Sci. Food Agric.* 69, 61–65.
45. ChiaLing, J., WenChing, K., Jao, and Ko, W.C. 2002. Utilization of cooking juice of young tuna processed into canned tuna as condiments: effect of enzymatic hydrolysis and membrane treatment. *Fish Sci.* 68, 1344–1351.
46. Lalasidis, G., Bostrom, S. and Sjoberg, L-B. 1978. Low molecular weight enzymatic fish protein hydrolysates: chemical composition and nutritive value. *J. Agric. Food Chem.* 26(3), 751–756.
47. Hoyle, N. and Merritt, J.H. 1994. Quality of fish protein hydrolysates from herring (*Clupea harengus*). *J. Food Sci.* 59, 76–79.
48. Chakrabarti, R. 1983. A method of debittering fish protein hydrolysate. *J. Food Sci. Technol.*, 20(4), 154–158.
49. Suzuki, T. 1981. *Fish and Krill Protein: Processing Technology*, Applied Science, London, UK.
50. Liu, C., Morioka, K., Itoh, Y. and Obatake, A. 2000. Contributions of lipid oxidation to bitterness and loss of free amino acids in the autolytic extract from fish wastes: effective utilization of fish wastes. *Fish. Sci.* 66, 343–348.
51. Atia, M. and Shekib, L. 1992. Preparation of fish protein hydrolysate from bolti frame (*Tilapia nilotica*) and evaluation of its chemical composition. *Assiut. J. Agric. Sci.* 23, 75–87.
52. Diniz, F.M. and Martin, D.M. 1996. Use of response surface methodology to describe the combined effects of pH, temperature and E/S ratio on the hydrolysis of dogfish (*Squalus acanthias*) muscle. *Int. J. Food Sci. Technol.* 31, 419–426.
53. Liaset, B., Lied, E. and Espe, M. 2000. Enzymatic hydrolysis of by-products from the fish-filleting industry; chemical characterisation and nutritional evaluation. *J. Sci. Food Agric.* 80(5), 581–589.
54. Abdul-Hamid, A., Bakar, J. and Bee, G.H. 2002. Nutritional quality of spray dried protein hydrolysate from black tilapia (*Oreochrombis mossambicus*). *Food Chem.* 78, 69–74.
55. Ballester, D., Yanez, E., Brunser, O., Stekel, A., Chadud, P., Castano, G. and Mockenberg, F. 1977. Safety evaluation of an enzymatic fish protein hydrolysate: 10 month feeding study and reproduction performance in rats. *J. Food Sci.* 42, 407–409.
56. Barrias, C. and Oliva-Teles, A. 2000. The use of locally produced fish meal and other dietary manipulations in practical diets for rainbow trout *Oncorhynchus mykiss* (Walbaum). *Aquaculture Res.* 31, 213–218.

57. Lian, P. and Lee, C.M. 2003. Characterization of squid hydrolysates for its potential as aquaculture feed ingredient. In: *Proceedings of the First Joint Trans-Atlantic Fisheries Technology Conference—TAFT 2003,* June 10–14, The Icelandic Fisheries Laboratories, Reykjavik, Iceland, pp. 379–380.
58. Wilson, R.P., Freeman, D.W. and Poe, W.E. 1984. Three types of catfish offal meals for channel catfish fingerlings. *Prog. Fish Cult.* 46, 126–132.
59. Oliva-Teles A., Cerqueira, A.L. and Goncalves, P. 1999. The utilization of diets containing high levels of fish protein hydrolysate by turbot (*Scophthalmus maximus*) juveniles. *Aquaculture* 179,195–201.
60. Cordova-Murueta, J.H. and Garcia-Carreno, F.L. 2002. Nutritive value of squid and hydrolyzed protein supplement in shrimp feed. *Aquaculture* 210, 371–384.
61. Gildberg, A., Bogwald, J., Johansen, A. and Stenberg, E. 1996. Isolation of acid peptide fractions from a fish protein hydrolysate with strong stimulatory effect on Atlantic salmon (*Salmo salar*) head kidney leucocytes. *Comp. Biochem. Physiol. B* 114(1), 97–101.
62. Bogwald, J., Dalmo, R.A., Leifson-McQueen, R., Stenberg, E., and Gildberg, A. 1996. The stimulatory effect of a muscle protein hydrolysate from Atlantic cod, *Gadus morhua* L., on Atlantic salmon, *Salmo salar* L., head kidney leucocytes. *Fish Shellfish Immunol.* 6, 3–16.
63. Gildberg, A.J. 2003. Enzymes and bioactive peptides from fish waste related to fish silage: fish feed and fish sauce production. In: *Proceedings of the First Joint Trans-Atlantic Fisheries Technology Conference – TAFT 2003,* June 10–14, The Icelandic Fisheries Laboratories, Reykjavik, Iceland, pp. 328–331.
64. Refstie, S., Olli, J.J. and Standal, H. 2004. Feed intake, growth and protein utilization by post-smolt Atlantic salmon (*Salmo salar*) in response to graded levels of fish. *Aquaculture* 239(1–4), 331–349.
65. Andrawulan, N. and Shetty, K. 1999. Influence of acetyl salicylic acid in combination with fish protein hydrolysates on hyperhydricity reduction and phenolic synthesis in oregano (*Origanum vulgare*) tissue cultures. *J. Food Biochem.* 23, 619–635.
66. Vattem, D.A. and Shetty, K. 2002. Solid-state production of phenolic antioxidants from cranberry pomace by *Rhizopus oligosporus. Food Biotechnol.* 16, 189–210.
67. Koeppen, B.M. and Stanton, B.A. 2000. Part VII renal systems, In: *Principles of Physiology,* 3rd ed., R.M. Berne RM and M.N. Levy, eds., Mosby, St. Louis, MO, pp. 408–483.
68. Kohama, Y., Oka, H., Kayamori, Y., Tsujikawa, K., Mimura, T., Nagase, Y. and Satake, M. 1991. Potent synthetic analogues of angiotensin-converting enzyme inhibitor derived from tuna muscle. *Agric. Biol. Chem.* 55(8), 2169–2170.
69. Ukeda, H., Matsuda, H., Osjima, K., Matufuji, H., Matsui, T. and Osjima, Y. 1992. Peptides from peptic hyrolysate of heated sardine meat that inhibit angiotensin I converting enzyme. *Nippon Nogeikagaku Kaishi* 65(8), 1223–1228.
70. Matsumura, N., Fujii, M., Takeda, Y. and Shimizu, T. 1993. Isolation and characterization of angiotensin I-converting enzyme inhibitory peptides derived from bonito bowels. *Biosci. Biotechnol. Biochem.* 57(10), 1743–1744.
71. Wako, Y., Ishikawa, S. and Muramoto, K. 1996. Angiotensin I-converting enzyme inhibitors in autolysates of squid liver and mantle muscle. *Biosci. Biotechnol. Biochem.* 60(8), 1353–1355.
72. Bordenave, S., Fruitier, I., Ballander, I., Sannier, F., Gildberg, A., Batista, I. and Piot, J.M. 2002. HPLC preparation of fish waste hydrolysate fractions. Effect on guinea pig ileum and ACE activity. *Prep. Biochem. Biotechnol.* 32, 65–77.

73. Jung, W.K., Mendis, E., Je, J.Y., Park, P.J., Son, B.W., Kim, H.C., Choi, Y.K. and Kim, S.K. 2006. Angiotensin I-converting enzyme inhibitory peptide from yellowfin sole (Limanda aspera) frame protein and its antihypertensive effect in spontaneously hypertensive rats. *Food Chem.* 94(1), 26–32.
74. Je, J-Y., Park, P-J., Kwon, J.Y. and Kim, S-K. 2004. A novel angiotensin I converting enzyme inhibitory peptide from Alaska pollack (*Theragra chalcogramma*) frame protein hydrolysate. *J. Agric. Food Chem.* 52, 7842–7845.
75. Fahmi, A., Morimura, S., Guo, H.C., Shigematsu, T., Kida, A. and Uemura, Y. 2004. Production of angiotensin I converting enzyme inhibitory peptides from sea bream scales. *Process Biochem.* 39, 1195–2000.
76. Cheung, H.S., Wang, H.L., Ondetti, M.A., Sabo, E.F. and Cushman, D.W. 1980. Binding of peptide substrates and inhibitors of angiotensin-converting enzyme: Importance of the COOH- terminal dipeptide sequence. *J. Biol. Chem.* 255, 401–407.
77. Wergedahl, H., Liaset, B., Gudbrandsen, O.A., Lied, E., Espe, M., Muna, Z., Mork, S. and Berge, R.K. 2004. Fish protein hydrolysate reduces plasma total cholesterol, increases the proportion of HDL cholesterol, and lowers acyl-CoA: cholesterol acyltransferase activity in liver of Zucker rats.
78. Chuang, W-L., Sun Pan, B. and Tsai, J-S. 2000. Inhibition of lipoxygenase and blood thinning effects of mackerel protein hydrolysate. *J. Food Biochem.* 24, 333–343.
79. Boukortt, F.O., Girard, A., Prost, J., Ait-Yahia, D., Bouchenak, M. and Belleville, J. 2004. Fish protein improves the total antioxidant status of streptozotocin-induced diabetes in spontaneously hypertensive rat. *Med. Sci. Monit.* 10, BR397–BR404.
80. Dorman, T., Bernard, L., Glaze, P., Hogan, J., Skinner, R., Nelson, D., Bowker, L. and Head, D. 1995. The effectiveness of *Garum armoricum* (stabilium) in reducing anxiety in college students. *J. Adv. Med.* 8, 193–200.
81. Le Poncin, M. 1996. Experimental study: stress and memory. *Eur. Neuropsychopharmacol.* 6, 110-P10-2.
82. Le Poncin, M. 1996. Nutrient presentation of cognitive and memory performances. *Eur. Neuropsychopharmacol.* 6, 187-P19-4.
83. Bernet, F., Montel, V., Noel, B. and Dupouy, J.P. 2000. Diazepam-like effects of a fish protein hydrolysate (Gabolysat PC60) on stress responsiveness of the rat pituitary-adrenal system and sympathoadrenal activity. *Psychopharmacology* 149, 34–40.
84. Thongthai, C. and Gildberg, A. 2005. Asian fish sauce as a nutritional source. In: *Asian Functional Foods*, J. Shi, C.T. Ho and F. Shahidi, eds., Marcel Dekker/CRC Press, Boca Raton, FL, pp. 21–72.
85. Fitzgerald, A.J., Raj, P.S., Marchbank, T., Taylor, G.W., Ghosh, S., Ritz, B.W. and Playford, R.J. 2005. Reparative properties of a commercial fish protein hydrolysate preparation. *Gut* 54(6), 775–781.
86. Rajapakse, N., Jung, W.K., Mendis, E., Moon, S.H. and Kim, S.K. 2005. A novel anticoagulant purified from fish protein hydrolysate inhibits factor XIIa and platelet aggregation. *Life Sci.* 76(22), 2607–2619.

10 Marine-Derived Protein Hydrolysates, Their Biological Activities and Potential as Functional Food Ingredients: ACE-Inhibitory Peptides Derived from Bonito

Hiroyuki Fujita and Masaaki Yoshikawa

CONTENTS

10.1 INTRODUCTION

Bioactive peptides showing beneficial effects on human health have been found in foods of marine origin. For example, anserine (β-alanyl 1-methyl-histidine) and carnosine (alanyl-histidine) found in bonito, tuna, and oyster show antioxidative activity [1]. Bioactive peptides showing antihypertensive [2–16], antioxidative [17], α-glucosidase-inhibitory [18], and hypocholesterolemic activities [19], etc., also have been found in enzymatic digests of proteins from marine organisms. However, biological activities of these digests have been studied mostly *in vitro*, and a few of them have been studied *in vivo*, especially in humans. Among them, antihypertensive peptides based on the inhibition of angiotensin I-converting enzyme (ACE) have been studied most extensively. In this chapter, the antihypertensive functions of ACE-inhibitory peptides derived from marine organisms are described.

10.2 ACE-INHIBITORY ACTIVITIES OF ENZYMATIC DIGESTS OF PROTEINS FROM MARINE ORGANISMS

ACE is a dipeptidyl carboxypeptidase, catalyzing the conversion of angiotensin I to angiotensin II, a strong vasopressor. ACE also inactivates bradykinin, a vasodilatory peptide (Figure 10.1). Therefore, inhibitors for ACE such as captopril show an antihypertensive effect.

ACE-inhibitory peptides have been found in enzymatic digest of many food proteins such as gelatin [2], casein [3], and zein [4]. The ACE-inhibitory activity has also been found in hydrolysate of protein of marine origin, such as tuna [5], dried bonito (Katsuobushi) [6], krill [7], sardine [8–10,16], salmon [11], *Porphyra yezoensis* (nori) [12], and *Undaria pinnatifida* (wakame) [13,14] (Table 10.1). Antihypertensive activities of some of these digests such as Katsuobushi, sardine, nori, and wakame have been confirmed in humans, and approved by the Japanese Ministry of Health, Labor and Welfare as "foods for specified health uses" (FOSHU).

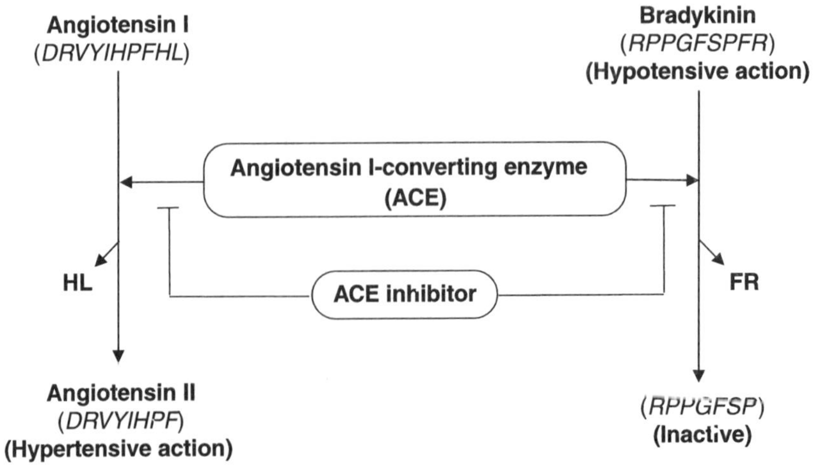

FIGURE 10.1 Mechanisms for antihypertensive effect of angiotensin I-converting enzyme inhibitors.

TABLE 10.1
ACE-Inhibitory Activities of Marine Protein Hydrolysates

Origin	Enzymes	ACE-Inhibitory Activity IC_{50} (μg/mL)	References
Dried bonito	Thermolysin	29	6
	Pepsinw	47	6
	Trypsin	161	6
	Chymotrypsin	117	6
	Trypsin+Chymotrypsin	175	6
	Pepsin→Trypsin	65	6
	Pepsin→Chymotrypsin	41	6
	Pepsin→Trypsin+Chymotrypsin	38	6
Sardine	Pepsin	620	8
	B. licheniformis protease	250	9
Salmon	Subtilisin	110	11
Nori (*Porphyra yezoensis*)	Pepsin	910	12
Wakame (*Undaria pinnatifida*)	Pepsin	250	13
	B. stearothermophillus protease	86	14
Krill	Pepsin→Trypsin	106	7

10.3 ANTIHYPERTENSIVE ACTIVITY OF THERMOLYSIN DIGEST OF DRIED BONITO (KATSUOBUSHI-OLIGOPEPTIDE)

Katsuobushi made from bonito muscle contains inosinate and its extract has been used as a seasoning material in Japan. We hydrolyzed water-extracted residue of Katsuobushi with various proteases since it is rich in proteins with almost no fishy odor. Enzymatic digests of Katsuobushi residue showed higher ACE-inhibitory activities than those from other proteins of marine origin (Table 10.2). Among them, the thermolysin digest showed the highest ACE-inhibitory activity (IC_{50} = 29 μg/mL), and the inhibitory activity was not impaired after digestion by gastrointestinal proteases (Table 10.3) [6]. Then, we tested the antihypertensive activity of the thermolysin digest of Katsuobushi (the Katsuobushi-oligopeptide) in spontaneously hypertensive rats (SHR) after single oral administration. It significantly reduced systolic blood pressure (SBP) after oral administration at a dose of 500 mg/kg [20]. We also tested the antihypertensive activity of the Katsuobushi-oligopeptide in SHR after long-term feeding. In SHR fed with 0.025% of the Katsuobushi-oligopeptide, which corresponds to 15 mg/kg/day, elevation of SBP was suppressed significantly (Figure 10.2). Thus, in long-term feeding, the minimum effective dose was lowered to about 1/33 that of single oral administration experiment [20]. The IC_{50} value for ACE of a Katsuobushi-oligopeptide preparation produced at industrial scale was about 50 μg/mL. The antihypertensive

TABLE 10.2
ACE-Inhibitory Peptides Derived from Marine Organisms

Peptide	Origin	IC_{50} (μM)	References
PTHIKWGD	Tuna	0.9	5
IY	Dried bonito	3.7	6
FQP	Dried bonito	12	6
LKPNM	Dried bonito	2.4	6
IWHHT	Dried bonito	5.8	6
ALPHA	Dried bonito	10	6
IKPLNY	Dried bonito	43	6
DYGLYP	Dried bonito	62	6
IVGRPRHQG	Dried bonito	6.2	6
KLKFV	Krill	30	7
LKVGGKGY	Sardine	9.2	8
MF	Sardine	44.7	9
RY	Sardine	51	9
MY	Sardine	193	9
LY	Sardine	38.5	9
YL	Sardine	122	9
VL	Sardine	43.7	9
KW	Sardine	1.63	9
GRP	Sardine	20	9
AKK	Sardine	3.13	9
RVY	Sardine	205.6	9
GWAP	Sardine	3.86	9
VY	Sardine	11	9
DW	Salmon	13	11
GIG	Salmon	30	11
LRY	Nori (*Porphyra yezoensis*)	5.06	12
MKY	Nori (*Porphyray ezoensis*)	7.26	12
AKYSY	Nori (*Porphyra yezoensis*)	1.52	12
YNKL	Wakame (*Undaria pinnatifida*)	21	13
AIYK	Wakame (*Undaria pinnatifida*)	213	13
YKYY	Wakame (*Undaria pinnatifida*)	64.2	13
KFYG	Wakame (*Undaria pinnatifida*)	90.5	13
KY	Wakame (*Undaria pinnatifida*)	38.7	14
KF	Wakame (*Undaria pinnatifida*)	56.2	14
FY	Wakame (*Undaria pinnatifida*)	59.9	14
KY	Wakame (*Undaria pinnatifida*)	48.4	14
VF	Wakame (*Undaria pinnatifida*)	62.7	14

activity of this preparation was confirmed in borderline or mildly hypertensive subjects [21]. The Katsuobushi-oligopeptide showed antihypertensive effect in hypertensive subjects at a dose of 3 g/day. Based on the antihypertensive activity and safety data, "peptide soup," containing the Katsuobushi-oligopeptide as the

TABLE 10.3
Stability of ACE-Inhibitory Activity of the Thermolysin Digest of Dried Bonito Against Gastrointestinal Proteases

Enzymes	IC_{50} (μg/mL)
None	29
Pepsin	22
Pepsin → Trypsin	26
Pepsin → Chymotrypsin	29
Pepsin → Trypsin + Chymotrypsin	26

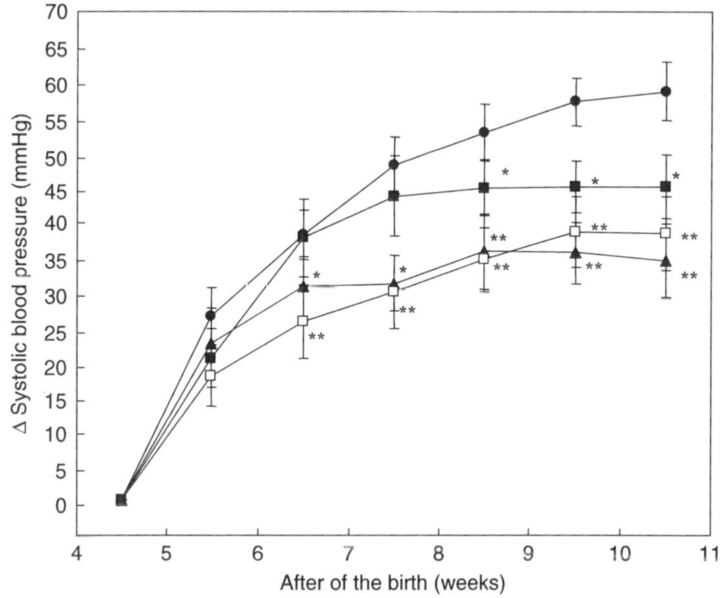

FIGURE 10.2 Antihypertensive effect of Katsuobushi-oligopeptide after long-time feeding in SHR. SHR of 4 weeks after birth were fed a standard chow containing 0% (●), 0.025% (■), 0.05% (□), and 1% (▲) Katsuobushi-oligopeptide (IC_{50} = 55 μg/mL), respectively. The daily intake of chow containing 0.025% the Katsuobushi-oligopeptide corresponds to 15 mg/kg/day. Difference in systolic blood pressure was expressed by mean ±S.E. Statistical analyses were done by student *t*-test (*$p<0.05$, **$p<0.01$).

active ingredient, as been approved by the Japanese Ministry of Health, Labor and Welfare in 1997 as "FOSHU."

By adding an ultrafiltration step, a Katsuobushi-oligopeptide with higher ACE-inhibitory (IC_{50} = 30 μg/mL) and antihypertensive activities was obtained [22]. Taste and flavor were also improved in this preparation. This strong-type Katsuobushi-oligopeptide was monitored against a placebo in a double-blind,

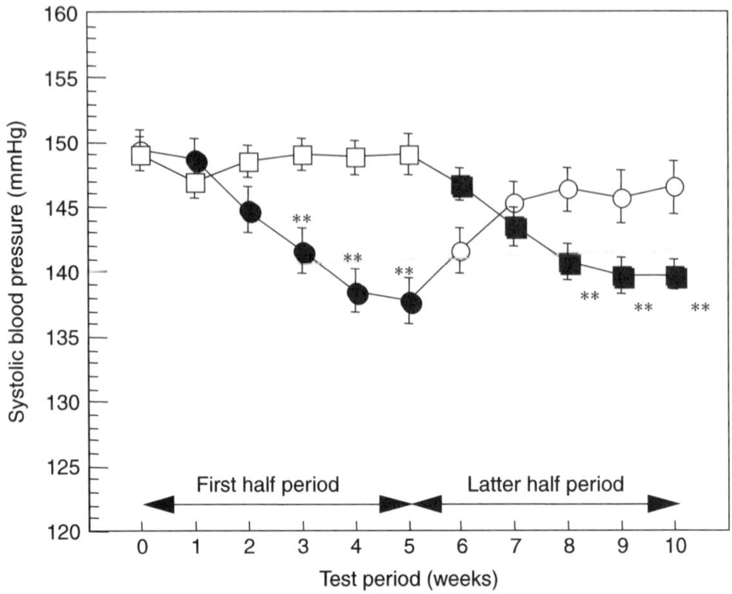

FIGURE 10.3 Antihypertensive effect of Katsuobushi-oligopeptide in borderline or mild hypertensive subjects. Thirty subjects (circles) ingested 1.5 g/day Katsuobushi-oligopeptide (IC_{50} = 31.5 μg/mL) and another 31 subjects ingested (squares) a placebo during the first 5-week period. In the second 5-week period, samples were reversed. Systolic blood pressure was monitored every week. The values are expressed as the mean ±S.E. Statistical analyses were done by Mann-Whitney *U* test (*$p < 0.05$, **$p < 0.01$).

randomized, crossover study, with 61 borderline or mildly hypertensive subjects (Figure 10.3). In the first group (30 subjects), who ingested the strong-type Katsuobushi-oligopeptide at a dose of 1.5 g/day for 5 weeks, the SBP was reduced significantly (11.7 mmHg) [22]. In the second group, SBP was reduced significantly (9.4 mmHg) after the crossover. The positively effective ratio (decreased SBP < −10 mmHg or DBP < −5 mmHg) was 64.3% in the first group and 61.3% in the second group [22]. This strong-type Katsuobushi-oligopeptide was approved as “FOSHU” effective at a dose of 1.5 g/day in 2001.

10.4 ISOLATION AND ANTIHYPERTENSIVE ACTIVITIES OF ACE-INHIBITORY PEPTIDES FROM THE KATSUOBUSHI-OLIGOPEPTIDE

We tried to isolate the ACE-inhibitory peptides in the Katsuobushi-oligopeptide. The Katsuobushi-oligopeptide was fractionated using octadecyl silica, phenyl silica, and cyanopropyl silica columns [6]. After three or four chromatography steps, eight potent ACE-inhibitory peptides IY, FQP, LKPNM, IWHHT, ALPHA, IKPLNY, DYGLYP, and IVGRPRHQG (IC_{50} = 3.7, 12, 2.4, 5.8, 10, 43, 62, and 6,2 μM, respectively) were isolated [6]. Among them, we tested antihypertensive

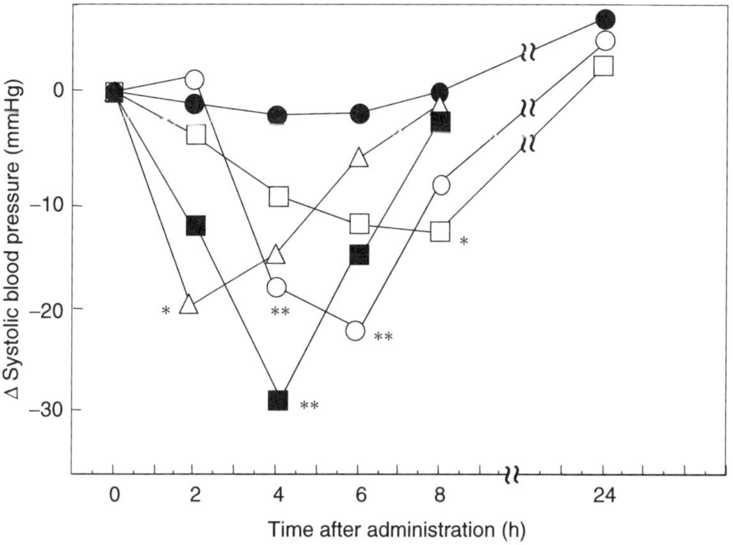

FIGURE 10.4 Antihypertensive activities of ACE-inhibitory peptides after single oral administration in SHR. SHR were administered saline (●), IY (△), IKW (■), LKPNM (○), and IVGRPRHQG (□) at a dose of 60 mg/kg. Changes in systolic blood pressure were expressed as mean ±S.E. (n = 5). Statistical analyses were done by student t test (*$p<0.05$, **$p<0.01$).

activities of IY, LKPNM, IWHHT, ALPHA, IKPLNY, and IVGRPRHQG. These peptides effectively lowered blood pressure after oral administration in SHR at a dose of 60 mg/kg.

The duration of blood pressure reduction by these ACE-inhibitory peptides derived from the Katsuobushi-oligopeptide differed depending on the number of amino acid residues. Dipeptide, such as IY, exhibited maximal blood pressure reduction 2 h after oral administration, while tripeptide such as IWH, pentapeptide such as LKPNM, and nonapeptide IVGRPRHQG showed maximum decrease of 4, 6, and 8 h after administration, respectively (Figure 10.4). The Katsuobushi-oligopeptide exhibited long-lasting antihypertensive activity since it contains ACE-inhibitory peptides of various maximally effective times.

10.5 CLASSIFICATION OF ACE-INHIBITORY PEPTIDES BY PREINCUBATION METHOD

In general, enzymatic digests of any proteins show an apparent ACE-inhibitory activity to a certain extent. This is why there are many ACE substrate peptides in enzymatic digest of food proteins based on its broad substrate specificity, and not only inhibitors but also substrates for ACE could inhibit the enzyme reaction. The inhibition of ACE reaction by its substrates is tentative since it is lost as the substrates are consumed. Therefore, substrate peptides for ACE fail to show antihypertensive activity *in vivo*. Thus, the most important thing to obtain peptides showing antihypertensive activity after oral administration is to select true ACE-inhibitory peptides from substrates [23,24]. True ACE-inhibitory peptides can be discriminated from substrates by preincubating peptides with ACE before measurement of ACE-inhibitory activity.

The IC_{50} values for ACE of individual peptides before and after preincubation, and their antihypertensive activities after oral administration are summarized in Table 10.4 [23,24]. The IC_{50} values of true inhibitors are unaltered, while those of substrates for ACE were altered by preincubation with ACE. ACE-inhibitory peptides isolated from thermolysin digest of dried bonito, such as IY, LW, and IKW, were true inhibitors since their IC_{50} values were not altered by preincubation. In contrast, peptides such as FKGRYYP from chicken muscle, FFGRCVSP and ERKIKVYL from ovalbumin were classified as substrates for ACE, since their potent ACE-inhibitory activities are almost lost after preincubation with ACE [23,24]. In fact, they failed to show antihypertensive activities [23].

Interestingly, LKPNM ($IC_{50} = 2.4$ μM) and IWHHT ($IC_{50} = 5.8$ μM) were activated to $IC_{50} = 0.76$ μM and $IC_{50} = 3.5$ μM, respectively, after preincubation with ACE (Table 10.4) [23,24]. LKPNM and IWHHT were converted to LKP and IWH, respectively, the true ACE inhibitors after preincubation. Therefore, these peptides were classified as prodrug-type inhibitor. Finally, LKPNM is activated eight times

TABLE 10.4
The Relationship between ACE-Inhibitory and Antihypertensive Activities of Peptides Derived from Various Proteins

		IC_{50} (μM)		Maximum Antihypertensive Effect[a]	
Peptides	**Origin**	**−Preincubation**	**+Preincubation**	**ΔmmHg**	**Hours after Administration**
Inhibitor type					
IY	Bonito	2.3	1.9	−19	2
LW	OVA	6.8	6.6	−22	2
IKW	Chicken	0.21	0.18	−17	4
Prodrug type					
LKPNM	Bonito	2.4	0.76	−23	6
LKP (active form)		0.32	0.32	−18	4
IWHHT	Bonito	5.8	3.5	−26	6
IWH (active form)		3.5	3.5	−30	4
Substrate type					
FKGRYYP	Chicken	5.8	34	0	
FFGRCVSP	OVA	0.4	4.6	0	
ERKIKVYL	OVA	1.2	6	0	

[a] Antihypertensive effects of peptides after oral administration in SHR at a dose of 60 mg/kg.

into LKP (IC_{50} = 0.3 μM) that belong to the most potent class among ACE-inhibitory peptides derived from food proteins [20,23,24]. The maximum decrease in blood pressure of LKP was observed 4 h after oral administration at a dose of 60 mg/kg. However, a maximum decrease in blood pressure of LKPNM was observed 6 h after oral administration at the same dose (Table 10.4). Thus, antihypertensive activity of LKPNM was more long lasting than with LKP [24]. This effect might be explained by the time required for the intestinal absorption of the larger peptide or for the enzymatic conversion of LKPNM into LKP by ACE *in vivo.*

Subsequently, we compared the antihypertensive activities of LKPNM and LKP with that of captopril (Figure 10.5). LKPNM, LKP, and captopril were orally administered at the same dose on molar base (25 μmol/kg). At this dosage, antihypertensive effect of LKP was maximal 2 h after the administration, while LKPNM and captopril showed their maximal effect after 4 h. The antihypertensive effect of LKP and LKPNM were 66.5 and 90.5% that of captopril on molar base. It should be noted that the antihypertensive effect of LKPNM was about two-third that of captopril although its ACE-inhibitory activity was less than 1/100 that of captopril (Table 10.5) [24]. These unique properties of LKPNM

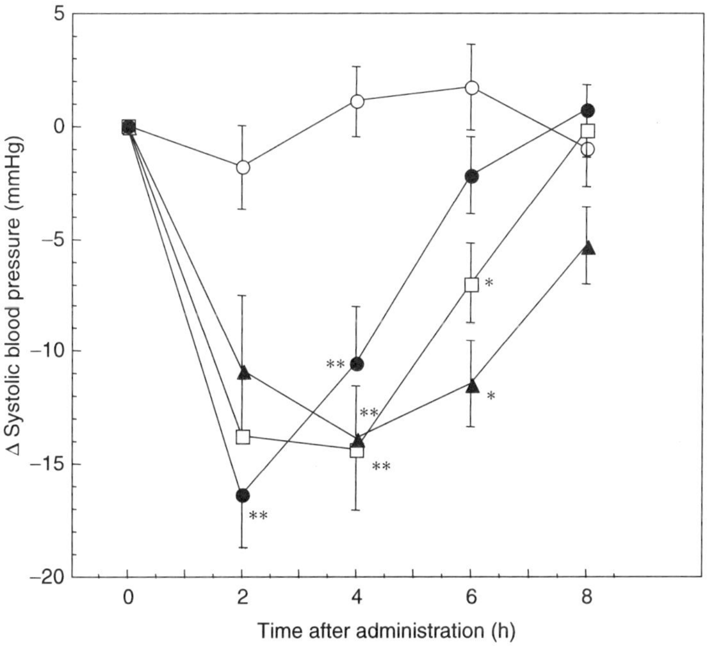

FIGURE 10.5 ACE-inhibitory and antihypertensive activities of captopril, LKPNM, and LKP. SHR were administered saline (○), LKPNM (▲), LKP (●), and captopril (□) at a dose of 25 μmol/kg (LKPNM, 15 mg/kg; captopril, 10 mg/kg; LKP, 9 mg/kg). Changes in systolic blood pressure were expressed as mean ±S.E. (n = 8). Statistical analyses were done by student t-test (*$p < 0.05$, **$p < 0.01$).

TABLE 10.5
ACE-Inhibitory and Antihypertensive Activities of Captopril and Peptides Derived from Dried Bonito

	ACE-Inhibitory Activity		Antihypertensive Activity		
				Relative Activity (%)	
Sample	IC_{50} (μM)	Relative Activity (%)	Dose[a] (m	By wt.	By mol.
Captopril	0.022	100	2.5	100	100
LKPNM	2.4	0.92	10.5	23.8	66.5
LKP	0.3	7.3	4.2	59.5	90.6

[a] Dose of sample to reduce SBP by 10 mmHg.

and LKP might be attributable to a higher affinity to the tissue owing to its basic properties, and slower excretion rate than captopril, a synthetic compound.

Thus, ACE-inhibitory peptides isolated from the Katsuobushi-oligopeptide are either inhibitor or prodrug-type peptides [23]. This may be the reason why the Katsuobushi-oligopeptide shows long-lasting antihypertensive activity in SHRs and human.

10.6 CONCLUSION

ACE-inhibitory peptides preventing hypertension are released from proteins of marine origin. To obtain a digest showing potent antihypertensive activities, the choice of the best combination of the substrate protein and protease is very important.

REFERENCES

1. Shiba, T. 1979. 5.3 Amino acid and peptides. In *Chemistry of Marine Products*, The Chemical Society of Japan Ed., Japan Scientific Society Press, Tokyo, p. 246.
2. Oshima, G., Shimabukuro, H. and Nagasawa, K. 1979. Peptide inhibitors of angiotensin I-converting enzyme in digests of gelatin by bacterial collagenase. *Biochem. Biophys. Acta* 566, 128.
3. Maruyama, S., Nakagomi, K., Tomizuka, N. and Suzuki, H. 1985. Angiotensin I-converting enzyme inhibitor derived from an enzymic hydrolysate of casein. II. Isolation and bradykinin-potentiating activity on the uterus and the ileum of rats. *Agric. Biol. Chem.* 49(5), 1405–1409.
4. Miyoshi, H., Ishikawa, H., Kaneko, T., Fukui, F., Tanaka, H. and Maruyama, S. 1991. Structures and activity of angiotensin-converting enzyme inhibitors in an alpha-zein hydrolysate. *Agric. Biol. Chem.* 55, 1313.
5. Kohama, Y., Matsumoto, S., Oka, H. and Suzuki, H. 1988. Isolation of angiotensin-converting enzyme inhibitor from tuna muscle. *Biochem. Biophys. Res. Commun.* 155, 332.

6. Yokoyama, K., Chiba, H. and Yoshikawa, M. 1992. Peptide inhibitors for angiotensin I-converting enzyme from thermolysin digest of dried bonito. *Biosci. Biotech. Biochem.* 56(10), 1541–1545.
7. Kawamura, Y., Sugimoto, T., Takane, T. and Satake, M. 1992. Physiologically active peptide motif in proteins: peptide inhibitors of ACE from the hydrolyzates of Antarctic krill muscle protein. *JARQ* 26(3), 210–213.
8. Suetsuna, K. and Osajima, K. 1986. The inhibitory activities against angiotensin I-converting enzyme of basic peptides originating from sardine and hairtail meat. *Nippon Suisan Gakkaishi* 52(11), 1981–1984.
9. Matsufuji, H., Matsui, T., Seki, E., Osajima, K. and Osajima, Y. 1994. Angiotensin I-converting enzyme inhibitory peptides in an alkaline protease hydrolyzate derived from sardine muscle. *Biosci. Biotechnol. Biochem.* 58, 2244.
10. Seki, E., Osajima, K., Matsufuji, H., Matsui, T. and Osajima, M. 1995. Val-Tyr, an angiotensin I converting enzyme inhibitor from sardines that have resistance to gastrointestinal proteases. *Nippon Nogei Kagaku Kaishi* 69(8), 1013–1020.
11. Ohta, T., Iwashita, A., Sasaki, S. and Hatano, M. 1997. Antihypertensive action of the orally administered protease hydrolysates of chum salmon head and their angiotensin I-converting enzyme inhibitory peptides. *Food Sci. Technol. Int. Tokyo*, 3, 339.
12. Suetsuna, K. 1998. Purification and identification of angiotensin I-converting enzyme inhibitors from the red alga *Porhyra yezoensis*. *J. Marine Biotech.* 6, 163.
13. Suetsuna, K. 2000. Identification of an antihypertensive peptide from digest of wakame (*Undaria pinnatifida*). *J. Nutr. Biochem.* 11, 450.
14. Suetsuna, K., Maekawa, K. and Chen, J.R. 2004. Antihypertensive effects of undaria pinnatifida (wakame) pepetide on blood pressure in spontaneously hypertensive rats. *J. Nutr. Biochem.* 15, 267.
15. Sato, M., Oba, T., Yamaguchi, T., Nakano,T., Kahara, T., Funayama, K., Kobayashi, A. and Nakano, T. 2002. Antihypertensive effects of hydrolysates of wakame (*Undaria pinnatifida*) and their angiotensin I-converting inhibitory activity. *Ann. Nutr. Metab.* 46, 259.
16. Kawasaki, T., Seki, E., Osajima, K., Yoshida, M., Asada, K., Matsui, T. and Osajima, M. 2000. Antihypertensive effect of valyl-tyrosine, a short chain peptide derived from sardine muscle hydrolyzate, on mild hypertensive subjects. *J. Hum. Hypertension* 14(8), 519–523.
17. Suetsuna, K. 2000. Antioxidant peptides from the protease digest of prawn (*Penaeus japonicus*) muscle. *Marine Biotechnol.* 2, 5.
18. Matsui, T., Yoshimoto, C., Osajima, K., Oki, T. and Osajima, Y. 1996. In vitro survey of alpha-glucosidase inhibitory food components. *Biosci. Biotechnol. Biochem.* 60, 2019.
19. Hagino, H. 2002. Physiological effects and application of peptides derived from Nori (laver). *Food Style* 21, 6(11), 69–72
20. Yoshikawa, M. and Fujita, H. 1994. Studies on the optimum conditions to utilize biologically active peptides derived from food proteins. In *Developments in Food Engineering*, Yano, T., Matsuno, R. and Nakamura, K. Ed., Blackie Academic and Professional, London, p. 1053.
21. Fujita, H. and Yoshikawa, M. 2004. Angiotensin I-converting enzyme inhibitory peptides derived from food protein. *Foods Food Ingredients J. Jpn* 209(8), 661–670.
22. Fujita, H., Yamagami, T. and Ohshima, K. 2001. Effects of an ACE-inhibitory agent, katsuobushi oligopeptide, in the spontaneouly hypertensive rat and in borderline and mildly hypertensive subjects. *Nutr. Res.* 21, 1149.

23. Fujita, H., Yokoyama, K. and Yoshikawa, M. 2000. Classification and antihypertensive activity of angiotensin I-converting enzyme inhibitory peptides derived from food proteins. *J. Food Sci.* 65, 564.
24. Fujita, H. and Yoshikawa, M. 1999. LKPNM: a prodrug-type ACE-inhibitory peptide derived from fish protein. *Immunopharmacology* 44, 123.

11 Marine Algal Constituents

Yvonne V. Yuan

CONTENTS

11.1 INTRODUCTION

Edible marine algae or seaweeds, sometimes referred to as sea vegetables, from the Protista taxonomies Phaeophyceae (brown [B]), Chlorophyceae (green [G]), and Rhodophyceae (red [R]) are ubiquitous in Pacific (e.g., Indonesia, Philippines, Maori, Hawaii) and Asian (e.g., Japan, China, Korea) diets, and are also present, albeit to a lesser extent, in waters of France, Iceland, Ireland, Norway, Wales, as well as the Canadian Maritime provinces and Maine in the United States. Traditionally, seaweeds have been incorporated into diets as fresh or blanched algae in salads, soups, or garnishes such as the *Caulerpa* spp. (G), *Sargassum* spp. (B), and *Porphyra spp.* (R) in the Philippines [1]; sushi components known as "Nori" or "Kim" (*Porphyra tenera* and *P. yezoensis*) in Japan and Korea, respectively; seasonings and condiments such as the *Laminaria* spp. (B), "Arame" (*Eisenia bicyclis*, B), "Hijiki"(*Hijikia fusiformis*, B), "Wakame" (*Undaria pinnatifida*, B) in Japan. Seaweeds are also used in soups such as "Hai dai" (*L. japonica*, B) in China; ingredients in stews, salads, or condiments such as "Limu Pālahalaha" (*Ulva fasciata*, G) in Hawaii; snack foods and ingredients in soups, salads, etc., such as the European/North American (NA) red alga "Dulse" (*Palmaria palmata*, R) [2]; as fresh or sundried ingredients for steaming, fermenting, or as a chewing gum such as "Karengo" (*Porphyra columbina*, R) by the Maori of New Zealand [3]; and even as an ingredient in baked goods such as "Laver" (*P. umbilicalis*, R) in laver bread in Wales. More recently, fresh and dried seaweeds have been enjoying a growing popularity in gourmet cuisine as side dishes, garnishes, and condiments on today's restaurant menus. Seaweeds also have a long history of use in the food industry as sources of gelling, thickening, and emulsifying agents, including alginates (alginic acid) from the brown kelps (e.g., *Macrocystis* and *Laminaria* spp.), carrageenans from the red alga *Chondrus crispus*, and agars from red algae such as *Gelidium*, *Gracilaria,* and *Gelidiella* spp.; in fact, a dessert unique to Prince Edward Island in the Canadian Maritimes, Seaweed Pie, is based on the thickening capacity of the native "Irish Moss" (*C. crispus*) harvested from island beaches.

In general, seaweeds are not known to play a large role in human nutrition other than as a rich source of minerals, in particular, iodine [4,5]. However, recent studies do indicate that various edible seaweeds can be good sources of protein (e.g., *P. palmata*; [6]; long-chain polyunsaturated fatty acids (PUFAs) such as eicosapentaenoic acid (e.g., *P. tenera* and *Porphyridium* spp., R) [7,8], as well as soluble and insoluble dietary fibers [9–11]. Interestingly, in 1990, the government of France published regulations on the usage of specific marine algae as vegetables, condiments, and as raw or semiprocessed ingredients in prepared and processed foods [12]; these included several brown kelps (i.e., *Ascophyllum nodosum*, *Fucus vesiculosus*, *F. serratus*, *Himanthalia elongata*, *Undaria pinnatifida*), green algae (i.e., *Ulva* and *Enteromorpha* spp.), and red algae (i.e., *P. umbilicalis*, *P. palmata*, *G. verrucosa*, *C. crispus*). In contrast, algae do have a relatively long history of use in animal feed and as a fodder supplement to provide additional vitamins, minerals, and trace elements to livestock diets [4]. Understandably, the use of algae in animal nutrition has been practiced primarily in coastal regions, particularly Scandinavia and Europe where brown kelps (e.g., *A. nodosum*, *Laminaria* spp., *Alaria esculenta*) and Rhodophyceae species have been used as adjuncts in the diets of poultry, swine, cattle, and sheep.

More recently, a growing area of inquiry has focused on marine algae and constituents as functional foods and nutraceuticals for potential beneficial health effects as sources of antioxidants and bioactives in reducing the risk for various diet-related chronic diseases such as atherosclerosis and hyperlipoproteinemia in cardiovascular disease (CVD) as well as breast and colon carcinogenesis. Indeed, much of the interest in marine algae as functional foods in chronic disease risk reduction is derived from epidemiological evidence relating the habitual consumption of seaweeds with reduced risk for CVD or various diet-related cancers between Asian and Western (i.e., European or NA) populations. For example, seaweed consumption by the Japanese ranges between 0.4 and 29.2 g/day by virtually 100% of the population [13], with the highest intakes attributed to the people of Okinawa [14], in contrast to the greatly reduced consumption of seaweed outside of Japan, for example, by 12–20% by the Japanese of Hawaii [15]. Very low to zero seaweed intakes can be expected in most Western populations. These seaweed consumption estimates represent a significant environmental (i.e., dietary) difference between populations known to differ in CVD rates: in Japan, the age-standardized mortality rates (ASMR) for coronary heart disease (CHD) are 43 and 22 per 100,000 in males and females, respectively [16], or approximately 25% the rate of CHD in NA. These statistics reflect the decline in CVD rates in Japan with increased economic wealth compared to Western countries. In China, the CHD mortality rates are approximately 50% of those of the West, albeit overall CVD mortality is similar to NA [16]. It is noteworthy that both these Asian cultures have traditional diets low in saturated fat and cholesterol, which manifests itself in low serum cholesterol levels, thereby affording protection against this CVD risk factor. With respect to diet-related cancers, the 1 year prevalence/100,000 for breast cancer in Japan and China were 42.2 and 13.1, respectively, compared to 125.9 and 106.2 in NA and Europe, respectively [17]; the corresponding statistics for prostate cancer in Japan and China were 10.4 and 0.7, respectively, compared to 117.2 and 53.1 in NA and Europe, respectively [17]. These population data are supported by a growing body of evidence elucidating the efficacy of seaweeds and their constituents in reducing the risk of chronic disease in animal models, such as mammary [18,19], intestinal [20,21], and skin carcinogenesis [22,23]; as well as hypocholesterolemic effects in rodent [24] and poultry [8] model studies.

This chapter will review marine algal macronutrient composition and the proposed mechanisms of their bioactivity in health and risk reduction of diet-related chronic diseases. An in-depth review of algal water- and lipid-soluble antioxidants has recently been published elsewhere [2]; however, the present chapter will review the polyphenol composition of marine algae. It is noteworthy that in some situations, increased consumption of marine algae can be detrimental, such as the concern with high levels of inorganic arsenic in "Hijiki" (*H. fusiformis*), which has been linked to gastrointestinal effects, anemia, and liver damage [25]; thus, this chapter will also address potential detrimental effects of increased marine algal consumption where appropriate.

11.2 PROXIMATE COMPOSITION OF MARINE ALGAE

In general, the nutritional value of the majority of edible marine algae is concentrated primarily in the ash (range 8.4–43.6% dry weight) and both soluble (range

0.2–59.7% dry weight) and insoluble dietary fiber fractions (range 5.3–52.3% dry weight; Table 11.1; [26–37]). The marine algae are not particularly rich in lipids (range 0.92–5.2% dry weight) with the exception of two Hawaiian samples of *Dictyota* spp. (B) containing 16.1 and 20.2% lipid on a dry weight basis [26]; albeit these lipids are enriched with the long-chain PUFAs, in particular C20:5ω3 and C22:6ω3 [7,27–30]. In contrast, while somewhat variable (range 4.9–37.8% protein dry weight), marine algal protein content does have potential value as an alternate plant protein source in the diet, particularly that of the Rhodophyceae such as the *Palmaria* and *Porphyra* spp. with protein levels between 13.7 and 37.8% dry weight [6,12,31,32]. Chlorophyceae and Phaeophyceae algal species are generally noted to have lower protein contents compared to Rhodophyceae, with the exceptions of *E. compressa* (G) at 32.4% protein dry weight [33] and *U. pinnatifida* (B), which has been reported to range between 11 and 24% protein on a dry weight basis [10,11,27,31,34].

It is clear from the data summarized in Table 11.1 that there is considerable variability in the proximate composition of the same algal species not only between studies conducted on samples from different geographical locations, but also within the same study on samples collected at different times of the year [6,32,38]. As photosynthetic organisms, the composition of marine algae can be seen to be

TABLE 11.1
Proximate Composition of a Variety of Edible Red, Green, and Brown Algae (% Dry Weight)

Algae	Crude Protein	Crude Lipid	Total Dietary Fiber	Soluble Dietary Fiber	Insoluble Dietary Fiber	Ash
Red						
Chondrus crispus	20.10		34.29	22.25	12.04	21.44
C. ocellatus	8.3	1.3		30.6		22.8
Eucheuma denticulatum	4.9	2.2		28.0		43.6
Gracilaria changgi	6.9	3.3	24.7			22.7
Hypnea charoides	18.4	1.48	50.3			22.8
H. japonica	19.0	1.42	53.2			22.1
Palmaria sp.	13.87	1.80				34.00
P. palmata			45.6	24.3	21.3	
P. palmata (W)	24.5					23.43
P. palmata (S)	13.7					15.97
Porphyra sp.	24.11	1.03				19.07
P. tenera	28.29		33.78	14.56	19.22	21.00
P. tenera	37.8	1.3	34.9	2.9	32.0	8.4
P. vietnamensis	16.5	4.4		30.5		25.2
P. yezoensis			59.4	23.1	36.3	
P. yezoensis			29.6	18.3	11.3	

(Continued)

TABLE 11.1
(Continued)

Algae	Crude Protein	Crude Lipid	Total Dietary Fiber	Soluble Dietary Fiber	Insoluble Dietary Fiber	Ash
Green						
Enteromorpha sp.			33.4	17.2	16.2	
E. compressa	32.4	4.2	28.5	0.2	28.2	17.1
E. flexuosa	7.9	5.6		39.9		23.2
E. intestinalis	11.4	5.2		22.2		29.2
Ulva fasciata	12.3	3.6		20.6		25.4
U. lactuca	7.06	1.64	55.4			21.3
U. lactuca			38.1	21.3	16.8	
Brown						
Dictyota acutiloba	12.0	16.1		5.9		28.9
D. sandvicensis	6.4	20.2		6.7		28.9
Eisenia bicyclis			74.6	59.7	14.9	
Fucus vesiculosus	6.19		50.09	9.8	40.29	30.12
Hijikia fusiformis			49.2	32.9	16.3	
H. fusiformis			60.6	22.9	37.7	
H. fusiformis			69.3	24.5	44.8	
Himanthalia elongata	5.46	0.97				26.78
H. elongata			32.7	25.7	7.0	
Laminaria digitata	10.70		36.12	9.15	26.98	37.60
L. japonica	8.1	1.9	30.8	3.3	27.5	32.5
L. japonica			41.1	7.1	34.0	
L. japonica			35.9	5.0	30.9	
L. ochroleuca	7.49	0.92				29.47
Sargassum hemiphyllum	10.1	3.04	62.9	9.00	52.3	19.6
S. echinocarpum	10.3	3.8		10.5		32.0
S. obtusifolium	13.0	2.6		12.3		28.9
Undaria pinnatifida	18.00	1.05				31.24
U. pinnatifida	18.5	2.1	51.0	11.5	39.4	26.9
U. pinnatifida	15.47		33.58	17.31	16.26	39.82
U. pinnatifida			35.3	30.0	5.3	

Note: W = winter/spring sample; S = summer/fall sample.

Source: Han, K-H., Lee, E-J., and Sung, M-K., *J. Food Sci. Nutr.*, 4(3), 180–183, 1999; Rupérez, P. and Saura-Calixto, F., *J. Eur. Food Res. Technol.*, 212, 349–354, 2001; McDermid, K.J. and Stuercke, B., *J. App. Phycol.*, 15, 513–524, 2003; Sánchez-Machado, D.I., López-Cervantes, J., López-Hernández, J., and Paseiro-Losada, P., *Food Chem.*, 85, 439–444, 2004; Norziah, M.H. and Ching, C.Y., *Food Chem.*, 68, 69–76, 2000; Chan, J.C.C., Cheung, P.C .K., and Ang, P.O., *J. Agric. Food Chem.*, 45, 3056–3059, 1997; Hagen-Rødde, R.S., Vårum, K.M., Laren B.A., and Myklestad, S.M., *Bot. Marina*, 47, 125–133, 2004; Wong, K.H. and Cheung, P.C.K., *Food Chem.*, 71, 475–482, 2000; Lahaye, M. J. *Sci. Food Agric.*, 54, 587–594, 1991; Lahaye, M., Michel, C., and Barry, J.L., *Food Chem.*, 47, 29–36,1993; Wang, W., Onnagawa, M., Yoshie, Y., and Suzuki, T., *Fish. Sci.*, 67, 1169–1173, 2001.

influenced by not only nutrient concentration in the seawater, but also water temperature and depth as well as the amount of photosynthetically active radiation (PAR; 400–700 nm) and more specifically, the amounts of UV-B (280–320 nm) and UV-A (320–400 nm) irradiation during growth, which are all dependent on geographical location and season [2,32,39]. Algae will be exposed to varying amounts of UV irradiation depending on their tidal zone habitat; for example, *P. palmata* (R) is classified as an upper sublittoral alga within the lower intertidal (i.e., between the average and lowest low tides) and upper subtidal (i.e., are never exposed to air during tidal fluctuations) zones, typically between depths of 0 and 3 m; *L. setchellii* (B) also inhabits low intertidal and upper subtidal zones; whereas *M. integrifolia* (B) and *N. leutkeana* (B) both inhabit low intertidal depths [40]. While the lower zonal limit for *M. integrifolia* is at depths of approximately 10 m, this alga typically inhabits depths between 1 and 4 m. Thus, when comparing the proximate compositions of marine algae, it is valuable to collect information about the harvest locations and collection dates.

11.2.1 Protein

Due to the fact that Rhodophyceae such as the *Palmaria* and *Porphyra* spp. have a long history of consumption in Asian, North American Maritime, and some European cultures as well as being noted as particularly good algal sources of protein, there has been a fair amount of interest in studying their proximate compositions, with an emphasis on protein content and quality, and how they are influenced by geographical location and harvest dates [6,32,38]. Studies evaluating *P. palmata* (dulse) collected from the French Brittany coast [6] and Northern and Southern Ireland; Bay of Biscay, Spain; and Trondheim, Norway [32] confirmed that the protein content was variable throughout the year, peaking with values between 21.9 and 24.5% dry weight during the winter/spring months spanning December through March, and dropping as low as 11.9–13.7% dry weight during the summer/fall months spanning May through November. Interestingly, these changes in protein content coincided with distinct changes in the amino acid profiles of these dulse samples, with marked decreases of up to four to six of the essential amino acids (His, Leu, Lys, Cys, Phe, Tyr, Thr) occurring from April to September [6]. Conversely, the nonessential amino acids Glu, Ser, and Ala were observed to be relatively higher in dulse samples harvested in late winter, spring, September, and October; but absent in July and November through January. It was thought that these seasonal fluctuations in algal amino acid content reflected changes in the types of proteins and thereby, enzymes present in the tissues. Moreover, a positive correlation between the availability of nitrogenous nutrients in seawater and protein content appears to exist with enhanced algal protein content in the winter and early spring when maximum nutrients are available, as well as geographic locations with higher concentrations of nitrogenous nutrients in the seawater such as Northern Ireland compared to locations in Southern Ireland, Spain, and Norway [32]. Support for the hypothesis that the intensity and availability of PAR UV irradiation during growth in the summer months affects algal metabolism and the profile of proteins present can be found in studies evaluating the influence of growth conditions on the endogenous antioxidant profile of seaweeds, particularly the mycosporine-like amino acids (MAAs) of Rhodophyceae and the phlorotannins of Phaeophyceae [2,39,40]. For example, when *P. palmata* or *Laminaria* spp. are exposed to high levels of UV-A and UV-B irradiation during growth, the synthesis of these endogenous antioxidant and sunscreen molecules is up-regulated. Indeed, the aromatic amino acids, Phe and Tyr, are key intermediates in the synthesis of phenolic acids and the MAAs via the shikimic acid pathway [2], which coincides with the decreased concentrations of Phe and Tyr in April through June and September reported by Galland-Irmouli and coworkers [6]. Also, Gly is a principal substituent conjugated with the MAA chromophore structures, potentially contributing to the reduced levels of this nonessential amino acid during the spring and summer months. Thus, there are likely multiple variables influencing not only the protein concentration, but also the protein/enzyme profile and amino acid composition of edible seaweeds, which need to be considered when comparing species and studies in the literature.

When evaluating the quality of a dietary protein, not only is the amino acid composition, and in particular the complement of essential amino acids, important, but so too is the digestibility of the protein, since it is this latter factor, which will determine whether the protein will support growth and maintenance *in vivo*. In general, Table 11.2 indicates that many of the edible Rhodophyceae, Chlorophyceae, and Phaeophyceae contain amino acid profiles with a good complement of essential amino acids ranging between 36.3 and 57.65% of total amino acids. The Rhodophyceae tended to contain a predominance of acidic versus basic amino acids such as in the case of *P. palmata* with 28.5 versus 7.8% of total amino acids [6]; *H. charoides* with 31.1 versus 14.7%; *H. japonica* with 29.7 versus 16.2%; and *G. changgi* with 17.3 versus 11.8% of total amino acids (Table 11.2); *P. tenera* was the exception with 15.4 acidic versus 24.2% basic amino acids. Similar to terrestrial plant proteins, marine algal proteins were limiting in more than one essential amino acid when compared to either hen's egg protein or the FAO/WHO reference pattern: Lys was limiting for *G. changgi* [28]; *P. palmata* [6] particularly when harvested in the summer months; *D. antarctica* [30] particularly in the fronds versus stem structures; *H. charoides*, *H. japonica,* and *U. lactuca* [41]. Additionally, *P. palmata* was limited in the sulfur-containing amino acids (Met and Cys) in general and limited in Thr in summer and early winter [6]; *D. antarctica* (stem) and *U. lactuca* were both limited in Ile and Leu [30]; *H. charoides*, *H. japonica,* and *U. lactuca* were all limited in the sulfur-containing amino acids in another study (Met and Cys) [41]. These data reflect the release of amino acids following acid hydrolyses *in vitro* and therefore would not necessarily reflect the bioavailability of these same amino acids *in vivo*. Moreover, *in vitro* and *in vivo* digestibility studies are highly variable in reporting marine algal protein digestibility depending on whether a crude algal powder versus a protein extract was studied, or individual proteolytic enzymes versus intestinal juice [6], a multienzyme preparation [41], or in sacco rumen incubation [42]. For example, when Galland-Irmouli and coworkers [6] evaluated the protein digestibility of

TABLE 11.2
Amino Acid Profiles of a Variety of Edible Red, Green, and Brown Algae (% of Total Amino Acids)

	Red Algae					Green Algae		Brown Algae	
Amino Acids	*Palmaria palmata*	*Hypnea charoides*	*Hypnea japonica*	*Gracilaria changgi*	*Porphyra tenera*	*Ulva lactuca*	*Ulva pertusa*	*Sargassum hemiphyllum*	*Durvillaea antarctica*
Essential amino acids									
His	0.50	0.83	1.15	3.17	1.52	1.35	4.96	1.50	9.83
Ile	3.7	4.22	4.80	4.88	4.34	4.12	4.34	4.10	4.59
Leu	7.1	7.52	7.09	6.06	9.44	7.47	8.55	7.01	7.91
Lys	3.3	4.22	4.65	2.76	4.88	4.77	5.58	5.03	6.64
Met	2.7	1.75	2.06	3.32	1.19	0.63	1.98	2.01	11.98
Cys	0.0	0.0	0.0	0.70	0.0	0.0	0.0	0.0	0.056
Phe	5.1	4.55	4.80	4.81	4.23	5.88	4.83	4.21	4.91
Tyr	3.4	3.14	3.58	1.56	2.60	3.74	1.73	22.6	2.33
Thr	3.6	5.21	5.11	6.61	4.34	6.38	3.84	4.57	3.34
Val	6.9	5.62	5.49	5.26	6.94	7.21	6.07	5.36	6.06
Total EAA	36.3	37.1	38.7	39.1	39.48	41.5	41.88	56.4	57.65
Nonessential amino acids									
Asp	18.5	17.6	16.6	6.66	7.59	14.3	8.05	9.99	9.76
Glu	9.9	13.5	13.1	10.6	7.81	11.3	8.55	11.4	13.79
hPro	2.3	–	–	–	–	–	–	–	–
Pro	1.8	3.80	3.96	9.33	6.94	4.70	4.96	3.91	0.004
Ser	6.3	5.04	5.26	ND	3.14	6.46	3.72	0.92	5.69
Gly	13.3	5.95	5.56	17.6	7.81	6.72	6.44	5.24	2.89
Ala	6.7	6.53	6.33	10.8	8.03	9.95	7.56	6.15	5.85
Arg	5.1	10.6	10.4	5.89	17.79	5.00	18.46	4.25	4.35

Note: – = not assayed.

Source: Norziah, M.H. and Ching, C.Y., *Food Chem.*, 68, 69–76, 2000; Chan, J.C.C., Cheung, P.C.K. and Ang, P.O., *J. Agric. Food Chem.*, 45, 3056–3059, 1997; Ortiz, J., Romero, N., Robert, P., Araya, J., Lopez-Hernández, J., Bozzo, C., Navarrete, E., Osorio, A. and Rios, A., *Food Chem.*, 99, 98–104, 2006; Fleurence, J., *Trends Food Sci. Technol.*, 10, 25–28, 1999.

P. palmata using a variety of *in vitro* methods, dulse protein was 56% digestible by pepsin–pancreatin in a digestion cell compared to casein when a crude algal powder was evaluated versus previous reports of 50 or 73% digestibility with pancreatin alone or pronase. Gel electrophoresis revealed that a dulse protein extract was highly digestible by trypsin and pronase, but less so by intestinal juice and chymotrypsin. Differences in digestibility were attributed to the potential presence of protease inhibitors in dulse as well as a general inhibitory effect of algal polysaccharides on enzyme access to peptide bond sites of algal proteins [6]. Further evidence of the potential detrimental influence of algal polysaccharide content on protein digestibility is evident from the work of Rupérez and Saura-Calixto [11] indicating that the available protein content of *F. vesiculosus* (B) and *L. digitata* (B) was reduced by the presence of resistant protein associated with insoluble dietary fiber accounting for up to 72 and 57% of total protein, respectively. The protein digestibility of Chlorophyceae and Phaeophyceae species is also potentially decreased by the greater polyphenolic content of these algae compared to Rhodophyceae; for example, Wong and Cheung [41] demonstrated a negative correlation between the *in vitro* protein digestibility and total phenolic contents of *H. charoides* (R) and *H. japonica* (R) compared to *U. lactuca* (G). When Ventura and Castañón [42] incubated ground *U. lactuca* in sacco within the cannulated rumen of Canarian goats, 85% of the protein appeared to be available for digestion, although degradable protein represented approximately 60% of available protein; thus, *U. lactuca* was described as a medium-quality forage for goat nutrition, similar to a medium-quality alfalfa hay.

In vivo studies with rats fed on diets containing marine algae are able to provide data on the significance of the chemical and *in vitro* investigations discussed above. Animals fed on AIN-93M-based diets containing 7% *P. tenera* (R) or *L. digitata* (B [43]), or 14.7% *P. tenera* or *U. pinnatifida* (B [44]), all exhibited similar food intakes, body weight gains and feed efficiency ratios (FER; weight gain [g]/feed intake [g]) compared to control rats fed on algae-free diets in the respective studies; albeit, the FER for 7% *L. digitata*–fed rats was slightly decreased compared to 7% *P. tenera*–fed counterparts. In this latter study, it was observed that the overall apparent digestibility of *P. tenera* (97.8%) and *L. digitata* (97.6%) diets were decreased compared to the control group (98.4% [43]). Overall, when rats were fed on 7% marine algal diets for 3 weeks, the diets supported growth well without changes in hepatic, kidney, or heart weights, but with a slight decrease in spleen weight in *L. digitata*–fed rats, thought to be related to the decreased dietary Na/K ratio for these animals [43]. Similarly, rats fed on 14.7% marine algal diets grew as well as control diet–fed counterparts despite decreases in the apparent digestibility of dietary protein from 92.6% with control rats to 86.1 and 86.2% with *P. tenera* and *U. pinnatifida*–fed rats, respectively [44]. These decreases could be attributed to increased fecal protein excretion from increased gut microflora protein or protein resistant to digestion associated with the inhibitory effects of dietary fiber on proteolytic enzyme activity discussed above. Thus, the protein efficiency ratios (weight gain [g]/protein intake [g]) of the marine algae–fed groups were increased compared to controls.

Taken together, these reports suggest that marine algae protein can support growth well when consumed as part of a mixed diet to offset slight decreases in apparent absorption of protein from diets, which can be attributed in part to the dietary fiber fractions of algae-supplemented diets.

11.2.2 Lipids

While the majority of marine algae have very low lipid contents, ranging from as low as 0.3% dry weight in *U. lactuca* harvested in Northern Chile [30] to 7.2% dry weight in *Caulerpa lentillifera* (G) harvested from Hilton Waikaloa, Hawaii [26], there are exceptions such as the edible Phaeophyceae *Dictyota* spp. from Hawaii, *D. acutiloba* and *D. sandvicensis* both commonly known as "Limu Alani" with lipid contents of 16.1 and 20.2% dry weight, respectively ([26]; Table 11.1). Thus, the energy content, determined by bomb calorimetry, of these algae is correspondingly low as well, ranging from as low as <1.0 kCal/g dry weight for certain Chlorophyceae to a maximum of 2.4–3.1 kCal/g for the *Dictyota* spp. above [26]. However, the fatty acid profiles of the marine algal lipid fractions can be good sources of the essential C18 PUFAs: 0.69–10.03 and 0.23–11.97% total fatty acids for C18:2ω6 and C18:3ω3, respectively (Table 11.3). Moreover, the marine algal lipid fractions are rich in the long-chain PUFAs associated most often with fish oils; particularly C20:5ω3 ranging from a low of 1.01% total fatty acids in the Chlorophyceae, *U. lactuca* to as high as 33.1% total fatty acids in the Rhodophyceae, *Gracilaria changgi* with intermediate amounts in most Phaeophyceae, as well as C22:6ω3 which was detected in *G. changgi* (12.9% total fatty acids), and *U. lactuca* (0.80% total fatty acids), but not other edible algae (Table 11.3). Among saturated fatty acids, C16:0 was the most predominant, particularly in the *Porphyra* (63.19% total fatty acids) and *Palmaria* spp. (45.44% total fatty acids) of the Rhodophyceae, whereas Chlorophyceae such as *U. lactuca* (14.00% total fatty acids) and most Pheophyceae (range from 16.51 to 36.0 total fatty acids) contained lower levels of C16:0. When Kaneda and Ando [7] investigated the distribution of fatty acids in tissue lipid fractions, the Rhodophyceae *P. tenera* exhibited a concentration of C16:0 in tissue phospholipids and triacylglycerols, whereas C20:5ω3 was primarily concentrated in the phospholipid fraction. For monounsaturated fatty acids (MUFAs), C18:1ω9 was the predominant fatty acid particularly in *G. changgi* and *U. lactuca* (21.9 and 27.43% of total fatty acids, respectively) followed by most Phaeophyceae (range from 6.79 to 19.96% of total fatty acids); marine algal lipids also contained significant amounts of C16:1ω7 ranging from 1.87 to 6.28% total fatty acids and C18:1ω7 in *Palmaria* and *Porphyra* spp. (2.08 and 1.29% of total fatty acids, respectively), as well as the Phaeophyceae, *Sargassum hemiphyllum* (0.91% total fatty acids). Lamberto and Ackman [45] identified and quantified two lesser-known *trans*-3-monoethylenic fatty acids in two Rhodophyceae species, *P. palmata* and *C. crispus*, harvested in Nova Scotia, Canada: *trans*-3-hexadecenoic acid (C16:1ω3) and *trans*-3-tetradecenoic acid (C14:1ω3). These two fatty acids are known to be involved in algal photosynthesis and are present at relatively low concentrations, particularly

C14:1ω3 at approximately 0.1% total fatty acids in both *P. palmata* and *C. crispus*, with C16:1ω3 at 2.4 and 0.9% total fatty acids, respectively [45]. In general, the predominant fatty acid class for Rhodophyceae was the saturates (>60% total fatty acids for *Palmaria* and *Porphyra* spp.), whereas PUFAs predominated with Phaeophyceae ranging from 36.33 and 69.11% total fatty acids; the proportions of saturates, MUFAs and PUFAs in Chlorophyceae, *U. lactuca* were fairly uniform at 33.12, 34.20, and 22.21, respectively (Table 11.3). Despite these differences in fatty acid classes between marine algal species, the ratios of ω6 to ω3 fatty acids were well below the recommended ratio of 4:1 for reducing the risk of CVD, various cancers, as well as inflammatory and autoimmune diseases in most Western populations whose habitual dietary intakes are closer to ratios of 15:1 to 16.7:1 owing to the preponderance of oilseed ω6 fatty acids in their diets [46]. Thus, marine algae with lipid ω6: ω3 ratios ranging from a low of 0.08 with *Palmaria* spp. to 1.57 in *U. lactuca* (Table 11.3) would appear to be an useful adjunct to Western diets to decrease dietary ω6: ω3 ratios in a bid to reduce chronic disease risks.

While the majority of marine algae available commercially and for retail sale are sun-dried after harvest, some products may be oven-dried at temperatures around 60°C, freeze-dried, or even heat-processed as a canned product, such as the Spanish Phaeophyceae, *Saccorhiza polyschides* and *Himanthalia elongata* [27]. Owing to the susceptibility of PUFAs to oxidation, one can expect differences in the fatty acid profiles of marine algae subjected to different heat-processing conditions. For example, Chan and coworkers [29] reported that the lipid fraction of freeze-dried *S. hemiphyllum* contained greater amounts of C16:1ω9, C18:1ω7, C18:3ω3, C18:4ω3, C20:4ω6, and C20:5ω6 as percentage of total fatty acids compared to sun-dried and oven-dried (60°C) samples of the same edible seaweed; however, there were no differences between the different seaweed drying treatments with respect to total crude lipid contents, which ranged from 3.04 to 4.42% dry weight. Similarly, Sánchez-Machado and coworkers [27] reported that sun-dried *H. elongata* exhibited greater amounts of C16:3ω4, C18:3ω3, C18:4ω3, C20:4ω3, and C20:5ω3 in the lipid fraction compared to a canned sample of *H. elongata*. These data were supported by the findings that the canned *H. elongata* also exhibited an enhanced level of saturated fatty acids and ω6:ω3 ratio than the sun-dried specimen, which contained correspondingly greater levels of PUFAs, particularly the ω3 PUFAs as described in Table 11.3.

Studies with Rhodophyceae, Chlorophyceae, and Phaeophyceae species fed to poultry and mammalian animal models have demonstrated that not only are marine algal lipid fatty acids able to significantly alter the fatty acid profile of egg yolk lipids from hens fed on 10% *Porphyridium* spp. [8], but also the hypocholesterolemic, and specifically the hypolipoproteinemic efficacy of algae-containing diets (i.e., Rhodophyceae, Phaeophyceae, and Chlorophyceae species) in rodent animal models of hypercholesterolemia [24,33,47]. When laying hens were fed on a standard poultry diet containing 3.5% fat supplemented with either 5 or 10% *Porphyridium* spp. for 10 days, blood cholesterol levels were decreased in hens fed on the 10% algae diet only [8]. There were also slight decreases in egg yolk cholesterol levels when hens were fed on 5 and 10% algal diets: 9.5 and

TABLE 11.3
Fatty Acid Profiles of a Variety of Edible Red, Green, and Brown Algae (% of Total Fatty Acids)

	Red Algae			Green Algae	Brown Algae			
Fatty acids	*Palmaria* sp.	*Porphyra* sp.	*G.changgi*	*U. lactuca*	*S. hemiphyllum*	*L. ochroleuca*	*U. pinnatifida*	*H. elongata*[a]
C12:0	ND	ND	NR	0.14	3.28	ND	ND	ND (ND)
C14:0	13.76	0.53	NR	1.14	3.83	4.97	3.17	5.85 (9.57)
C16:0	45.44	63.19	26.0	14.00	36.0	28.51	16.51	32.53 (36.73)
C16:1ω9	ND	ND	NR	0.69	0.80	ND	ND	ND (ND)
C16:1ω7	5.26	6.22	NR	1.87	6.28	5.62	3.70	2.79 (3.00)
C16:2ω4	–	tr	NR	1.03	ND	–	tr	tr (tr)
C16:3ω4	1.20	1.56	NR	ND	ND	0.87	2.31	4.38 (0.06)
C18:0	1.28	1.23	NR	8.39	1.52	0.34	0.69	0.68 (0.59)
C18:1ω9	3.13	6.70	21.9	27.43	11.0	13.62	6.79	19.96 (22.64)
C18:1ω7	2.08	1.29	NR	–	0.91	–	–	– (–)
C18:2ω6	0.69	1.17	NR	10.03	4.51	6.79	6.23	4.39 (5.80)
C18:3ω3	0.59	0.23	NR	4.38	6.84	5.15	11.97	8.79 (6.77)
C18:4ω3	0.74	0.24	NR	0.41	5.94	10.77	22.60	3.53 (1.94)
C20:1ω9	0.20	4.70	NR	4.21	ND	–	–	– (–)
C20:4ω6	1.45	6.80	NR	0.34	10.4	14.20	15.87	10.69 (9.78)

C20:4ω3	0.14	0.07	NR	ND	ND	0.54	0.70	0.88 (0.35)
C20:5ω3	24.05	6.03	33.1	1.01	8.64	8.62	9.43	5.50 (2.77)
C22:6ω3	ND	ND	12.9	0.80	ND	ND	ND	ND (ND)
C24:0	ND	ND	NR	9.45	ND	ND	ND	ND (ND)
Saturated FAs	60.48	64.95	26.0	33.12	44.6	33.82	20.39	39.06 (46.89)
MUFAs	10.67	18.91	21.9	34.20	18.99	19.23	10.50	22.75 (25.64)
PUFAs	28.86	16.10	46.0	22.21	36.33	46.94	69.11	38.16 (27.47)
PUFAs ω6	2.14	7.97		10.37	14.91	20.99	22.10	15.08 (15.58)
PUFAs ω3	25.52	7.20	46.0	6.60	21.42	25.08	44.70	18.70 (11.83)
Ratio ω6:ω3	0.08	1.11		1.57	0.70	0.84	0.49	0.81 (1.32)

[a] Values in brackets are for a canned processed sample. NR = not reported; ND = not detected.

Source: Sánchez-Machado, D.I., López-Cervantes, J., López-Hernández, J. and Paseiro-Losada, P., *Food Chem.*, 85, 439–444, 2004; Norziah, M.H. and Ching, C.Y., *Food Chem.*, 68, 69–76, 2000; Chan, J.C.C., Cheung, P.C.K. and Ang, P.O., *J. Agric. Food Chem.*, 45, 3056–3059, 1997; Ortiz, J., Romero, N., Robert, P., Araya, J., Lopez-Hernández, J., Bozzo, C., Navarrete, E., Osorio, A. and Rios, A., *Food Chem.*, 99, 98–104, 2006.

10 mg cholesterol/g yolk, respectively, compared to 12.5 mg cholesterol/g yolk with the control diet, albeit the decreases were not significant. Interestingly, despite a 10% decrease in feed consumption among hens fed on algal diets, there were no differences in hen body weight, numbers of eggs laid, or egg weights among groups of hens. The eggs of hens fed on algal diets were also observed to have a darker color, attributed to the carotenoid content of the algae (8). In a short-term animal feeding study (25 days), rats fed on diets containing 2% cholesterol, 7% corn oil, and 5% of the following algae: *Ulva* spp. (G), *H. charoides* (R), *Colpomenia sinuosa* (B), and *S. hemiphyllum* (B), did not exhibit any differences among food intakes, body weight gained, or hepatic and spleen weights among treatment groups [24]. Effects of algal diets on plasma hypercholesterolemia were varied: *S. hemiphyllum* and *C. sinuosa* both increased plasma total cholesterol (TC) compared to *Ulva* spp. and *H. charoides*; *C. sinuosa* was the only alga, which was hypercholesterolemic compared to the cholesterol-fed control group; none of the other algae-fed groups were significantly different from each other for plasma TC. Interestingly, not only did the *C. sinuosa* diet increase LDL cholesterol levels compared to the control and *H. charoides* groups, but also HDL cholesterol [24]. These results could be partially attributed to the increased level of hepatic cholesterol exhibited by *C. sinuosa*–fed animals compared to the control animals. From this short-term study, only *H. charoides* exhibited potentially beneficial, albeit insignificant effects, on plasma hyperlipoproteinemia with a small decrease in LDL, concomitant with a slight increase in HDL cholesterol. These results were in contrast to an earlier report by Kaneda and coworkers [33], in which rats fed on 1% cholesterol with 0.25% bile salts, 5% cottonseed oil, and 5% *P. tenera* (R), *E. compressa* (G), or *L. japonica* (B) over 10 weeks, exhibited marked decreases in plasma cholesterol fractions. Total and free cholesterol levels were reduced in animals fed on *E. compressa* and *P. tenera*, but not in *L. japonica*–fed counterparts. Interestingly, these decreases in *P. tenera*–fed animals coincided with increased total and free hepatic cholesterol levels to balance the hypocholesterolemic efficacy of this diet [33]. Hepatic cholesterol fractions were not influenced in *E. compressa*- or *L. japonica*–fed animals, while, hepatic lipid levels appeared to increase these latter two groups compared to the others. More recently, in an attempt to isolate the hypolipidemic effect of edible Rhodophyceae, Chlorophyceae, and Phaeophyceae species, Ara and coworkers [47] treated normal rats fed a standard chow diet, or rats fed a hypercholesterolemic diet (5% coconut oil, 1% cholesterol with 0.5% cholic acid) with ethanol extracts from *Solieria robusta* (R), *Caulerpa racemosa* (G), *Iyengaria stellata* (B), *C. sinuosa* (B), or *Spatoglossum asperum* (B). All extracts decreased plasma total cholesterol (by 11.2–20.0%) and triacylglyercol levels (by 16.66–33.33%) except *S. robusta* in the normal diet group. In these same animals, plasma LDL cholesterol levels were decreased in all algae-treated groups by 32.8–51.38%, whereas plasma HDL cholesterol levels were increased by 9.75–45.14% in all groups, except *I. stellata* [47]. In contrast, all algal extracts reduced plasma total cholesterol (by 11.2–37.38%), triacylglycerol (by 14.1–28.62%), LDL cholesterol (by 15.59–78.72%), and increased plasma HDL cholesterol (by 13.71–66.24%) in rats fed on the hypercholesterolemic diet.

In this latter study, the hypolipidemic efficacy of edible marine algae was restricted to the ethanol-soluble fractions, removing the effects of soluble and insoluble dietary fibers, and thereby suggests that these extracts may modulate LDL cholesterol metabolism in algae-treated animals [47].

11.2.3 Sterols

The few investigations of marine algal sterol composition have focused on the Rhodophyceae, *P. palmata* harvested from the Isle of Wight, UK [48] and Newfoundland, Canada [38]. Cholesterol (cholestan-5-en-3β-ol) and desmosterol (3β-cholesta-5,24-dien-3-ol or 24-dehydrocholesterol) accounted for approximately 85% of total sterol content in *P. palmata*; the remaining sterols identified were 24-methylene cholesterol, β-sitosterol and fucosterol (24,28-methylene cholesterol; [38]; for structures, see below). Indeed, 24-methylene cholesterol is an intermediate in the conversion of desmosterol into fucosterol, which can then be converted into β-sitosterol [49]. The desmosterol content of *P. palmata* was observed to vary throughout the year with maxima of 70 mg/100 g dry weight in November and May and minima of 20 mg/100 g dry weight in March through April, reaching as low as 1 mg/100 g dry weight in June through August [38]. Conversely, *P. palmata* cholesterol content remained fairly constant throughout the year at approximately ≤1 mg/100 g dry wt. It was also suggested that algal desmosterol content may vary not only seasonally, but also with geographic location and thus, likely UV irradiation exposure during growth. Other workers identified 7-oxo-desmosterol in a methanol extract of *P. palmata,* which had been fractionated with *n*-hexane [48].

It is noteworthy that desmosterol is a C24 unsaturated cholesterol precursor late in the cholesterol biosynthetic pathway via desmosterol reductase. In fact, recent *in vitro* studies indicate that desmosterol binds to liver X receptors (LXR), which regulate the expression of multiple genes involved with sterol transport and fatty acid biosynthesis [50]. The mechanisms related to desmosterol-mediated activation of LXR and reduced expression of LDL receptors and HMG-CoA reductase in Chinese Hamster Ovary (CHO)-7 cells in this latter study, likely played a role in the hypocholesterolemic effects observed with marine algae above. Thus, there appears to be some interest in finding rich sources of desmosterol, such as the Rhodophyceae *P. palmata*, to isolate pharmaceutic or nutraceutical ingredients to potentially modulate cholesterol metabolism to reduce CVD risk factors [38].

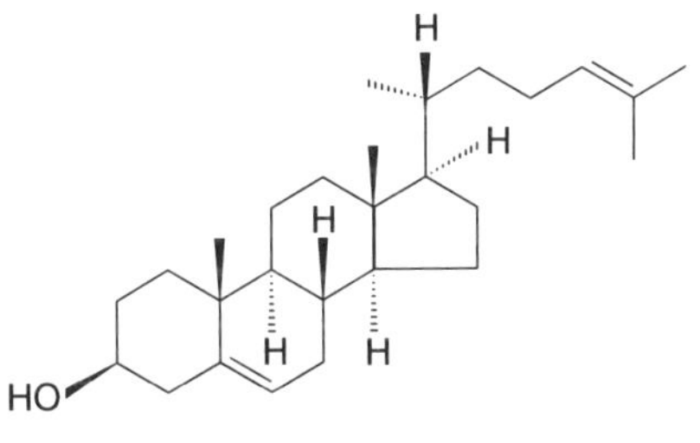

Cholesterol

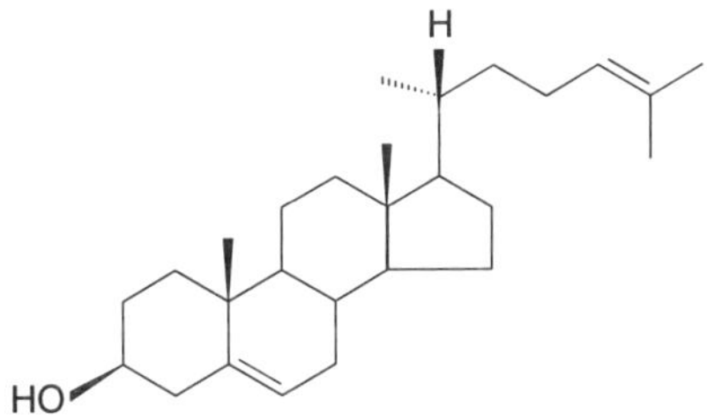

Desmosterol

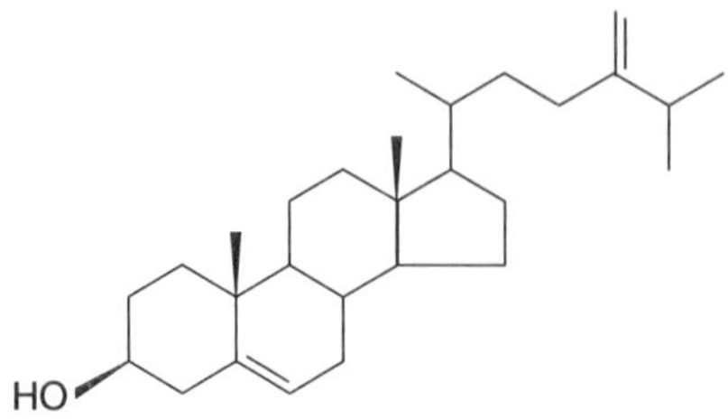

24-Methylene cholesterol

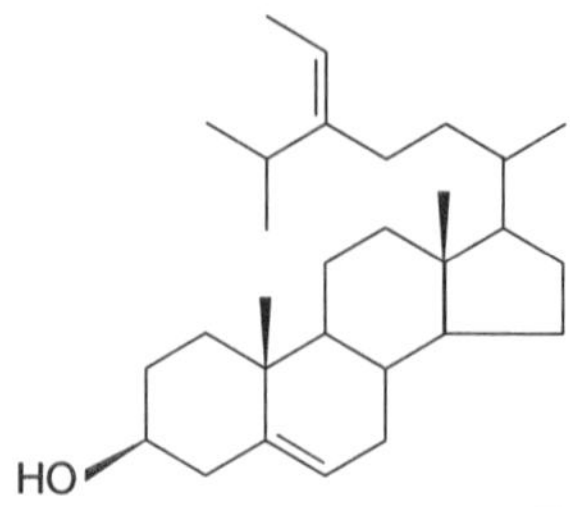

Fucosterol

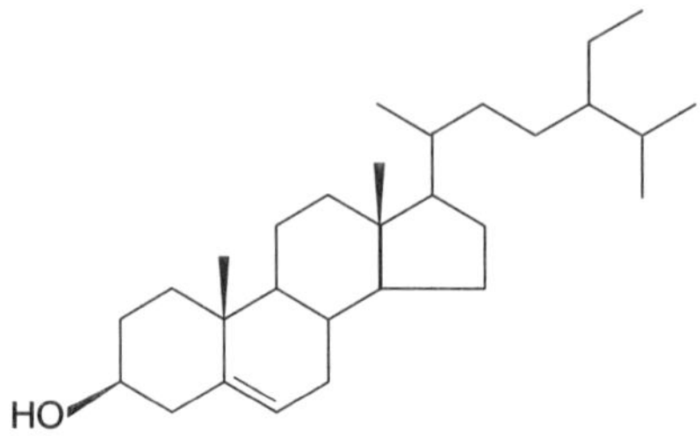

β-Sitosterol

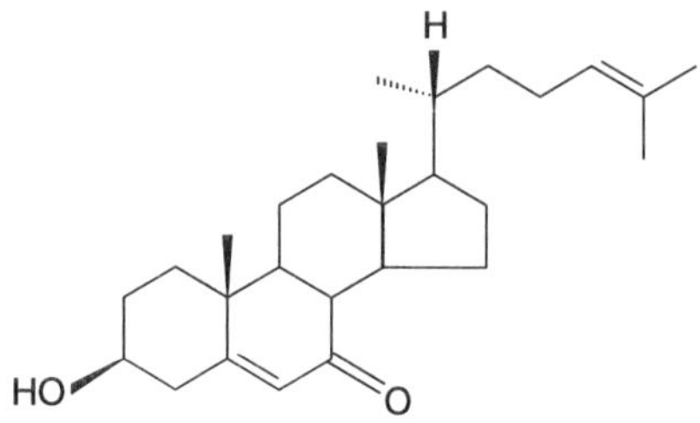

7-Oxo-desmosterol

11.2.4 Polysaccharides

11.2.4.1 Total Dietary Fiber

Current understanding and definitions of dietary fiber have resulted in an expansion of the terminology to include not only nonstarch polysaccharides associated with plant cell wall components (i.e., cellulose, hemicellulose, lignin, pectins, gums, and waxes), which are resistant to human alimentary digestive enzymes, but also oligosaccharides, resistant starches, and resistant proteins along with bound constituents such as polyphenols [9]. Soluble dietary fiber (SDF) is defined as being dispersible in water, thereby giving rise to viscous gels in the gastrointestinal tract and comprises xyloglucans, galactomannan hemicelluloses, β-glucans, pectins, gums, mucilages and thus, the majority of marine algal polysaccharide structures discussed below. The bioactivity of soluble algal dietary fibers is associated with viscosity-mediated effects such as reductions in plasma cholesterol and glucose levels [9,51]. In contrast, insoluble dietary fiber (IDF) is not dispersible in water, but does contribute to fecal bulking and comprises cellulose, lignin, arabinoxylan hemicelluloses, and resistant starch. Thus, one of the main biological effects of IDF is reduced intestinal transit time. It is noteworthy however, that both SDF and IDF can bind water or mineral ions in their matrices; thus, much of the interest in marine algal polysaccharides and total dietary fiber (TDF) is related to water-holding capacity (WHC) as well as metal ion– and bile acid–binding bioactivities of these components [9,10,36,37,52]. Moreover, marine algal SDF that reach the bowel can be susceptible to fermentation by the colonic microflora to yield short-chain fatty acids and thereby, alteration to the bowel pH [36].

Table 11.1 indicates that the proportions of SDF (ranging from 8.3 to 85.0% of TDF) and IDF (ranging from 20.0 to 98.9% of TDF) were highly variable between Rhodophyceae, Chlorophyceae, and Phaeophyceae classes, species, as well as the same algae between and within individual studies, albeit IDF appeared to predominate. The major polysaccharides of marine algae include the alginates, fucans, and laminarans of Phaeophyceae, agaran and agaropectin from the Rhodophyceae, *Gracilaria* and *Gelidium* spp., as well as the κ-, λ-, and ι-carrageenans from Rhodophyceae such as *C. crispus*. Alginates comprise the major polyuronide matrix component of *Phaeophyceae* spp. and consist of alternating sequences of β-(1,4)-D-mannuronic acid and α-(1,4)-L-guluronic acid with 20–30 uronic acid residues.

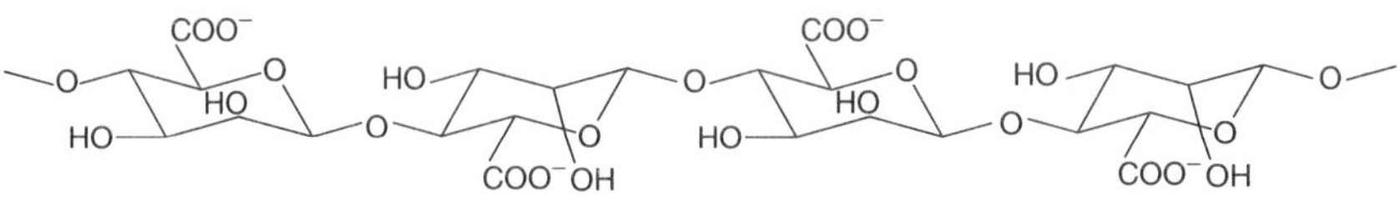

Poly (D-mannuronic acid) segment of alginate

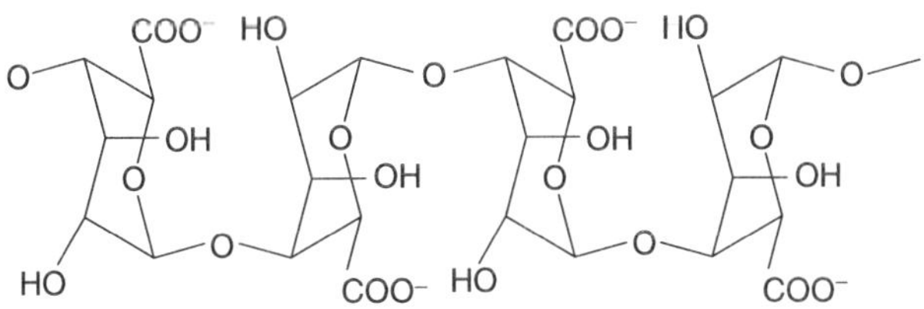

Poly (L-guluronic acid) segment of alginate

Fucans are a heterogeneous group of polysaccharides made up of the fucoidans, xylofucoglycuronans, and glycouronogalactofucans: fucoidans comprise (1,2)-α-L-fucose-4-sulfate with branching or sulfate esters on C3 with small amounts of D-xylose, galactose, mannose, and uronic acids; xylofucoglycouronans (also referred to as ascophyllans) contain a backbone of poly-β-(1,4)-D-mannuronic acid with branches of 3-*O*-D-xylosyl-L-fucose-4-sulfate or uronic acid; whereas glycouronogalactofucans consist of linear chains of (1,4)-D-galactose with branching at C5 with L-fucosyl-3-sulfate or uronic acid [9]. Laminarans represent a polysaccharide reserve in brown kelps and consist of (1,3)-β-D-glucose and (1,6)-β-D-glucose polymers with mannitol end groups.

Agaran is based on β-D-galactosyl with (1,4) links to 3,6-anhydro-α-L-galactopyranosyl units, whereas agaropectin also contains 5–10% sulfate ester groups plus D-glucuronic acid residues and pyruvate ester.

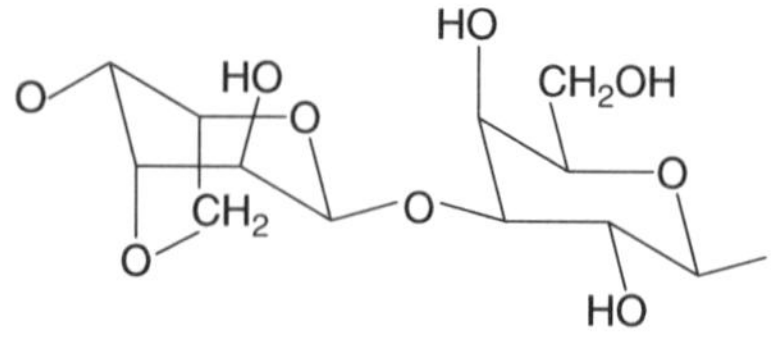

Agaran

The carrageenans are sulfated linear galactans with α-(1,3)-galactose alternating with β-(1,4,3,6)-anhydro-D-galactose with varying amounts of sulfate ester substitution among the three main species of κ-, λ-, and ι-carrageenan:

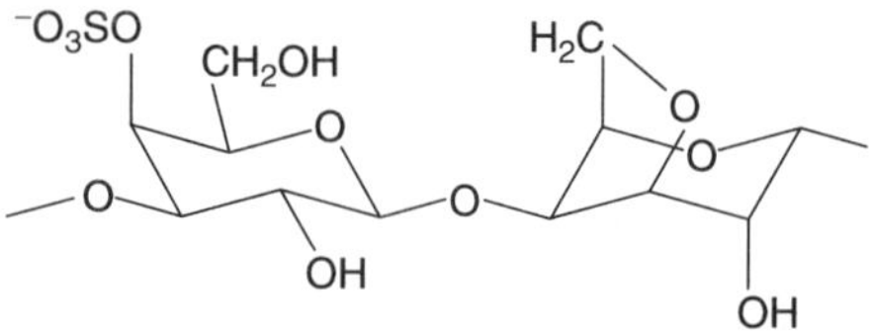

κ-Carrageenan

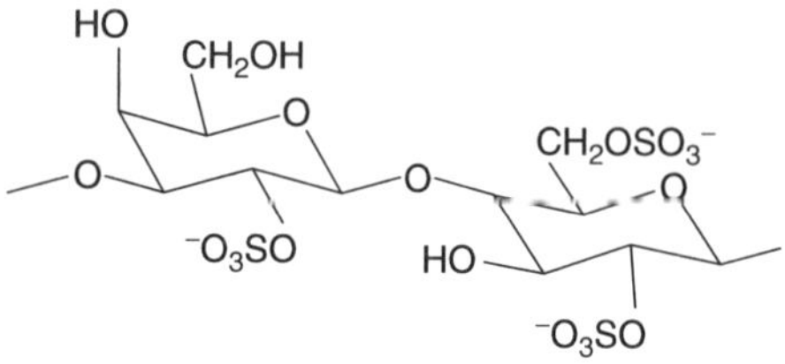

λ-Carrageenan

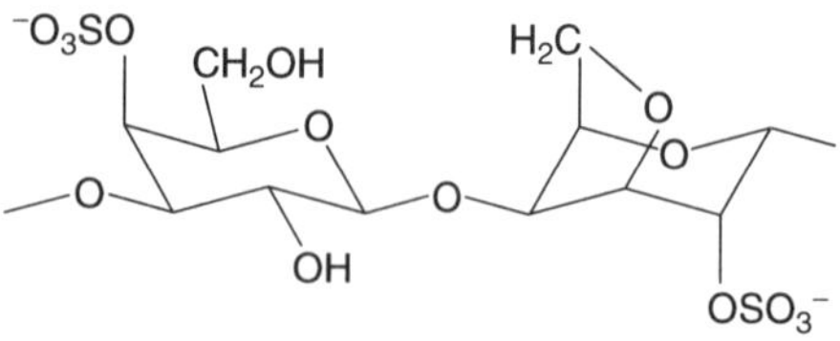

τ-Carrageenan

Xylans are the principle matrix and cell wall polysaccharide of the Rhodophyceae *P. palmata* consisting of linear xylans with mixed β-(1,4) and β-(1,3) linkages with the degree of polymerization varying between 40 and 114 [36].

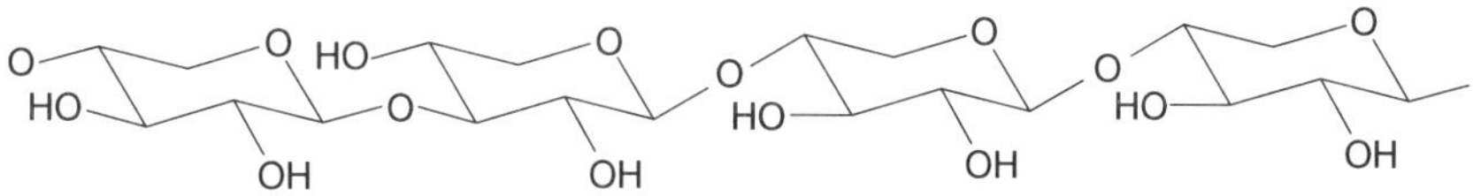

Xylan

However, Phaeophyceae cell wall material is predominantly cellulose. The reserve polysaccharide of Rhodophyceae is known as floridean starch [10,12,36]. Chlorophyceae are noted to contain varying amounts of starch, cellulose, xylans, mannans, polysaccharides with sulfate and uronic acid substituents as well as rhamnose, xylose, galactose, and arabinose groups [9].

The chemical compositions of the marine algal polysaccharide fractions discussed above were confirmed by the analysis of neutral sugars and uronic acid in the SDF and IDF hydrolysates of selected edible commercial algae from Spain [11]. For brown kelps (*F. vesiculosus*, *L. digitata,* and *U. pinnatifida*), the SDF fractions contained between 1.3 and 4.9% uronic acids from alginates compared to only trace amounts (<0.6%) in Rhodophyceae (*C. crispus* and *P. tenera*). The SDF neutral sugar profile of *F. vesiculosus* included (in descending order, % dry weight) fucose, glucose, arabinose, galactose, and xylose; for *L. digitata* they were fucose, galactose, mannose, xylose, rhamnose, arabinose with trace amounts of glucose; for *U. pinnatifida* they were fucose, galactose, xylose, glucose, mannose, arabinose, and rhamnose; for *C. crispus* they were galactose, xylose, 6-*O*-methyl-galactose, mannose, glucose, rhamnose, and arabinose; and for *P. tenera* they were galactose, 3,6-anhydrogalactose, mannose, xylose, glucose, rhamnose, and arabinose [11]. Thus, SDF for brown kelps were confirmed to contain not only alginates, but also laminaran and sulfated fucoidans, whereas those of the red algae were confirmed to contain agar and carrageenans. However, the IDF neutral sugar profile of *F. vesiculosus* included fucose, arabinose, xylose, mannose, and galactose with traces of glucose; *L. digitata* included glucose, fucose, mannose, galactose, xylose, arabinose, and rhamnose; *U. pinnatifida* included glucose, arabinose, galactose, mannose, xylose, and rhamnose; *C. crispus* included galactose, glucose, xylose, mannose, 6-*O*-methyl-galactose, arabinose, and rhamnose; and that of *P. tenera* included mannose, xylose, galactose, 3,6-anhydrogalactose, glucose, and arabinose. Thus, IDF for brown kelps was comprised of not only cellulose and other bound polysaccharides, but also residual insolubles referred to as "Klason Lignin" associated with resistant bound protein; while that of red algae was composed of insoluble mannans and xylans [11]. When Lahaye and coworkers [36] elucidated the neutral sugar profile of *P. palmata*, the SDF fraction was composed of a majority of xylose derived from xylans and trace amounts of mannose, galactose, and glucose, thought to be derived from glycoproteins and floridoside (2-*O*-glycero-α-D-galactopyranoside). In contrast, the IDF of *P. palmata* was composed of a majority of xylose and small amounts of glucose and galactose, reflecting the xylan content as well as a small amount of cellulose, common to red algae [36].

11.2.4.2 Water-Holding Capacity

When Rupérez and Saura-Calixto [11] evaluated the WHC of freeze-dried and milled Spanish brown and red algae, Phaeophyceae *L. japonica* and *U. pinnatifida* exhibited the greatest swelling capacities at 9.82 and 10.53 mL/g dry weight compared to *F. vesiculosus* (5.77 mL/g dry weight) and the red algae *P. tenera* and *C. crispus* with 6.08 and 5.87 mL/g dry weight, respectively. The WHC of the same algae was similar with values of 10.33 and 10.96 g H_2O/g dry weight for *L. digitata* and *U. pinnatifida*, respectively, compared to *F. vesiculosus* with 5.48 g H_2O/g dry weight and the red algae *P. tenera* and *C. crispus* with 5.19 and 7.29 g H_2O/g dry weight, respectively [11]. These results may reflect the alginate and thereby uronic acid content of the brown kelps; for example, as representative examples of SDF, brown kelp alginates have been characterized as forming strong gels in the presence of excess calcium ions, and indeed, can form gels in the stomach to slow gastric emptying [9]. Interestingly, guluronic acid–rich alginates have greater water solubility compared to mannuronic acid–rich alginates. Thus, it is the guluronic acid–rich alginates that may potentially be responsible for any decreased food intake in kelp-fed animals. However, the greater WHC of *C. crispus* compared to its Rhodophyceae counterpart *P. tenera* likely reflects the carrageenan content of the former, versus the xylan content of the latter red alga. Among the carrageenans, κ-carrageenan is noted for its strong gels in the presence of potassium ions; ι-carrageenan for its elastic gels in the presence of calcium ions; whereas, λ-carrageenan is nongelling, but adds viscosity to solutions.

The results detailed above were in contrast to the WHC of Phaeophyceae, Rhodophyceae, and Chlorophyceae obtained commercially in Korea [10] where *U. pinnatifida* exhibited the greatest increase in weight of 1310% followed by *P. tenera* (943%), *L. japonica* (854%), and *E. compressa* (816%), which could potentially be attributed to the greater amounts of TDF and SDF in *U. pinnatifida* (51.0 and 11.5% dry weight, respectively) compared to *P. tenera* (34.9 and 2.9%), *L. digitata* (30.8 and 3.3%) and *E. compressa* (28.5 and 0.2%). However, the viscosity of 20% suspension of these same powdered algae was greatest for *L. japonica* at 39.7 cP, followed by *U. pinnatifida* (23.4 cP), *P. tenera* (5.0 cP), and *E. compressa* (4.0 cP). Unfortunately, the varying results among studies are difficult to compare because of the existing differences in sample preparation and measurement. The WHC of commercial ground and freeze-dried *P. palmata* powder from France was 4.3 and 4.7 g H_2O/g powder at pH 3.0 and 7.3 (20°C), respectively [36]. Interestingly, the viscosity of the *P. palmata* SDF fraction was 49.4 mL/g in 155 mM NaCl at 37°C, which was very low and therefore predicted to be unlikely to influence blood glucose or cholesterol levels *in vivo*.

11.2.4.3 Fermentation

The SDF fraction of *P. palmata* containing soluble xylans was highly fermentable when incubated with a human fecal inoculum over a 6 h period [36]. At the conclusion of the fermentation, the pH was reduced to 6.7 in the presence of 107 mM short-chain fatty acids representing a C2:C3:C4 molar ratio of

58:28:14. Thus, approximately 96% of the xylose had been fermented over the 6 h incubation [36].

11.2.4.4 Ion Exchange Capacity

The cation exchange capacity of selected Spanish edible brown kelp and red algae was greatest in *U. pinnatifida*, followed in descending order by *F. vesiculosus*, *L. digitata*, *C. crispus*, and *P. tenera*, thereby reflecting the greater amount and number of charged groups (i.e., uronic acid and sulfate substituents) in the brown kelps versus the red algae [11].

11.2.4.5 Bile Acid Binding

Binding of bile acids to selected commercial dried and ground Japanese marine algae varied between 1 (cholate and chenodeoxycholate) and 2° (deoxycholate) bile acids as well as SDF and IDF among Phaeophyceae and Rhodophyceae [37]. SDF fractions from *U. pinnatifida, H. fusiformis, L. japonica,* and *P. yezoensis* bound 1–24 μmol sodium cholate/g seaweed; 4–57 μmol sodium chenodeoxycholate/g seaweed; 2–31 μmol sodium deoxycholate/g seaweed; whereas IDF from these same algae bound 7–27 μmol sodium cholate/g seaweed; 42–121 μmol sodium chenodeoxycholate/g seaweed; 29–141 μmol sodium deoxycholate/g seaweed, respectively [37]. Thus, binding of cholate to seaweeds was the lowest, followed by chenodeoxycholate and then deoxycholate. It is noteworthy that while algal IDF fractions bound the most bile acids owing to the greater concentration of this dietary fiber fraction (*U. pinnatifida*, 61.1–71.3%; *H. fusiformis*, 37.7–44.8%; *L. japonica*, 34.0–30.9%; *P. tenera*, 36.3–11.3% dry weight), overall, SDF fractions exhibited a greater binding ability based on fiber dry matter. When purified samples of various algal SDFs were incubated with bile acids, high-viscosity alginate was most effective in binding cholate versus low-viscosity alginate, agar, and carrageenan; agar and carrageenan bound more chenodeoxycholate, and high-viscosity alginate the least; all four SDFs were equally effective in binding large amounts of deoxycholate [37]. If these *in vitro* data can be translated to the *in vivo* situation, marine algal SDF and IDF can then promote fecal bile acid excretion and thereby influence cholesterol metabolism to reduce hypercholesterolemia.

11.2.4.6 Cholesterolemic Effects

Rabbits fed on cholesterol-enriched diets for two months and administered a laminaran sulfate (extracted from *L. cloustoni*) preparation subcutaneously (7.5 mg/kg body weight) daily, exhibited decreased intimal thickening and decreased intimal lipid in the small and large coronary arteries [53]. Aortic lesions of rabbits also exhibited both decreased intimal thickening and lipid deposits. The laminaran sulfate preparation contained β-(1,3) linked glucose units with an average molecular weight of 4000 Da with 45% sulfate esters. Serum cholesterol levels were highly variable between animals, particularly since both male and female animals were studied; however, female rabbits did exhibit decreased serum cholesterol levels with laminaran treatment associated with their more severe hypercholesterolemia compared to male animals [53]. It is noteworthy that rabbits treated with laminaran sulfate also exhibited weight loss during the study, potentially related to WHC and viscosity effects in the gastrointestinal trac and thereby, potentially decreased feed intake. When rats were fed on a hyperlipidemic diet (1% cholesterol with 0.5% cholic acid and 5% coconut oil) and given a polysaccharide extract isolated from *F. vesiculosus* in the drinking water (10 mg/kg body weight) for 12 days, rats treated with the *F. vesiculosus* polysaccharide extract exhibited an increase in HDL cholesterol, albeit there were no differences in plasma triacylglycerols, total or LDL cholesterol levels of these animals [54]. In a short-term study with chow-fed rats, which had been fasted prior to *F. vesiculosus* polysaccharide extract treatment (2.5, 5.0, or 10 mg/kg body weight), animals exhibited dose response reductions in plasma total and LDL cholesterol levels, albeit these differences were only significant in the 10 mg/kg body weight group; plasma HDL cholesterol was not affected by the *F. vesiculosus* polysaccharide extract treatment [54]. Interestingly, plasma triacylglycerol levels were actually increased in rats treated with the lowest dose of *F. vesiculosus* polysaccharide extract. Taken together, these results suggest that the *F. vesiculosus* polysaccharide extract mediated effects via inhibition of hepatic cholesterol synthesis and may be efficacious in the treatment of hypercholesterolemia since the normal group of rats did demonstrate decreases in plasma total and LDL cholesterol [54]. Thus, polysaccharide extracts from Phaeophyceae such as *Laminaria* and *Fucus* spp. may prove efficacious in reducing hypercholesterolemia in clinical trials if results similar to these animal studies can be demonstrated in humans.

11.2.4.7 Glycemic Effects

When Rhodophyceae *P. tenera* at a 3 g level was administered with white bread to healthy subjects and the postprandial glycemic response compared to that of white bread alone, the areas under the glycemic response curves (AUC) were 1795.95 and 2638.01 mg glucose min/dL, respectively [51]. Thus, the glycemic index for the *P. tenera* and white bread treatment was calculated at 68.08%. The *P. tenera* and white bread treatment was efficacious in decreasing the sharp blood glucose peaks from 30 to 60 min postingestion observed when white bread alone was consumed by subjects. Similarly, when the *P. tenera* and white bread was subjected to *in vitro* starch hydrolysis, 63.2% of starch was hydrolyzed after 90 min compared to 76% for white bread alone; thus, the degree of starch hydrolysis in the former sample was clearly decreased compared to the latter sample [51]. Taken together, these data suggest that the SDF of *P. tenera* likely played a role in delaying the gastric emptying time of subjects consuming the *P. tenera* and white bread treatment.

11.2.4.8 Carcinogenesis Effects

Protective effects of dietary Pheophyceae and Rhodophyceae have been reported in several rodent models of carcinogenesis including the reduction of

1,2-dimethylhydrazine-induced intestinal tumors [21], and inhibition of 7,12-dimethylbenz[α]anthracene (DMBA)-induced mammary tumors [18,19] in rats fed diets containing 0.4–5.0% powdered *Laminariales* spp. or the red alga *P. tenera*; as well as reduced growth of implanted sarcoma-180 cells in mice when fed diets containing 2.0% of various kelps or *P. tenera* [22]. More recently, Lee and Sung [20] reported that 15% dietary *L. japonica* reduced colon aberrant crypts and aberrant crypt foci in rats treated with azoxymethane. One hypothesis for the anticarcinogenic effects of these algae was thought to be the moderate sulfated ester content of the hot water–extracted polysaccharide fractions from *P. tenera* and the various kelps [19,20,22]. However, a protective effect of algal sulfated polysaccharides could not be consistently demonstrated between several species of kelps or the red alga *P. tenera*, compared to control diets in animal model studies of mammary and intestinal carcinogenesis [18,21]. Other workers have reported that the anticarcinogenicity of dietary kelps may be associated with the inhibition of hyaluronidase (EC.3.2.1.35) by brown algal phlorotannins such as phloroglucinol, eckol, phlorofucofuroeckol, and dieckol [55]. Hyaluronidase is thought to play a role in carcinogenesis through the depolymerization of hyaluronic acid in the extracellular matrix of connective tissues and organs and thus contribute to the metastasis of cancers [55]. Clearly, the anticarcinogenic effects of dietary algal species are associated with more than one constituent (e.g., dietary fiber, polyphenols) and mechanism; particularly in light of the report from Reddy and coworkers [56] that diets containing 10% of the Pheophyceae *Laminaria angustata* fed to male F344 rats administered the intestinal carcinogen azoxymethane enhanced the incidence (percentage of animals with colon tumors) and multiplicity of colon adenomas (number of tumors/animal), as well as the size of colon tumors (those larger than 0.25 cm in diameter) in kelp-fed rats compared to the control diet. Interestingly, these workers observed decreases in fecal cholesterol in the kelp-fed group (1.22 mg/g dry feces) compared to control animals (2.31 mg/g dry feces), albeit excretion of other neutral sterols (coprostanol, coprostanone, and cholesterol) was not different between these two groups; similarly, excretion of bile acids (cholic acid, chenodeoxycholic acid, deoxycholic acid, etc.) was not affected by the 10% kelp diet compared to the control. While these results are contrary to those of others discussed above, the colon cancer-promoting effects of the 10% *L. angustata* diet may have been because of irritation of the colon mucosa or possibly attributable to genetic and metabolic differences in the rat strain used by these workers.

11.2.5 Vitamins

11.2.5.1 Vitamin C

Due to the fact that the majority of edible marine algae are sun-dried, freeze-dried, canned, or even subjected to toasting or roasting before consumption as in the case of the Japanese or Korean "Nori" and "Kim" (*P. tenera* or *P. yezoensis*), the vitamin content of these plant tissues is both highly variable and even minimal for those compounds, which are highly susceptible to oxidation, such as the reducing agent

L-ascorbic acid [26] (Table 11.4). Indeed several vitamins, including ascorbic acid, the carotenoids and the tocopherols, have a role in protecting marine algal tissues from oxidative stress and potential for tissue damage associated with not only tidal fluctuations, but also exposure to UV-A and UV-B wavelengths depending on the season, water depth, and turbidity [39]. A fresh specimen of the Rhodophyceae *Stictosiphonia arbuscula*, harvested in Brighton Beach, Otago, New Zealand, contained approximately 2.5–2.8 μmol ascorbate/g wet weight [57]. Fresh *P. palmata* has been reported to contain between 220 and 520 μg ascorbic acid/g wet weight when harvested from Spitsbergen, Norway, and unknown locations in France and elsewhere [58,59], similar to *G. changgi* (R) from Ban Merbok, Kedah, on the west coast of Malaysia (Table 11.4 [28]). Fresh specimens of Phaeophyceae such as the *Laminaria* spp. harvested from Spitsbergen, Norway, exhibited lower (170 μg/g wet wt) or only trace amounts of ascorbic acid. It is noteworthy that air- and sun-drying, as well as storage of *P. palmata* is known to reduce the vitamin C content measured in samples [59]. For example, when McDermid and Stuercke [26] surveyed 22 species of Rhodophyceae, Chlorophyceae, and Phaeophyceae algae harvested from various locations in the Hawaiian Islands, the vitamin C content of the oven-dried samples (60°C) was minimal with detectable levels reported in only four samples: between 1300 and 3000 μg/g dry weight in the Chlorophyceae, *Enteromorpha flexuosa*, *Monostroma oxyspermum*, *U. fasciata* (Table 11.4), and 2000 μg/g dry weight in the Rhodophyceae *Eucheuma denticulatum*; neither of the Phaeophyceae tested (*Sargassum echinocarpum* and *S. obtusifolium*) exhibited detectable amounts of vitamin C. The vitamin C content of the Phaeophyceae *Ascophyllum nodosum* harvested from an unknown location has been reported to range between 550 and 1650 μg/g dry weight [4]. Therefore, marine algal levels of L-ascorbic acid are likely substantial only prior to harvest and in fresh specimens, and are dramatically decreased in the dehydrated and thermally processed products available through retail to the consumer.

11.2.5.2 Carotenoids

The lipophilic carotenoid and tocopherol vitamins are known to be associated with the oxidative status of marine algal tissues to combat cell membrane damage against dessication from the daily tidal fluctuations as well as the photooxidative stress associated with UV exposure during growth. Within marine algal tissues, chlorophylls are the major photosynthetic pigments; however, carotenoids are also active in a secondary capacity; moreover, algal tissue levels of the carotenoids have been reported to vary seasonally depending upon UV irradiation exposure levels [59,60]. For example, levels of α- and β-carotenes in *P. palmata* appear to peak during the spring and summer months (April to September) ranging between 110 and 420 μg/g dry weight in samples harvested from North Berwick, Scotland; Helgoland, Germany; and unknown locations in Norway, in comparison to wintertime (January) at 37 μg/g dry weight [59], as supporting evidence of a protective role for carotenoids against oxidative stress. Tissue levels of β-carotene in algae harvested from different locations in the Hawaiian Islands

TABLE 11.4
Vitamin Content of a Variety of Edible Red, Green, and Brown Algae

	Red Algae				Green Algae		Brown Algae	
Vitamin	*Chondrus ocellatus*	*Gracilaria parvispora*	*G. changgi*	*Porphyra vietnamensis*	*Ulva lactuca*	*U. fasciata*	*Durvillaea antarctica*	*Sargassum echinocarpum*
Vitamin C mg/100 g wet wt.			28.5					
Vitamin C mg/g dry wt.	ND	ND		ND		2.2		ND
β-carotene mg/100 g dry wt			5.2					
β-carotene IU/g dry wt.	30	ND		430		180		97
Niacin mg/g dry wt.	0.06	ND		ND		ND		0.09
Riboflavin mg/g dry wt.	ND	0.006		ND		0.010		ND
Thiamine mg/g dry wt.	0.09	ND		ND		ND		ND
Tocols mg/kg lipid					1071.4		1112.9	
α-tocopherol					9.3		179.4	
β-tocopherol					14.3		16.5	
γ-tocopherol					25.8		19.4	
δ-tocopherol					25.3		245.9	
α-tocotrienol					33.2		ND	
γ-tocotrienol					963.5		651.7	

Note: ND = not detected.

Source: McDermid, K.J. and Stuercke, B., *J. App. Phycol.*, 15, 513–524, 2003; Norziah, M.H. and Ching, C.Y., *Food Chem.*, 68, 69–76, 2000; Ortiz, J., Romero, N., Robert, P., Araya, J., Lopez-Hernández, J., Bozzo, C., Navarrete, E., Osorio, A. and Rios, A., *Food Chem.*, 99, 98–104, 2006.

varied widely depending on class, in the range of 27–180, 60–97, and 15–430 IU/g dry weight in Chlorophyceae, Pheophyceae, and Rhodophyceae, respectively [26]. The Rhodophyceae *G. changgi* harvested in Malaysia was reported to contain 52 μg/g dry weight (Table 11.4 [28]).

Analytical parameters that will have an impact on the carotenoid levels reported in the literature include whether the data reflect total carotenoids, the separate identification of α- and β-carotenes, or whether other carotenoids have also been determined in the marine algae, such as lutein, zeaxanthin, and fucoxanthin [48,59,61]. For example, it is noted that α- and β-carotene may account for only 1/3–1/8 of total carotenoids in the Rhodophyceae *P. palmata* [59], in addition to lutein at 240 μg/g dry weight in a sample harvested from an unknown location in Norway [59]; albeit lutein is devoid of provitamin A activity. Similarly, neither zeaxanthin nor fucoxanthin are vitamin A precursors in the diet.

11.2.5.3 Tocopherols

Similar to the carotenoids above, marine algal tissue levels of the tocopherols vary seasonally with low concentrations during winter and spring between 22 and 35 μg/g dry weight in *P. palmata* and trace levels in *L. digitata*, compared to summer and fall levels of approximately 139 μg/g dry weight in *P. palmata* and 2% dry weight in *L. digitata* [59,62]. The brown kelp *Ascophyllum nodosum* exhibited between 260 and 450 μg tocopherols/g dry weight [4]. The tocopherol homolog composition of various Phaeophyceae, *L. digitata* (0.005 mg/g extract for both α- and γ-homologs), *H. elongata* (0.052 and 0.038 mg/g for α- and γ-isomers), *F. vesiculosus* (4.3, 1.7, 2.2 mg/g for α-, γ-, and δ-homologs), *F. serratus* (2.23, 0.8, 1.1 mg/g for α-, γ-, and δ-homologs), and *A. nodosum* (3.42, 1.2, 1.7 mg/g for α-, γ-, and δ homologs) indicated that the biologically active form of vitamin E, α-tocopherol predominated in these algae [62]. In contrast, the γ- and δ-homologs predominated in the Chlorophyceae *U. lactuca*, while the δ- and α-tocopherols were predominant in the Phaeophyceae *D. antarctica* (Table 11.4 [30]). These same workers also reported substantial amounts of tocotrienols in these algae.

11.2.5.4 B Vitamins

The B vitamin composition of Rhodophyceae, Chlorophyceae, and Phaeophyceae, summarized in Table 11.4, suggests that oven-dried (60°C) marine algae are not good sources of niacin, riboflavin, or thiamine [26]. Watanabe and coworkers [63] demonstrated that commercial samples of *Enteromorpha* spp. (G) and *Porphyra* spp. (R) contained 63.58 and 32.26 μg of biologically active vitamin B_{12}/100 g dry weight, and that the hydroxo- and cyano-forms (i.e., noncoenzyme forms) of the vitamin predominated in these algae tissues. More recently, Rodríguez-Bernaldo de Quirós and coworkers [64] reported that the total folate content (expressed as folic acid) of selected Phaeophyceae and Rhodophyceae was as follows: 161.59, 159.11, 149.61, 99.42, 71.71, 66.5, and 61.40 μg/100 g dry weight for *L. ochroleuca* (B), *Palmaria* spp., *U. pinnatifida* (B), *H. elongata* (B), *U. pinnatifida* (canned),

Saccorhiza polychides (R), and *Porphyra* spp., respectively. The most abundant form of folate was 5-methyltetrahydrofolate for *U. pinnatifida*, *L. ochroleuca*, *Palmaria* spp., and *S. polychides*, whereas 5-formyltetrahydrofolate predominated in *H. elongata* samples and folic acid in *Porphyra* spp.

11.2.6 Minerals

From Table 11.1, the ash and thereby, mineral contents of the various edible marine algae can be considerable, ranging from 8.4 to 43.6% dry weight in Rhodophyceae; 17.1–29.2% dry weight in Chlorophyceae, and 19.6–39.82% dry weight in Pheophyceae. It is noteworthy that similar to the protein content discussed above, marine algal ash content has also been observed to vary seasonally [32,65](Table 11.1). Hagen-Rødde and coworkers [32] reported that *P. palmata* harvested during the winter months (i.e., January through April) exhibited ash contents between approximately 22.5 and 28% dry weight, in contrast to ash contents of approximately 15 to 18% dry weight during the summer (i.e., June through October). Individual mineral concentrations also displayed slight variations in concentration seasonally in *P. palmata,* although a consistent pattern among the different minerals with Ca varying between 2.40 and 7.94 mg/g dry weight; Mg varying between 2.31 and 3.18 mg/g dry weight; Na varying between 11.17 and 22.47 mg/g dry weight; K varying between 47.68 and 86.61 mg/g dry weight; Sr varying between 0.031 and 0.062 mg/g dry weight; and Br varying between 0.16 and 0.34 mg/g dry weight, could not be discerned [32]. A certain proportion of the seasonal variation in algal mineral contents can be expected to be a function of the growth patterns of these marine species, with increased concentrations of minerals during the winter months characterized by slow to minimal growth of plant tissues, and decreased concentrations of minerals during the summer months upon resumption of rapid growth of plant tissues [65]. Moreover, the seasonal variation in marine algal mineral constituents can also reflect the variation in available PAR, specifically UV irradiation, and thereby, photosynthetic pigments in algal tissues, between seasons; for example, Rhodophyceae, Chlorophyceae, and Pheaophyceae exhibited greater levels of chlorophyll a, which contains Mg within the porphyrin structure, in early June (ice cover on seawater) compared to markedly lower levels in mid- to late-July (after sea-ice breakup) in algae harvested from Spitsbergen, Norway [60]. Indeed, *P. palmata* exhibited greater levels of Mg in samples harvested in the spring month of April, compared to the fall harvest in October [32](Table 11.5). Also, Rhodophyceae such as *Porphya* spp., *Gelidium* spp., and *Pterocladia* spp. are noted to contain a variety of halogenated compounds including brominated mono- and dihydroxy C_6-C_1, C_6-C_2 and C_6-C_3 phenols, primarily 2,4,6-tribromophenol, which are thought to have a role in modulating oxidative stress attributed to the phenolic structure of these molecules [66–68]. For example, Phaeophyceae such as *Sargassum* spp. contain mono-, di-, and tribromophenols with isoprenoid substituents, primarily 2,4,6-tribromophenol and 2,4-dibromophenol [67]; whereas the Rhodophyceae *Polysiphonia urceolata* contained four bromophenols: 5-bromo-3, 4-dihydroxybenzaldehyde,

TABLE 11.5
Mineral Composition of a Variety of Edible Red, Green, and Brown Algae (mg/g Dry Weight)

Algae	Na	K	Ca	Mg	Fe[a]	Zn[a]	Cu[a]	Cd[a]	Mn	Sr	Br	Pb[a]	I
Red													
Chondrus ocellatus		22.6	4.4	9.2	142	284	39		0.070				
Eucheuma denticulatum		124.0	4.5	7.6	112	7	2		0.009				
Gracilaria coronopifolia		221.6	1.8	3.4	136	42	2		0.057				
G. parvispora		160.0	3.8	4.9	198	8	3		0.048				
Halymenia formosa		46.0	5.3	12.5	66	22	4		0.011				
Palmaria palmata (S)	20.15	86.61	4.82	3.18					0.062		0.16		
P. palmata (F)	15.93	47.68	7.94	2.81					0.050		0.19		
P. palmata	17–25	70–90	5.6–12	1.7–5.0	150–1400								0.1–1.0
Porphyra sp.													0.043
P. tenera					110	31	6.3		32.3				0.017
P. vietnamensis		39.7	2.9	7.8	154	11	7		0.041				
Green													
Caulerpa lentillifera		7.0	9.5	16.5	167	17	6		0.010				
Enteromorpha sp.													0.064
E. flexuosa		16.0	7.4	11.7	104	6	3		0.005				
Monostroma oxyspermum		31.4	5.8	13.6	142	32	28		0.010				
Ulva spp.	9.0–59	7.3–10.3	8.6–56	20–52	60–1000								0.02–0.25
U. fasciata		28.7	4.7	21.9	86	9	5		0.012				
Brown													
Alaria marginata					60	23.6	0.5		3.79				0.151
Dictyota acutiloba		72.6	10.3	13.6	438	16	5		0.012				
D. sandvicensis		55.7	18.1	9.1	608	13	5		0.021				
Eisenia bicyclis					80	14	3.3		13.8				0.600

Fucus vesiculosus						14.0	2.16	3.30		0.249	
F. vesiculosus					520	24	1.4		27.9		0.732
Hijikia fusiformis					30	4	0.9		5.77		0.436
Laminaria japonica					80	13	0.5		6.79		2.11
L. saccharina					40	8.5	0.5		3.04		0.238
L. setchellii					10	22.3	0.5		3.67		1.070
Laminaria spp.	9–60	13–106	5.0–30	5.0–20	40–800						2.0–10.0
Macrocystis integrifolia					110	6.9	0.5		9.96		0.240
Nereocystis leutkeana					20	10	0.9		5.55		0.734
Sargassum echinocarpum		95.0	13.1	11.6	92	7	11		0.006		
S. hemiphyllum	9.51	44.7	22.4	9.89	20.8	1.6	tr		0.002		0.0002
S. obtusifolium		79.0	15.0	9.3	129	16	9		0.015		
Undaria pinnatifida					40	13	1.1		6.46		0.102
U. pinnatifida	16–70	55–63	11–30	10–30	80–400						0.25

[a] µg/g dry wt.

Note: S = spring harvest, April; F = fall harvest, October.

Source: van Netten, C., Hoption Cann, S.A., Morley, D.R., and van Netten, J.P., *Sci. Total Environ.*, 255, 169–175, 2000; Mabeau, S. and Fleurence, J., *Trends Food Sci. Technol.*, 4,103–107, 1993; McDermid, K.J. and Stuercke, B., *J. App. Phycol.*, 15, 513–524, 2003; Chan, J.C.C., Cheung, P.C.K., and Ang, P.O., *J. Agric. Food Chem.*, 45, 3056–3059, 1997; Hagen-Rødde, R.S., Vårum, K.M., Laren B.A., and Myklestad, S.M., *Bot. Marina*, 47, 125–133, 2004; Watanabe, F., Takenaka, S., Katsura, H., Zakir Hussain Massumder, S.A.M., Abe, K., Tamura, Y., and Nakano, Y., *J. Agric. Food Chem.*, 47, 2341–2343, 1999; Riget, F., Johansen, P., and Asmund, G., *Mar. Poll. Bull* 30(6), 409–413, 1995.

5-bromo-3,4-dihydroxybenzyl alcohol, 3,5-dibromo-4-hydroxybenzyl alcohol, and 3-bromo-4-hydroxybenzyl alcohol [66]. In a follow-up study, Fujimoto and coworkers [68] determined that 5-bromo-3,4-dihydroxybenzaldehyde comprised 75% of the total bromophenols in *P. urceolata* with substantial amounts of 5-bromo-3,4-dihydroxybenzyl alcohol present as well. Indeed, *P. palmata* harvested in the spring (April) exhibited lower amounts of Br (0.16 mg/g dry weight) compared to the fall (October) with 0.19 mg Br/g dry weight (Table 11.5 [32]).

Interestingly, while the amount of salt (Na^+, K^+, Cl^-) in the seawater was not thought to contribute to the increase in *P. palmata* ash during the winter [32], there does appear to be a temporal nature in the concentrations of certain heavy metals in seawater [65]. For example, approximately 80% of the dissolved Cd in seawater is derived from elimination processes of phytoplankton, in that Cd is taken up by these organisms during the summer months and released during the winter [65]. Thus, while marine algae are considered good sources of various essential minerals, particularly iodine, there can also be concern about the bioaccumulation of heavy metals such as Cd, Pb, Hg, and As in edible seaweeds (Table 11.5 [5,65]). In its 1990 regulations, the French government outlined the acceptable limits for several minerals of concern: inorganic As ≤3.0 mg/kg dry matter; Pb ≤5.0 mg/kg; Cd ≤0.5 mg/kg; Sn ≤5.0 mg/kg; Hg ≤0.1 mg/kg; and I ≤5.0 mg/kg dry matter [12]. It is noteworthy that the Phaeophyceae *F. vesiculosus,* an edible seaweed, is often used as a bioindicator to monitor trace metal pollution owing to its capacity for elemental accumulation [65]. Samples of *F. vesiculosus* harvested from a West Greenland fjord thought to be free from pollutant inputs exhibited Cd, Cu, Pb, and Zn concentrations close to the background values of samples from other regions of West Greenland without known metal pollution [65](Table 11.5). Seasonal variation in metal concentrations for Cd was between 2.25 and 4.81 μg/g dry weight; Cu was between 1.30 and 3.30 μg/g; Pb was between 0.047 and 0.700 μg/g; while that for Zn was between 5.71 and 25.4 μg/g dry weight attributed in part to growth patterns of the algae as well as temporal changes in metal availability such as Cd above. When van Netten and coworkers [5] investigated the elemental composition of a variety of imported and locally harvested marine algae (west coast of Vancouver Island, [Bamfield], British Columbia, Canada) six out of eight imported algae from Japan and Norway (*U. pinnatifida, H. fusiformis, P. tenera* × 2, *L. japonica, F. vesiculosus*) exhibited greater Hg concentrations (0.24–1.08 μg/g dry weight) than local algae (*N. leutkeana, M. integrifolia, A. marginata, L. saccharina, L. setchellii*) at 0.05 μg/g dry weight, the detection limit of analysis. Lead does not appear to be a widespread contaminant in most edible algae with levels ranging between the assay detection limit of 0.01 and 0.57 μg/g dry weight [5,65](Table 11.5); it is thought that where some Pb pollution does occur, it can often be traced to the use of lead weights in the fishing industry. Similarly, Zn in seawater can be attributable to the use of zinc in products used on marine vessels to protect against the corrosion of brass fittings [5]. These same workers demonstrated the presence of arsenic in both imported algae from Japan and Norway (ranging between 20 and 88 μg/g dry weight) as well as local algae (ranging between 33.1 and

79 μg/g dry weight); however, they did not distinguish between the organic and inorganic forms of this metal. This information is vital since the organic forms of As are considered relatively nontoxic [25], whereas inorganic forms of As such as As_2O_3 can cause acute toxicity manifested as gastrointestinal tract inflammation leading to diarrhea and vomiting and eventually multiorgan failure as well as fever, emaciation, irritability, and hair loss; or chronic toxicity manifested in skin lesions, nerve damage, skin cancer, and diseases of the blood vessels. Thus, following the release of a Consumer Advisory in 2001 by the Canadian Food Inspection Agency [25] that consumers should avoid the consumption of "Hijiki" seaweed (*H. fusiformis*, B) owing to the elevated levels of inorganic As found in this type of seaweed, the Food Standards Agency (FSA) of the United Kingdom conducted a survey of imported edible algae to determine the concentrations of total and inorganic arsenic in selected Phaeophyceae and Rhodophyceae [69]. It is noteworthy that while the As content of samples of *P. tenera* was between 18.2 and 31.9 mg/kg dry weight; *L. digitata* contained between 18.9 and 75.2 mg/kg; *U. pinnatifida* contained between 29.2 and 41.9 mg/kg; *E. bicyclis* contained between 27.9 and 32.3 mg/kg; and *H. fusiformis* contained between 94.6 and 134 mg/kg dry weight; only *H. fusiformis* was found to contain measurable amounts of inorganic As (above the analytic detection limit of 0.3 mg/kg dry weight) accounting for 70% of the total As content [69]. However, in all cases except for *P. tenera*, the edible algae above are subjected to soaking in water during preparation, which does help to remove a portion of the total and inorganic As to somewhat lessen the exposure. Nevertheless, the current advisories are to avoid consumption of "Hijiki" to lessen the risk of exposure to this known carcinogen.

Iodine content of the edible marine algae summarized in Table 11.5 is highly variable, ranging between 0.017 and 1.0 mg/g dry weight for Rhodophyceae, 0.02 and 0.25 mg/g dry weight for Chlorophyceae, and 0.0002 and 10.0 mg/g dry weight for Phaeophyceae. When imported algae from Japan and Norway were compared to local varieties harvested from the west coast of Vancouver Island (Bamfield), British Columbia, Canada, imported *L. japonica* contained a higher iodine content (2.11 mg/g dry weight) than local *Laminaria* spp. such as *L. saccharina* (0.238 mg/g dry weight) and *L. setchellii* [5](1.07 mg/g dry weight) (Table 11.5). Again, similar to the discussion above with As, true consumption of iodine by consumers will depend upon the method of preparation, particularly for the majority of Phaeophyceae, which are soaked in water prior to consumption.

11.2.7 Polyphenols

As secondary metabolites, polyphenols include not only the flavonoids such as flavones, flavonols, flavanones, flavononols, chalcones, and flavan-3-ols, but also the lignans, lignins, tocopherols (discussed above), tannins, and the related marine polyglucinol polymers (i.e., phlorotannins), plus phenolic acids. Flavonoids are products of the shikimic acid pathway via the aromatic amino acid precursors, Phe and Tyr, giving rise to C_6-C_3 cinnamic acid derivatives. Cinnamic acid can then be converted into flavonoids, isoflavones, or various phenolic acids via *p*-coumaric, caffeic, and ferulic acids. Thus, flavonoids are characterized by a C_6-C_3-C_6 skeleton, that is, two aromatic rings bridged by an aliphatic chain, which is often condensed into a pyran or furan ring; these compounds most often occur as glycosides in plant tissues, which are then deglycosylated during digestion. The diversity of these compounds is also reflected in the potentially bioactive antioxidant mechanisms involved, including scavenging of reactive oxygen species, quenching free radicals, as well as chelating transition metal ions.

When the flavan-3-ol profiles of Japanese Phaeophyceae, Chlorophyceae, and Rhodophyceae were investigated, catechin was detected in only two Chlorophyceae species: *Acetabularia ryukyuensis* (3.33 mg/g dry weight) and *Tydemaniz expeditionis* (0.25 mg/g dry weight [70]). Catechin was detected in most Phaeophyceae species including *E. bicyclis* (1.24 mg/g dry weight), although only trace amounts were observed in *L. religiosa* and none in *U. pinnatifida* and *H. fusiformis*; as well as most Rhodophyceae species including *P. yezoensis* (0.036 mg/g dry weight). The catechin epimer, epicatechin (EC), was detected in only five Japanese algae: *A. ryukyuensis* (0.50 mg/g dry weight), *E. bicyclis* (3.86 mg/g dry weight), *Sargassum muticum* (2.62 mg/g dry weight), *Gelidium elegans* (0.13 mg/g dry weight), and *Chondrococcus hornemannii* (0.36 mg/g dry weight). In contrast, the corresponding gallate esters were not common: catechin gallate was detected in only *G. elegans* (0.07 mg/g dry weight); epicatechin gallate (ECG) in *E. bicyclis* (0.29 mg/g dry weight), and *S. muticum* (0.12 mg/g dry weight). Epigallocatechin (EGC) was detected in only one Chlorophyceae species: *T. expeditionis* (0.35 mg/g dry weight); most Phaeophyceae including *E. bicyclis* (4.77 mg/g dry weight), *H. fusiformis* (3.77 mg/g dry weight), albeit not in *U. pinnatifida* or *L. religiosa*; a few Rhodophyceae such as *C. hornemannii* (16.0 mg/g dry weight), but not in *P. yezoensis* [70]. Epigallocatechin gallate (EGCG) was not detected in any of the Chlorophyceae investigated, but was detected in three Phaeophyceae: *E. bicyclis* (0.28 mg/g dry weight), *Padina arborescens* (0.68 mg/g dry weight), and *P. minor* (0.49 mg/g dry weight); and in three Rhodophyceae: *P. yezoensis* (0.032 mg/g dry weight), *Gracilaria texorii* (0.024 mg/g dry weight), and *Gracukarua asiatica* (0.018 mg/g dry weight). Thus, the Japanese Chlorophyceae are relatively poor sources of the flavan-3-ols compared to the Phaeophyceae and Rhodophyceae species; moreover, the flavan-3-ol gallate esters are not common among the Japanese edible algae.

Further work investigating the polyphenol profile of edible algae focused on the hydroxylated cinnamic acid derivatives, flavonols, flavanones, and their glycosides [71,72]. Two Chlorophyceae not currently used for consumption, *Halimeda macroloba* and *H. opuntia* were reported to be rich in EGC: 28.0 and 12.7 mg/g dry weight, respectively; catechol: 1.88 and 0.38 mg/g dry weight; the flavonols myricetin: 0.41 and 0.15 mg/g dry weight, and morin: 0.43 and 0.23 mg/g dry weight, respectively [71]. *H. macroloba* also contained caffeic acid (0.085 mg/g dry weight) and the flavanone glycoside hesperidin (hesperitin-7-rhamnoglucoside; 0.14 mg/g dry weight). While most Japanese edible Phaeophyceae, Chlorophyceae, and Rhodophyceae were found to be rich in the flavonol morin: between 0.26 and 2.47 mg/g dry weight, only two algae contained myricetin, the

Phaeophyceae *Tubinaria ornata* (0.35 mg/g dry weight) and the Rhodophyceae *Chondrus verruscosus* (0.27 mg/g dry weight; [72]). Japanese Pheophyceae, Chlorophyceae, and Rhodophyceae were rich in catechol, ranging between 0.24 mg/g dry weight in *L. religiosa* and a high of 77.7 mg/g dry weight in *Caulerpa serrulata*. However, caffeic acid was detected in only one Chlorophyceae: *A. ryukyuensis* (0.32 m/g dry weight); two Phaeophyceae: *U. pinnatifida* (0.054 mg/g dry weight) and *Ishige okamurae* (0.15 mg/g dry weight); and most Rhodophyceae including *P. yezoensis* (0.047 mg/g dry weightt), *G. elegans* (0.13 mg/g dry weight), and *G. texorii* (0.17 mg/g dry weight). The flavonol glycoside rutin was particularly rich in most Rhodophyceae, including *P. yezoensis* (11.4 mg/g dry weight), *G. elegans* (23.2 mg/g dry weight), and *G. texorii* (30.0 mg/g dry weight); but was detected in only three Chlorophyceae: *A. ryukyuensis* (26.9 mg/g dry weight), *Monostroma nitidum* (2.70 mg/g dry weight), and *C. serrulata* (3.37 mg/g dry weight), and three Phaeophyceae: *U. pinnatifida* (0.46 mg/g dry weight), *E. cava* (2.73 mg/g dry weight), and *P. arborescens* (1.00 mg/g dry weight [72]). In contrast, the flavonol glycoside quercitrin was absent from Chlorophyceae and Rhodophyceae, and detected in only two Phaeophyceae: *U. pinnatifida* (0.20 mg/g dry weight) and *P. arborescens* (0.47 mg/g dry weight). The flavanone glycoside hesperidin was particularly rich in most Japanese Chlorophyceae, Pheophyceae, and Rhodophyceae including *A. ryukyuensis* (117 mg/g dry weight), *E. bicyclis* (6.93 mg/g dry weight), and *P. yezoensis* (51.3 mg/g dry weight), respectively [72].

A few commercially available Japanese edible algae have been reported to yield mammalian lignans following a human fecal fermentation *in vitro*: *U. pinnatifida* yielded 1.84 μg enterolactone/g dry weight and 10.83 μg enterodiol/g dry weight; *H. fusiformis* yielded 2.97 μg enterolactone/g dry weight and 4.32 μg enterodiol/g dry weight [73]. Thus, these Pheaophyceae likely contain the plant lignan secoisolariciresinol diglucoside (SDG) and possibly matairesinol (MAT). The plant lignan SDG is initially hydrolyzed to yield the aglycone plant lignan SECO (R-(R*', R*)-2,3-*bis*(4-OH-3-methoxy phenyl)-CH_3-1, 4-butanediol), which is then dehydroxylated and demethylated to yield the mammalian lignan enterodiol (2,3-*bis*(3-OH phenyl) methylbutane-1,4-diol), which is subsequently oxidized to form enterolactone (EL; *trans*-dihydro-3,4-*bis*(3-OH phenyl) CH_3-γ-butyrolactone) by colonic facultative aerobes (*Clostridia* spp.). However, the plant lignan MAT can be dehydroxylated and demethylated to yield EL directly.

The brominated mono- and dihydroxy C_6-C_1, C_6-C_2 and C_6-C_3 phenols, known as bromophenols, which are unique to Rhodophyceae such as *Porphya* spp., *Gelidium* spp., and *Pterocladia* spp. are discussed above.

11.3 CONCLUSIONS AND FUTURE WORK

It is clear that edible Rhodophyceae, Chlorophyceae, and Phaeophyceae species can contribute to a healthful diet, particularly those of red algae with a high protein content (*Porphyra* and *Palmaria* spp.); as well as the red and brown algae with high concentrations of C20:5ω3 and C22:6ω3 in their lipids. Moreover, most edible algae can potentially contribute to a healthful diet through contributions to TDF in general, or SDF for the beneficial effects to intestinal transit time and cholesterol metabolism, as well as the rich mineral composition of these marine algae. In contrast, the efficacy of marine algae as functional foods or nutraceuticals needs to be further investigated with *in vivo* studies and eventually clinical trials to elucidate the mechanisms of observed beneficial effects. Producers and importers of marine algal products need to evaluate their safety and composition with respect to heavy metal contamination (i.e., Hg, Cd, and inorganic As) to ensure confidence in their products. Moreover, because marine algae are photosynthetically active organisms and algal composition can vary seasonally and with harvest location, this information needs to be recorded and reported in the literature to allow for easier comparison among studies.

ACKNOWLEDGMENTS

The work of the author was funded by the Natural Sciences and Engineering Research Council of Canada (NSERC), Faculty of Community Services of Ryerson University, Ontario Workstudy Program and was conducted with the help of postdoctoral fellows, technicians, and many student research assistants.

REFERENCES

1. Trono, G.C. Jr. 1999. Diversity of the seaweed flora of the Philippines and its utilization. *Hydrobiologia* 398/399, 1–6.
2. Yuan, Y.V. 2007. Antioxidants from edible seaweeds. In: *Antioxidant Measurement and Applications*, Shahidi, F. and Ho, C.-T. (eds). Washington D.C.: ACS Oxford Press, pp. 268–301.
3. Cambie, R.C. and Ferguson, L.R. 2003. Potential functional foods in the traditional Maori diet. *Mutation Res.* 523–524, 109–117.
4. Indergaard, M. and Minsaas, J. 1991. Animal and human nutrition. In: *Seaweed Resources in Europe: Uses and Potential.* Toronto: John Wiley & Sons, pp. 21–64.
5. van Netten, C., Hoption Cann, S.A., Morley, D.R. and van Netten, J.P. 2000. Elemental and radioactive analysis of commercially available seaweed. *Sci. Total Environ.* 255, 169–175.
6. Galland-Irmouli, A.V., Fleurence, J., Lamghari, R., Luçon, M., Rouxel, C., Barbaroux, O., Bronowicki, J.P., Villaume, C. and Guéant, J.L. 1999. Nutritional value of proteins from edible seaweed *Palmaria palmata* (Dulse). *J. Nutr. Biochem.* 10, 353–359.
7. Kaneda, T. and Andom, H. 1971. Component lipids of purple laver and their antioxygenic activity. *Proc. Int. Seaweeds Symp.* 7, 553–557.
8. Ginzberg, A., Cohen, M., Sod-Moriah, U.A., Shany, S., Rosenshtrauch, A. and Arad, S. 2000. Chickens fed with biomass of the red microalga *Porphyridium* spp. have reduced blood cholesterol level and modified fatty acid composition in egg yolk. *J. Appl. Phycol.* 12, 325–330.
9. Jiménez-Escrig, A. and Sánchez-Muniz, F.J. 2000. Dietary fiber from edible seaweeds: chemical structure, physicochemical properties and effects on cholesterol metabolism. *J. Nutr. Res.* 20(4), 585–598.
10. Han, K-H., Lee, E-J. and Sung, M-K. 1999. Physical characteristics and antioxidative capacity of major seaweeds. *J. Food Sci. Nutr.* 4(3), 180–183.

11. Rupérez, P. and Saura-Calixto, F. 2001. Dietary fiber and physiochemical properties of edible Spanish seaweeds. *J. Eur. Food Res. Technol.* 212, 349–354.
12. Mabeau, S. and Fleurence, J. 1993. Seaweed in food products: biochemical and nutritional aspects. *Trends Food Sci. Technol.* 4, 103–107.
13. Nagata, C., Shimizu, H., Takami, R., Hayashi, M., Takeda, N. and Yasuda, K. 2003. Association of blood pressure with intake of soy products and other food groups in Japanese men and women. *Prev. Med.* 36(6), 692–697.
14. Sho, H. 2001. History and characteristics of Okinawan longevity food. *Asia Pac. J. Clin. Nutr.* 10(2), 159–164.
15. Teas, J. 1981. The consumption of seaweed as a protective factor in the etiology of breast cancer. *Med. Hypoth.* 7, 601–613.
16. Yusuf, S., Reddy, S., Ôunpuu, S. and Anand, S. 2001. Global burden of cardiovascular diseases. *Circulation* 104, 2855–2864.
17. Pisani, P., Bray, F. and Parkin, D.M. 2002. Estimates of the world-wide prevalence of cancer for 25 sites in the adult population. *Int. J. Cancer* 97, 72–81.
18. Yamamoto, I., Maruyama, H. and Moriguchi, M. 1987. The effect of dietary seaweeds on 7,12-dimethylbenz[α]anthracene-induced mammary tumorigenesis in rats. *Cancer Lett.* 35, 109–118.
19. Teas, J., Harbison, M.L. and Gelman, R.S. 1984. Dietary seaweed (*Laminaria*) and mammary carcinogenesis in rats. *Cancer Res.* 44, 2758–2761.
20. Lee, E-J. and Sung, M-K. 2003. Chemoprevention of azoxymethane-induced rat colon carcinogenesis by seatangle, a fiber-rich seaweed. *Plant Foods Hum. Nutr.* 58, 1–8.
21. Yamamoto, I. and Maruyama, H. 1985. Effect of dietary seaweed preparations on 1,2-dimethylhydrazine-induced intestinal carcinogenesis in rats. *Cancer Lett.* 26, 241–251.
22. Yamamoto, I., Maruyama, H., Takahasi, M. and Komiyama, K. 1986. The effect of dietary or intraperitoneally injected seaweed preparations on the growth of sarcoma-180 cells subcutaneously implanted into mice. *Cancer Lett.* 30, 125–131.
23. Higashi-Okai, K., Otani, S. and Okai, Y. 1999. Potent suppressive effect of a Japanese edible seaweed, *Enteromorpha prolifera* (Sujiao-nori) on initiation and promotion phases of chemically induced mouse skin tumorigenesis. *Cancer Lett.* 140, 21–25.
24. Wong, K.H., Sam, S.W., Cheung, P.C.K. and Ang, P.O. Jr. 1999. Changes in lipid profiles of rats fed with seaweed-based diets. *J. Nutr. Res.* 19(10), 1519–1527.
25. Canadian Food Inspection Agency 2001. *Consumer Advisory: Inorganic Arsenic and Hijiki Consumption*. Ottawa: CFIA, pp. 1–2.
26. McDermid, K.J. and Stuercke, B. 2003. Nutritional composition of edible Hawaiian seaweeds. *J. App. Phycol.* 15, 513–524.
27. Sánchez-Machado, D.I., López-Cervantes, J., López-Hernández, J. and Paseiro-Losada, P. 2004. Fatty acids, total lipid, protein and ash contents of processed edible seaweeds. *Food Chem.* 85, 439–444.
28. Norziah, M.H. and Ching, C.Y. 2000. Nutritional composition of edible seaweed *Gracilaria changgi*. *Food Chem.* 68, 69–76.
29. Chan, J.C.C., Cheung, P.C.K. and Ang, P.O. 1997. Comparative studies on the effect of three drying methods on the nutritional composition of seaweed *Sargassum hemiphyllum*. *J. Agric. Food Chem.* 45, 3056–3059.
30. Ortiz, J., Romero, N., Robert, P., Araya, J., Lopez-Hernández, J., Bozzo, C., Navarrete, E., Osorio, A. and Rios, A. 2006. Dietary fiber, amino acid, fatty acid and tocopherol contents of the edible seaweeds *Ulva lactuca* and *Durvillaea antarctica*. *Food Chem.* 99, 98–104.

31. Fleurence, J. 1999. Seaweed proteins: biochemical, nutritional aspects and potential uses. *Trends Food Sci. Technol.* 10, 25–28.
32. Hagen-Rødde, R.S., Vårum, K.M., Laren B.A. and Myklestad, S.M. 2004. Seasonal and geographical variation in the chemical composition of the red alga *Palmaria palmata* (L.) Kuntze. *Bot. Marina* 47, 125–133.
33. Kaneda, T., Tokuda, S. and Arai, K. 1963. Studies on the effects of marine products on cholesterol metabolism-I. The effects of edible seaweeds. *Bull Jpn. Soc. Sci. Fish.* 29(11), 1020–1023.
34. Wong, K.H. and Cheung, P.C.K. 2000. Nutritional evaluation of some subtropical red and green seaweeds Part 1—proximate composition, amino acid profiles and some physico-chemical properties. *Food Chem.* 71, 475–482.
35. Lahaye, M. 1991. Marine algae as sources of dietary fibers: determination of soluble and insoluble dietary fiber content in some 'sea vegetables'. *J. Sci. Food Agric.* 54, 587–594.
36. Lahaye, M., Michel, C. and Barry, J.L. 1993. Chemical, physiochemical and in-vitro fermentation characteristics of dietary fibers from *Palmaria palmata* (L.) Kuntze. *Food Chem.* 47, 29–36.
37. Wang, W., Onnagawa, M., Yoshie, Y. and Suzuki, T. 2001. Binding of bile salts to soluble and insoluble dietary fibers of seaweeds. *Fish. Sci.* 67, 1169–1173.
38. Idler, D.R. and Atkinson, B. 1976. Seasonal variation in the desmosterol content of dulse (*Rhodymenia palmata*) from Newfoundland waters. *Comp. Biochem. Physiol.* 53B, 517–519.
39. Yuan, Y.V., Carrington, M.F. and Walsh, N.A. 2005. Extracts from dulse (*Palmaria palmata*) are effective antioxidants and inhibitors of cell proliferation *in vitro*. *Food Chem. Toxicol.* 43, 1073–1081.
40. Yuan, Y.V. and Walsh, N.A. 2006. Antioxidant and antiproliferative activities of extracts from a variety of edible seaweeds. *Food Chem. Toxicol.* 44, 1144–1150.
41. Wong, K.H. and Cheung, P.C.K. 2001. Nutritional evaluation of some subtropical red and green seaweeds Part 2. *In vitro* protein digestibility and amino acid profiles of protein concentrates. *Food Chem.* 72, 11–17.
42. Ventura, M.R. and Castañón, J.I.R. 1998. The nutritive value of seaweed (*Ulva lactuca*) for goats. *Small Ruminant Res.* 29, 325–327.
43. Bocanegra, A., Nieto, A., Blas, B. and Sánchez-Muniz, F.J. 2003. Diets containing a high percentage of Nori or Konbu algae are well-accepted and efficiently utilized by growing rats but induce different degrees of histological changes in the liver and bowel. *Food Chem. Toxicol.* 41, 1473–1480.
44. Urbano, M.G. and Goñi, I. 2002. Bioavailability of nutrients in rats fed on edible seaweeds, Nori (*Porphyra ternera)* and Wakame (*Undaria pinnatifida*), as a source of dietary fiber. *Food Chem.* 76, 281–286.
45. Lamberto, M. and Ackman, R.G. 1994. Confirmation by gas chromatography/mass spectrometry of two unusual *trans*-3-monethylenic fatty acids from the Nova Scotian seaweeds *Palmaria palmata* and *Chondrus crispus*. *Lipids* 29(6), 441–444.
46. Simopoulos, A.P. 2002. The importance of the ratio of omega-6/omega-3 essential fatty acids. *Biomed. Pharmacother.* 56(8), 365–379.
47. Ara, J., Sultana, V., Qasim, R. and Ahmad, V.U. 2002. Hypolipidaemic activity of seaweed from Karachi coast. *Phytother. Res.* 16, 479–483.
48. Ma, Y.C., Blunden, G., Barwell, C.J. and Yang, M-H. 1995. 7-Oxo-desmosterol from *Palmaria palmata*. *Botanica Marina* 38, 133–134.
49. Patterson, G.W. and Karlander, E.P. 1967. Fucosterol reduction to clionasterol *in vivo* by *Chlorella ellipsoidea*. *Plant Physiol.* 42, 1651–1652.

50. Yang, C., McDonald, J.G., Patel, A., Zhang, Y., Umetani, M., Xu, F., Mangelsdorf, D.J., Westover, E.J., Covey, D.F., Cohen, J.C. and Hobbs, H.H. 2006. Sterol intermediates from cholesterol biosynthetic pathway as LXR ligands. *J. Biol. Chem.* online, July 20.
51. Goñi, I., Valdivieso, L. and Garcia-Alonso, A. 2000. Nori seaweed consumption modifies glycemic response in healthy volunteers. *Nutr. Res.* 20(10), 1367–1375.
52. Mishra, V.K., Ooraikul, B. and Temelli, F. 1996. Physical characterization and water sorption of freeze dried dulse *Palmaria palmata* powder. *J. Food Proc. Preserv.* 20, 25–39.
53. Besterman, E.M.M. 1970. Effects of laminarin sulphate on experimental atherosclerosis and on serum lipids in rabbits during long-term intermittent cholesterol feeding. *Atherosclerosis* 12, 85–96.
54. Vázquez-Freire, M.J., Lamela, M. and Calleja, J.M. 1996. Hypolipidaemic activity of a polysaccharide extract from *Fucus vesiculosus* L. *Phytother. Res.* 10, 647–650.
55. Shibata, T., Fujimoto, K., Nagayama, K., Yamaguchi, K. and Nakamura, T. 2002. Inhibitory activity of brown algal phlorotannins against hyaluronidase. *Int. J. Food Sci. Technol.* 37, 703–709.
56. Reddy, B.S., Numoto, S. and Choi, C-I. 1985. Effect of dietary *Laminaria angustata* (brown seaweed) on azoxymethane-induced intestinal carcinogenesis in male F344 rats. *Nutr. Cancer* 7, 59–64.
57. Burritt, D.J., Larkindale, J. and Hurd, C.L. 2002. Antioxidant metabolism in the intertidal red seaweed *Stictosiphonia arbuscula* following dessication. *Planta* 215, 829–838.
58. Aguilera, J., Dummermuth, A., Karsten, U., Schriek, R. and Wiencke, C. 2002. Enzymatic defences against photooxidative stress induced by ultraviolet radiation in Arctic marine macroalgae. *Polar Biol.* 25, 432–441.
59. Morgan, K.C., Wright, J.L.C. and Simpson, F.J. 1980. Review of chemical constituents of the red alga *Palmaria palmata* (dulse). *Econ Bot.* 34(1), 27–50.
60. Aguilera, J., Bischof, K., Karsten, U. and Hanelt, D. 2002. Seasonal variation in ecophysiological patterns in macroalgae from an Arctic fjord. II. Pigment accumulation and biochemical defence systems against high light stress. *Marine Biol.* 140, 1087–1095.
61. Yan, X., Chuda, Y., Suzuki, M. and Nagata, T. 1999. Fucoxanthin as the major antioxidant in *Hijikia fusiformis*, a common edible seaweed. *Biosci. Biotechnol. Biochem.* 63(3), 605–607.
62. Le Tutour, B., Benslimane, F., Gouleau, M.P., Gouygou, J.P., Saadan, B. and Quemeneur, F. 1998. Antioxidant and pro-oxidant activities of the brown algae, *Laminaria digitata, Himanthalia elongata, Fucus vesiculosus, Fucus serratus* and *Ascophyllum nodosum*. *J. Appl. Phycol.* 10, 121–129.
63. Watanabe, F., Takenaka, S., Katsura, H., Zakir Hussain Massumder, S.A.M., Abe, K., Tamura, Y. and Nakano, Y. 1999. Dried green and purple lavers (nori) contain substantial amounts of biologically active vitamin B12 but less of dietary iodine relative to other edible seaweeds. *J. Agric. Food Chem.* 47, 2341–2343.
64. Rodríguez-Bernaldo de Quirós, A., Castro de Ron, C., López-Hernández, J. and Lage-Yusty, M.A. 2004. Determination of folates in seaweeds by high-performance liquid chromatography. *J. Chromatogr.* 1032, 135–139.
65. Riget, F., Johansen, P. and Asmund, G. 1995. Natural seasonal variation of cadmium, copper, lead and zinc in brown seaweed (*Fucus vesiculosus*). *Mar. Poll. Bull* 30(6), 409–413.
66. Fujimoto, K. and Kaneda, T. 1984. Separation of antioxygenic (antioxidant) compounds from marine algae. *Hydrobiologia* 116/117, 111–113.

67. Clifford, M.N. 2000. Miscellaneous phenols in foods and beverages—nature, occurrence and dietary burden. *J. Sci. Food Agric.* 80, 1126–1137.
68. Fujimoto, K., Ohmura, H. and Kaneda, T. 1985. Screening for antioxygenic compounds in marine algae and bromophenols as effective principles in a red alga *Polysiphonia ulceolate. Bull. Jpn. Soc. Sci. Fish.* 51(7), 1139–1143.
69. Food Standards Agency UK 2004. *Food Survey Information Sheet: Arsenic in seaweed.* London: FSA, pp. 1–4.
70. Yoshie, Y., Wang, W., Petillo, D. and Suzuki, T. 2000. Distribution of catechins in Japanese seaweeds. *Fish. Sci.* 66, 998–1000.
71. Yoshie, Y., Wang, W., Hsich, Y-P. and Suzuki, T. 2002. Compositional difference of phenolic compounds between two seaweeds, *Halimeda* spp. *J. Tokyo Univ. Fish.* 88, 21–24.
72. Yoshie-Stark, Y., Hsieh, Y-P. and Suzuki, T. 2003. Distribution of flavonoids and related compounds from seaweeds in Japan. *J. Tokyo Univ. Fish.* 89, 1–6.
73. Thompson, L.U., Robb, P., Serraino, M. and Cheung, F. 1991. Mammalian lignan production from various foods. *Nutr. Cancer* 16, 43–52.

12 Beneficial Health Effects of Seaweed Carotenoid, Fucoxanthin

Kazuo Miyashita and Masashi Hosokawa

CONTENTS

12.1 INTRODUCTION

Seaweeds are macroscopic marine algae that form a part of the staple diet in Japan and Korea. They are often referred to as seaweeds or marine macroalgae with some authors referring to them as sea vegetables. They have been harvested for centuries by many countries, especially in the Asian continent, for several uses. Seaweed is traditionally consumed in the Far East countries and in Hawaiian Islands, while in the West, the principal uses of seaweeds are as sources of phycocolloids, thickening, and gelling agents for various industrial applications including uses in foods.

Chemical composition of seaweed varies with species, habitats, maturity, and environmental conditions. In comparison with land vegetables, edible seaweeds are potentially good sources of nonstarch polysaccharides, minerals, trace elements, and certain vitamins [1]. The undigested polysaccharides of seaweed can form important source of dietary fiber. They also contain some valuable nutrients and chemicals that can be used as functional foods and nutraceuticals by humans.

Seaweeds have been studied for a long time for production of industrially important polysaccharides like agar, carrageenan, and fucoidan, among others. However, they have not been looked upon as sources of lipids as these are found in relatively small quantities [2]. The occurrence of polyunsaturated fatty acids (PUFAs), mainly n-3 fatty acids, that have been found to have significant health effects, is a unique feature of marine lipids. Unlike marine animals such as fish, marine plants have not been looked upon as important substrates for further processing of their lipids because of the relatively small amounts of lipids present in them. In contrast, the important cardioprotective effect along with strong effects against skin diseases of PUFAs found in marine lipids make these unique lipids more interesting. In spite of their low lipid content, fatty acids of marine seaweeds have attracted considerable interest among researchers for their nutritional value to other marine organisms, occurrence of bioactive compounds, and potential medical applications [3–8]. Several researchers have reported the occurrence of bioactive conjugated fatty acids from red seaweeds that have potential biomedical applications [3,6,9–13].

Interest in seaweed lipids has been on the rise owing to the recognition of important bioactive molecules such as conjugated fatty acids and pigments (carotenoids) that have profound physiological effects in the treatment of tumors and other cancer-related problems [14–21]. In addition, seaweeds have been considered as delicacies in some Asian countries such as Japan and Korea. However, in other countries, they are not important as culinary items, although many of them are being used in the production of phycocolloids like agar, carrageenan, and alginate, among others. Since, lipids would be wasted as a major by-product during the production of these phycocolloids, it would be worthwhile to utilize them as sources of nutritious lipids and other compounds that can have profound health benefits.

The seaweed lipids mainly consist of glycolipids, phospholipids, and neutral lipids apart from containing carotenoids as the major pigment. Fucoxanthin is the characteristic pigment of brown seaweeds (phaeophyceae), which are the largest occurring group among seaweeds. Carotenoids of terrestrial origin have been thoroughly reviewed with respect to their occurrence, biological functions, and possible health benefits. However, relatively much less attention has been paid to physiological effects or beneficial applications of aquatic plant carotenoids.

12.2 FUCOXANTHIN

Fucoxanthin is a xanthophyll that contains allenic bond and two epoxy groups (Figure 12.1). It is one of the most abundant carotenoids accounting for >10% of estimated total natural production of carotenoids. Fucoxanthinol is a major metabolite from fucoxanthin. Fucoxanthin and fucoxanthinol have been reported to occur in brown seaweeds [22–24]. Apart from fucoxanthinol, various metabolites arising from fucoxanthin (like apo-9′-fucoxanthinone, apo-13′-fucoxanthinone)

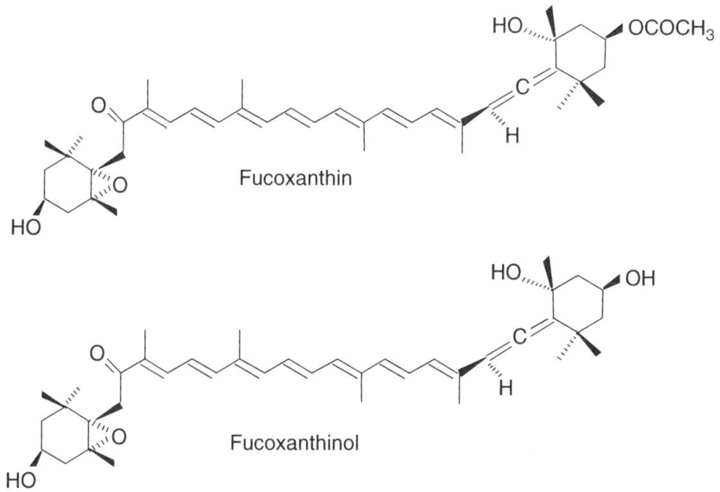

FIGURE 12.1 Structure of fucoxanthin and fucoxanthinol.

have also been reported in some cultured edible brown seaweeds [24]. Several geometrical isomers (*cis* isomers) of fucoxanthin also exist in nature.

Fucoxanthin content in seaweeds exhibits seasonal variation [23] and also varies depending on the life cycle of the seaweeds [24] indicating the possible biological significance of this pigment in seaweeds. In addition, it has also been conclusively shown that freshwater fish do not accumulate fucoxanthin, although their natural feed (i.e., aquatic insects) contains fucoxanthin [25]. Studies involving quantification of fucoxanthin in different brown seaweeds, both wild and cultured, are limited. However, an effort is made to summarize the available data with regard to quantification in the published works. The content of fucoxanthin/fucoxanthinol reported in several brown seaweeds, based on the published literature, is summarized in Table 12.1. Fucoxanthin, when present in the thallus of seaweeds, was found to be quite stable in the presence of organic ingredients apart from surviving the drying process and storage at ambient temperature, although fucoxanthin in pure form is susceptible to oxidation [24]. Effective stabilization method will therefore be necessary to apply fucoxanthin in nutraceuticals and formulation of functional foods.

Seaweeds, as mentioned earlier, are being used for producing commercial polysaccharides in many countries. These polysaccharides mainly find use as an ingredient in food. A large amount of waste is discarded during the production of polysaccharides or other water-soluble compounds from seaweed. This waste has been found to be rich in lipids. Wakame (*Undaria pinnatifida*) is one of the most popular edible seaweed in Japan and has been used in many seafood products. Wakame lipid–rich fraction is obtained after the extraction of polysaccharide and proteins in a laboratory scale. Wakame lipids were found to contain 5–10% fucoxanthin apart from containing polar lipids such as glycolipids. Purified

TABLE 12.1
Total Carotenoid (mg/g) and Fucoxanthin/Fucoxanthinol (mg/g) Content in Selected Brown Seaweeds

No.	Species	Total Carotenoid	Fucoxanthin	Fucoxanthinol
1	*Undaria pinnatifida* (young thallus)	–	0.32	–
2	*U. pinnatifida* (commercial-dried)	–	0.33	–
3	*U. pinnatifida* (female gametophyte)	–	1.64	–
4	*U. pinnatifida* (male gametophyte)	–	2.67	–
5	*Scytosiphon lamentaria* (young thallus)	–	0.24	–
6	*S. lamentaria* (germlings)	–	0.56	–
7	*Petalonia binghamiae* (young thallus)	–	0.43	–
8	*P. binghamiae* (germlings)	–	0.58	–
9	*Laminaria religiosa* (young thallus)	–	0.24	–
10	*Ecklonia radiata*	6.85	1.65	0.24
11	*Carphophyllum maschalocarpum*	6.21	1.17	–
12	*C. plumosum*	5.68	1.44	0.41
13	*Harmosira banksii* (coastal rocks)	6.48	–	–
14	*H. banksii* (mangroves)	3.75	–	–
15	*Cystophora retroflexa*	4.71	0.46	0.62
16	*Sargassum sinclairii*	9.79	0.54	–
17	*Fucus serratus*	0.80	0.56	–

Note: Content on wet weight basis for species from nos. 1 to 9 and on dry weight basis for species listed in nos. 10–16.

Source: Nos. 1–9 from Mori, K., Ooi, T., Hiraoka, M., Oka, N., Hamada, H., Tamura, M. and Kusumi, T., *Mar. Drugs*, 2, 63–72, 2004; nos. 10–16 from Czeczuga, B. and Taylor, F.J., *Biochem. Syst. Ecol.*, 15, 5–8, 1987; no. 17 from Haugan, J. and Liaeen-Jensen, S., *Biochem. Syst. Ecol.*, 22, 31–41, 1994.

fucoxanthin (purity>95%) is obtained by column chromatographic separation from fucoxanthin-containing Wakame lipids. Most laboratories employ different methods of extraction and isolation protocols in case of seaweed carotenoids. Readers can also refer to the method employed by Mori et al. [24] for isolation and purification of fucoxanthin and its metabolites. Typical separation procedure for fucoxanthin from brown seaweed is shown in Figure 12.2.

12.3 ANTICANCER EFFECT

12.3.1 Inhibitory Effect of Fucoxanthin on Cancer Cells

In a study screening the antiproliferative activity of seafood extracts on tumor cells, fucoxanthin from the brown algae, *U. pinnatifida,* was found to be the active principle [15]. In another study dealing with the antiproliferative activity of fucoxanthin HL-60 cells [26], fucoxanthin activity was higher than that of

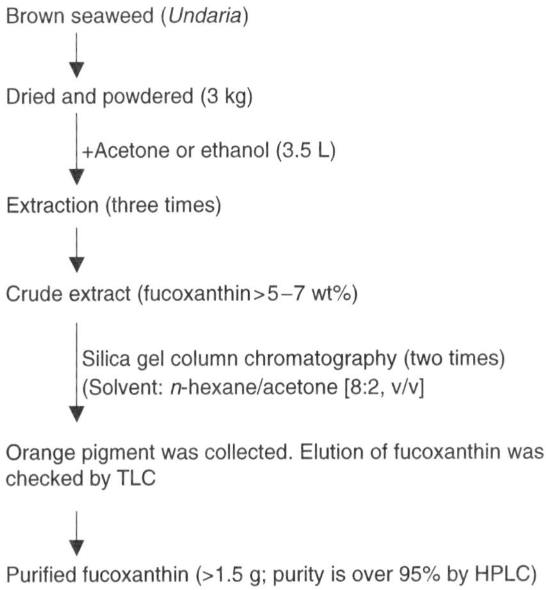

FIGURE 12.2 Typical procedure for fucoxanthin separation from the brown seaweed, *Undaria pinnatifida*.

β-carotene (Figure 12.3). Fucoxanthin completely inhibited the proliferation of the cancer cells at concentrations as low as 22.6 μM. Further, when HL-60 cells were treated with fucoxanthin for 24 h, viable cell numbers decreased in a dose-dependent manner. Cell proliferation is the key to promoting and further progression of carcinogenesis [27]. In an investigation on the apoptosis-inducing activity of fucoxanthin, a DNA ladder, which is a characteristic feature of apoptotic cells, was clearly visible in HL-60 cells treated with 22.6 μM fucoxanthin for 12 h. Similar results have been obtained with camptothecin, which is known to be a strong apoptosis-inducing agent [28]. After 24 h incubation, the DNA ladder became clearer, even though the concentration of fucoxanthin was reduced to 11.3 or 4.5 μM. In contrast, DNA fragmentation was not observed in HL-60 cells treated with 22.6 μM β-carotene or retinoic acid. The enrichment factor also increased with fucoxanthin treatment. The increase in enrichment factor was found to be time-dependent and reached a plateau after an incubation period of 24 h. Some apoptosis-inducing agents are known to arrest a specific cell phase [29]. The doubling time of HL-60 cells is approximately 24 h. This suggests that fucoxanthin affects cell cycle. However, viable cell numbers decreased gradually until 96 h after fucoxanthin addition. Therefore, mechanisms other than apoptosis induction by fucoxanthin may be speculated to be involved in the antiproliferative activity of fucoxanthin on HL-60 cells.

The strong inhibitory effect of fucoxanthin has also been confirmed using human prostate cancer cells [15]. In their study, the effect of 15 kinds of

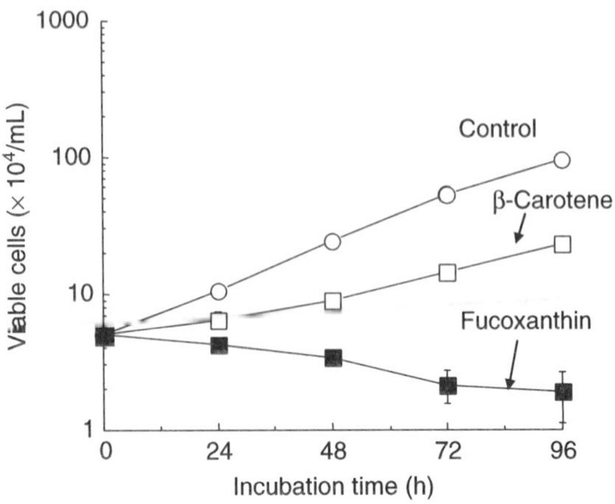

FIGURE 12.3 Comparison of the inhibitory effects of fucoxanthin and β-carotene on viability of HL-60 cells. HL-60 cells (5×10^4 cells/mL) were incubated with 22.6 μM fucoxanthin or β-carotene. Results are shown as means ± S.D. ($n = 3$). (Adapted from Hosokawa, M., Wanezaki, S., Miyauchi, K., Kurihara, H., Kohno, H., Kawabata, J., Odashima, S., and Takahashi, K., *Food Sci. Technol. Res.*, 5, 243–246, 1999. With permission.)

carotenoids (phytoene, phytofluene, ξ-carotene, lycopene, α-carotene, β-carotene, β-cryptoxanthin, canthaxanthin, astaxanthin, capsanthin, lutein, zeaxanthin, vio-axanthin, neoxanthin, and fucoxanthin) present in foodstuffs was evaluated on the growth of human prostate cancer cell lines (PC-3, DU 145, and LNCap). Among the carotenoids evaluated, neoxanthin and fucoxanthin were reported to cause a remarkable reduction in the growth of prostate cancer cells. DNA fragmentation, indicated by *in situ* TUNEL (terminal deoxynucleotidyl transferase-mediated dUTP nick end labeling), revealed that these two carotenoids apparently reduced the cell viability by inducing apoptosis. TUNEL is acknowledged as the method of choice in rapid identification and quantification of the apoptotic cell fraction in cultured cell preparations [30]. Although other acyclic carotenoids such as phytofluene, β-carotene, and lycopene also significantly reduced cell viability, the effect was lower as compared to neoxanthin and fucoxanthin. Further, other carotenoids did not affect the growth of the prostate cancer cells.

12.3.2 Mechanism

In their study on the effect of fucoxanthin on the viability of human colon cancer cells Caco-2, Hosakawa et al. [17] revealed that after 72 h of incubation with 7.6 μM of fucoxanthin, the number of viable cells decreased by 39% compared to the control (Figure 12.4). In contrast, incubation with 7.6 μM β-carotene and astaxanthin did not alter the viability of Caco-2 cells. Fucoxanthin has also been reported to reduce the viability of other human colon cancer cell lines (DLD-1 and HT-29 cells) in a dose- and time-dependent manner (Table 12.2). Caco-2 cells were more sensitive to fucoxanthin followed by DLD-1 and HT-29 cells.

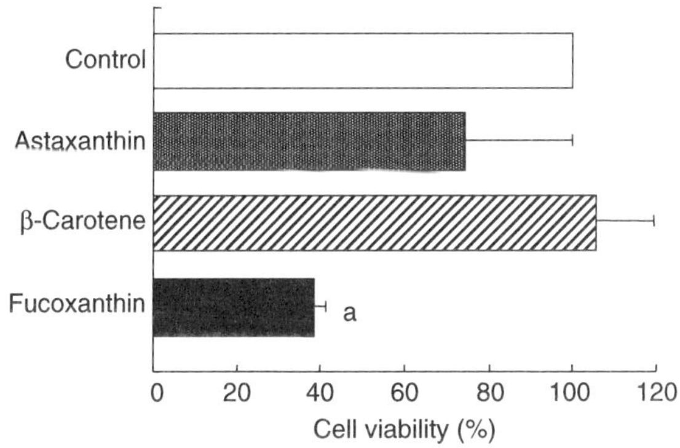

FIGURE 12.4 Comparison of viability of Caco-2 cells incubated with fucoxanthin, astaxanthin, or β-carotene. Caco-2 cells were incubated with 7.6 μM of each carotenoid for 72 h. Cell viability was measured by WST-1 assay. Values are means ± SD ($n = 3$–4). a: significant difference from control ($P < 0.01$). (Adapted from Hosokawa, M., Kudo, M., Maeda, H., Kohno, H., Tanaka, T., and Miyashita, K., *Biochim. Biophys. Acta*, 1675, 113–119, 2004. With permission.)

TABLE 12.2
Effect of Fucoxanthin on Cell Viability of Human Colon Cancer Cells

	24 h Incubation		72 h Incubation	
Colon Cancer Cells	**7.6 μM**	**15.2 μM**	**7.6 μM**	**15.2 μM**
Caco-2	$79.6 + 2.0^{a,c}$	$56.1 + 4.9^{b}$	$36.8 + 4.3^{a}$	$14.8 + 0.4^{c}$
HT-29	$93.1 + 2.9^{a}$	$82.2 + 0.5^{a,c}$	$93.7 + 1.1^{b}$	$50.8 + 0.6^{d}$
DLD-1	$88.5 + 1.5^{a}$	$70.9 + 4.4^{c}$	$103.8 + 5.2^{b}$	$29.4 + 2.3^{a}$

Note: The values with different letters (a, b, c, d) were significantly different from each other ($P < 0.05$).

The morphological evidence in Caco-2 cells treated with 11.6 μM fucoxanthin for 72 h indicated a diminished size and rounded shape [17]. Also, the cell membrane had shrunk with a condensed cytoplasm. In other words, the morphological appearance of Caco-2 cells treated with fucoxanthin had the properties observed in apoptotic cells. The cellular DNA fragmentation (measured as an indicator of apoptosis with quantitative sandwich ELISA using an antihistone antibody and an anti-DNA antibody) showed a dose-dependent increase in the case of Caco-2 cells treated with 11.3–22.6 μM fucoxanthin. However, β-carotene and astaxanthin did not induce DNA fragmentation in Caco-2 cells at a concentration of 7.6 μM, even after 48 h incubation, although fucoxanthin induced DNA fragmentation at the same concentration (Figure 12.5).

Fucoxanthin suppresses the level of Bcl-2 protein, which is responsible for suppression of programmed cell death as a survival factor [31,32]. The Bcl-2

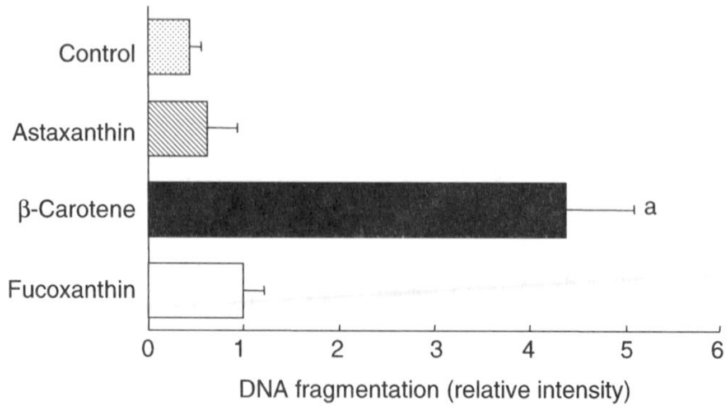

FIGURE 12.5 DNA fragmentation in Caco-2 cells incubated with fucoxanthin, astaxanthin, or β-carotene. Caco-2 cells were incubated in cultured medium containing 7.6 μM of each carotenoid for 48 h. Values are means ± SD ($n = 3$). a: significant difference from control ($P < 0.01$). (Adapted from Hosokawa, M., Kudo, M., Maeda, H., Kohno, H., Tanaka, T., and Miyashita, K., *Biochim. Biophys. Acta*, 1675, 113–119, 2004. With permission.)

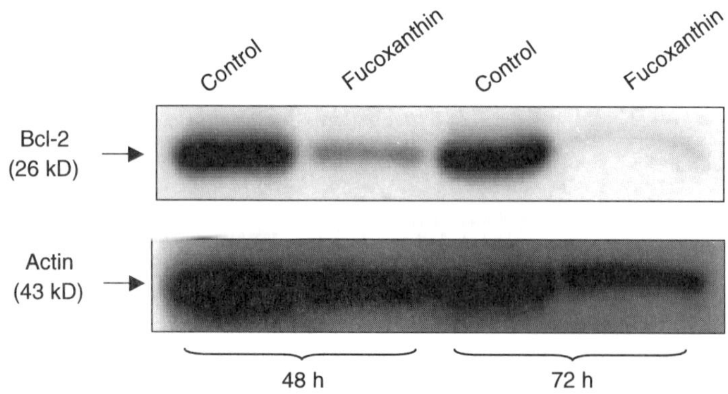

FIGURE 12.6 Expression of Bcl-2 protein in Caco-2 cells treated with fucoxanthin. Caco-2 cells were incubated in cultured medium containing 22.6 μM fucoxanthin for 48 or 72 h. Cellular protein was extracted, and levels of Bcl-2 protein were detected using Western blot analysis. (Adapted from Hosokawa, M., Kudo, M., Maeda, H., Kohno, H., Tanaka, T., and Miyashita, K., *Biochim. Biophys. Acta*, 1675, 113–119, 2004. With permission.)

protein levels in Caco-2 cells treated with fucoxanthin has been reported to decrease remarkably at concentrations as low as 22.6 μM (Figure 12.5) [17]. As can be observed from the Figure 12.6, after 72 h of incubation, the relative expression level of Bcl-2 protein compared to β-actin was less than 20% that of the control cells [17]. This is indicative of the fact that down-regulation of Bcl-2 protein may contribute to fucoxanthin-induced apoptosis in Caco-2 cells.

Caspases are known to play an important role in inducing apoptosis [33]. DNA fragmentation induced by fucoxanthin has been reported to be partially inhibited by a caspase inhibitor Z-VAD-fmk [17]. However, since Z-VAD-fmk diminished DNA fragmentation by only 40%, it is suggested that the apoptosis signaling in Caco-2 cell by fucoxanthin is mediated both by caspase-dependent and -independent pathways. Further, fucoxanthin may also regulate the redox signals, and then facilitate the progression of apoptosis through Bcl-2 protein suppression, and the caspase-dependent and -independent pathway. The combination effect of fucoxanthin and troglitazone on the reduction of Caco-2 cell viability has also been demonstrated [17]. Troglitazone is known to inhibit cell growth and induce apoptosis through the activation of PPARγ [34–38]. Oral administration of troglitazone has been reported to inhibit the early stage of colon tumorigenesis [39,40]. In their study, Hosakwa et al. [17] found that preincubation of Caco-2 cells with fucoxanthin remarkably enhanced the effect of troglitazone. Hence, the combined action of PPARγ ligand and fucoxanthin may provide new perspectives in developing novel approaches in the chemoprevention of cancer.

12.4 ANTIOBESITY

12.4.1 General Aspects

Obesity is defined as accumulation of body fat. Especially, the accumulation of fat around the internal organs is a major risk factor causing many kinds of diseases. This is because when the fat cell differentiates and accumulates the fat into the cell, the cell secretes various bioactive components, adipo-cytokines. These adipo-cytokines induce various health problems such as diabetes, hypertension, and hyperlipemia. These problems have been regarded as metabolic syndrome. Therefore, safe and effective antiobesity component has been keenly expected to be found from food materials.

Some dietary components have been reported to show an antiobesity effect. Based on the nature of fat digestion and absorption, diacylglycerols (DG) with a 1,3-configuration [41,42] and medium-chain triacylglycerols (MCT) [43] have been used for the prevention of obesity. Several studies have demonstrated that conjugated linoleic acid (CLA) reduces body fat accumulation in growing animals but not all CLA isomers contributed to this effect equally [44–46]. Among CLA isomers, *trans*(t)10, *cis*(c)12-18:2 induced body fat loss. The reported mechanism of this CLA action includes stimulation of lipolysis, reduction of lipid synthesis, and direct action on adipocytes [47,48]. Caffeine is a naturally consumed substance that is widely present in beverages such as coffee and tea. It has thermogenic properties and increases the metabolic rate in humans [49–53]. This effect can be explained by the stimulation of the secretion of catecholamine such as noradrenaline from the nerve endings. Noradrenaline stimulates β_3-adrenergic receptor (β_3-AR) and then induces promotion of energy expenditure through uncoupling protein 1 (UCP1) expression in brown adipose tissue (BAT) [54–57] (Figure 12.7). Capsaicin, the major pungent principle of red pepper, also up-regulates UCP1 in BAT by release of catecholamine

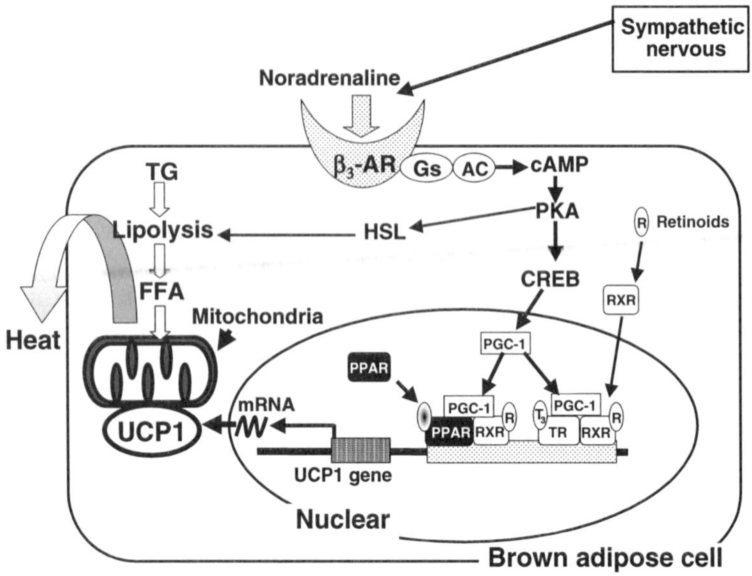

FIGURE 12.7 Possible mechanism for up-regulation of UCP1 in BAT.

such as noradrenaline [58–61]. Green tea extract is reported to increase energy expenditure and fat oxidation in humans [62]. The tea extract contains caffeine and catechin. Epigallocatechin gallate (EGCG), the main tea catechin, promotes fat oxidation and decreases fat synthesis, but does not activate β_3-adrenergic receptor [63]. Antiobesity activity of green tea extract will be attributed to both effects of UCP1 up-regulation by caffeine and of lipid metabolism control by catechin.

Effect of eicosapentaenoic (EPA) and docosahexaenoic acid (DHA) on lipid metabolism would be strongly correlated to antiobesity effect of marine lipids. Some publications have described reducing effect of fish oil on abdominal fat pad [64,65]. The same result was obtained by Kawada et al. [66]. They also found that the expression of uncoupling protein (UCP1) in interscapular brown adipose tissue (BAT) was significantly higher in the fish oil diet–fed rats compared to that in the lard-fed group. In BAT mitochondria, substrate oxidation is poorly coupled to ATP synthesis because of the presence of UCP1 (Figure 12.7), thereby leading to energy dissipation, that is, heat production. Kawada et al. [66] suggested that the intake of PUFA found in fish oil such as EPA and DHA causes UCP induction and enhancement of thermogenesis, resulting in suppression of the excessive growth of abdominal fat pads. PUFA from vegetable oils also suppressed the excessive accumulation of adipose tissue, as compared to animal fats [67,68]. However, the activity of PUFA from vegetable oils was less than those of EPA and DHA from fish oil [66]. Another study supports the previous results in terms of the higher antiobesity effect of EPA and DHA in fish oils [69].

12.4.2 UCP1

Although antiobesity compounds render their activity by different molecular mechanisms, adaptive thermogenesis through UCP1 expression is most important. Therefore, a great deal of interest has been focused on the physiological roles of UCP families (UCP1, UCP2, and UCP3) [70,71]. UCP are found in BAT (UCP1, UCP2, and UCP3), white adipose tissue (WAT) (UCP2), skeletal muscle (UCP2 and UCP3), and brain (UCP4 and UCP5) [71,72]. UCP2 and UCP3 are expressed in tissues besides BAT and, thus, are candidates to influence energy efficiency and expenditure [71]. Since metabolic rate, metabolic efficiency, and obesity are integrated properties of the whole animal, researchers have produced mice lacking UCP2 [72] and UCP3 [73,74]. However, despite the lack of UCP2 or UCP3, no consistent phenotypic abnormality was observed in the knockout mice. They were not obese and had normal thermogenesis. These results suggest that UCP2 and UCP3 are not a major determinant of metabolic rate in normal condition, but rather, have other functions [71,72,75–80]. Indeed, unexpected physiological or pathological implications of the UCP2 and UCP3 function (such as possible UCP2 involvement in diabetes and in apoptosis) could also be implicated [71]. UCP3 has a diminished thermogenic response to the drug 3,4-methylene-dioxymethamphetamine or MDMA [80]. Apart from UCP2 and UCP3, it is certain that UCP1 can potentially reduce excess abdominal fat [81].

UCP1 is a dimeric protein present in the inner mitochondrial membrane of BAT, and it dissipates the pH gradient generated by oxidative phospholyration, releasing chemical energy as heat. UCP1 is exclusively expressed in BAT, where the gene expression is increased by cold, adrenergic stimulation, β_3-agonists, retinoids, and thyroid hormone [82]. Thermogenic activity of BAT is dependent on UCP1 expression level controlled by the sympathetic nervous system via noradrenaline [81,83–85] (Figure 12.7). As a consequence of noradrenaline binding to the adipocyte plasma membrane, protein kinase (PKA) is expressed, and then, cyclic AMP response element binding protein (CREB) and hormone-sensitive lipase (HSL) are expressed. HSL stimulates lipolysis and the free fatty acids liberated serve as substrate in BAT thermognesis [85]. Free fatty acids also act as cytosolic second messengers, which activate UCP1 as PPARγ ligand. The same activity is expected in dietary PUFAs [86]. As shown in Figure 12.7, UCP1 expression in BAT controlled by sympathetic nervous system via noradrenaline and expressions of PPARγ, RXR, and TR. Antiobesity effect is shown by regulation of these molecular actions.

12.4.3 Up-Regulation of UCP1 in WAT by Fucoxanthin

As revealed above, it is certain that UCP1 is a key molecule for antiobesity. UCP1 is usually expressed only in BAT; however, adult humans have very little BAT and most of the fat is stored in WAT. Considered as breakthrough discoveries for an ideal therapy of obesity, regulation of UCP1 expression in tissues other than BAT by food constituents would be important. UCP1, usually expressed only in BAT, has also been found in WAT of mice overexpressing *Foxc2*, a winged helix gene, with a change in steady-state levels of several WAT- and BAT-derived mRNAs [87]. This result suggests the possibility of UCP1 expression in WAT, which would be an increasingly attractive target for the development of antiobesity therapies. As the key molecular components become defined, screening for food constituent that increase energy dissipation is becoming a more attainable goal. From this viewpoint, the antiobesity effect of edible seaweed carotenoid, fucoxanthin, is very interesting, as its activity depends on the protein and gene expressions of UCP1 in WAT [88].

In the study on antiobesity effect of the seaweed carotenoid, fucoxanthin, lipids were separated from edible seaweed, Wakame (*U. pinnatifida*), one of the most popular edible seaweeds in Japan and Korea. *Undaria* lipids, containing 10% fucoxanthin, reduced significantly the weight of WAT (comprising perirenal and epididymal abdominal adipose tissues) of both rats and mice (Figure 12.8). Furthermore, body weight of mice fed 2% *Undaria* lipid was significantly ($P < 0.05$) lower than that of the control, although there was no significant difference in the mean daily intake of diet between the two groups. To confirm the active component of *Undaria* lipids, fucoxanthin and *Undaria* glycolipids, the main fraction of the lipids, was administered to obese KK-Ay mice. Dietary effects of both fractions on WAT weight of obese mice is shown in Figure 12.9. The WAT weight of fucoxanthin-fed mice was significantly lower than that of control mice. However, there was no difference in WAT weight of mice fed *Undaria* glycolipids and

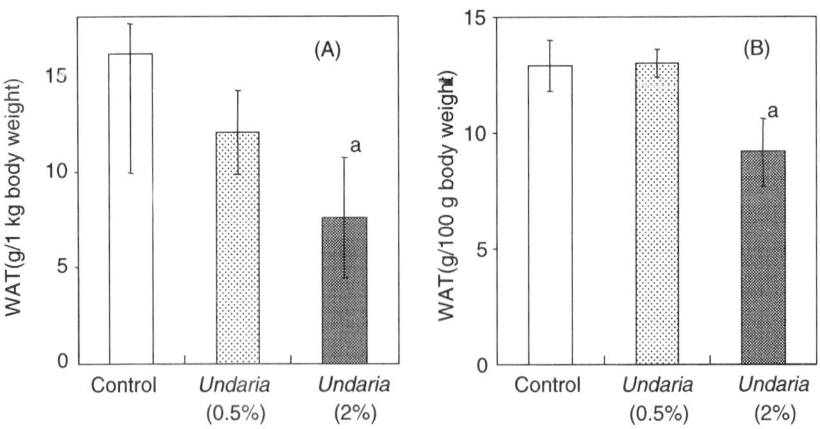

FIGURE 12.8 Weight of WAT of rats (A) and mice (B) fed *Undaria* lipids and control diet. a: significant different from control ($P < 0.01$). A diet was prepared according to the recommendation of American Institute of Nutrition (AIN-93G). The dietary fats for rats were 7% soybean oil (control), 6.5% soybean oil + 0.5% *Undaria* lipids, and 5% soybean oil + 2% *Undaria* lipids. Those for mice were 13% soybean oil (control), 12.5% soybean oil + 0.5% *Undaria* lipids, 11% soybean oil + 2% *Undaria* lipids. (Adapted from Maeda, H., Hosokawa, M., Sashima, T., Funayama, K., and Miyashita, K., *Biochim. Biophys. Res. Commun.*, 332, 392–397, 2005. With permission.)

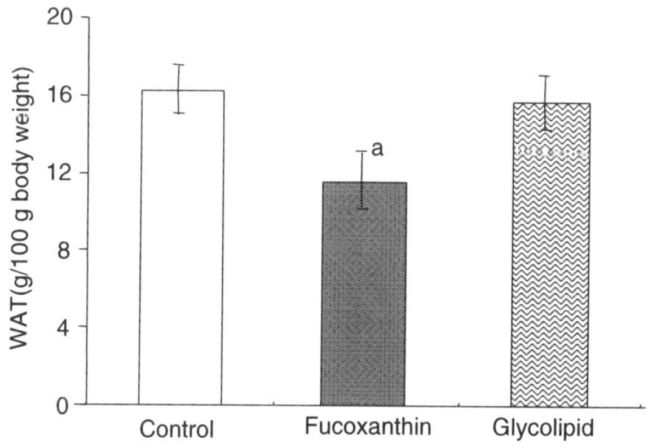

FIGURE 12.9 Weight of WAT of mice fed fucoxanthin, *Undaria* glycolipids, and control diet. a: significant difference from control ($P < 0.01$). The dietary fats were 13% soybean oil (control), 12.6% soybean oil + 0.4% fucoxanthin, and 11.2% soybean oil + 1.8% *Undaria* glycolipid. (Adapted from Maeda, H., Hosokawa, M., Sashima, T., Funayama, K., and Miyashita, K., *Biochim. Biophys. Res. Commun.*, 332, 392–397, 2005. With permission.)

control diet. This result indicates that fucoxanthin is the active component in the *Undaria* lipids resulting in the antiobesity effect.

BAT is implicated as an important site of facultative energy expenditure in small rodents because of its capacity for uncoupled mitochondrial respiration. This has led to speculation that BAT normally functions to prevent obesity. In 2.0% *Undaria* lipids–fed mice, BAT weight was significantly greater than that in the control mice. However, there was no significant difference in UCP1 expression among three different dietary groups. Thus, the decrease in abdominal fat pad weight found in *Undaria* lipids–fed mice could not be explained only by energy expenditure in BAT mitochondria by UCP1. As shown in Figure 12.10, UCP1 expression was found in WAT of *Undaria* lipids–fed mice, although there was little expression in that of control mice. Expression of UCP1 mRNA was also found in WAT of *Undaria* lipids–fed mice, but little expression in that of the control (Figure 12.10). UCP1 is known as a specific protein, which induces fat oxidation and conversion of the energy to heat. Therefore, the decrease in WAT weight of *Undaria* lipids–fed mice would be because of the adaptive thermogenesis through UCP1 expression in WAT. UCP1 expression in WAT was also found in fucoxanthin-fed mice, but little expression of UCP1 was found in WAT of mice fed *Undaria* glycolipids and control diets. This result confirmed the antiobesity activity of the seaweed carotenoid, fucoxanthin, through up-regulation of UCP1 expression in WAT.

The finding that fucoxanthin induces both protein and mRNA expressions of UCP1 in WAT will give a clue for new dietary antiobesity therapy. An enormous amount of data has been collected on thermogenesis in BAT through UCP1 expression. However, there had been no information on UCP1 expression in WAT induced by a diet component until the above report had appeared. An excessive

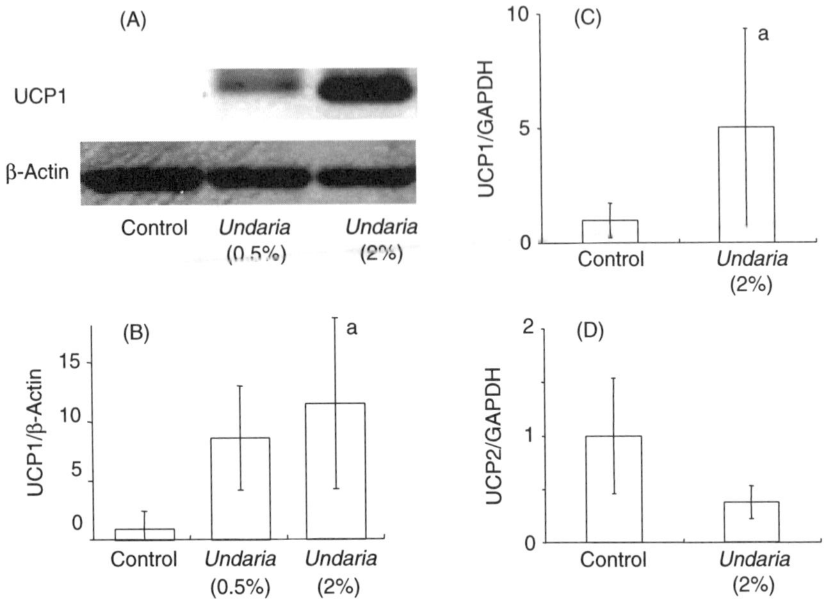

FIGURE 12.10 UCP1 and UCP2 expressions in WAT of mice fed *Undaria* lipids and control diet. (A) Western blot analysis of UCP1. (B) UCP1 protein expression. (C) UCP1 mRNA expression. (D) UCP2 mRNA expression. a: significant different from control ($P < 0.05$). (Adapted from Maeda, H., Hosokawa, M., Sashima, T., Funayama, K., and Miyashita, K., *Biochim. Biophys. Res. Commun.*, 332, 392–397, 2005. With permission.)

accumulation of fat in WAT induces some diseases such as type II diabetes. Direct heat production by fat oxidation in WAT, therefore, will reduce risk of these diseases in humans.

An enhancer element of UCP1 gene promotes transcription that is both BAT-selective and responsive to β-adrenergic stimulation through camp [75]. This complex enhancer element has putative binding sites for the thyroid hormone receptor, retinoic acid receptor, and peroxisome proliferators–activated receptor-γ (PPARγ). The study of gene expression in BAT is of special interest because this has been regarded as the only tissue in the mammalian body that functions exclusively as a thermogenic organ. Treatment with β-carotene and α-carotene promoted UCP1 expression in BAT. This effect of carotenoids could be explained by their conversion to retinoic acid [89]. However, fucoxanthin has no potency as vitamin A. Furthermore, there was no significant difference in UCP1 expression in BAT of *Undaria* lipids–fed mice as compared to that of the control mice, although BAT of 2% *Undaria* lipids–fed mice was significantly greater than that of the control. The significant decrease in WAT weight of rats and mice fed *Undaria* lipids diet (Figure 12.8) would be due to energy dissipation via the generation of heat by UCP1 expression in the WAT. This may be due to a specific activity of fucoxanthin or its relative carotenoids.

12.4.4 Reducing Effect of Fucoxanthin and Fucoxanthinol on Adipocyte Differentiation

Fucoxanthin has a unique structure including an allenic bond and a 5,6-monoepoxide in its molecule (Figure 12.1). It is a major marine carotenoid found in edible seaweeds such as *Undaria pinnatifida, Hijikia fusiformis*, and *Sargassum fulvellum*. In Southeast Asian countries, some seaweeds containing fucoxanthin are often used as a food source. Fucoxanthin is easily converted into fucoxanthinol (Figure 12.1) in the human intestinal cells and in mice [90], suggesting that the active form of fucoxanthin in biological system would be fucoxanthinol.

As shown in Figure 12.11, both fucoxanthin and fucoxanthinol inhibited intercellular lipid accumulation during adipocyte differentiation of 3T3-L1 cells.

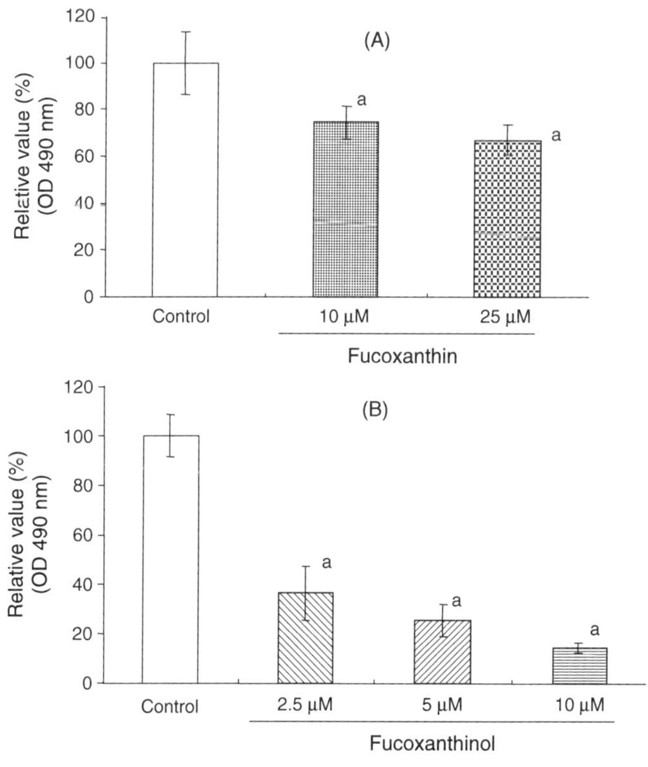

FIGURE 12.11 Effect of fucoxanthin (A) and fucoxanthinol (B) on lipid accumulation of 3T3-L1 cells during adipocyte differentiation. 3T3-L1 cells were treated with fucoxanthin or fucoxnthinol in differentiation medium for 120 h. The intercellular lipid accumulation was determined by Oil Red-O staining. The values ($n = 3$) are expressed as absorbance at 490 nm. a: significant different from control ($P < 0.01$). (Adapted from Maeda, H., Hosokawa, M., Sashima, T., Takahashi, N., Kawada, T., and Miyashita, K., *Int. J. Mol. Med.*, 18, 147–152, 2006. With permission.)

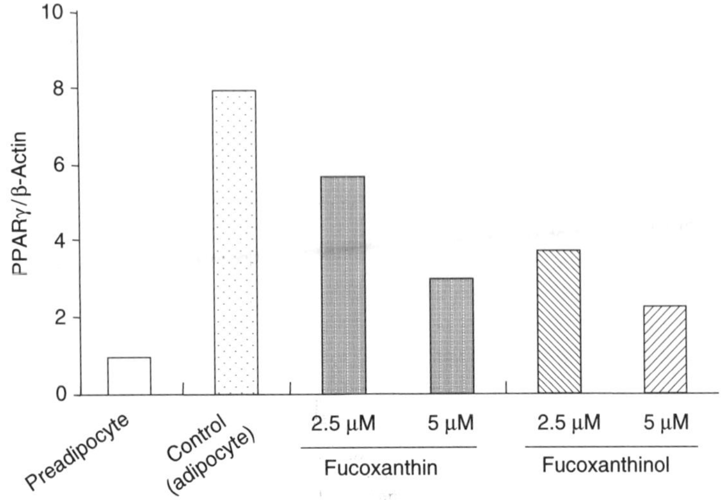

FIGURE 12.12 Expression of PPARγ in 3T3-L1 cells treated with fucoxanthin and fucoxanthinol. The PPARγ protein expression level was normalized to the β-actin level and expressed as the value relative to preadipocyte PPAR levels. (Adapted from Maeda, H., Hosokawa, M., Sashima, T., Takahashi, N., Kawada, T., and Miyashita, K., *Int. J. Mol. Med.*, 18, 147–152, 2006. With permission.)

Fucoxanthin and fucoxanthinol also decrease glycerol-3-phosphate dehydrogenase activity, an indicator of adipocyte differentiation [91]. The effects of fucoxanthinol were stronger than those of fucoxanthin. When 3T3-L1 cells treated with fucoxanthin and fucoxanthinol, PPARγ, a regulator of adipogenic gene expression, was down-regulated by both carotenoids in a dose-dependent manner [91] (Figure 12.12). These results suggest that fucoxanthin and fucoxanthinol inhibit the adipocyte differentiation of 3T3-L1 cells through down-regulation of PPARγ and fucoxanthinol would be an active compound for the antiobesity effect of fucoxanthin.

It is noteworthy that PPARγ levels were down-regulated in 3T3-L1 cells treated with fucoxanthin and fucoxanthinol. PPARγ has an important role in the early stages of 3T3-L1 cell differentiation, because it is a nuclear transcription factor that regulates adipogenic gene expression. Regulation of PPARγ would be one of the expected mechanisms underlying the antiobesity effect of dietary fucoxanthin. Catechin [92], sterols [93], tannic acid [94], phenolic lipids [95], and red yeast rice extracts [96] also inhibit 3T3-L1 differentiation. Retinoids inhibit the early stage of differentiation of 3T3-L1 cells [97]. Some of these compounds, however, have low absorption and have not been fully investigated for their antiobesity effects *in vivo*. Because fucoxanthin accumulates as metabolites, including fucoxanthinol, in WAT of mice, dietary fucoxanthin might be a useful natural compound for the prevention of obesity.

12.5 ANTI-INFLAMMATORY EFFECT

Several marine carotenoids, including fucoxanthin and astaxanthin, have been shown to have anti-inflammatory potency [21,98]. Endotoxin-induced uveitis (EIU), an ocular inflammation, is an acute anterior segment intraocular inflammation that can be induced by injection of lipopolysaccharides [99,100]. EIU is suggested to serve as a model for certain types of human uveitis like the one associated with seronegative arthritis, where a Gram-negative bacteria is presumed to play a role in pathogenesis [101]. In a study dealing with the effect of fucoxanthin on lipopolysaccharide-induced EIU in a rat model system [21], it was reported that fucoxanthin obtained from *Undaria* lipids significantly suppressed the development of EIU. Further, the effect exhibited by a unit quantity (10 ppm) of fucoxanthin was similar to that by a similar amount of the steroid prednisolone.

Generation of nitric oxide (NO) by inducible nitric oxide synthase (iNOS) is one of the major responses in the eye to lipopolysaccharide injection during EIU [102,103]. Apart from this, lipopolysaccharides and proinflammation cytokines induce iNOS in macrophages, neutrophils, and endothelial cells that can further contribute to increase in nitric oxide in the eye during EIU; and, activation of iNOS is strongly associated with the development of inflammation in EIU [104]. Fucoxanthin exhibits a dose-dependent antiocular inflammatory effect on EIU, as evidenced by a decreased NO concentration in an *in vitro* study involving lipopoly saccaride (LPS)-induced RAW264.7 (a mouse macrophage cell line) cells, as well as an *in vitro* study involving rats [21]. It is well documented that NO is produced by bacterial LPS or cytokines plays a major role in endotoxemia and inflammatory conditions [105]. The possible mechanism of antiocular inflammatory effect exhibited by fucoxanthin is the suppression of NO production in the eye through inhibition of the iNOS enzyme and thus exhibits a therapeutic effect in the treatment of inflammation.

REFERENCES

1. Wong, K.H. and Cheung, P.C.K. 2000. Nutritional evaluation of some subtropical red and green seaweeds part 1—proximate composition, amino acid profiles and some physico chemical properties. *Food Chem.* 71, 475–482.
2. Ramavat, B.K., Pazdro, K., Kosakowska, A. and Pempkowiak, J. 1997. Fatty acid content in seaweeds from the Baltic Sea and the Indian ocean. *Oceanologia* 39, 279–287.
3. Lopez, A. and Gerwick, W.H. 1987. Two new icosapentaenoic acids from temperate sea weed *Ptilota filicina. J. Agardh. Lipids* 22, 190–194.
4. Dembitsky, V.M., Rezanka, T. and Rozensvet, O.A. 1993. Lipid composition of three macrophytes from the Caspian sea. *Phytochemistry* 33, 1015–1019.
5. Jones, A.L. and Harwood, J.L. 1993. The component fatty acids of some marine algal lipids. *Phytochemistry* 11, 1423–1432.
6. Wise, M.L., Hamberg, M. and Gerwick, W.H. 1994. Biosynthesis of conjugated triene containing fatty acids by a novel isomerase from the red marine alga *Ptilota filicina. Biochemistry* 33, 15223–15232.

7. Vaskovsky, V.E., Khotimchenko, S.V., Xia, B. and Hefang, L. 1996. Polar lipids and fatty acids of some marine macrophytes from the yellow sea. *Phytochemistry* 42, 1347–1356.
8. Kamenarska, Z., Yalcin, F.N., Ersoz, T., Calis, I., Stefanov, K. and Popov, S. 2002. Chemical composition of *Cytoseira crinita* Bory from the Eastern Mediterranean. *Z. Naturforsch.* 57c, 584–590.
9. Burgess, J.R., de la Rosa, R.I., Jacobs, R.S. and Butler, A. 1991. A new eicosap entaenoic acid formed from arachidonic acid in the coralline red algae *Bosiella orbingniana. Lipids* 26, 162–165.
10. Hamberg, M. 1992. Metabolism of 6,9,12-octadecatrienoic acid in the red alga *Lithothamnion corallioides*: mechanism of formation of a conjugated tetraene fatty acid, *Biochem. Biophys. Res. Commun.* 188, 1220–1227.
11. Mikhailova, M.V., Bemis, D.L., Wise, M.L., Gerwick, W.H., Norris, J.N. and Jacobs, R.S. 1995. Structure and biosynthesis of novel conjugated polyene fatty acids from the marine green alga *Anadyomene stellata. Lipids* 30, 583–589.
12. Bhasakar, N., Kinami, T., Miyashita, K., Park, S.-B., Endo, Y. and Fujimoto, K. 2004. Occurrence of conjugated polyenoic fatty acids in seaweeds from the Indian ocean. *Z. Naturforsch.* 59C, 310–314.
13. Bhasakar, N., Hosokawa, M. and Miyashita, K. 2004. Growth inhibition of human pro-myelocytic leukemia (HL-60) cells by lipid extracts of marine alga *Sargassum marginatum* (Fucales, Phaeophyta) harvested off Goa (west coast of India) with special reference to fatty acid composition. *Indian J. Mar. Sci.* 33, 355–360.
14. Suzuki, R., Noguchi, R., Ota, T., Abe, M., Miyashita, K. and Kawada, T. 2001. Cytotoxic effect of conjugated trienoic fatty acids on mouse tumor and human monocytic leukemia cells. *Lipids* 36, 477–482.
15. Kotake-Nara, E., Kushiro, M., Zhang, H., Sugawara, T., Miyashita, K. and Nagao, A. 2001. Carotenoids affect proliferation of human prostate cancer cells. *J. Nutr.* 131, 3303–3306.
16. Kohno, H., Suzuki, R., Noguchi, R., Hosokawa, M., Miyashita, K. and Tanaka, T. 2002. Dietary conjugated linolenic acid inhibits azoxymethane-induced colonic aberrant crypt foci in rats. *Jpn. J. Cancer Res.* 93, 133–142.
17. Hosokawa, M., Kudo, M., Maeda, H., Kohno, H., Tanaka, T. and Miyashita, K. 2004. Fucoxanthin induces apoptosis and enhances the antiproliferative effect of the PPARγ ligand, troglitazone, on colon cancer cells. *Biochim. Biophys. Acta* 1675, 113–119.
18. Kohno, H., Yasui, Y., Suzuki, R., Hosokawa, M., Miyashita, K. and Tanaka, T. 2004. Dietary seed oil rich in conjugated linolenic acid from bitter melon inhibits azoxymethane-induced rat colon carcinogenesis through elevation of colonic PPARγ expression and alteration of lipid composition. *Int. J. Cancer* 110, 896–901.
19. Kohno, H., Suzuki, R., Yasui, Y., Hosokawa, M., Miyashita, K. and Tanaka, T. 2004. Pomegranate seed oil rich in conjugated linolenic acid suppresses chemically induced colon carcinogenesis in rats. *Cancer Sci.* 95, 481–486.
20. Yasui, Y., Hosokawa, M., Sahara, T., Suzuki, R., Ohgiya, S., Kohno, H., Tanaka, T. and Miyashita, K. 2005. Bitter gourd seed fatty acid rich in 9c,11t,13t-conjugated linolenic acid induces apoptosis and up-regulates the GADD45, p53 and PPARγ in human colon cancer Caco-2 cells. *Prostag. Leukotr. Ess.* 73, 212–219.
21. Shiratori, K., Ohgami, K., Ilieva, I., Jin, X.-H., Koyama, Y., Miyashita, K., Yoshida, K., Kase, S. and Ohno, S. 2005. Effects of fucoxanthin on lipopolysaccharide-induced inflammation in vitro and in vivo. *Exp. Eye Res.* 81, 271–277.
22. Czeczuga, B. and Taylor, F.J. 1987. Carotenoid content in some species of the brown and red algae from the coastal area of New Zealand. *Biochem. Syst. Ecol.* 15, 5–8.

23. Haugan, J. and Liaeen-Jensen, S. 1994. Algal carotenoids 54.: carotenoids from brown algae (Phaeophyceae). *Biochem. Syst. Ecol.* 22, 31–41.
24. Mori, K., Ooi, T., Hiraoka, M., Oka, N., Hamada, H., Tamura, M. and Kusumi, T. 2004. Fucoxanthin and its metabolites in edible brown algae cultivated in deep seawater. *Mar. Drugs* 2, 63–72.
25. Matsuno, T., Ohkubo, M., Toriiminami, Y., Tsushima, M., Sakaguchi, S., Minami, T. and Maoka, T. 1999. Carotenoids in food chain between freshwater fish and aquatic insects. *Comp. Biochem. Physiol. Part B*, 24, 341–345.
26. Hosokawa, M., Wanezaki, S., Miyauchi, K., Kurihara, H., Kohno, H., Kawabata, J., Odashima, S. and Takahashi, K. 1999. Apoptosis-inducing effect of fucoxanthin on human leukemia cell HL-60. *Food Sci. Technol. Res.* 5, 243–246.
27. Compton, M.M. 1992. A biochemical hallmark of apoptosis: internucleosomal degradation of the genome. *Cancer Metastasis Rev.* 11, 105–119.
28. Solary, B.E., Bertrand, R., Kohn K.W. and Pommier, Y. 1993. Differential induction of apoptosis in undifferentiated and differentiated HL-60 cells by DNA topoisomerase I and II inhibitors. *Blood* 81, 1359–1368.
29. Bino, G.D., Skierski, J.S. and Darzynkiewicz, Z. 1991. The concentration-dependent diversity of effects of DNA topoisomerase I and II inhibitors on the cell cycle of HL-60 cells. *Exp. Cell Res.* 195, 485–491.
30. Negoescu, A., Lorimier, P., Labot-Moleur, F., Azoti, L., Robert, C., Guillermet, C., Brambilla, C. and Brambilla, E. 1997. TUNEL: improvement and evaluation of method for *in-situ* apoptotic cell identification. *Biochemica* 2, 12–17.
31. Hockenbery, D., Nunez, G., Milliman, C., Schreiber, R.D. and Korsmeyer, S. 1990. Bcl-2 is an inner mitochondrial membrane protein that blocks programmed cell death. *Nature* 348, 334–336.
32. Levy, P., Robin, H., Bertrand, F., Kornprobst, M. and Capeau, J. 2003. Butyrate-treated colonic Caco-2 cells exhibit defective integrin-mediated signaling together with increased apoptosis and differentiation. *J. Cell Physiol.* 197, 336–347.
33. Satomi, Y., Tokuda, H., Fujii, H., Shimidzu, N., Tanaka, Y. and Nishino, H. 1996. Anti-tumor-promoting activity of fucoxanthin, a natural carotenoid. *J. Kyoto Pref. Univ. Med.* 105, 739–743.
34. Brockman, J.A., Gupta, T.A. and Dubois, R.N. 1998. Activation of PPAR gamma leads to inhibition of anchorage-independent growth of human colorectal cancer cells. *Gastroenterology* 115, 1049–1055.
35. Kitamura, S., Miyazaki, Y., Shinomura, Y., Kondo, S., Kanayama, S. and Matsuzawa, Y. 1999. Peroxisome proliferator-activated receptor γ induces growth arrest and differentiation markers of human colon cancer cells. *Jpn. J. Cancer Res.* 90, 75–80.
36. Yang, W.L. and Frucht, H. 2001. Activation of the PPAR pathway induces apoptosis and COX-2 inhibition in HT-29 human colon cancer cells. *Carcinogenesis* 22, 1379–1383.
37. Shimada, T., Kojima, K., Yoshiura, K., Hiraishi, H. and Terano, A. 2002. Characteristics of the peroxisome proliferator activated receptor γ (PPARγ) ligand induced apoptosis in colon cancer cells. *Gut* 50, 658–664.
38. Chen, G.G., Lee, J.F., Wang, S.H., Chan, U.P., Ip, P.C. and Lau, W.Y. 2002. Apoptosis induced by activation of peroxisome-proliferator activated receptor-gamma is associated with Bcl-2 and NF-kappaB in human colon cancer. *Life Sci.* 70, 2631–2646.
39. Tanaka, T., Kohno, H., Yoshitani, S., Takashima, S., Okumura, A., Murakami, A. and Hosokawa, M. 2001. Ligands for peroxisome proliferators-activated receptors alpha and gamma inhibit chemically induced colitis and formation of aberrant crypt foci in rats. *Cancer Res.* 61, 2424–2428.

40. Kohno, H., Yoshitani, S., Takashima, S., Okumura, A., Yamaguchi, N. and Tanaka, T. 2001. Troglitazone, a ligand for peroxisome proliferators-activated receptor γ, inhibits chemically induced aberrant crypt foci in rats. *Jpn. J. Cancer Res.* 92, 396–403.
41. Flickinger, B.D. and Matsuo, N. 2003. Nutritional charecteristics of DAG oil. *Lipids* 38, 129–132.
42. Flickinger, B.D. 2006. Diacyglycerols (DAGs) and their mode of action, in *Nutraceuticals and Specialty Lipids and Their Co-Products*, Shahidi, F., Ed., CRC Press, Boca Raton, FL, pp. 181–186.
43. Che Man, Y.B. and Manaf, A.A. 2006. Medium-chain triacylglycerols, in *Nutraceuticals and Specialty Lipids and Their Co-Products*, Shahidi, F. Ed., CRC Press, Boca Raton, FL, pp. 27–56.
44. Atkinson, R.A. 1999. Conjugated linoleic acid for altering body composition and treating obesity, in *Advances in Conjugated Linoleic Acid Research*, Vol. 1, Yurawecz, M.P., Mossoba, M.M., Kramer, J.K.G., Pariza, M.W., and Nelson, G.J., Eds., AOCS Press, Champaign, IL, pp. 348–353.
45. Keim, N.L. 2003. Conjugated linoleic acid for altering body composition and treating obesity, in *Advances in Conjugated Linoleic Acid Research*, Vol. 2, Sébédio, J.-L., Christie, W.W., and Adlof, R., Eds., AOCS Press, Champaign, IL, pp. 316–324.
46. Watkins, B.A. and Li, Y. 2006. Conjugated linoleic acids (CLAs): Food, nutrition, and health, in *Nutraceuticals and Specialty Lipids and Their Co-Products*, Shahidi, F., Ed., CRC Press, Boca Raton, FL, pp. 187–200.
47. Miner, J.L., Cederberg, C.A., Nielson, M.K., Chen, X. and Baile, C.A. 2001. Conjugated linoleic acid (CLA), body fat, and apoptosis. *Obes. Res.* 9, 129–134.
48. Tsuboyama-Kasaoka, N., Takahashi, M., Tanemura, K., Kim, H.J., Tange, T., Okuyama, H., Kasai, M., Ikemoto, S. and Ezaki, O. 2000. Conjugated linoleic acid supplementation reduces adipose tissue by apoptosis and develops lipodystrophy in mice. *Diabetes* 49, 1534–1542.
49. Miller, D.S., Stock, M.J. and Stuart, J.A. 1974. The effect of caffeine and carnitine on oxygen consumption of fed and fasted subjects. *Proc. Nutr. Soc.* 33, A28–A29.
50. Acheson, K.J., Zahorska-Markiewicz, B., Pittet, P., Anantharman, K. and Jéquier, E. 1980. Caffeine and coffee. Their influence on metabolic rate and substrate utilization in normal weight and obese individuals. *Am. J. Clin. Nutr.* 33, 989–997.
51. Yoshida, T., Sakane, N., Umekawa, T. and Kondo, M. 1994. Relationship between basal metabolic rate, thermogenic response to caffeine, and body weight loss following combined low calorie and exercise treatment in obese women. *Int. Obes. Relat. Metab. Disord.* 18, 345–350.
52. Astrup, A. 2000. Thermogenic drugs as a strategy for treatment of obesity. *Endocrine* 13, 207–212.
53. Molnar, D., Torok, K., Erhardt, E. and Jeges, S. 2000. Safety and efficacy of treatment with an ephedrine/caffeine mixture. The first double-blind placebo-controlled pilot study in adolescents. *Int. Obes. Relat. Metab. Disord.* 24, 1573–1578.
54. Bellet, S., Roman, L., Decastro, O., Kim, K.E. and Kershbaum, A. 1969. Effect of coffee ingestion on catecholamine release. *Metabolism* 18, 288–291.
55. Berkowitz, B.A. and Spector, S. 1971. Effect of caffeine and theophylline on peripheral catecholamines. *Eur. J. Pharmacol.* 13, 193–197.
56. Yoshida, K., Yoshida, T., Kamanaru, K., Hiraoka, N. and Kondo, M. 1990. Caffeine activates brown adipose tissue thermogenesis and metabolic rate in mice. *J. Nutr. Sci. Vitaminol.* 36, 173–178.

57. Kogure, A., Sakane, N., Takakura, Y., Umekawa, T., Yoshida, K., Nishino, H., Yamamoto, T., Kawada, T., Yoshikawa, T. and Yoshida, T. 2002. Effects of caffeine on the uncoupling protein family in obese yellow KK mice. *Clin. Exp. Pharm. Phys.* 29, 391–394.
58. Kawada, T., Hagihara, K. and Iwai, K. 1986. Effects of capsaicin on lipid metabolism in rats fed a high fat diet. *J. Nutr.* 116, 1272–1278.
59. Kawada, T., Sakabe, S., Aoki, N., Watanabe, T., Higeta, K. and Iwai, K. 1991. Intake of sweeteners and pungent ingredients increases the thermogenin content in brown adipose tissue of rat. *J. Agric. Food Chem.* 39, 651–654.
60. Watanabe, T., Kawada, T., Kato, T., Harada, T. and Iwai, K. 1994. Effects of capsaicin analogs on adrenal catecholamine secretion in rats. *Life Sci.* 54, 369–374.
61. Masuda, Y., Haramizu, S., Oki, K., Ohnuki, K., Watanabe, T., Yazawa, S., Kawada, T., Hashizume, S. and Fushiki, T. 2003. Upregulation of uncoupling proteins by oral administration of capsiate, a nonpungent capsaicin analog. *J. Appl. Phys.* 95, 2408–2415.
62. Duloo, A.G., Duret, C., Rohrer, D., Girardier, L., Mensi, N., Fathi, M., Chantre, P. and Vandermander, J. 1999. Efficacy of a green tea extract rich in catechin polyphenols and caffeine in increasing 24-h energy expenditure and fat oxidation in humans. *Am. J. Clin. Nutr.* 70, 1040–1045.
63. Klaus, S., Pültz, S., Thöne-reineke, C. and Wolfram, S. 2005. Epigallocatechin gallate attenuates diet-induced obesity in mice by decreasing energy absorption and increasing fat oxidation. *Int. J. Obesity* 19, 615–623.
64. Parrish, C.C., Pathy, D.A. and Angel, A. 1990. Dietary fish oils limit adipose tissue hypertrophy in rats. *Metabolism* 39, 217–219.
65. Couet, C., Delarue, J., Ritz, P., Antoine, J.-M. and Lamisse, F. 1997. Effect of dietary fish oil on body fat mass and basal fat oxidation in healthy adults. *Int. J. Obesity* 21, 637–643.
66. Kawada, T., Kayahashi, S., Hida, Y., Koga, K., Nadachi, Y. and Fushiki, T. 1998. Fish (bonito) oil supplementation enhances the expression of uncoupling protein in brown adipose tissue of rat. *J. Agric. Food Chem.* 46, 1225–1227.
67. Shimomura, Y., Tamura, T. and Suzuki, M. 1990. Less body fat accumulation in rats fed a safflower oil diet than in rats fed a beef tallow diet. *J. Nutr.* 120, 1291–1296.
68. Okuno, M., Kajiwara, K., Imai, S., Kobayashi, T., Honma, N., Maki, T., Suruga, K., Goda, T., Takase, S., Muto, Y. and Noriwaki, H. 1997. Perilla oil prevents the excessive growth of visceral adipose tissue in rats by down-regulating adipocyte differentiation. *J. Nutr.* 127, 1752–1757.
69. Toyoshima, K., Noguchi, R., Hosokawa, M., Fukunaga, K., Nishiyama, T., Takahashi, R. and Miyashita, K. 2004. Separation of sardine oil without heating from surimi waste and its effect on lipid metabolism in rats. *J. Agric. Food Chem.* 52, 2372–2375.
70. Dulloo, A.G. and Samec, S. 2001. Uncoupling proteins: their roles in adaptive thermogenesis and substrate metabolism reconsidered. *British J. Nutr.* 86, 123–139.
71. Ježek, P. 2002. Possible physiological roles of mitochondrial uncoupling proteins-UCP*n*. *Int. J. Biochem. Cell Biol.* 34, 1190–1206.
72. Arsenijevic, D., Onuma, H., Pecqueur, C., Raimbault, S., Manning, B.S., Miroux, B., Couplan, E., Alves-Guerra, M.-C., Goubern, M., Surwit, R., Bouillaud, F., Richard, D., Collins, S. and Ricquier, D. 2000. Disruption of the uncoupling protein-2 gene in mice reveals a role in immunity and reactive oxygen species production. *Nat. Genet.* 26, 435–439.

73. Gong, D.-W., Monemdjou, S., Garvilova, O., Leon, L.R., Marcus-Samuels, B., Chou, C.J., Everett, C., Kozak, L.P., Li, C., Deng, C., Harper, M.-E. and Reitman, M.L. 2000. Lack of obesity and normal response to fasting and thyroid hormone in mice lacking uncoupling protein-3. *J. Biol. Chem.* 275, 16251–16257.
74. Vidal-Puig, A.J., Grujic, D., Zhang, C.-Y., Hagen, T., Boss, O., Ido, Y., Szczepanik, A., Wade, J., Mootha, V., Cortright, R., Muoio, D.M. and Lowell, B.B. 2000. Energy metabolism in uncoupling protein 3 gene knockout mice. *J. Biol. Chem.* 275, 16258–16266.
75. Lowell, B.B. and Spiegelman, B.M. 2000. Towards a molecular understanding of adaptive thermogenesis. *Nature* 404, 652–660.
76. Harper, M.-E., Dent, R.M., Bezaire, V., Antoniou, A., Gauthier, A., Monemdjou, S. and McPherson, R. 2001. UCP3 and its putative function: consistencies and controversies. *Biochem. Soc. Trans.* 29, 768–773.
77. Rodríquez, A.M., Roca, P. and Palou, A. 2002. Synergic effect of overweight and cold on uncoupling proteins expression, a role of α2/β3 adrenergic receptor balance. *Eur. J. Phys.* 444, 484–490.
78. Schrauwen, P. and Hesselink, M. 2002. UCP2 and UCP3 in muscle controlling body metabolism. *J. Exp. Biol.* 205, 2275–2285.
79. Ricquier, D. 2005. Respiration uncoupling and metabolism in the control of energy expenditure. *Pro. Nutri. Soc.* 64, 47–52.
80. Mills, E.M., Banks, M.L., Sprague, J.E. and Finkel, T. 2003. Uncoupling the agony from ecstasy. *Nature* 426, 403–404.
81. Nedergaard, J., Golozoubova, V., Matthias, A., Asadi, A., Jacobsson, A. and Cannon, B. 2001. UCP1: the only protein able to mediate adaptive non-shivering thermogenesis and metabolic inefficiency. *Biochim. Biophys. Acta* 1504, 82–106.
82. Silva, J.E. and Rabelo, R. 1997. Regulation of the uncoupling protein gene expression. *Eur. J. Endocrinol.* 136, 251–264.
83. Del Mar Gonzalez-Barroso, M., Ricquier, D. and Cassard-Doulcier, A.-M. 2000. The human uncoupling protein-1 gene (UCP1): present status and perspectives in obesity research. *Obesity Rev.* 1, 61–72.
84. Argyropoulos, G. and Harper, M.-L. 2002. Molecular biology of thermoregulation. Invited review: uncoupling proteins and thermoregulation. *J. Appl. Physiol.* 92, 2187–2198.
85. Mozo, J., Emre, Y., Bouillaud, F., Ricquier, D. and Criscuolo, F. 2005. Thermoregulation: what role for UCPs in mammals and birds? *Biosci. Reports* 25, 227–249.
86. Clarke, S.D. 2000. Polyunsaturated fatty acid regulation of gene transcription: a mechanism to improve energy balance and insulin resistance. *Br. J. Nutr.* 83, 59–66.
87. Cederberg, A., Grønning, L.M., Ahrén, B. Taskén, K., Carlsson, P. and Enerbäck, S. 2001. *FOXC2* is a winged helix gene that counteracts obesity, hypertriglyceridemia, and diet-induced insulin resistance. *Cell* 106, 563–573.
88. Maeda, H., Hosokawa, M., Sashima, T., Funayama, K. and Miyashita, K. 2005. Fucoxanthin from edible seaweed, *Undaria pinnatifida*, shows antiobesity effect through UCP1 expression in white adipose tissues. *Biochim. Biophys. Res. Commun.* 332, 392–397.
89. Serra, F., Bonet, M.L., Puigserver, P., Oliver, J. and Palou, A. 1999. Stimulation of uncoupling protein 1 expression in brown adipocytes by naturally occurring carotenoids. *Int. J. Obesity* 23, 650–655.
90. Sugawara, T., Baskaran, V., Tsuzuki, W. and Nagao, A. 2002. Brown algae fucoxanthin is hydrolyzed to fucoxanthinol during absorbtion by Caco-2 human intestinal cells and mice. *J. Nutr.* 132, 946–951.

91. Maeda, H., Hosokawa, M., Sashima, T., Takahashi, N., Kawada, T. and Miyashita, K. 2006. Fucoxanthin and its metabolite, fucoxanthinol, suppress adipocyte differentiation in 3T3-L1 cells. *Int. J. Mol. Med.* 18, 147–152.
92. Paul, A.G. 2001. The roles of PPARs in adipocyte differentiation. *Prog. Lipid Res.* 40, 269–281.
93. Furuyashiki, T., Nagayasu, H., Aoki, Y., Bessho, H., Hashimoto, T., Kanazawa, K. and Ashida, H. 2004. Tea catechin suppresses adipocyte differentiation accompanied by down-regulation of PPAR gamma2 and C/EBPalpha in 3T3-L1 cells. *Biosci. Biotechnol. Biochem.* 68, 2353–2359.
94. Awad, A.B., Begdache, L.A. and Fink, C.S. 2000. Effect of sterols and fatty acids on growth and triglyceride accumulation in 3T3-L1 cells. *J. Nutr. Biochem.* 11, 153–158.
95. Liu, X., Kim, J.K., Li, Y., Liu, F. and Chen, X. 2005. Tannic acid stimulates glucose transport and inhibits adipocyte differentiation in 3T3-L1 cells. *J. Nutr.* 135, 165–171.
96. Rejman, J. and Kozubek, A. 2004. Inhibitory effect of natural phenolic lipids upon NAD-dependent dehydrogenases and on triglyceride accumulation in 3T3-L1 cells in culture. *J. Agric. Food Chem.* 52, 246–250.
97. Jeon, T., Hwang, S.G., Hirai, S., Matsui, T., Yano, H., Kawada, T., Lim, B.O. and Park, D.K. 2004. Red yeast rice extracts suppress adipogenesis by down-regulating adipogenic transcription factors and gene expression in 3T3-L1 cells. *Life Sci.* 75, 3195–3203.
98. Ohgami, K., Shiratori, K., Kotake, S., Nishida, T., Mizuki, N., Yazawa, K. and Ohno, S. 2003. Effects of astaxanthin on lipopolysaccharide-induced inflammation in vitro and in vivo. *Invest. Opthulmol. Vis. Sci.* 44, 2694–2701.
99. Hikita, N., Chan, C.C., Whitcup, S.M., Nussenblatt, R.B. and Mochizuki, M. 1995. Effects of tropical FK506 on endotoxin-induced uveitis (EIU) in the Lewis rat. *Curr. Eye Res.* 14, 209–214.
100. Suzuma, I., Mandai, M., Suzuma, K., Ishida, K., Tojo, S.J. and Honda, Y. 1998. Contribution of E-selectin to cellular infiltration during endotoxin-induced uveitis. *Invest. Optholmol. Vis. Sci.* 39, 1620–1630.
101. Rosenbaum, J.T., McDevitt, H.O., Guss, R.B. and Egbert, P.R. 1980. Endotoxin-induced uveitis in rats as a model for human disease. *Nature* 286, 611–613.
102. Goureau, O., Bellot, J., Thillaye, B., Courtois, Y. and de Kozak, Y. 1995. Increased nitric oxide production in endotoxin-induced uveitis. Reduction of uveitis by an inhibitor of nitric oxide synthase. *J. Immunol.* 154, 6518–6523.
103. McMenamin, P.G. and Crewe, J.M. 1997. Cellular localisation and dynamics of nitric oxide synthase expression in the rat anterior segment during endotoxin-induced uveitis. *Exp. Eye Res.* 65, 157–164.
104. Mandai, M., Yoshimura, N., Yoshida, M., Iwaki, M. and Honda, Y. 1994. The role of nitric oxide synthase expression in the rat anterior segment during endotoxin-induced uveitis. *Exp. Eye Res.* 65, 157–164.
105. Bellot, J.L., Palmero, M., Garcia-Cabanes, C., Espi, R., Hariton, C. and Orts, A. 1996. Additive effect of nitric oxide and prostaglandin-E2 synthesis inhibitors in endotoxin induced uveitis in the rabbit. *Inflamm. Res.* 45, 203–208.

13 The Production and Health Benefits of Astaxanthin

Miguel Olaizola

CONTENTS

13.1 ASTAXANTHIN

Astaxanthin is a fat-soluble xanthophyll, an oxygenated derivative of carotenoids ($C_{40}H_{52}O_4$, FW 596.8 Da). It is closely related to carotenoids such as β-carotene, zeaxanthin, and lutein. Like other carotenoids, the astaxanthin molecule consists of a long chain of carbon atoms held together by conjugated double bonds, the chromophore. Thus, carotenoids absorb light resulting in colorful compounds of different shades of yellow, orange, and red. Two ionone rings, one at each end of the astaxanthin chromophore, modify its optical and chemical characteristics. Also, the molecular environment of astaxanthin (e.g., carotenoproteins in lobster carapace) affect the optical characteristics of the chromophore [1].

13.1.1 Chemistry of Astaxanthin

The conjugated double-bond chain that forms the backbone of the astaxanthin molecule indicates the possibility of geometrical isomers of astaxanthin (cis or trans). All *trans* carotenoid isomers are usually the more stable ones and thus the most abundant in nature. The presence of chiral centers at the 3 and 3′ carbon atoms on the rings allows for the presence of three different stereoisomers: 3S,3′S; 3S,3′R; and 3R,3′R (Figure 13.1). The 3S,3′S is the most common form found in nature [2].

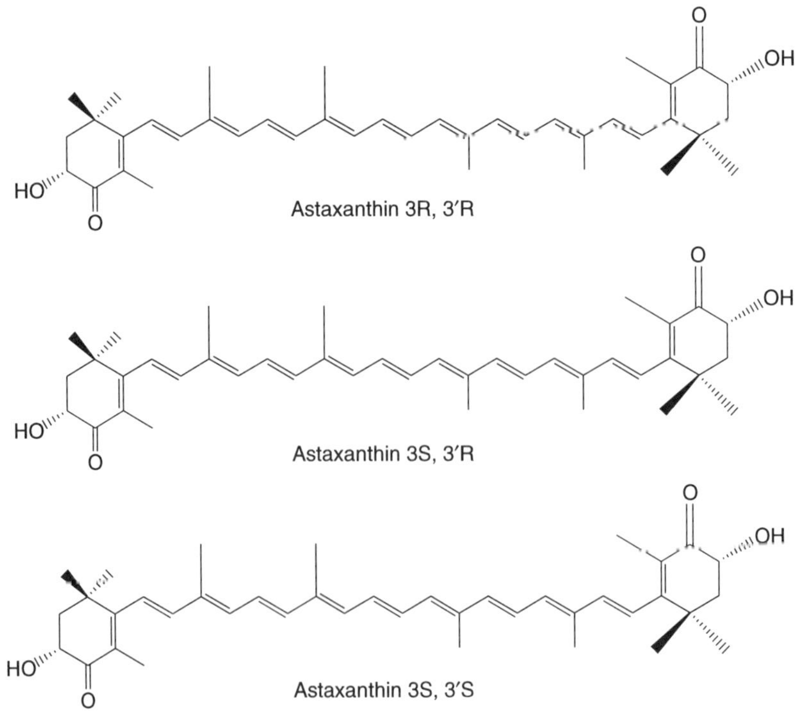

FIGURE 13.1 Astaxanthin stereoisomers.

The hydroxyl groups on the 3 and 3′ carbon atoms also give astaxanthin the ability to be esterified. Esterification of astaxanthin affects its absorbability and transport in the organism since it changes its polarity: free astaxanthin is more polar than astaxanthin monoester, which is more polar than astaxanthin diester. Astaxanthin can also be found bound to proteins; the change in the molecular environment causes a shift in the absorbance of astaxanthin such that the carotenoprotein complex may look green and blue instead of red or dark orange such as in lobsters [3].

Carotenoids are potent biological antioxidants, that quench singlet oxygen and other reactive species, by absorbing the excited energy of singlet oxygen onto the carotenoid chain, leading to the degradation of the carotenoid molecule, but preventing other molecules or tissues from being damaged. They also can prevent the chain reaction production of free radicals initiated by the degradation of polyunsaturated fatty acids, which can dramatically accelerate the degradation of lipid membranes [4,5]. Astaxanthin shows very good capability at protecting membranous phospholipids and other lipids against peroxidation [6,7]. Several studies have shown that astaxanthin has a stronger antioxidant activity than vitamin E (up to 500 times [8]) and other carotenoids [9].

Keeping in mind astaxanthin's molecular structure, lipophilic along the polyene chain but somewhat polar at the rings, is useful in understanding its transport into and availability to organisms as well as its location in the organism (see below).

13.1.2 Astaxanthin in Nature

Astaxanthin can be synthesized *de novo* only by some bacteria (such as *Agrobacterium aurantiacum*), fungi (such as *Xanthophyllomyces dendrorhous—Phaffia*), and microalgae (such as *Haematococcus pluvialis*). Animals obtain astaxanthin from their diets either by directly ingesting astaxanthin-rich prey or by converting other ingested carotenoids into astaxanthin. In aquatic environments, carotenoids obtained from microalgae are transformed into astaxanthin by zooplankton, which make it available to higher trophic levels [10]. At least one study has suggested that differences in nutrient and phytoplankton and zooplankton population dynamics may affect the antioxidant protection of higher trophic levels [11].

Thus, astaxanthin can be naturally found in widely consumed seafood such as salmon and marine invertebrates such as krill, lobster, and shrimp. Astaxanthin is also found in birds like flamingo, quail, and other species [12,13]. Free astaxanthin is particularly sensitive to oxidation. Consequently, it is usually found either conjugated to proteins, like in salmon muscle or lobster exoskeleton, or esterified with one or two fatty acids (monoester and diester forms), which stabilizes the molecule.

Of the three organisms listed as primary producers of astaxanthin, *Haematococcus* (Figure 13.2) is considered the most promising source of natural astaxanthin at industrial scale. In the case of *Agrobacterium*, astaxanthin is not even the most abundant carotenoid in the organism [14]. In the case of

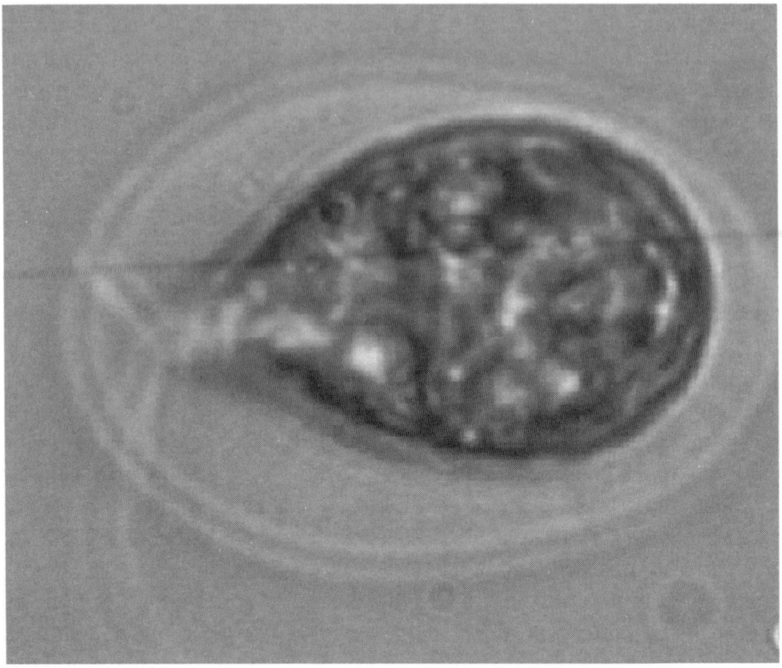

FIGURE 13.2 *Haematococcus pluvialis* cell in the flagellated state. The cell is about 10 μm long.

Phaffia, isolated strains accumulate only as much as 0.3% of the dry weight as astaxanthin and extraction of the pigment from the biomass is capital intensive [15]. Commercially produced *Haematococcus*, however, can accumulate upward of 3% of its dry weight as astaxanthin [16] (Figure 13.3). Astaxanthin can also be extracted from animals such as krill and crawfish, but the concentration in the raw material is low.

13.1.3 Astaxanthin Accumulation in *Haematococcus*

Haematococcus accumulates astaxanthin in oil droplets in the cytoplasm following environmental changes nonconducive to growth [17]. Other changes that accompany or precede astaxanthin accumulation include changes in the cell wall [18] and level of other enzyme-mediated defenses [19].

Over the past decades, starting with Droop [20], many authors have described the culture conditions that induce or accelerate the accumulation of astaxanthin by *Haematococcus* (e.g., light, nutrients, salt, acetate, temperature). Early results were interpreted to indicate that cessation of cell division was necessary for astaxanthin accumulation, as well as the production of reactive oxygen species (ROS) and accumulation of lipids (summarized in Ref. 21). It is now believed that synthesis and accumulation of astaxanthin from β-carotene involves the action

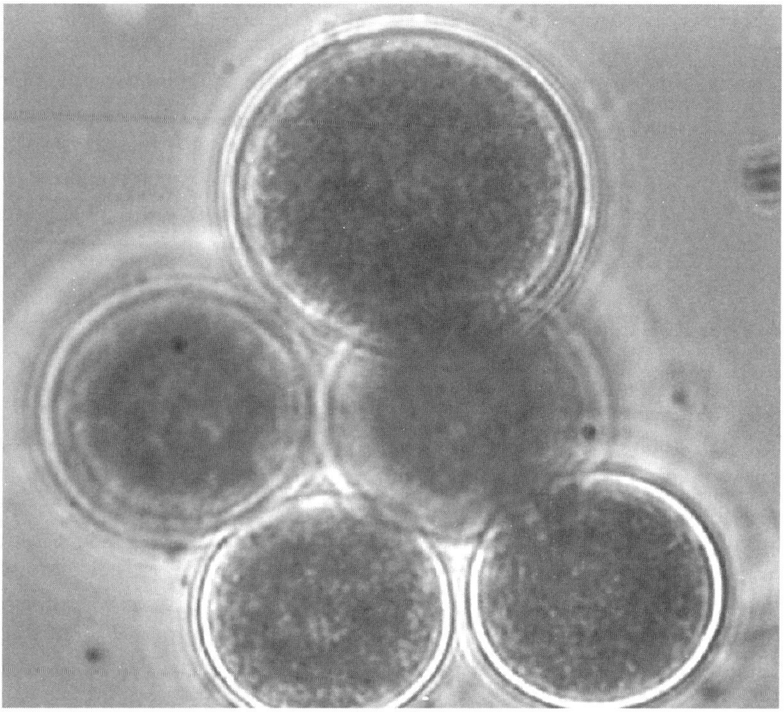

FIGURE 13.3 Encysted *Haematococcus pluvialis* cells. The large cell (top) is about 35 μm long.

of two enzymes, β-carotene ketolase (β-carotene → canthaxanthin) and carotene hydroxylase (canthaxanthin → astaxanthin). Interestingly, although β-carotene is manufactured in the chloroplast, astaxanthin accumulates in lipid vesicles in the cytoplasm [17]. It has been shown that β-carotene ketolase exists in these lipid vesicles supporting the suggestion that β-carotene (or other precursors) might be transported from the chloroplast to the lipid vesicles where the pigment is transformed into astaxanthin [22]. Similarly, it has been shown that fatty acid synthesis and lipid accumulation accompanies, and may be required for, astaxanthin esters accumulation in *Haematococcus* [23,24]. Recently, it has been determined that carotenoid and astaxanthin biosynthetic genes up-regulate during induction following environmental stresses such as high light. It was found that increased levels of transcription correlated with the redox state of the photosynthetic electron transport system [25].

As mentioned above, *Haematococcus* accumulates astaxanthin as part of a suite of responses to environmental insult, possibly translated in the cell as a change in the redox state of the photosynthetic electron transport system. What is, then, the function of the astaxanthin in the cytoplasm? Droop [20] suggested a possible role for astaxanthin as a storage material following observations on the accumulation and disappearance of astaxanthin during reddening and greening.

Evidence has also been presented to support a light screening function [26]. It appears that astaxanthin also has an antioxidant function [27]. This would explain why astaxanthin seems to protect *Haematococcus* lipids against peroxidation *in vivo* [28]. It has also been suggested that astaxanthin accumulation in globules surrounding the nucleus may also indicate a physicochemical barrier to oxidative damage in nuclear material [26,28]. However, it has also been suggested that astaxanthin *per se* is itself the result of a photoprotective process, not the protective agent [29].

13.2 PRODUCTION OF ASTAXANTHIN FROM *HAEMATOCOCCUS PLUVIALIS*

Although *Haematococcus* may be induced to accumulate astaxanthin indoors under heterotrophic and mixotrophic growth conditions, successful commercial producers take advantage of abundant sunlight outdoors. There are a handful of commercial producers of *Haematococcus* astaxanthin today that have developed the strategies necessary to produce large quantities of biomass, induce astaxanthin accumulation, harvest and process the reddened biomass, and formulate the final product [30]. The production strategies must take into consideration the physiological characteristics of *Haematococcus* at the different stages in the production process.

13.2.1 Biomass Production

The culture conditions that favor *Haematococcus* biomass production are incompatible with the stresses needed to induce astaxanthin accumulation (described earlier). Thus, the strategy to produce commercial quantities of astaxanthin-rich biomass is two-phased. The objective of the first phase is to produce large quantities of green biomass under conditions (nutrient sufficiency, and light, pH, and temperature control) conducive to fast growth. Not surprisingly, those growth conditions are amenable to a number of other microalgae that can contaminate and overtake the culture. Thus, commercial production of *Haematococcus* has been dependent on the development of large-scale enclosed photobioreactors (see e.g., Figure 13.4; see also Ref. 30).

13.2.2 Accumulation of Astaxanthin

Haematococcus accumulates astaxanthin following changes in the culture conditions from growth-promoting to stress promoting (see Section 13.1.3). One can envision that the easiest way to accomplish this is to significantly dilute the green culture. Dilution accomplishes two things immediately. On the one hand, the concentration of nutrients decreases, inducing nutrient stress. At the same time, the average irradiance per cell in culture increases significantly. This plus the addition of other induction factors (e.g., acetate, Fe, salt, temperature, etc.) creates the proper environment for quick accumulation of astaxanthin by *Haematococcus*.

FIGURE 13.4 The Mera growth module, an enclosed photobioreactor used for the production of *Haematococcus pluvialis* biomass.

While some producers use enclosed photobioreactors for the astaxanthin accumulation phase, open pond systems can also be used (Figure 13.5). Open ponds are quite adequate for the carotenogenesis phase. First, culture conditions in reddening cultures are very poor for growth (designed to induce astaxanthin accumulation in *Haematococcus*) even for contaminant microalgae. Second, under these conditions, *Haematococcus* develops into fully reddened cysts within 5–6 days; too short for microalgal contaminants to affect the system.

13.2.3 Harvesting

Besides accumulating astaxanthin, the reddening *Haematococcus* cells lose their flagella and become significantly larger (up to 50 μm versus 10–15 μm for the green flagellates) and heavier [31]. Since the cells are denser than the medium, harvesting can be easily accomplished by gravitational settling and with the use of centrifuges [30].

13.2.4 Biomass Processing

A further characteristic of the reddening *Haematococcus* cells is the formation of a very tough cell wall designed to protect the cell from further environmental insult. The wall resists digestion by animals (including humans) and solvent extraction [32]. Thus, the next step in astaxanthin production is the breaking up of the cell wall.

FIGURE 13.5 Open ponds used by Mera Pharmaceuticals for reddening *Haematococcus pluvialis* biomass.

Ball mills can be used to rupture *Haematococcus* cell walls as well as high-pressure homogenizers at commercial scale [30]. Our own experience has shown that solvent and enzyme pretreatments may help in this process (unpublished data; see also Ref. 32). However, depending on the application, such chemical aids may not be advisable if residues may be found in the final product.

After the cells are disrupted, the water content of the biomass is carefully reduced. Since the astaxanthin molecules are no longer protected by the cyst structure, they are prone to degradation. Thus, the mass of broken cells must be protected from exposure to oxygen, light, and high temperature to the extent possible. Standard spray driers, that reach temperatures near 200°C can reduce the yield of astaxanthin significantly (20–40%, unpublished observation). Commercial producers have developed technologies, which limit the exposure of the biomass to oxygen and use lower temperatures [30].

13.2.5 End Products and Formulation

For animal husbandry applications, the dried biomass can be directly incorporated into animal diets. For human nutraceutical applications, the astaxanthin is usually extracted from the biomass using nonpetrochemical solvents such as edible oils or supercritical CO_2. The extract obtained is then diluted to the desired concentration with other oils or ingredients. Finally, the astaxanthin extract is encapsulated according to the desired dosage, usually in gelatin capsules or beadlets.

13.3 ASTAXANTHIN APPLICATIONS IN ANIMAL HUSBANDRY

In the organisms in which it is found, astaxanthin has several physiological functions from coloration (communication, sexual selection) to reproductive and developmental success (see below). These functions may be related to astaxanthin's function as an antioxidant (see above). Normal metabolism produces free radicals (e.g., hydroxyl and peroxyl radicals) and ROS (e.g., singlet oxygen). Physiological and environmental stresses can enhance the production of these agents. These reactive agents can themselves damage DNA, proteins, and lipid membranes. As will be seen below, the presence of astaxanthin in animal tissues protects the organism from these chemical insults.

Astaxanthin, thus, has two types of applications in animal husbandry: as a source of natural color and as a nutritional supplement because of its physiological effects.

13.3.1 COLORATION

Appearance (e.g., coloration) of farmed animal products affects consumer acceptability. As was noted above, animals cannot synthesize astaxanthin, they must obtain it from their diet. In the case of farmed animals, the pigment must be included in the feed since these animals do not have much opportunity to forage. The market for feed pigments is estimated at several hundred million US dollars.

Salmonid feed (salmon and trout) is probably the largest market for astaxanthin at this time. Today, most salmon is farmed and is fed artificial diets. To maintain coloration of the fish, the diets must incorporate carotenoids such as astaxanthin [33]. Red sea bream (red porgy) is another valuable fish appreciated for its coloration. Astaxanthin must be added to the diet, otherwise the captive fish have only a fraction of the astaxanthin of wild fish [34]. Astaxanthin is also effective in maintaining the coloration of ornamental fishes [35].

Shrimp and prawns also benefit from astaxanthin supplementation [36]. The pigment deposits in the exoskeleton (carotenoproteins) and turns red during cooking, giving the shrimp a consumer-desired orange or red hue. Other animals, such as chickens, also benefit from astaxanthin in their diet (yolk, muscle, and skin coloration [13]).

13.3.2 PHYSIOLOGICAL BENEFITS

The benefits of astaxanthin supplementation in fish include improved survivability, growth, photoprotection, fertility, egg quality, protection from cataracts, and the function of a provitamin [37–40]. There are also some indications that astaxanthin supplementation is beneficial for growth and survival of shrimp as well as resistance to environmental stresses [41].

There is little information available on the physiological effects of astaxanthin supplementation on chickens, but it seems that it may positively affect growth performance and offer resistance to certain bacterial infections [42]. Furthermore, positive effects of astaxanthin on immune response mechanisms in aquatic invertebrates have been demonstrated [43].

Continuing research efforts are resulting in the development of new applications for astaxanthin in animal husbandry. According to AstaReal (formerly Astacarotene, http:/www.astareal.com/), astaxanthin supplementation increases the fertility of horses, pigs, and fur animals (see also Ref. 44). According to Ito et al. [45], astaxanthin may enhance the effect of antistress agents on farm animals.

13.4 ASTAXANTHIN APPLICATIONS IN HUMAN HEALTH

The results of recent survey among users of a commercial astaxanthin supplement indicated that an improvement in 85% of health conditions with an inflammation component reportedly resulted from *Haematococcus* astaxanthin supplementation. Significantly, users reported that astaxanthin was as effective as, or more effective than, either prescription or over-the-counter anti-inflammatory drugs [46]. There is a dearth of human clinical data that could corroborate the experiences reported by those users of *Haematococcus* astaxanthin. However, in the following sections we review evidence, mainly from *in vitro* and animal models that support the conclusions reached by the survey results.

13.4.1 SAFETY AND BIOAVAILABILITY

It can be assumed that astaxanthin has been part of the human diet since people started consuming seafood. Still, its use as a supplement is recent. Mera Pharmaceuticals conducted the first human safety study with *Haematococcus* astaxanthin that we are aware of [47]. In this study, human subjects were given two different astaxanthin daily doses, 3.85 or 19.25 mg. The results of complete medical examinations before, during, and after the trial, including extensive urine and blood analysis, found no ill effects or toxicity from ingestion of the supplement. A more recent safety study [48] corroborated the safety of *Haematococcus* astaxanthin for humans. Other studies on animal models (summarized in Ref. 47) support the conclusion that *Haematococcus* astaxanthin consumption does not appear to possess health risks.

The steps to accomplish dietary carotenoid assimilation include the destruction of the food matrix, solubilization and emulsification of the carotenoids with other lipids, formation of micelles with bile salts, deesterification, absorption through the intestinal mucosa, loading onto lipoproteins, and transfer to chylomicrons (reviewed in Ref. 49). Uptake studies on humans have shown that the largest fraction of astaxanthin in the serum is associated with the very low-density lipoprotein fraction (VLDL) and that there appears to be selective enrichment during uptake [50,51]. More recent work has shown that presence of fats and emulsifiers may significantly enhance the bioavailability of astaxanthin supplements [52].

Once in the blood stream, astaxanthin may be transported throughout the organism, including the nervous system since it has been shown to cross the blood–brain barrier [53].

13.4.2 Antioxidant Activity of Astaxanthin

Astaxanthin is a very potent antioxidant. Several studies have shown that astaxanthin's antioxidant activity can be several folds higher than those of, for example, β-carotene and vitamin E [8,54], and that it is very capable of protecting membrane phospholipids and other lipids against peroxidation [55–59]. One reason for astaxanthin's superiority in protecting cellular membranes against oxidation may derive from its polyene chain and terminal rings, which allow the molecule to protect both the inner and outer membrane surfaces [60] and its effect on membrane permeability [61]. It is believed that astaxanthin's antioxidant activity may play a significant role in conditions that may be either triggered or aggravated by oxidative damage such as UV-light damage, cancer, inflammation, and ulcerous infections. The powerful antioxidant activity may also play a role in astaxanthin's enhancement of immune responses, liver function and eye, joint, prostate, and heart health (see below).

13.4.3 Photoprotective Properties of Astaxanthin

The photoprotective activity of astaxanthin appears to be directly related to its antioxidant properties. UV light can lead to production of singlet oxygen and free radicals and photooxidative damage of lipids and tissues [62,63]. In nature, carotenoids are often found in tissues directly exposed to light. O'Connor and O'Brien [64] demonstrated that astaxanthin could be more effective than β-carotene and lutein at preventing UV-light photooxidation of lipids by a factor of up to 200 and 1000 folds, respectively. In humans, the eyes and skin are the organs with highest exposure to UV light, and are the organs where astaxanthin photoprotection might prove to be most beneficial.

Age-related macular degeneration (AMD) and age-related cataracts are two leading causes of visual impairment that appear to be related to light-induced oxidative damage in the eye. In the case of cataracts, free radicals impair the crystalline proteins in the lens and damage the enzymes that remove damaged proteins [65] while in the case of AMD, peroxidation of membranes likely leads to the death of photoreceptor cells in the retina [66]. Perhaps not surprisingly, high intake of dietary carotenoids has been associated with reduced risk for cataracts and AMD [67,68]. While astaxanthin has not been isolated in the human eye, it is found in the eye or eye parts of a number of animals [12]. Furthermore, an animal study has demonstrated that astaxanthin is capable of crossing the blood–brain barrier, depositing in the retina of mammals, and offering protection against UV-light photodamage if included in the diet [53]. Furthermore, astaxanthin was shown to protect human lens epithelial cells against ultraviolet-b radiation (UVB) insult [69].

Skin exposure to sunlight may result in light-induced oxidation, inflammation, immunosuppression, aging, and even carcinogenesis of skin cells. Several studies have shown that dietary antioxidants can reduce photo-induced damages [70–72]. As was mentioned earlier, astaxanthin is believed to protect against photooxidation in those tissues in which it is found, such as the skin of salmon and red seabream, salmon eggs, and *Haematococcus* [33,37] and was found to protect retinal photoprotectors against UV-light injury in rats [53]. Considering that astaxanthin has a stronger *in vitro* protective effect against UV-induced photooxidation than β-carotene and lutein [64] it seems reasonable that astaxanthin has an excellent potential as an oral sun protectant as well as to treat other skin conditions [73].

13.4.4 Anti-Inflammatory Properties of Astaxanthin

Another condition associated with oxidative damage is inflammation. ROS may aggravate the inflammation that accompanies various forms of asthma [74,75], exercise-induced muscle damage [76], and ulcers [77]. Inflammation has also been recently implicated in the onset of AMD [78]. Also, it has been recently hypothesized that decreased inflammation early in life may have led to increases in human life span [79]. Furthermore, inflammation has been implicated in the progression of cancers (see summary by Marx [80]).

Kurashige et al. [8] were the first to identify an anti-inflammatory role for astaxanthin. They injected an inflammation agent in the paw of experimental rats, fed either natural astaxanthin, vitamin E, or no antioxidant (control treatment), and measured swelling of the paw. Only the astaxanthin-fed rats showed a significant decrease in inflammation when compared to the control. A more recent rat study [81] showed a dose-dependent anti-inflammatory effect of astaxanthin by the suppression of nitric oxide, prostaglandin E2, and tumor necrosis factor. Studies on peripheral blood mononuclear cells from asthmatics have shown that astaxanthin can be as effective as commonly used antihistamines at suppressing T-cell activation [82]. The authors concluded that astaxanthin might have a role in novel antiasthmatic formulations.

Dietary astaxanthin also helps fight symptoms of ulcer disease from *Helicobacter pylori*, which among other effects leads to inflammation of gastric tissues [77]. Bennedsen et al. [83] have shown that astaxanthin supplementation in the diet of mice is able to reduce symptoms of gastric inflammation and is associated with shifts in the inflammation response (change in the activity of T-lymphocytes), consistent with the reduced symptoms. A more recent study [84] found that astaxanthin can protect against gastric lesions (ulcers) induced by the use of nonsteroid anti-inflammatory drugs such as naxopren. Finally, a rat study found that astaxanthin supplementation accumulated in heart and skeletal muscle and that it reduced the amount of exercise-induced inflammation markers [85].

13.4.5 Immunomodulating Properties of Astaxanthin

A recent review on the action of carotenoids in general on the immune response noted that carotenoids without vitamin A activity, such as astaxanthin, can be as active (an even more active) than β-carotene [86]. Specifically, astaxanthin has been shown to significantly influence immune function in a number of *in vitro* and *in vivo* assays using animal models. A significant amount of this work has been carried out by Jyonouchi and colleagues at the University of Minnesota. For example, they have found that astaxanthin influences antibody production by mouse spleen cells stimulated with sheep red blood cells [87] and can partially restore decreased humoral immune responses in old mice [88]. Studies on human blood cells *in vitro* demonstrated that astaxanthin can enhance immunoglobulin production [89].

Astaxanthin has also been found effective in preventing the development of symptoms in autoimmune-prone mice [90]. It has also been suggested that astaxanthin's immunomodulating functions may be related to its antitumor activity [91]. The immunomodulating capacity of astaxanthin has been found to be superior to that of β-carotene and canthaxanthin [92].

13.4.6 Astaxanthin and Cardiovascular Health

It has been recognized for some time that oxidative damage plays a central role in the development of degenerative diseases, including cardiovascular disease (CVD), one of the major causes of morbidity and mortality in Western society [93,94].

Atherosclerosis is the principal cause of CVD (and cerebrovascular disease). Initially, atherosclerosis consists of fatty streaks containing "foam" cells, which develop into fibrous plaques. As other cells, cell debris, lipids, cholesterol, and calcium accumulate, plaque deposits are formed, which can ultimately limit or block blood flow. The whole chain of events is triggered by uptake of oxidized low-density lipoprotein (LDL, the "bad" cholesterol) by macrophages [95]. In principle, protection from oxidation by LDL could prevent the trigger that results in atherosclerosis.

Astaxanthin is carried by VLDL, LDL, and HDL (high-density lipoprotein) in human blood. A study in which human subjects consumed daily dosages as low as 3.6 mg astaxanthin per day for two consecutive weeks demonstrated that astaxanthin protects LDL-cholesterol against oxidation [96], suggesting that astaxanthin supplementation may lower the risk of CVD. Astaxanthin has also been found to decrease the degree of macrophage infiltration of atherosclerotic plaques as well as apoptosis in an animal model [97], which resulted in plaque stability.

Blood levels of HDL (the "good" cholesterol) are indicative of protection against atherosclerosis [98]. According to one study, astaxanthin supplementation was correlated with an increase in HDL levels [99], suggesting that astaxanthin may also offer protection from CVD by modifying the ratio of LDL and HDL.

Astaxanthin can also protect the heart directly in mammals. First, it accumulates in the heart muscle and, following exercise, astaxanthin appears to lower the level of exercise-induced oxidative damage to lipids and DNA in heart tissue [85]. Second, two studies showed that astaxanthin reduced infarct size following an induced heart attack [100,101]. Thus, astaxanthin supplementation may be beneficial for those people with enhanced risk for heart attacks. Finally, in at least one animal model, it was shown that astaxanthin might have a role in the reduction of high blood pressure and the delay in the incidence of stroke [102].

Astaxanthin may also be beneficial to cardiovascular health by reducing the inflammation (see above) associated with the development of coronary heart disease. C-reactive proteins (CRP) are nonglycosylated proteins produced by hepatocytes and normally present only in trace amounts in the blood [103]. In the acute phase response to inflammatory stimuli such as infection, tissue injury, damage to blood vessels, strenuous exercise, and lung irritants, the blood levels of CRP increase dramatically. Elevated blood levels of CRP are believed to indicate "smoldering" low-grade inflammation in arteries of patients without known coronary risk factors [104–106].

13.4.7 Astaxanthin and Neurodegenerative Diseases

Intense metabolic activity, high levels of unsaturated fats and iron, and rich irrigation with blood vessels make the nervous system (brain, spinal cord, and peripheral nerves) very susceptible to oxidative damage [107,108]. Oxidative stress is also a suspected factor in the pathogenesis of major neurodegenerative diseases. These include Parkinson's disease [109–111], Alzheimer's disease [112,113], and amyotrophic lateral sclerosis (ALS, "Lou Gehrig's disease") [114–116] as well as in cases of stroke, trauma, and seizures [108].

It has been suggested that Alzheimer's disease may be linked to diet, with reduced risk associated with diets high in antioxidants [117,118]. A number of *in vitro* studies have shown that dietary antioxidants, vitamins, and carotenoids can protect nervous tissue from damage by oxidative stress [119–121]. In an experimental model of diabetes-caused neurovascular dysfunction, β-carotene was found to protect cells most effectively, followed by vitamins E and C [119]. A number of *in vivo* and clinical studies have provided evidence that dietary supplementation with lipid-soluble antioxidants can help fight neurological diseases.

One report demonstrated that the rate at which Parkinson's disease progressed to the point when the patient required treatment with levodopa was slowed by 2.5 years in patients given large doses of vitamin C and synthetic vitamin E [122]. Another study found that the risk for Parkinson's disease was lower for subjects who had higher dietary intakes of antioxidants, particularly vitamin E [123]. The same group reported that a low dietary intake of β-carotene was associated with impaired cognitive function in a group of people aged 55–95. Further, it was found that patients suffering from Parkinson's disease had consumed less of the small-molecule antioxidants β-carotene and vitamin C than did nonsufferers of the disease, implying that dietary antioxidants do play a protective role in this disease [124].

About 20% of familial ALS cases are associated with a mutation in the gene for copper/zinc superoxide dismutase, an important antioxidant enzyme, and *in vitro* experiments demonstrated that expression of the mutant enzyme in neuronal cells caused cell death, which could be prevented by antioxidant small molecules such as glutathione and vitamin E [125].

The above-mentioned study with rats fed natural astaxanthin [53] demonstrated that astaxanthin is able to cross the blood–brain barrier in mammals and can extend its antioxidant benefits beyond that barrier. Astaxanthin, which has a stronger antioxidant activity than β-carotene or vitamin E, is therefore an excellent candidate for testing in Alzheimer's disease and other neurological diseases. To our knowledge, no clinical study involving astaxanthin and Alzheimer's has been reported at this time. However, a recent mouse study has shown what appears to be a neuroprotective effect of high astaxanthin doses [102]. Mice were subjected to transient cerebral ischemia (two vessel occlusion, 20 min). Those mice that had been pretreated with astaxanthin (1 h before the ischemia) performed significantly better in the Morris water maze. The authors concluded that the effect might have been caused by the antioxidant activity of astaxanthin on ischemia-induced free radicals, which resulted in the reduced impairment of spatial memory in the mice.

13.4.8 Anticancer Activity of Astaxanthin

In the mid-1990s, several animal studies demonstrated that dietary astaxanthin could protect mammals from cancer following exposure to known carcinogens (e.g., urinary bladder, colon, and mouth [126–128]). Later, it was shown that dietary astaxanthin could reduce the growth of mammary tumors [129].

Oxidative damage of tissues may lead to cancerous growth. Thus, one can expect that enhancing the levels of antioxidants in body tissues may lead to protection from cancer. It does appear, however, that the anticancer activity of astaxanthin goes beyond its function as an antioxidant (see above).

It is, however, not clear what the mode of action of astaxanthin is. Jyonouchi et al. [91] found that astaxanthin fed mice had enhanced immunological responses against transplanted fibrosarcoma cells resulting in smaller tumors. Kurihara et al. [130] suggested that astaxanthin improves immunological antitumor responses by inhibiting stress-induced lipid peroxidation. However, it has been shown that astaxanthin may offer protection from prostate cancer by inhibiting the enzyme responsible for prostate growth [131]. Astaxanthin has also been shown to induce xenobiotic-metabolizing enzymes in rat liver, a process that may help prevent carcinogenesis [132]. Similarly, Jewell and O'Brien [133] demonstrated that astaxanthin was also able to induce xenobiotic metabolizing enzymes in the lung and kidney. It has also been suggested that carotenoids' role in cell communications at cell-gap junctions (by up-regulating gene expression) may play a role in slowing the growth of cancer cells [134].

For an exhaustive review of investigations into the anticancer activity of astaxanthin in cell and animal models as well as the possible modes of action, the reader is directed to consult Dore [135].

13.5 ASTAXANTHIN MARKETS AND FUTURE OF ASTAXANTHIN

As can be deduced from the material reviewed above, there are two main markets, at the present time, for *Haematoccocus* astaxanthin: as a feed additive and as a human nutraceutical. Astaxanthin has applications in feeds for animals for which color appearance is important for consumer acceptability. The largest consumer of astaxanthin today is the salmon feed industry. However, the largest producers of astaxanthin in the world market today produce astaxanthin via chemical synthesis (BASF, Hoffman-La Roche) at significantly lower costs than is possible using *Haematococcus* at present [30]. In some applications (koi, chickens, sea bream), *Haematococcus* astaxanthin commands a premium price over synthetic astaxanthin because of enhanced pigment deposition or regulatory requirements. It can be expected that as the public at large becomes educated to the fact that most salmon available today is fed an artificial pigment, desire for a naturally pigmented product will increase the demand, and price, of *Haematococcus* astaxanthin. Thus, for *Haematococcus* astaxanthin to compete in the feed market, two things must happen. First, production costs need to be lowered: by improving the production technology (higher productivity, higher final astaxanthin content of the biomass) and expanding production to locales with lower costs. Second, astaxanthin producers need to ensure that the public becomes educated regarding the source of pigmentation in food and that they demand naturally pigmented food products.

In this review, we have made the point that *Haematococcus* astaxanthin supplementation is a practical strategy in protecting the body from oxidative damage, which has repercussions in a number of health conditions. Thus, a market for nutraceutical astaxanthin has emerged. This market is very attractive to *Haematococcus* astaxanthin producers since the price obtainable is significantly higher than in feed applications. The size of the nutraceutical astaxanthin market is expected to reach a size of several hundred million US dollars within a decade [30]. To realize this potential, however, *Haematoccocus* producers and marketers need to invest in human clinical trials that will demonstrate the presumed benefits of astaxanthin supplementation.

REFERENCES

1. Britton, G. 1995. UV/visible spectroscopy, in *Carotenoids Volume 1B: Spectroscopy*, Britton, G., Liaaen-Jensen, S., and Pfander, H., Eds., Birkhäuser Verlag, Basel, Chapter 2.
2. Weedon, B.C.L. and Moss, G.P. 1995. Structure and nomenclature, in *Carotenoids Volume 1A: Isolation and Analysis*, Britton, G., Liaaen-Jensen, S., and Pfander, H., Eds., Birkhäuser Verlag, Basel, Chapter 3.
3. Tlusty, M. and Hyland, C. 2005. Astyaxanthin deposition in the cuticle of juvenile American lobster (*Homarus americanus*): implications for phenotypic and genotypic coloration. *Mar. Biol.* 147, 113.

4. Tinkler, J.H., Böhm, F., Schalch, W. and Truscott, T.G. 1994. Dietary carotenoids protect human cells from damage. *J. Photochem. Photobiol.* 26, 283.
5. Mortensen, A., Skibsted, L.H., Sampson, J., Rice-Evans, C. and Everett, S.A. 1997. Comparative mechanisms and rates of free radical scavenging by carotenoid antioxidants. *FEBS Lett.* 418, 91.
6. Lim, B., Nagao, A., Terao, J., Tanaka, K., Suzuki, T. and Takama, K. 1992. Antioxidant activity of xanthophylls on peroxyl radical-mediated phospholipid peroxidation. *Biochim. Biophys. Acta* 1126, 178.
7. Naguib, Y.M.A. 2000. Antioxidant acitivities of astaxanthin and related carotenoids. *J. Agric. Food Chem.* 48, 1158.
8. Kurashige, M., Okimasu, E., Inoue, M. and Utsumi, K. 1990. Inhibition of oxidative injury of biological membranes by astaxanthin. *Physiol. Chem. Phys. Med. NMR*, 22, 27.
9. Shimidzu, N., Goto, M. and Miki, W. 1996. Carotenoids as singlet oxygen quenchers in marine organisms. *Fish. Sci.* 62, 134.
10. Andersson, M., Nieuwerburgh, L.V. and Snoeijs, P. 2003. Pigment transfer from phytoplankton to zooplankton with emphasis on astaxanthin production in the Baltic Sea food web. *Mar. Ecol. Progr. Ser.* 254, 213.
11. Nieuwerburgh, L.V., Wänstrand, I., Liu, J. and Snoeijs, P. 2005. Astaxanthin production in marine pelagic copepeods grazing on two different phytoplankton diets. *J. Sea Res.* 53, 147.
12. Egeland E.S. 1993. Carotenoids in combs of capercaillie (*Tetrao urogallus*) fed defined diets, *Poult. Sci.* 72, 747.
13. Inborr, J. 1998. *Haematococcus* the poultry pigmentor. *Feed Mix* 6, 31.
14. Yokoyama, A. and Miki, W. 1995. Composition and presumed biosynthetic pathway of carotenoids in the astaxanthin-producing bacterium *Agrobacterium aurantiacum*. *FEMS Microbiol. Lett.* 128, 139.
15. An, G.-H. and Choi, E.-S. 2003. Preparation of the red yeast, *Xanthophyllomyces dendrorhous*, as feed additive with increased availability of astaxanthin. *Biotech. Lett.* 25, 767.
16. Olaizola, M. 2000. Commercial production of astaxanthin from *Haematococcus pluvialis* using 25,000-liter outdoor photobioreactors. *J. Appl. Phycol.* 12, 499.
17. Santos, M.F. and Mesquita, J.F. 1984. Ultrastructural study of *Haematococcus lacustris* (Girod.) Rostafinski (Volvocales) I. Some aspects of carotenogenesis. *Cytologia* 49, 215.
18. Wang, S.B., Hu, Q., Sommerfeld, M. and Chen, F. 2004. Cell wall proteomics of the green alga *Haematococcus pluvialis* (Chlorophyceae). *Proteomics* 4, 692.
19. Wang, S.B., Chen, F., Sommerfeld, M. and Hu, Q. 2004. Proteomic analysis of molecular response to oxidative stress by the green alga *Haematococcus pluvialis* (Chlorophyceae). *Planta* 220, 17.
20. Droop, M.R. 1954. Conditions governing haematochrome formation and loss in the alga *Haematocuccus pluvialis* flotow. *Arch. Mikrobiol.* 20, 391.
21. Boussiba, S. 2000. Carotenogenesis in the green alga *Haematococcus pluvialis*: cellular physiology and stress response. *Physiologia Plantarum* 108, 111.
22. Grünewald, K. and Hagen, C. 2001. β-carotene is the intermediate exported from the chloroplast during accumulation of secondary carotenoids in *Haematococcus pluvialis*. *J. Appl. Phycol.* 13, 69.
23. Schoefs, B., Rmiki, N.-E., Rachadi, J. and Lemoine, Y. 2001. Astaxanthin accumulation in *Haematococcus* requires a cytochrome P450 hydroxylase and an active synthesis of fatty acids. *FEBS Lett.* 500, 125.

24. Zhekisheva, M., Boussiba, S., Khozin-Goldberg, I., Zarka, A. and Cohen, Z. 2002. Accumulation of oleic acid in *Haematococcus pluvialis* (Chlorophyceae) under nitrogen starvation or high light stress is correlated with that of astaxanthin esters. *J. Phycol.* 38, 325.
25. Steinbrenner, J. and Linden, H. 2003. Light induction of carotenoid biosynthesis genes in the green alga *Haematococcus pluvialis*: regulation by photosynthetic redox control. *Plant Mol. Biol.* 52, 343.
26. Wang, B., Zarka, A., Trebts, A. and Boussiba, S. 2003. Astaxanthin accumulation in *Haematococcus pluvialis* (Chlorophyceae) as an active photoprotective process under high irradiance. *J. Phycol.* 39, 1116.
27. Kobayashi, M. 2000. *In vivo* antioxidant role of astaxanthin under oxidative stress in the green alga *Haematococcus pluvialis. Appl. Microbiol. Biotechnol.* 54, 550.
28. Hagen, C., Braune, W. and Greulich, F. 1993. Functional aspects of secondary carotenoids in *Haematococcus lacustris* [Girod] Rostafinski (Volvocales) IV. Protection from photodynamic damage. *J. Photochem. Photobiol. B: Biol.* 20, 153.
29. Fan, L., Vonshak, A., Zarka, A. and Boussiba, S. 1998. Does astaxanthin protect *Haematococcus* against light damage? *Z. Naturforsch.* 53c, 93.
30. Olaizola, M. and Huntley, M.E. 2003. Recent advances in commercial production of astaxanthin from microalgae, in *Recent Advances in Marine Biotechnology. Volume 9. Biomaterials and Bioprocessing*, Fingerman, M. and Nagabhushaman, R., Eds., Science Publishers, New Hampshire, 143.
31. Kobayashi, M., Katsuragi, T. and Tani, Y. 2001. Enlarged and astaxanthin-accumulating cyst cells of the green alga *Haematococcus pluvialis. J. Biosc. Bioeng.* 92, 565.
32. Mendes-Pinto, M.M., Raposo, M.F.J., Bowen, J., Young, A.J. and Morais, R. 2001. Evaluation of different cell disruption processes on encysted cells of *Haematococcus pluvialis*: effects on astaxanthin recovery and implications for bioavailability. *J. Appl. Phycol.* 13, 19.
33. Torissen O.J., Hardy, R.W. and Shearer, K. 1989. Pigmentation of salmonids—carotenoid deposition and metabolism. *CRC Cr. Rev. Aq. Sci.* 1, 209.
34. Kalinowski, C.T., Robaina, L.E., Fernández-Palacios, H., Schuchardt, D. and Izquierdo, M.S. 2005. Effect of different carotenoid sources and their dietary levels on red porgy (*Pagrus pagrus*) growth and skin colour. *Aquaculture* 244, 223.
35. Gouveia, L., Rema, P., Pereira, O. and Empis, J. 2003. Colouring ornamental fish (*Cyprinus carpio* and *Carassius auratus*) with microalgal biomass. *Aquaculture Nut.* 9, 123.
36. Yamada, S., Tanaka, Y., Sameshima, M. and Ito, Y. 1990. Pigmentation of prawn (*Penaeus japonicus*) with carotenoids. I. Effect of dietary astaxanthin, beta-carotene and canthaxanthin on pigmentation. *Aquaculture* 87, 323.
37. Meyers S.P. 1993. The biological role of astaxanthin in salmonids and other aquatic species, presented at First International Symposium of Natural Colors for Foods, Nutraceuticals, Beverages and Confectionary. Amherst, November 7–10.
38. Christiansen, R., Lie, O. and Torissen, O.J. 1995. Growth and survival of Atlantic salmon, *Salmo salar* L., fed different dietary levels of astaxanthin. First feeding fry. *Aquaculture Nutr.* 1, 189.
39. Vasallo-Agius, R., Imaizumi, H., Watanabe, T., Yamazaki, T., Satoh, S. and Kiron, V. 2001. The influence of astaxanthin supplemented dry pellets on spawning of striped jack. *Fish. Sci.* 67, 260.
40. Waagbø, R., Hamrel, K., Bjerkås, E., Berge, R., Wathne, E., Lie, Ø. and Torstensen, B. 2003. Cataract formation in Atlantic salmon, *Salmo salar* L., smolt relative to dietary pro- and antioxidants and lipid level. *J. Fish Dis.* 26, 213.

41. Chien, Y.-H., Pan, C.H. and Hunter, B. 2003. The resistance to physical stress by *Penaeus monodon* fed diets supplemented with astaxanthin. *Aquaculture* 216, 177.
42. Waldenstedt, L., Inborr, J., Hanhsson, I. and Elwinger, K. 2003. Effects of astaxanthin-rich algal meal (*Haematococcus pluvialis*) on growth performance, caecal campylobacter and clostridial counts and tissue astaxanthin concentration of broiler chickens. *A. Feed Sci. Technol.* 108, 119.
43. Kawakami, T., Tsushima, M., Katabami, Y., Mine, M., Ishida, A. and Matsuno, T. 1998. Effect of B,B-carotene, B-echinenone, astaxanthin, fucoxanthin, vitamin A and vitamin E on the biological defense of sea urchin *Pseudocentrotus depressus. J. Exp. Mar. Biol. Ecol.* 226, 165.
44. Hansen, K.B., Tauson, A.H. and Inborr, J. 2001. Effect of supplementation with the antioxidant astaxanthin on reproduction, pre-weaning growth performance of kits and daily milk intake in mink. *J. Reprod. Fertil. Suppl.* 57, 331.
45. Ito, S., Ogata, E. and Yamada, M. 2003. United States Patent #5,937,790, 1999.
46. Guerin, Huntley, M., and Olaizola, M., *Haematococcus* astaxanthin: applications for human health and nutrition. *Trends Biotech.* 21, 210.
47. Mera Pharmaceuticals. 1999. *Haematococcus pluvialis* and astaxanthin safety for human consumption. Technical Report TR.3005.001 available at http://www.astaxanthin.org/humansafety.htm.
48. Spiller, G.A. and Dewell, A. 2003. Safety of an astaxanthin-rich *Haematococcus pluvialis* algal extract: a randomized clinical trial. *J. Med. Food* 6, 51.
49. Furr, H.C. and Clark, R.M. 1977. Intestinal absorption and tissue distribution of carotenoids. *Nutr. Biochem.* 8, 364.
50. Østerlie, M., Bjerkeng, B. and Liaaen-Jensen, S. 2000. Plasma appearance and distribution of astaxanthin *E/Z* and *R/S* isomers in plasma lipoproteins of men after single dose administration of astaxanthin. *J. Nutr. Biochem.* 11, 482.
51. Coral-Hinostroza, G.N., Ytrestoyl, T., Ruyter, B. and Bjerkeng, B. 2004. Plasma appearance of unesterified astaxanthin geometrical E/Z and optical R/S isomers in men given single doses of a mixture of optical 3 and 3′R/S isomers of astaxanthin fatty acyl diesters. *Comp. Biochem. Physiol. C. Toxicol. Pharmacol.* 139, 99.
52. Odeberg, J.M., Lignell, Å., Len Pattersson, A. and Höglund, P. 2003. Oral bioavailability of the antioxidant astaxanthin in humans is enhanced by incorporation of lipid based formulations. *E. J. Pharma. Sci.* 19, 299.
53. Tso, M.O.M. and Lam, T.-T. 1996. United States Patent #5,527,533.
54. Shimidzu, N., Goto, M. and Miki, W. 1996. Carotenoids as singlet oxygen quenchers in marine organisms. *Fish. Sci.* 62, 134.
55. Terao, J. 1989. Antioxidant activity of beta-carotene-related carotenoids in solution. *Lipids* 24, 659.
56. Miki, W. 1991. Biological functions and activities of animal carotenoids. *Appl. Chem.* 63, 141.
57. Palozza, P. and Krinsky, N.I. 1992. Astaxanthin and canthaxanthin are potent antioxidants in a membrane model. *Arch. Biochem. Biophys.* 297, 291.
58. Lim, B., Nagao, A., Terao, J., Tanaka, K., Suzuki, T. and Takama, K. 1992. Antioxidant activity of xanthophylls on peroxyl radical-mediated phospholipid peroxidation. *Biochim. Biophys. Acta* 1126, 178.
59. Naguib, Y.M.A. 2000. Antioxidant acitivities of astaxanthin and related carotenoids. *J. Agric. Food Chem.* 48, 1150.
60. Goto, S., Kogure, K., Abe, K., Kimata, Y., Kitahama, K., Yamashita, E. and Terada, H. 2001. Efficient radical trapping at the surface and inside the phospholipid membrane is responsible for highly potent antiperoxidative activity of the carotenoid astaxanthin. *Biochim. Biophys. Acta* 1512, 251.

61. Barros, M.P., Pinto, E., Colepicolo, P. and Pedersén, M. 2001. Astaxanthin and peridinin inhibit oxidative damage in Fe^{2+}-loaded liposomes: Scavenging oxyradicals or changing membrane permeability? *Biochem. Biophys. Res. Comm.* 288, 225.
62. Mc Vean, M., Kramer-Stickland, K. and Liebler, D.C. 1999. Oxidants and antioxidants in ultraviolet-induced nonmelanoma skin cancer, in *Antioxidant Status, Diet, Nutrition, and Health*, Papas, A.M., Ed., CRC Press, Boca Raton, FL, Chapter 18.
63. Nogushi, N and Niki, E. 1999. Chemistry of active oxygen species and antioxidants, in *Antioxidant Status, Diet, Nutrition, and Health*, Papas, A.M., Ed., CRC Press, Boca Raton, FL, Chapter 1.
64. O'Connor, I. and O'Brien, N. 1998. Modulation of UVA light-induced oxidative stress by beta-carotene, lutein and astaxanthin in cultured fibroblasts. *J. Dermatol. Sci.* 16, 226.
65. Head, K.A. 2001. Natural therapies for ocular disorders, part two: cataracts and glaucoma. *Altern. Med. Rev.* 6, 141.
66. Gerster, H. 1991. Review: antioxidant protection of the ageing macula. *Age Ageing* 20, 60.
67. Seddon, J.M., Ajani, U.A., Sperduto, R.D., Hiller, R., Blair, N., Burton, T.C., Farber, M.D., Gragoudas, E.S., Haller, J., Miller, D.T., Yannuzzi, L.A. and Willett, W. 1994. Dietary carotenoids, vitamins A, C, and E, and advanced age-related macular degeneration. *J. Am. Med. Assoc.* 272, 1413.
68. Lyle, B.J., Mares-Perlman, J.A., Klein, B.E., Klein, R. and Greger, J.L. 1999. Antioxidant intake and risk of incident age-related nuclear cataracts in the Beaver Dam Eye Study. *Am. J. Epidemiol.* 149, 801.
69. Chitchumroonchokchai, C., Bomser, J.A., Glamm, J.E. and Failla, M.L. 2004. Xanthophylls and alpha-tocopherol decrease UVB-induced lipid peroxidation and stress signaling in human lens epithelial cells. *J. Nutr.* 234, 3225.
70. Fuchs, J. 1998. Potentials and limitations of the natural antioxidants RRR-alpha-tocopherol, L-ascorbic acid and beta-carotene in cutaneous photoprotection. *Free Radical Bio. Med.* 25, 848.
71. Lee, J., Jiang, S., Levine, N. and Watson, R.R. 2000. Carotenoid supplementation reduces erythema in human skin after simulated solar radiation exposure. *P. Soc. Exp. Biol. Med.* 223, 170.
72. Stahl, W., Heinrich, U., Jungmann, H., Sies, H. and Tronnier, H. 2000. Carotenoids and carotenoids plus vitamin E protect against ultraviolet light-induced erythema in humans. *Am. J. Clin. Nutr.* 71, 795.
73. Lorenz, R.T. 2002. United States Patent #6,433,025.
74. Casillas, A., Hiura, T., Li, N. and Nel, A.E.T. 1999. Enhancement of allergic inflammation by diesel exhaust particles: permissive role of reactive oxygen species. *Ann. Allerg. Asthma Im.* 83, 624.
75. Comhair, S., Bhathena, P.R., Dweik, R.A., Kavuru, M. and Erzurum, S.C. 2000. Rapid loss of superoxide dismutase activity during antigen-induced asthmatic response. *Lancet* 355, 624.
76. Dekkers, J., van Doornen, L.J. and Kemper, H.C. 1996. The role of antioxidant vitamins and enzymes in the prevention of exercise-induced muscle damage. *Sports Med.* 21, 213.
77. Alejung, P. and Wadstroem, T. 1998. World Patent #9,837,874.
78. Haines, J.L., Hauser, M.A., Schmidt, S., Scott, W.K., Olson, L.M., Gallins, P., Spencer, K.L., Kwan, S.Y., Noureddine, M., Gilbert, J.R., Schnetz-Boutaud, N., Agarwal, A., Postel, E.A. and Pericak-Vance, M.A. 2005. Complement factor H variant increases the risk of age-related macular degeneration. *Science* 308, 419.

79. Finch, C.E. and Crimmins, E.M. 2004. Inflammatory exposure and historical changes in human life-spans. *Science* 305, 1736.
80. Marx, J. 2004. Inflammation and cancer: the link grows stronger. *Science* 306, 966.
81. Ohgami, K., Shiratori, K., Kotake, S., Nishida, T., Mizuki, N., Yazawa, K. and Ohno, S. 2003. Effects of astaxanthin on lipopolysaccharide-induced inflammation *in vitro* and *in vivo*. *Invest. Opthalmol. Vis. Sci.* 44, 2694.
82. Mahmoud, F.F., Haines, D.D., Abul, H.T., Abal, A.T., Onadeko, B.O. and Wise, J.A. 2004. *In vitro* effects of astaxanthin combined with ginkgolide B on T lymphocyte activation in peripheral blood mononuclear cells from asthmatic subjects. *J. Pharmacol. Sci.* 94, 129.
83. Bennedsen, M., Wang, X., Willen, R., Wadstrom, T. and Andersen, L.P. 1999. Treatment of *H. pylori* infected mice with antioxidant astaxanthin reduces gastric inflammation, bacterial load and modulates cytokine release by splenocytes. *Immunol. Lett.* 70, 185.
84. Kim, J,-H., Kim, Y.-S., Song, G.-G., Park, J.-J. and Chang, H.-I. 2005. Protective effects of astaxanthin on naproxen-induced gastric antral ulceration in rats. *Eur. J. Pharmacol.* 514, 53.
85. Aoi, W., Naito, Y., Sakuma, K., Kuchide, M., Tokuda, H., Maoka, T., Toyokuni, S., Oka, S., Yasuhara, M. and Yoshikawa, T. 2003. Astaxanthin limits exercise-induced skeletal and cardiac muscle damage in mice. *Antiox. Redox Sign.* 5, 139.
86. Chew, B.P. and Park, J.S. 2004. Carotenoid action on the immune response. *J. Nutr.* 134, 257S.
87. Jyonouchi, H., Hill, R.J., Tomita, Y. and Good, R.A. 1991. Studies of immunomodulating actions of carotenoids. I. Effects of β-carotene and astaxanthin on murine lymphocyte functions and cell surface marker expression *in vitro* culture system. *Nutr. Cancer* 16, 93.
88. Jyonouchi, H., Zhang, L., Gross, M. and Tomita, Y. 1994. Immunomodulating actions of carotenoids: Enhancement of in vivo and in vitro antibody production to T-dependent antigens. *Nutr. Cancer* 21, 47.
89. Jyonouchi, H., Sun, S. and Gross, M. 1995. Effect of carotenoids on in vitro immunoglobulin production by human peripheral blood mononuclear cells: astaxanthin, a carotenoid without vitamin A activity, enhances in vitro immunoglobulin production in response to a T-dependent stimulant and antigen. *Nutr. Cancer* 23, 171.
90. Tomita, Y., Jyonouchi, H., Engelman, R.W., Day, N.K. and Good, R.A. 1993. Preventive action of carotenoids on the development of lymphadenopathy and proteinuria in MRL-lpr/lpr mice. *Autoimmunity* 16, 95.
91. Jyonouchi, H., Sun, S., Iijima, K. and Gross, M.D. 2000. Antitumor activity of astaxanthin and its mode of action. *Nutr. Cancer* 36, 59.
92. Okai, Y. and Higashi-Okai, K. 1996. Possible immunomodulating activities of carotenoids in in vitro cell culture experiments. *Int. J. Immunopharmacol.* 18, 753.
93. Ames, B.N., Shigenaga, M.K. and Hagen, T.M. 1993. Oxidants, antioxidants and the degenerative diseases of aging. *Proc. Natl. Acad. Sci. USA* 90, 7915.
94. Tapiero, H., Townsend, D.M. and Tew, K.D. 2004. The role of carotenoids in the prevention of human pathologies. *Biomed. Pharmacoth.* 58, 100.
95. Goldstein, J.L., Ho, Y.K., Basu, S.K. and Brown, M.S. 1979. Binding site on macrophages that mediates uptake and degradation of acetylated low density lipoprotein, producing massive cholesterol deposition. *Proc. Natl. Acad. Sci. USA* 76, 333.
96. Iwamoto, T., Hosoda, K., Hirano, R., Kurata, H., Matsumoto, A., Miki, W., Kamiyama, M., Ikatura, H., Yamamoto, S. and Kondo, K. 2000. Inhibition of low-density lipoprotein oxidation by astaxanthin. *J. Atheroscler. Thromb.* 7, 216.

97. Li, W., Hellsten, A., Jacobsson, L.S., Blomqvist, H.M., Olsson, A.G. and Yuan, X.M. 2004. Alpha-tocopherol and astaxanthin decrease macrophage infiltration, apoptosis and vulnerability in atheroma of hyperlipidaemic rabbits. *J. Mol. Cell. Cardiol.* 37, 969.
98. Drexel, H., Amann, F.W., Beran, J., Rentsch, K., Candinas, R., Muntwyler, J., Luethy, A., Gasser, T. and Follath, F. 1994. Plasma triglycerides and three lipoprotein cholesterol fractions are independent predictors of the extent of coronary atherosclerosis. *Circulation* 90, 2230.
99. Murillo, E. 1992. Efecto hipercolesterolémico de la cantaxantina y la astaxantina en ratas. *Arch. Latinoam. Nutr.* 42, 409.
100. Gross, G. and Lockwood, S.F. 2004. Cardioprotection and myocardial salvage by a disodium disuccinate astaxanthin derivative (CardaxTM). *Life Sci.* 75, 215.
101. Gross, G. and Lockwood, S.F. 2005. Acute and chronic administration of disodium disuccinate astaxanthin (Cardax TM) produces marked cardioprotection in dog hearts. *Mol. Cell. Biochem.* 272, 221.
102. Hussein, G., Nakamura, M., Zhao, Q., Iguchi, T., Goto, H., Sankawa, U. and Watanabe, H. 2005. Antihypertensive and neuroprotective effects of astaxanthin in experimental animals. *Biol. Pharm. Null.* 28, 47.
103. Kushner, I. 1990. C-reactive protein and the acute phase response. *Hosp. Pract.* 25, 13.
104. Danesh, J. 1999. Smoldering arteries? Low-grade inflammation and coronary heart disease. *J. Am. Med. Assoc.* 282, 2169.
105. Tracy, R.P. 1999. Inflammation markers and coronary heart disease. *Curr. Opin. Lipidol.* 10, 435.
106. Ridker, P.M. 1992. Novel risk factors for coronary disease. *Adv. Int. Med.* 45, 391.
107. Halliwell, B. 1992. Reactive oxygen species and the central nervous system. *J. Neurochem.* 59, 1609.
108. Fachinetti, F., Dawson, V.L. and Dawson, T.M. 1998. Free radicals as mediators of neuronal injury. *Cell. Mol. Neurobiol.* 18, 667.
109. Fahn, S. and Cohen, G. 1992. The oxidant stress hypothesis in Parkinson's disease: evidence supporting it. *Ann. Neurol.* 32, 804.
110. Borlongan, C., Kanning, K., Poulos, S.G., Freeman, T.B., Cahill, D.W. and Sanberg, P.R. 1996. Free radical damage and oxidative stress in Huntington's disease. *J. Fla. Med. Assoc.* 83, 335.
111. Ebadi, M., Srinivasan, S.K. and Baxi, M.D. 1996. Oxidative stress and antioxidant therapy in Parkinson's disease. *Prog. Neurobiol.* 48, 1.
112. Behl, C. 1999. Alzheimer's disease and oxidative stress: implications for novel therapeutic approaches. *Prog. Neurobiol.* 57, 301.
113. Markesbery, W.R. and Carney, J. M. 1999. Oxidative alterations in Alzheimer's disease. *Brain Pathol.* 9, 133.
114. Simonian, N. and Coyle, J. 1996. Oxidative stress in neurodegenerative diseases. *Annu. Rev. Pharmacol. Toxicol.* 36, 83.
115. Ferrante, R., Browne, S.E., Shinobu, L.A., Bowling, A.C., Balk, M.J., MacGarvey, U., Kowall, N.W., Brown, R.H. Jr. and Beal, M.F. 1997. Evidence of increased oxidative damage in both sporadic and familial amyotrophic lateral sclerosis. *J. Neurochem.* 69, 2064.
116. Hall, E., Andrus, P.K., Oostveen, J.A., Fleck, T.J. and Gurney, M.E. 1998. Relationship of oxygen radical-induced lipid peroxidative damage to disease onset and progression in a transgenic model of familial ALS. *J. Neurosci. Res.* 53, 66.
117. Grant, W.B. 1997. Dietary links to Alzheimer's disease. *Alz. Dis. Rev.* 2, 42.

118. Praticó, D. 2002. Alzheimer's disease and oxygen radicals: new insights. *Biochem. Pharmacol.* 15, 563.
119. Cotter, M.A., Love, A., Watt, M.J., Cameron, N.E. and Dines, K.C. 1995. Effects of natural free radical scavengers on peripheral nerve and neurovascular function in diabetic rats. *Diabetologia* 38, 1285.
120. Tagami, M., Yamagata, K., Ikeda, K., Nara, Y., Fujino, H., Kubota, A., Numano, F. and Yamori, Y. 1998. Vitamin E prevents apoptosis in cortical neurons during hypoxia and oxygen reperfusion. *Lab. Invest.* 78, 1415.
121. Mitchell, J., Paiva, M. and Heaton, M.B. 1999. Vitamin E and beta-carotene protect against ethanol combined with ischemia in an embryonic rat hippocampal culture model of fetal alcohol syndrome. *Neurosci. Lett.* 263, 189.
122. Fahn, S. 1991. An open trial of high-dosage antioxidants in early Parkinson's disease. *Am. J. Clin. Nutr.* 53, 380S.
123. de Rijk, M., Breteler, M.M., den Breeijen, J.H., Launer, L.J., Grobbee, D.E., van der Meche, F.G. and Hofman, A. 1997. Dietary antioxidants and Parkinson disease. The Rotterdam Study. *Arch. Neurol.* 54, 762.
124. Hellenbrand, W., Boeing, H., Robra, B.P., Seidler, A., Vieregge, P., Nischan, P., Joerg, J., Oertel, W.H., Schneider, E. and Ulm, G. 1996. Diet and Parkinson's disease. II: a possible role for the past intake of specific nutrients. Results from a self-administered food-frequency questionnaire in a case-control study. *Neurology* 47, 644.
125. Ghadge, G., Lee, J.P., Bindokas, V.P., Jordan, J., Ma, L., Miller, R.J. and Roos, R.P. 1997. Mutant superoxide dismutase-1-linked familial amyotrophic lateral sclerosis: molecular mechanisms of neuronal death and protection. *J. Neurosci.* 17, 8756.
126. Tanaka, T., Morishita, Y., Suzui, M., Kojima, T., Okumura, A. and Mori, H. 1994. Chemoprevention of mouse urinary bladder carcinogenesis by the naturally occurring carotenoid astaxanthin. *Carcinogenesis* 15, 15.
127. Tanaka, T., Kawamori, T., Ohnishi, M., Makita, H., Mori, H., Satoh, K. and Hara, A. 1995. Suppression of azomethane-induced rat colon carcinogenesis by dietary administration of naturally occurring xanthophylls astaxanthin and canthaxanthin during the postinitiation phase. *Carcinogenesis* 16, 2957.
128. Tanaka, T., Makita, H., Ohnishi, M., Mori, H., Satoh, K. and Hara, A. 1995. Chemoprevention of rat oral carcinogenesis by naturally occurring xanthophylls, astaxanthin and canthaxanthin. *Cancer Res.* 55, 4059.
129. Chew, B.P., Park, J.S., Wong, M.W. and Wong, T.S. 1999. A comparison of the anticancer activities of dietary b-carotene, canthaxanthin and astaxanthin in mice *in vivo*. *Anticancer Res.* 19, 1849.
130. Kurihara, H., Koda, H., Asami, S., Kiso, Y. and Tanaka, T. 2002. Contribution of the antioxidative property of astaxanthin to its protective effect on the promotion of cancer metastasis in mice treated with restraint stress. *Life Sci.* 70, 2509.
131. Andersson, M. 2001. United States Patent #6,277,417.
132. Gradelet, S., Astorg, P., Leclerc, J., Chevalier, J., Vernevaut, M.F. and Siess, M.H. 1996. Effects of canthaxanthin, astaxanthin, lycopene and lutein on liver xenobiotic-metabolizing enzymes in the rat. *Xenobiotica* 26, 49.
133. Jewell, C. and O'Brien, N. 1999. Effect of dietary supplementation with carotenoids on xenobiotic metabolizing enzymes in the liver, lung, kidney and small intestine of the rat. *Br. J. Nutr.* 81, 235.
134. Bertram, J.S. 1999. Carotenoids and gene regulation. *Nutr. Rev.* 57, 182.
135. Dore, J.E. 2005. Astaxanthin and cancer chemoprevention, in *Phytopharmaceuticals in Cancer Chemoprevention*, Bagchi, D. and Preuss, H.G., Eds., CRC Press, Boca Raton, FL, Chapter 34.

14 Marine Algae and Polysaccharides with Therapeutic Applications

J. Helen Fitton, Mohammad R. Irhimeh, and Jane Teas

CONTENTS

14.1 INTRODUCTION

Human use of marine macroalgae and microalgae for both food and medicine has taken place the world over for millennia [1]. Archaeological studies in South America revealed that marine algae were being used as "medicine" in approximately 12000 BC [2]. Ancient texts suggest that seaweeds were used for ailments as diverse as snakebite, lung diseases, and gout.

Algae and algal extracts are used in present-day medicine, and play a significant part in "ethnic" medicine. Therapeutically active agents found in marine algae have untapped potential in functional foods, nutraceuticals, and pharmaceuticals. Noteworthy activity includes profound antiviral properties, lipid-lowering activity, and anticancer activity. Interestingly, many of these effects can be observed with oral intake.

Most of the world's annual seaweed harvest is used to produce the algal hydrocolloids alginate, agar, and carrageenan. These bulk commodities are used largely as viscosity-modifying agents in foods and pharmaceuticals.

This short review describes the therapeutic benefits of marine algae from both historic and present-day perspectives. Macroalgae are the focus of our discussions, with some references to the more commonly ingested microalgae.

14.2 THE COMMERCIAL USES OF MARINE ALGAE

There are thousands of species of marine algae, which represent over 90% of all marine plants. They are classified into major groups: brown (Phaeophyta), red (Rhodophyta), or green (Chlorophyta), determined by their pigment contents [3]. Algae may be macrophytes (seaweeds) or microalgae such as *Spirulina* (*Arthrospira platensis*). Marine algae thrive in well-lit, relatively shallow areas of the sea, where they can photosynthesize and absorb nutrients. The growth rates and composition are dependent upon environmental factors. Perennial macroalgae such as *Macrocystis* are some of the largest plants on Earth. Macroalgae chelate minerals and are often rich in trace elements and iodine.

Seaweeds are a rich source of polysaccharides that are used largely as bulk commodities in food and pharmaceutical industries. The main commercial phycocolloids are agar, alginate, and carrageenan. The production and uses of these are well covered in the FAO documents prepared by Dennis McHugh and colleagues [4]. At the end of the last century, the annual seaweed harvest was above 2 million tonnes (dry weight) with a value in excess of US$ 6 billion with more than 8 million tonnes of wet seaweed used annually. Of the 221 species of seaweed in use, 145 species are for food and 101 species for phycocolloid production [5].

Microalgae that tolerate high salt conditions, surviving in estuarine areas and salt-laden lakes, may also be considered "marine algae." Both *Spirulina* (*Arthrospira platensis*) and *Dunaliella salina* are cultured. The latter yields high concentration of vitamin A and lipids, whereas the former is cultivated largely for use as a health food [6]. *Spirulina* is also harvested from naturally occurring sources for food use [7].

14.3 TYPES OF POLYSACCHARIDES FOUND IN MARINE ALGAE

The structure characteristics of some well-known polysaccharides are detailed in Table 14.1. Agar is extracted from the "agarophytes," such as *Gelidium*, *Gracilaria*, and *Phyllophora* by soaking them in water to remove foreign matter, and then heating with water, which causes the agar to dissolve in the water. After filtration and cooling, a 1% agar gel is formed. This is broken up, washed to remove soluble salts, and the water is removed from the gel, either by a freeze–thaw process or by squeezing it out using pressure. The product is dried in hot air, and then milled [4].

Similarly, carrageenan can be obtained by hot-water extraction, although this method is relatively expensive compared to a less refined version popular today. In the latter method, the seaweed is treated with alkali and water, leaving the carrageenan and other insoluble matter behind. The insoluble residue is carrageenan and about 15% cellulose, which is sold as semirefined carrageenan.

The first source of carrageenan was *Chondrus crispus*, or "Irish moss" (often used in traditional set milk puddings). Today, most carrageenan is extracted from *Kappaphycus alvarezii* (also called *Eucheuma cottonii*) and *Eucheuma denticulatum*. Some South American species used are *Gigartina skottsbergii*, *Sarcothalia crispata,* and *Mazzaella laminaroides* [4].

Carrageenan has three major commercially useful fractions, kappa, lambda, and iota carrageenan. It is possible to find both agarans and carrageenans in the same alga [8]. Each species yields a blend of fractions. *C. crispus* yields a mixture of kappa and lambda, *Kappaphycus* yields mainly kappa, whereas *E. denticulatum* yields mainly iota [4].

The characteristics of the different carrageenans are as follows:

- Iota forms clear, freeze–thaw stable gels with calcium salts.
- Kappa forms strong, opaque gels with potassium salts.
- Lambda forms no gels, but creates viscosity suitable for suspensions.

Alginate or "algin" is extracted from brown macroalgae such as *Ascophyllum*, *Durvillaea*, *Ecklonia*, *Laminaria*, *Lessonia,* and *Macrocystis.* Sodium alginate is produced using alkali extraction technique. The brown seaweed source material is treated with cross-linking agent to immobilize polyphenol fractions. The material is then acid rinsed to eliminate fucoidan and laminarin fractions and then alkali treated, which turns the alginates into a gelatinous mass. This material is then treated with acid to extract it from the cellulosic mass. Sodium alginate, and smaller quantities of alginic acid and the calcium, ammonium, and potassium salts, and an ester, propylene glycol alginate are also produced. Recent publications by Hernadez-Carmona et al. [9,10] detail the process for extraction.

Polysaccharides with sulfate hemiester groups attached to sugar units are found in the form of "fucoidan" in Phaeophyceae, as "galactans" (agar and carrageenans) in Rhodophyceae, and as "arabinogalactans" with lesser amounts of other sugars in Chlorophyceae. The fucoidans encompass three different

TABLE 14.1
Types of Polysaccharides Found in Marine Algae

Extract	Polysaccharide Components	Sugars	Sulfate	Isolated From	Uses (Other Than Food)
Agar	Agarose (60%)	D-galactose, 3,6 L-anhydro-galactose	No	Rhodophyta e.g., *Gracilaria, Phyllophora*	Bacterial culture media, wound-healing absorbents, bulk-forming laxative
	Agaropectin (40%)	D-galactose, 3,6 L-anhydro-galactose D-guluronic acid	Yes 3.5–9.7% sulfate		
Carrageenan	Kappa	3,6-anhydro-α-D-galactose	24% alkali stable ester sulfate	Rhodophyta *Chondrus crispus Gigartina, Furcellaria Eucheuma cottonii*	Antiviral topical agent
	Lambda	Mostly D-galactose 3,6-anhydro-α-D-galactose, small amount of L-galactose	35% ester sulfate		
	Iota	3,6-anhydro-α-D-galactose			
Alginic acid	Blocky or dispersed mannuronic and guluronic acid polymers	Guluronic and mannuronic acids	No	Phaeophycaea, *Durvillea, Macrocystis*	Wound dressing Drug delivery Dental mold making

Furcelleran	Carrageenan-type polysaccharide	43–45% D-galactose, 30% 3,6,anhydro D-galactose, some xylose	20% ester sulfate	*Furcellaria fastigiata*	Approved as a food additive, as for carrageenan
Fucoidan	Fucose rich branched polysaccharide	Fucose rich	Up to 38% ester sulfate	Phaeophyceae e.g., *Laminaria digitata*, *Fucus vesiculosis*	Functional food, nutraceutical
Ascophyllan	Also called glucuronoxylofucan sulfate	25% fucose, 26% xylose, 19% sodium uronate	13% sulfate	*Ascophyllyum nododum*	As for fucoidan
Laminarin	Beta glucan 1,3 linked	Glucose	Yes	Phaeophyceae	No commercial use
Polyuronides from green algae	Highly branched polysaccharides Structure unit 4-O-β-D-glucuronosyl-L-rhamnose present in all the mucilages from different species	D-glucuronic acid, L-rhamnose, D-xylose, D-glucose	20% ester sulfate on rhamnose, xylose	*Ulva lactuca*, *Enteromorpha compressa*, and *Spongomorpha arcta*	Nutraceuticals

subclasses of sulfated fucose–rich polysaccharides according to their origin: fucoidans, ascophyllans, and sargassan [11].

Investigation of the sulfated polysaccharides from green algae reveals water-soluble polysaccharides composed of a variety of sugars in variable molar ratios.

14.4 HISTORY OF ALGAE IN MEDICINAL APPLICATIONS

The earliest records of the use of marine algae in medicine were uncovered by Jack Rossen and Tom Dillehay at Monte Verde in Chile, a site that dates back to approximately 12000 BC. Four species were found at the site: *Sargassum*, *Gracilaria*, *Porphyra*, and *Durvillea*. *Sargassum* was found wrapped around a plant called boldo, which has hallucinogenic qualities. These sargassum/boldo packets were apparently chewed into mouth-shaped "cuds," and were found in a wishbone-shaped healer's hut [2]. *Sargassum* must have been transported from some distance away, as it grows much further to the north.

Traditional use of algae as food and medicine by indigenous people is well recognized. North American native people prize *Porphyra* for its medicinal food value [1,12]. Coastal harvesting and transport of macroalgae inland for food use occurs today in South- and North America, preventing iodine deficiency. *Porphyra abbottae* is a culturally important species used by the First Peoples of coastal British Colombia, Canada. It is said to be a "health food" and used to alleviate indigestion or heartburn, as a laxative and can be used as an antiseptic poultice for a deep cut or swelling, or even a broken collarbone [12].

The UNESCO courier reports on the present-day Chilean harvest: "Near Temuco and Chiloé, in the central regions of Chile and in the very heart of Araucana, they gather 'cochayuyo' and 'luche' (*Durvillaea antarctica* and *Porphyra*), with which they make bread and cakes, or cook with mutton" [1,13]. Island populations in the Caribbean enjoy "healthy" or reputedly aphrodisiac milk–based seaweed drinks. Brie Cocos [14] noted that in Belize, boiled *Euchuema isoforme* is used to make a much prized concoction with condensed milk, nutmeg, cinnamon, and a dash of brandy or rum.

In Northern Europe, Ireland, and the United Kingdom, marine algae has long been valued as a food for people and livestock, and sometimes used "medicinally." Home-produced remedies include cough medicines and strengthening milk–based jellies made from *Chondrus crispus* (Irish moss). In Victorian times, macroalgae-based remedies were popular as dieting aids. Various orally ingestible concoctions of brown algae were also used to treat arthritis and rheumatism, and topical applications of "seapod liniment" or bruised algae were recommended as poultices for scrofula (mycobacterium infection of the skin), sprains and bruises, and embrocations for "the limbs of rickety children" [15]. Today, brown algae are still included in supplements designed to aid weight loss [16]. Many of the medicinal applications described by Mrs. Grieve in the 1901 classic "A modern herbal" were probably developed from earlier texts, such as Gerard's *On the History of Plants,* published in 1633 [17]. This in turn, borrowed from ancient texts such as that by Pliny the elder who describes uses for macroalgae [18]. In Book XXVI, Chapter 66, he discusses a seaweed treatment for gout: *But it is the phycos thalassion, or sea-weed, more particularly, that is so excellent a remedy for the gout . . . Used before it becomes dry, it is efficacious as a topical application not only for gout, but for all diseases of the joints. There are three kinds of it; one with a broad leaf, another with a longer leaf of a reddish hue, and a third with a crisped leaf, and used in Crete for dyeing cloths. All these kinds have similar properties; and we find Nicander prescribing them in wine as an antidote to the venom of serpents even.*

Pliny's reference to the Greek philosopher Nicander's treatise "Theriaca" is a curious echo of recent research indicating the potential of fucoidans as snakebite enzyme inhibitors [19]. He also mentions red seaweed as a treatment for the sting of a scorpion. Interestingly, there is traditional use of seaweed for snakebite by the Guguyalanji tribe in Northern Queensland, Australia [20].

Throughout the texts of Grieve, Gerard, and Pliny, there are references to the use of particular seaweeds as antiparasitic remedies. These references are to algae such as "Corsican moss" that contain kainic acid, a potent antihelminthic. Today, kainic acid and a related chemical, domoic acid, are extracted from macroalgae for use as neuroexcitory agents [21]. Tseng [22] also refers to the use of similar antihelminthic algae in present-day China.

Antitumor properties of seaweeds are also mentioned in Gerard and Pliny. Gerard comments: "Lungwoort (probably Fucus) is much commended of the learned physitions of our time against the diseases of the lungs, especially for the inflammations and ulcers of the same, being brought into pouder and drunke with water." These observations correlate with today's research findings in animals, in which tumor arrest or delay is seen with oral, i.v., or i.p delivery of macroalgal extracts [23–28].

In traditional Chinese medicine, diet is considered necessary for the prevention of disease, and superior to treatment. Kelp (large brown marine algae) is considered to be a "Han" or cold food, in a system that considers foods to be cold, hot, neutral, or strengthening [29]. There are many Chinese medicines that include macroalgae. An example is "concoction of the Jade flask" (a reference to the shape of the neck in goiter), which includes *Ecklonia* and *Sargassum*. The use of macroalgae in traditional medicines is common in present-day China [22].

Another traditional community that still makes use of seaweed are the Maori people of New Zealand. They used *Durvillea* (Rimu) and *Porphyra* (karengo) as treatments for goiter prevention and as laxatives, in addition to food purposes. The tender tips of *Durvillea*, roasted, were considered useful as antiparasitic agents. In addition, Riley comments *the author has been told of the use, to good effect, of a large blade of kelp to wrap around a Maori child suffering from burns* [30].

14.5 CONTEMPORARY USE OF MACROALGAE IN MEDICINE

Whole macroalgae are only used in two applications in conventional medicine. Iodine supplementation using dried kelp granules is sometimes suggested for patients with suspected deficiencies. Iodine levels vary widely between macroalgae, and there are some potential risks in ingesting high doses of kelps, although the commonly eaten *Porphyra* and *Undaria* are relatively low in iodine [31]. The second application is in surgery. Specially prepared stipes from *Laminaria* are used to dilate the cervix in gynaecological procedures. This long-used procedure is still in vogue, as its slow swelling does not damage the cervix [32].

Radiation medicine has made use of macroalgae to inhibit radioactive iodine uptake, and assist in the chelation and elimination of radioisotopes [33]. Sodium alginate from *Sargassum siliquastrum* was a potent agent for reducing strontium absorption in clinical tests, when added to bread at 6% level, while alginate syrup was more suitable for emergencies because of its rapid action [34].

Agars are commonly used as microbiological culture mediums for identifying infectious agents, and also in proprietary laxatives. Alginates are also used in constipation remedies. The hydrocolloids act as bulking agents, and may stimulate mucus production. In a rat model, carrageenan and sodium alginate, but not cellulose, increased colonic mucus [35]. Alginic acid is also a component of medicines designed to block acid reflux. Alginic acid forms a water-swollen raft on the top of the stomach contents, preventing acidic erosion of the oesophagus [36].

Tonneson reviewed the uses of purified alginates. More than 200 different alginate grades and a number of alginate salts are available [37]. Controlled-release drug applications use alginate, because it gels differentially, depending on pH. At low pH, such as in the stomach, alginic acid forms a high-viscosity "acid gel." Ionotropic gels are formed when alginate gels in the presence of calcium ions. These gels create a diffusion barrier to encapsulated drugs, extending delivery times. Calcium alginate dressings are used in wound-dressing applications for absorbing wound fluids. They are less painful to change, and popular for use in ulcers [38].

A niche application for *Corallina officinalis*, a calcified species once used in Europe as a vermifuge, is found in bone-defect filling materials, sold as "Algipore biomaterial" [39].

14.6 DRUGS IN DEVELOPMENT FROM MARINE ALGAE

There are a variety of smaller molecules found in some marine algae that have useful pharmacological properties. These molecules are then synthesized and tested for their safety and efficacy as drugs in a clinical setting. Examples include the depsipeptide Kahalalide F, which was discovered in the green marine alga *Bryopsis* and has clinically assessed anticancer properties [40]. Furanones from *Delisea pulchra* have disrupting effects on the accumulation of bacterial biofilms, and are being investigated for their ability to inhibit the buildup of *Pseudomonas* lung infections [41]. Other examples of such druglike molecules have been discovered in marine algae [42].

14.7 DISEASE INCIDENCE IN ALGAL CONSUMERS

The populations of Japan and Korea consume macroalgae every day, averaging 6.6 g/day in 1995 in Korea [43]. Cancer has an overall lower incidence in Japan. Breast cancer is typically 10 times less prevalent in postmenopausal Japanese women than in Western women, and dietary macroalgae have been implicated as a potential protective agent [44,45]. Macroalgae are also part of the macrobiotic diet, which may have some benefits to cancer sufferers [46].

Realizing the potential value of dietary algae as antiviral agents, we recently proposed algal consumption as one unifying characteristic of countries with anomalously low rates of HIV [47]. In a pilot study in which HIV positive patients ingested 5 g/day for 3 weeks of either *Undaria* or *Spirulina*, viral loads were reduced by up to 76%, and CD4 also rose [48]. Given the large percentage of HIV positive people in the world who will never have access to drugs, dietary supplementation with algae may be a realistic, helpful treatment.

14.8 BIOLOGICAL ACTIVITIES AND APPLICATIONS OF ALGAE

Marine algae possess antiviral and antimicrobial properties in addition to antitumor and immune activity. Sulfated polysaccharide extracts display anticlotting, enzyme inhibitory, and growth factor modulation activity, in addition to being potent selectin blockers.

14.8.1 Tumor Inhibition and Immune Modulation Activity

Macroalgae contain several potentially "anticancer" constituents. Whole macroalgae, macroalgae soaked water, and macroalgae extracts, all inhibit tumor development in solid tumors and leukemias in animal models [23–28]. These inhibitory effects are often mediated by either brown marine algae or their acid-soluble polysaccharides, fucoidans. The commonly eaten Japanese Kombu (*Laminaria*), wakame (*Undaria*), and Mozuku (*Cladosiphon*) all have substantial activity, although the mechanism is uncertain. Algal administration seems to enhance innate immunity, which includes an increase in the Th1 cytokine profile (IFN gamma) and increased NK cell activity [49,50]. Research by Shimuzu demonstrated an overall increase in cytotoxic T cells in mice fed on a high-molecular-weight fucoidan ($2–3 \times 10^5$) from *Cladosiphon okamuranus*, at a level of 5% in the diet. In this model, lower-molecular-weight fucoidans from the same source had no effect [51].

Other components that may exert anticancer effects include iodine, which causes apoptosis in cancer cells. Additionally, omega-3 fatty acids such as stearidonic acid and hexadecatetraenoic acid are found in edible marine algae such as *Undaria* and *Ulva* up to 40% of total fatty acids [52]. Algal galactolipids found in *Undaria*, *Laminaria*, *Porhphyra*, and *Cladosiphon* include high concentrations of the telomerase inhibitor sulfoquinovosyldiacylglyceride (SQDG) [53,54]. Fucoxanthins, the carotenoids found in algae, are metabolized to fucoxanthinol and thus may exert anticancer effects [55]. Glutathione, an antioxidant, is a constituent of all macroalgae (*Sargassum thumbergeii* and *Ishige okamurai* contain 1482 and 3082 mg glutathione/100 g) [56]. Fucan fractions, alginates, and polyphloroglucinols isolated from brown macroalgae have enzyme inhibitory properties against hyaluronidase, heparanases, phospholipase A2, and tyrosine kinase, which may also contribute to the anticancer activity [57–61]. Cyclic and noncyclic nucleotides with potential therapeutic value have been identified in *Porphyra umbilicalis* [62].

There have been a number of human clinical studies involving marine algae. In 1968, a fucoidan preparation known as Algosol T 128 was used to treat leukemia [63]. *Spirulina* ingestion was shown to inhibit oral cancer [64]. Decreased allergic responses were observed with an alginic acid oligosaccharide, which suppressed IgE production by inducing IL-12 production [65]. Similarly, in a mouse model, *Cladosiphon* fucoidan downregulated IL6 (A Th2 cytokine) and ameliorated colitis [66]. Interestingly, algal polysaccharides have immune stimulatory effects on plants. Sulfated oligo-fucan fractions elicited "systemic acquired resistance," and the accumulation of salicylic acid in tobacco plants [67].

14.8.2 Direct Antipathogenic Activity

All marine algae contain sulfated polysaccharides. Carrageenans, fucoidans, and sulfated rhamnogalactans have inhibitory effects on entry of enveloped viruses (such as herpes and HIV) into cells. Some other algal fractions have virucidal and enzyme inhibitory activity, or the ability to inhibit syncytium formation [68–74]. Pilot studies with *Undaria* showed inhibitory effects on herpes infections [75]. Lambda carrageenan preparations (Carraguard™) are being tested as vaginal microbiocides [76]. Antibacterial activity of marine algae is partly attributable to iodine, but some polysaccharides prevent bacterial adhesion. Funoran extracted from *Gloiopeltis furcata* inhibited the adherence of dental plaque [77]. Fucoidan extracts from *Cladosiphon* (Okinawan "Mozuku") inhibit adhesion of the ulcer-causing bacterium *Helicobacter.* Human trials in which subjects consumed 1.5–4.5 mg/kg/day of *Cladosiphon* fucoidan provided relief for nonulcer dyspepsia over 2 weeks [78]. Marine algal lectins provide antibacterial activity against marine *Vibrios* [79], and algal-derived furanones inhibit the ability to form biofilms, which may be clinically useful in a number of applications [80].

14.8.3 Anti-Inflammatory Activity

Fucoidans are excellent selectin blocking agents, and have been used experimentally to reduce postischemic leukocyte influx, or so-called "reperfusion injury" [81]. Fucoidans also have a protective effect on kidney function in animal models, when orally delivered [82]. Carrageenans may be macrophage toxic, and act as inflammation initiators in animal models [83]. Some carrageenan fractions are also anti-inflammatory [84]. Nori (*Porphyra*)-fed rats had a higher incidence of submucosal edema than the Konbu (*Laminaria*)-fed rats. However, this may have been related to the higher Na/K ratio, rather than any intrinsic inflammatory activity in the Nori diet [85].

14.8.4 Hypertension, Serum Lipids, and Sugar Metabolism

Marine algae come close to being an ideal food to reduce the potential for developing ischemic and cardiovascular diseases. They are nutrient dense, with protein yields exceeding 30% in some cases, containing all essential amino acids, omega-3 lipids and plenty of soluble fiber [1]. In addition, commonly eaten species contain a variety of lipid-modulating, blood pressure–lowering, and glucose metabolism modifying components. Ethnic communities benefit from the continuing tradition of algal consumption, which counteracts the atherogenic qualities of some Western foods [12].

In the 1950s, there was considerable interest in the antitumor and lipid-modifying properties of sulfated polysaccharides from seaweeds. Heparin was known to clear "lipemia," but had undesirable anticoagulant qualities. Besterman and Evans carried out clinical trials with two relatively low sulfate laminarins (0.62 and 0.37 sulfate groups per glucose) produced by the British pharmaceutical company, Boots Ltd., in 12 patients with ishemic heart disease. The laminarins had low anticoagulant activity (1.4 and 1.3 IU/mg) and had low toxicity in guinea pigs in contrast with a high sulfated laminarin. The patients were administered 100 mg i.v. or i.m. Two patients had a fatty meal prior to the injection, and in these patients the visible turbity of the serum was reduced. Only the lower sulfated laminarin was active, and no effects were seen via intramuscular injection [86]. Despite the apparent success of this trial, in which lipemia-clearing activity without the anticoagulant effects was observed, there were no further published trials.

Since that time, algae and algal fractions delivered orally as well as i.v. have shown considerable activity in moderating serum lipid levels. Ingestion of carrageenans lowers serum lipid and cholesterol levels [87]. Synergistic effects on serum lipids were observed when *Undaria* and fish oils were combined as part of an experimental diet [88]. *Undaria* fucoidan fractions delivered by i.v. resulted in rapid clearance of serum lipids [89]. Up to 500 mg/kg polysaccharides from *Ulva pertusa* delivered orally to mice decreased plasma total cholesterol, low density lipoprotein (LDL), triacylglycerols and markedly increased high density lipoprotein (HDL) [90].

Sterols, which have dietary cholesterol modulating activities, are also found in fairly high concentrations in marine algae [91]. The predominant sterol was fucosterol in brown seaweeds (83–97% of total sterol content; 662–2320 μg/g dry weight), and desmosterol in red seaweeds (87–93% of total sterol content; 187–337 μg/g dry weight). Fucosterol has additional antidiabetic properties, which may explain the use of high fucosterol seaweeds in traditional diabetes remedies. Fucosterol from *Pelvetia siliquosa* was administered at 300 mg/kg in epinephrine-induced diabetic rats. It caused an inhibition of blood glucose level and glycogen degradation [92].

Peptide fractions of *Undaria* contain angiotensin converting enzyme (ACE) inhibitory activity [93,94]. The inhibition of development of cerebrovascular diseases in stroke-prone spontaneously hypertensive rats was observed with an *Undaria*-rich diet [95,96]. The polysaccharide components of macroalgae modify the uptake of glucose in the gut [97]. Brown marine macroalgae also contain phloroglucinols such as "eckol" and "dieckol" in *Eisenia bicyclis*, which modify glucose metabolism. These substances demonstrated inhibition of aldose reductase and glycation [98].

Finally, the inhibition of vascular smooth muscle proliferation by fucoidans may assist in the arrest or reversal of atherogenesis. Following binding to the cell surface, heparins or high MW fucoidan undergo internalization by receptor-mediated endocytosis and subsequent breakdown. The largest fucoidan fractions remain virtually intact, and have the highest antiproliferative activity [99]. The antimitogenic action of fucoidan differs from that of heparin, being effective even in "heparin-resistant" vascular smooth muscle cells [100].

14.8.5 Effects on Blood

14.8.5.1 Anticoagulants and Antithrombotic Effects

Fucoidans are intensively researched as heparin replacements. Fucoidans are plant sourced (as opposed to porcine gut or lung mucosae for heparin), have good potential antithrombotic qualities, and complement inhibitory qualities. These are well reviewed by Berteau and Blondin [101,102]. One attractive feature of algal-derived fucoidan fractions is their potential stability as compared to heparin.

Fucoidans appear to act on specific parts of the clotting cascade (antithrombin II or to potentiate heparin cofactor II) in similar ways to endogenous sulfated polysaccharides such as heparan sulfate and dermatan sulfate. The MW, patterns of sulfation, and sugar composition are critical to the activity of the fraction. Oversulfated fucoidan had a potent fibrinolytic (clot dissolving) activity, by stimulating the action of tissue plasminogen activator. It also inhibited hyaluronic acid–mediated enhancement of clot formation [103].

14.8.5.2 Stem Cell Modulation

Hemopoetic stem cells (CD34+ cells) give rise to the different lineages of immune cells, and possibly to other tissue cell types. It is sometimes desirable to mobilize greater numbers of these cells from the bone marrow stroma into the peripheral blood for harvesting and later engrafting. A cytokine called granulocyte macrophage stimulating factor (GCSF) is often used for this purpose, but it had been noted that heparin had some mobilizing effects. In a clinical study, conventional heparin anticoagulation resulted in a 2.49-fold increase in circulating CD34+ HPCs [104]. Unfractionated *Fucus* fucoidan was observed to have a potent, long-lasting mobilizing effect when injected into mice or primates [105,106]. It is thought that heparin and fucoidan mobilize stem cells by displacing growth factor called SDF-1 into the serum, thus creating a chemotracting gradient for CXCR4, the cell surface ligand for SDF-1. This was confirmed in a study, which demonstrated increased plasma SDF1 levels in an ischemic hindlimb revascularization model using a low-molecular-weight fucoidan fraction [107]. Pilot clinical studies with orally delivered *Undaria fucoidan* showed a large increase in the expression of CXCR4 on CD34+ cells [108].

14.8.6 Uptake and Toxicity of Marine Algal Polysaccharides

Much of this review has considered the immunomodulating, antitumor, and antiviral effects of algal polysaccharides. Although these materials have a high molecular weight, they exert effects after oral administration. It is not known whether absorption or uptake by the gut lymphatic system occurs. However, other high-molecular-weight sulfated polysaccharides such as chondroitin sulfate are absorbed whole in small amounts in the small intestine [109]. There were no toxicological changes observed in rats given up to 300 mg/kg orally of fucoidan from *Laminaria japonica*. Anticoagulant effects were observed at doses of 900–2500 mg/kg, but no other signs of toxicity were observed. The composition of the fucoidan was fucose 28%, sulfate 29%, fucose:galactose ratio 1:0.24, with a molecular weight of 189,000 [110]. No side effects were reported in human clinical studies with algae [108].

Gut permeability is affected by disease. Tight junctions between gut epithelium break down in a variety of disease states, such as irritable bowel syndrome and viral infections. Posttransplant cytomegalovirus (even subclinical infections)-infected patients demonstrated increased gut permeability [111]. HIV-infected patients who have progressed to AIDS and have diarrhea also have increased bowel permeability [112]. Gut permeability is also different in infants. This presents an easy route for HIV infection in breast-feeding infants of HIV-infected women, but paradoxically, it would also present a route for high-molecular-weight HIV inhibitory substances like fucoidans.

The uptake of high MW substances, such as alginates, by Peyers patches (part of the gut-associated lymphatic system) is now being exploited to develop orally delivered peptide vaccines. Ease of delivery, increased safety, and economic benefits make oral delivery of peptide vaccines highly desirable. However, peptides are easily degraded in the gastrointestinal tract, and lack the size to produce an immune response. Commercial vaccine development based on this technology is in progress [113].

14.9 REGULATORY STATUS AND SAFETY

Agar-agar and alginic acid and its salts were granted the GRAS (generally recognized as safe) status by US Food and Drug Administration (FDA). Carrageenan, furcelleran, and their salts, as well as propylene glycol alginate are also in the legal food additive list permitted for direct addition to food for human consumption. Some brown and red algae species are also included in FDAs GRAS list. Carrageenan and processed *Eucheuma* seaweed (a form of carrageenan with a higher cellulose content) have a Joint FAO/WHO Expert Committee on Food Additives (JEFCA) recommended group allowable daily intake (ADI) of "not specified" [114].

All marine algae sequester ions readily, and depending on their polysaccharide content, and the local environment, heavy metal loading may occur. Marine algae contain arsenic as the potentially toxic "inorganic arsenic" and a greater part of the possibly benign, "organic arsenic" [115]. The WHO has established a "provisional tolerable weekly intake" of 15 μg inorganic arsenic per week per kilogram of body weight.

Hijiki, a commonly eaten Japanese seaweed, has a naturally high inorganic arsenic content, and is banned in many Western countries [116], and some Porphyra products from China had excessive arsenic levels [117]. The toxicity of seaweed-derived arsenic is, to some extent, tested by animals that survive on seaweed. These include North Ronaldsey sheep [118] that forage almost exclusively on beachcast *Laminaria digitata*, and a species of deer that has a high intake of seaweed [119].

Excessive iodine intake could potentially induce thyrotoxicosis. In a recent study of 12 different species of macroalgae, iodine content ranged from 16 mcg/g in Nori (*Porphyra tenera*) to over 8165 mcg/g in kelp granules made from *Laminaria digitata*. Relatively low iodine–containing species included *Undaria* and *Sargassum* [31]. The maximum tolerated dose of iodine is 1000 mcg/day.

There have been some concerns about radioactivity in macroalgae in the northern hemisphere after nuclear accidents have occurred. A survey of Canadian, Japanese, and European sources of macroalgae found traces of cesium-137 in a product from Norway and radium-226 was found in a product from Japan [120].

Although marine algae contain sulfated polysaccharides known to inhibit clotting, no reports of clotting inhibition are reported in the literature subsequent to ingestion by humans. Marine algae also contain vitamin K (phylloquinone), which is a "procoagulant." One report exists of a 33-year-old woman eating Nori (*Porphyra*), averaging 18.8 mcg/100 g product, which interfered with postoperative warfarin therapy (an anticlotting agent) [121].

14.10 NEW HORIZONS FOR MARINE ALGAE

As outlined in this review, marine algae have great potential for further development as products in the nutraceutical, functional food, and pharmaceutical markets. Patent activity in this area has increased in the last few years, and several novel products based on macroalgae have entered the market. Red marine algae can be found in several nutraceuticals aimed at ameliorating herpes infections. A particularly novel application is found in "Plaque–Off™," an orally administered macroalgal supplement aimed at decreasing the deposition of calculus, in both people and their pets. Fucoidan extracts are well accepted in Japan and are included in mainstream functional foods such as yogurts and fruit juices. South Pacific–derived fucoidan-based liquid preparations are a popular nutraceutical offering in the United States.

The development of new products in the pharmaceutical area will depend on the ability to isolate well-characterized uniform polysaccharides with defined activity. Heterogeneous fucoidan preparations, such as that used for selectin blocking in many experimental systems [122], will gradually become less attractive as biological tools.

REFFRENCES

1. Aaronson, S. 2000. Algae. Volume I, Chapter II.C.I In: *The Cambridge World History of Food*, Kiple K and Ornelas KC, Eds., Cambridge University Press, Cambridge, UK, pp. 231–249.
2. Dillehay, T.D. 1997. *Monte Verde—A Late Pleistocene Settlement in Chile. Vol. 2 The Archaeological Context and Interpretation*. Smithsonian Institution Press, Washington, DC, pp. 307–350.

3. Guiry, M.D. and Nic Dhonncha, E. 2005. *AlgaeBase Version 2.1.* World-wide electronic publication, National University of Ireland, Galway. http://www.algaebase.org; accessed on 22 January 2005.
4. McHugh, D.J. 2003. A guide to the seaweed industry. FAO Fisheries Technical Papers—T441.
5. Zemke-White, W. and Ohno M. 1999. World seaweed utilisation: an end-of-century summary. *J. Appl. Phycol.* 11, 369–376.
6. Hejazi, M.A. and Wijffels, R.H. 2004. Milking of microalgae. *Trends Biotechnol.* 22, 189–194.
7. Abdulqader, G., Barsanti, L. and Tredici, M. 2000. Harvest of *Arthrospira platensis* from Lake Kossorom (Chad) and its household usage among the Kanembu. *J. Appl. Phycol.* 12, 493–498.
8. Estevez, J.M., Ciancia, M. and Cerezo, A.S. 2004. The system of galactans of the red seaweed, *Kappaphycus alvarezii*, with emphasis on its minor constituents. *Carbohydr. Res.* 339, 2575–2592.
9. Hernandez-Carmona, G., McHugh, D.J. and Lopez-Gutierrez, F. 1999. Pilot plant scale extraction of alginates from *Macrocystis pyrifera.* 2. Studies on extraction conditions and methods of separating the alkaline-insoluble residue. *J. Appl. Phycol.* 11, 493–502.
10. Hernandez-Carmona, G., McHugh, D.J., Arvizu-Higuera, D.L. and Rodriguez-Montesinos, Y.E. 1999. Pilot plant scale extraction of alginates from *Macrocystis pyrifera* 4. Conversion of alginic acid to sodium alginate, drying and milling. *J. App. Phycol.* 14, 445–451.
11. Kloareg, B. and Quatrano, R.S. 1988. Structure of the cell walls of marine algae and ecophysiological functions of the matrix polysaccharides. *Oceanogr. Mar. Biol. Ann. Rev.* 26, 259–315.
12. Turner, N.J. 2003. The ethnobotany of edible seaweed (*Porphyra abbottae* and related species; Rhodophyta: Bangiales) and its use by First Nations on the Pacific Coast of Canada. *Can. J. Bot.* 81(4), 283–293.
13. Boukhari, S. 1998. From Chile, a $50-million crop. *UNESCO Courier,* August.
14. Cokos, B. 2002. Got seaweed? *Belizean Journeys Newsletter.* http://www.belizeanjourneys.com/features/seaweed/newsletter.html.
15. Grieve, M. 1971. *A Modern Herbal.* New York: Dover Publications.
16. Egger, G., Cameron-Smith, D. and Stanton, R. 1999. The effectiveness of popular, non-prescription weight loss supplements. *Med. J. Aust.* 171, 604–608.
17. Gerard, J. 1975. *The Herbal or General History of Plants.* Originally published in 1633 Lib 3. NewYork: Dover Publications, pp. 1566–1615.
18. Pliny the Elder 1855. *The Natural History.* Translated by John Bostock, M.D., F.R.S. H.T. Riley, Esq., B.A. London: Taylor and Francis.
19. Angulo, Y. and Lomonte, B. 2003. Inhibitory effect of fucoidan on the activities of crotaline snake venom myotoxic phospholipases A_2. *Biochem. Pharmacol.* 66, 1993–2000.
20. Marchant, G. 1999. Masters in their own tepees—people and tourism—aboriginal tourism in Australia and Canada. *UNESCO Courier,* July.
21. Blunden, G. 1997. Biologically active compounds from marine organisms. *Pest. Sci.* 51, 483–486.
22. Tseng, C. 2001. Algal biotechnology industries and research activities in China. *J. Appl. Phycol.* 13, 375–380.
23. Funahashi, H., Imai, T., Mase, T., Sekiya, M., Yokoi, K., Hayashi, H., Shibata, A., Hayashi, T., Nishikawa, M., Suda, N., Hibi, Y., Mizuno, Y., Tsukamura, K., Hayakawa, A. and Tanuma, S. 2001. Seaweed prevents breast cancer? *Jpn. J. Cancer Res.* 92, 483–487.

24. Furusawa, E. and Furusawa, S. 1989. Anticancer potential of Viva-Natural, a dietary seaweed extract, on Lewis lung carcinoma in comparison with chemical immunomodulators and on cyclosporine-accelerated AKR leukemia. *Oncology* 46, 343–348.
25. Furusawa, E., Furusawa, S. and Chou, S.C. 1991. Antileukemic activity of Viva-Natural, a dietary seaweed extract, on Rauscher murine leukemia in comparison with anti-HIV agents, azidothymidine, dextran sulfate and pentosan polysulphate. *Cancer Lett.* 56, 197–205.
26. Itoh, H., Noda, H., Amano, H., Cun Zhuaug, Mizuno, T. and Itoh, H. 1993. Antitumor activity and immunological properties of marine algal polysaccharides, especially fucoidan, prepared from *Sargassum thunbergii* of Phaeophyceae. *Anticancer Res.* 13(6A), 2045–2052.
27. Yamatoto, I., Maruyama, H., Takahashi, M. and Komiyama, K. 1986. The effect of dietary or intraperitoneally injected seaweed preparations on the growth of sarcoma-180 cells subcutaneously implanted into mice. *Cancer Lett.* 30, 125–131.
28. Ohigashi, H., Sakai, Y., Yamaguchi, K., Umezaki, I. and Koshimizu, K. 1992. Possible anti-tumor promoting properties of marine algae and *in vivo* activity of Wakame seaweed extract. *Biosci. Biotechnol. Biochem.* 56, 994–995.
29. Ho, Zhi-chien. 1993. Principles of diet therapy in ancient Chinese medicine: "Huang Di Nei Jing." *Asia Pacific J. Clin. Nutr.* 2, 91–95.
30. Riley, M. 1997. *Maori Healing and Herbal.* Viking Sevenseas, Paraparaumu, New Zealand.
31. Teas, J., Pino, S., Critchley, A. and Braverman, L.E. 2004. Variability of iodine content in common commercially available edible seaweeds. *Thyroid* 14, 836–841.
32. Killick, S.R., Vaughan Williams, C.A. and Elstein, M. 1985. A comparison of prostaglandin E2 pessaries and laminaria tents for ripening the cervix before termination of pregnancy. *Br. J. Obstet. Gynaecol.* 92, 518–521.
33. Hesp, R. and Ramsbottom, B. 1965. Radiobiology—effects of sodium alginate in inhibiting uptake of radiostrontium by the human body. *Nature* 208, 1341–1342.
34. Gong, Y.F., Huang, Z.J., Qiang, M.Y., Lan, F.X., Bai, G.A., Mao, Y.X., Ma, X.P. and Zhang, F.G. 1991. Suppression of radioactive strontium absorption by sodium alginate in animals and human subjects. *Biomed. Environ. Sci.* 4, 273–282.
35. Shimotoyodome, A., Meguro, S., Hase, T., Tokimitsu, I. and Sakata, T. 2001. Sulfated polysaccharides, but not cellulose, increase colonic mucus in rats with loperamide-induced constipation digestive diseases and sciences. *Dig. Dis. Sci.* 46, 1482–1489.
36. Mandel, K.G., Daggy, B.P., Brodie, D.A. and Jacoby, H.I. 2000. Review article: alginate-raft formulations in the treatment of heartburn and acid reflux. *Aliment Pharmacol. Ther.* 14, 669–690.
37. Tonnesen, H.H. and Karlsen, J. 2002. Alginate in drug delivery systems. *Drug Dev. Ind. Pharm.* 28(6), 621–630.
38. Gilchrist, T. and Martin, A.M. 1983. Wound treatment with Sorbsan—an alginate fibre dressing. *Biomaterials* 4(4), 317–320.
39. Schopper, C., Moser, D., Sabbas, A., Lagogiannis, G., Spassova, E., König, F., Donath, K. and Ewers, R. 2003. The fluorohydroxyapatite (FHA) FRIOS Algipore is a suitable biomaterial for the reconstruction of severely atrophic human maxillae. *Clin. Oral Implants Res.* 14, 743–749.
40. López-Macià, A., Jiménez, J.C., Royo, M., Giralt, E. and Albericio, F. 2001. Synthesis and structure determination of kahalalide F (1,2). *J. Am. Chem. Soc.* 123, 11398–11401.

41. Wu, H., Song, Z., Hentzer, M., Andersen, J.B., Molin, S., Givskov, M. and Hoiby, N. 2004. Synthetic furanones inhibit quorum-sensing and enhance bacterial clearance in *Pseudomonas aeruginosa* lung infection in mice. *J Antimicrob. Chemother.* 53, 1054–1061.
42. Mayer, A.M. and Gustafson, K.R. 2004. Marine pharmacology in 2001-2: antitumor and cytotoxic compounds. *Eur. J. Cancer* 40, 2676–2704.
43. Kim, S., Moon, S. and Popkin, B. 2000. The nutrition transition in South Korea. *Am. J. Clin. Nutr.* 71, 44–53.
44. Adami, H.O., Signorello, L.B. and Trichpoulos, D. 1998. Towards an understanding of breast cancer etiology. *Cancer Biol. Semin.* 183, 255–262.
45. Teas, J. 1981. The consumption of seaweed as a protective factor in the etiology of breast cancer. *Med. Hypotheses* 7, 601–613.
46. kushi, L.H., Cunningham, J.E., Hebert, J.R., Lerman, R.H., Bandera, E.V. and Teas, J. 2001. The macrobiotic diet in cancer. *J.Nutr.* 131 (11 Suppl), 3056S–3064S.
47. Teas, J., Hebert, J.R., Fitton, J.H. and Zimba, P.V. 2004. Algae—a poor man's HAART? *Med. Hypotheses* 62, 507–510.
48. Teas, J. Patterson, K. and Royer, J. 2005. Could dietary algae protect against HIV progression? *HIV Nutrition Update* 9(3), 6–18.
49. Maruyama H, Tamauchi, H., Hashimoto, M. and Nakano, T. 2003. Antitumor activity and immune response of Mekabu fucoidan extracted from sporophyll of *Undaria pinnatifida. In Vivo.* 17, 245–250.
50. Mao, T.K., Van de Water, J. and Gershwin, M.E. 2005. Effects of a *Spirulina*-based dietary supplement on cytokine production from allergic rhinitis patients. *J. Med. Food.* 8, 27–30.
51. Shimizu, J., Wada-Funada, U., Mano, H., Matahira, U., Kawaguchi, M. and Wada, M. 2005. Proportion of murine cytotoxic T-cell is increased by high-molecular weight fucoidan extracted from Okinawa Mozuku (*Cladosiphon okamuranus*). *J. Health Sci.* 51, 394–397.
52. Ishihara, K., Murata, M., Kaneniwa, M., Saito, H., Komatsu, W. and Shinohara, K. 2000. Purification of stearidonic acid (18:4(n-3)) and hexadecatetraenoic acid (16:4(n-3)) from algal fatty acid with lipase and medium pressure liquid chromatography. *Biosci. Biotechnol. Biochem.* 64, 2454–2457.
53. Terasaki, M. and Itabashi, Y. 2003. Glycerolipid acyl hydrolase activity in the brown alga *Cladosiphon okamuranus* Tokida. *Biosci. Biotechnol. Biochem.* 67, 1986–1989.
54. Khotimchenko, S.V. 2003. The fatty acid composition of glycolipids of marine macrophytes. *Russ. J. Mar. Biol.* 29, 126–128.
55. Eitsuka, T., Nakagawa, K., Igarashi, M. and Miyazawa, T. 2004. Telomerase inhibition by sulfoquinovosyldiacylglycerol from edible purple laver (*Porphyra yezoensis*). *Cancer Lett.* 212, 15–20.
56. Sugawara, T., Baskaran, V., Tsuzuki, W. and Nagao, A. 2002. Brown algae fucoxanthin is hydrolyzed to fucoxanthinol during absorption by Caco-2 human intestinal cells and mice. *J. Nutr.* 132, 946–951.
57. Kakinuma, M., Park, C.S. and Amano, H. 2001. Distribution of free L-cysteine and glutathione in seaweeds. *Fish. Sci.* 67, 194–196.
58. Katsube, T., Yamasaki, Y., Iwamoto, M. and Oka, S. 2003. Hyaluronidase-inhibiting polysaccharide isolated and purified from hot water extract of *Sporophyll* of *Undaria pinnatifida. Food Sci. Technol. Res.* 9, 25–29.
59. Parish, C.R., Coombe, D.R., Jakobsen, K.B., Bennett, F.A. and Underwood, P.A. 1987. Evidence that sulphated polysaccharides inhibit tumour metastasis by blocking tumour cell derived heparanases. *Int. J. Cancer* 40, 511–518.

60. Shibata, T., Nagayama, K., Tanaka, R., Yamaguchi, K. and Nakamura, T. 2003. Inhibitory effects of brown algal phlorotannins on secretory phospholipase A_2s, lipoxygenases and cyclooxygenases. *J. Appl. Phycol.* 15, 61–66.
61. Wessels, M., Konig, G. and Wright, A. 1999. A new tyrosine kinase inhibitor from the marine brown alga *Stypopodim zonale*. *J Nat. Prod.* 62, 927–930.
62. Newton, R.P., Kingston, E.E. and Overton, A. 1995. Identification of novel nucleotides found in the red seaweed *Porphyra umbilicalis*. *Rapid Commun. Mass Spectrom.* 9, 305–311.
63. Claudio, F. and Stendardo, B. 1968. Contributo clinico sperimentale sull'uso di un fitocolloide in oncologia. *Minerva Medica* 3617–3622.
64. Mathew, B., Rengaswamy, S., Nair, P.P., Cherian, V., Thara, S., Padmavathy, A.B., Sreedevi, A.N., Madhavan, K. and Shnan, N. 1995. Evaluation of chemoprevention of oral cancer with *Spirulina fusiformis*. *Nutr. Cancer* 24, 197–202.
65. Yoshida, T., Hirano, A., Wada, H., Takahashi, K. and Hattori, M. 2004. Alginic acid oligosaccharide suppresses Th2 development and IgE production by inducing IL-12 production. *Int. Arch. Allergy Immunol.* 133, 239–247.
66. Matsumoto, S., Nagaoka, M., Hara, T., Kimura-Takagi, I., Mistuyama, T. and Ueyama, S. 2004. Fucoidan derived from *Cladosiphon okamuranus* Tokida ameliorates murine chronic colitis through the down-regulation of interleukin-6 production on colonic epithelial cells. *Clin. Exp. Immunol.* 136, 432–439.
67. Klarzynski, O., Descamps, V., Plesse, B., Yvin, J.C., Kloareg, B. and Fritig, B. 2003. Sulfated fucan oligosaccharides elicit defense responses in tobacco and local and systemic resistance against tobacco mosaic virus. *Mol. Plant Microbe Interact.* 16(2), 115–122.
68. Thompson, K.D. and Dragar, C. 2004. Antiviral activity of *Undaria pinnatifida* against herpes simplex virus. *Phytother. Res.* 18, 551–555.
69. Hudson, J.B., Kim, J.H., Lee, M.K., Dewreede, R.E. and Hong, Y.K. 1999. Antiviral compounds in extracts of Korean seaweeds; evidence for multiple activities. *J. Appl. Phycol.* 10, 427–434.
70. Ponce, N.M.A., Pujol, C.A., Damonte, E.B., Flores, M.L. and Storitz, C.A. 2003. Fucoidans from the brown seaweed *Adenocystis utricularis*: extraction methods, antiviral activity and structural studies. *Carbohydr. Res.* 338, 153–165.
71. Pujol, C.A., Esteves, J.M., Carlucci, M.J., Ciancia, M., Cerezo, A.S. and Damonte, E.B. 2002. Novel DL-galactan hybrids from the red seaweed *Gymnogongrus torulosus* are potent inhibitors of herpes simplex virus and dengue virus. *Antivir. Chem. Chemother.* 13, 83–89.
72. Schaeffer, D.J. and Krylov, V.S. 2000. Anti HIV activity of extracts and compounds from algae and cyanobacteria. *Ecotoxicol. Environ. Safety* 45, 208–227.
73. Witvrouw, M. and de Clercq, E. 1997. Sulfated polysaccharides extracted from sea algae as potential anti-viral drugs. *Gen. Pharmacol.* 29, 497–511.
74. Romanos, M., Andrada-Serpa, M.J., Dos, S., Ribeiro, A., Yoneshique-Valentin, Y., Costa, S.S. and Wiqq, M.D. 2002. Inhibitory effect of extracts of Brazilian marine algae on human T-cell lymphotropic virus type 1 (HTLV-1)-induced syncytium formation in vitro. *Cancer Invest.* 20, 46–54.
75. Cooper, R., Dragar, C., Elliot, K., Fitton, J.H., Godwin, J. and Thompson, K. 2002. GFS, a preparation of Tasmanian *Undaria pinnatifida* is associated with healing and inhibition of reactivation of herpes. *BMC Complement Altern. Med.* 2, 11.
76. Zeitlin, L. and Whaley, K.J. 2002. Microbicides for preventing transmission of genital herpes. *Herpes* 9(1), 4–9.

77. Sato, S., Yoshinuma, N., Ito, K., Tokumoto, T., Takiguchi, T., Suzuki, Y. and Murai, S. 1998. The inhibitory effect of funoran and eucalyptus extract-containing chewing gum on plaque formation. *J. Oral Sci.* 40, 115–117.
78. Nagaoka, M., Shibata, H., Kimura-Takagi, I., Hashimoto, S., Aiyama, R., Ueyama, S. and Yokokura, T. 2000. Anti-ulcer effects and biological activities of polysaccharides from marine algae. *Biofactors* 12, 267–274.
79. Liao, W.R., Lin, J.Y., Shieh, W.Y., Jeng, W.L. and Huang, R. 2003. Antibiotic activity of lectins from marine algae against marine vibrios. *J. Ind. Microbiol. Biotechnol.* 30, 433–439.
80. Wu, H., Song, Z., Hentzer, M., Andersen, J.B., Molin, S., Givskov, M. and Hoiby, N. 2004. Synthetic furanones inhibit quorum-sensing and enhance bacterial clearance in *Pseudomonas aeruginosa* lung infection in mice. *J. Antimicrob. Chemother.* 53, 1054–1061.
81. Barrabes, J.A., Garcia-Dorado, D., Mirabet, M., Iinserte, J., Agullo, L., Soriano, B., Massaguer, A., Padilla, F., Lidon, R.M. and Soler-Soler, J. 2005. Antagonism of selectin function attenuates microvascular platelet deposition and platelet-mediated myocardial injury after transient ischemia. *J. Am. Coll. Cardiol.* 45, 293–299.
82. Zhang, Q., Li, Z., Xu, Z., Niu, X. and Zhang, H. 2003. Effects of fucoidan on chronic renal failure in rats. *Planta Med.* 69, 537–541.
83. Sugawara, I., Ishizaka, S. and Moller, G. 1982. Carrageenans, highly sulfated polysaccharides and macrophage-toxic agents: newly found human T lymphocyte mitogens. *Immunobiology* 163(5), 527–38.
84. Matsui, M.S., Muizzuddin, N., Arad, S. and Marenus, K. 2003. Sulfated polysaccharides from red microalgae have antiinflammatory properties in vitro and in vivo. *Appl. Biochem. Biotechnol.* 104, 13–22.
85. Bocanegra, A., Nieto, A., Bias, B. and Sanchez-Muniz, F.J. 2003. Diets containing a high percentage of Nori or Konbu algae are well-accepted and efficiently utilised by growing rats but induce different degrees of histological changes in the liver and bowel. *Food Chem. Toxicol.* 41, 1473–1480.
86. Besterman, E.M. and Evans, J. 1957. Antilipaemic agent without anticoagulant action. *Br. Med. J.* 51(5014), 310–312.
87. Panlasigui LN, Baello, O.Q., Dimatangal, J.M. and Dumelod, B.D. 2003. Blood cholesterol and lipid-lowering effects of carrageenan on human volunteers. *Asia Pac. J. Clin. Nutr.* 12, 209–214.
88. Murata, M., Sano, Y., Ishihara, K. and Uchida, M. 2002. Dietary fish oil and *Undaria pinnatifida* (wakame) synergistically decrease rat serum and liver triacylglycerol. *J. Nutr.* 132, 742–747.
89. Mori, H., Kamei, H., Nishide, E. and Nisizawa, K. 1982. Sugar constituents of some sulphated polysaccharides from the sporophylls of Wakame (*Undaria pinnatifida*) and their biological activities. In: *Marine Algae in Pharamaceutical Science*, vol 2, Walter de Gruyter, New York, pp. 109–121.
90. Yu, P., Li, N., Liu, X., Zhou, G., Zhang, Q. and Li, P. 2003. Antihyperlipidemic effects of different molecular weight sulfated polysaccharides from *Ulva pertusa* (Chlorophyta). *Pharmacol. Res.* 48, 543–549.
91. Sachez-Machado, D.I., Lopez-Hernandez, J., Paseiro-Losada, P. and Lopez-Cervantes, J. 2004. An HPLC method for the quantification of sterols in edible seaweeds. *Biomed. Chromatogr.* 18, 183–190.
92. Yeon, S.L., Kuk, H.S. Kim, B.K. and Sanghyun, L. 2004. Anti-diabetic activities of fucosterol from *Pelvetia siliquosa*. *Arch. Pharm. Res.* 27, 1120–1122.

93. Girard, J.P., Marion, C., Liutkus, M., Boucard, M., Rechencq, E., Vidal, J.P. and Rossi, J.C. 1988. Hypotensive constituents of marine algae; 1. Pharmacological studies of laminine. *Planta Med.* 54, 193–196.
94. Sato, M., Oba, T., Yamaguchi, T., Nakano, T., Kahara, T., Funayama, K., Kobayashi, A. and Nakano, T. 2000. Antihypertensive effects of hydrolysates of Wakame (*Undaria pinnatifida*) and their angiotensin-I-converting enzyme inhibitory activity. *Ann. Nutr. Metab.* 46, 259–267.
95. Ikeda, K., Kitamura, A., Machida, H., Watanabe, M., Negishi, H., Hiraoka, J. and Nakano, T. 2003. Effect of *Undaria pinnatifida* (Wakame) on the development of cerebrovascular diseases in stroke-prone spontaneously hypertensive rats. *Clin. Exp. Pharmacol. Physiol.* 30, 44–48.
96. Yamori, Y., Nara, Y., Tsubouchi, T., Sogawa, Y., Ikeda, K. and Horie, R. 1986. Dietary prevention of stroke and its mechanisms in stroke-prone spontaneously hypertensive rats—preventive effect of dietary fibre and palmitoleic acid. *J. Hypertension* 4, 499–452.
97. Hoebler, C., Guillon, F., Darcy-Vrillon, B., Vaugelade, P., Lahaye, M., Worthington, E., Duee, P.H. and Barry, J.L. 2000. Supplementation of pig diet with algal fibre changes the chemical and physicochemical characteristics of digesta. *J. Sci. Food Agric.* 80, 1357–1364.
98. Okada, Y., Ishimaru, A., Suzuki, R. and Okuyama, T. 2004. A new phloroglucinol derivative from the brown alga *Eisenia bicyclis*: potential for the effective treatment of diabetic complications. *J. Nat. Prod.* 67, 103–105.
99. Logeart, D., Prigent-Richard, S., Boisson-Vidal, C., Chaubet, F., Durand, P., Jozefonvicz, J. and Letourneur, D. 1997. Fucans, sulfated polysaccharides extracted from brown seaweeds, inhibit vascular smooth muscle cell proliferation. II. Degradation and molecular weight effect. *Eur. J. Cell. Biol.* 74, 385–390.
100. Patel, M.K., Mulloy, B., Gallagher, K., O'Brien, L. and Hughes, A.D. 2002. The antimitogenic action of the sulphated polysaccharide fucoidan differs from heparin in human vascular smooth muscle cells. *Thromb. Haemost.* 87, 149–154.
101. Berteau, O. and Mulloy, B. 2003. Sulfated fucans, fresh perspectives: structures, functions, and biological properties of sulfated fucans and an overview of enzymes active toward this class of polysaccharide. *Glycobiology* 13, 29–40.
102. Blondin, C., Chaubet, F., Nardella, A., Jozefonvicz, J. and Sinquin, C. 1996. Relationships between chemical characteristics and anticomplementary activity of fucans. *Biomaterials* 17, 597–603.
103. Soeda, S., Sakaguchi, S., Shimeno, H. and Nagamatsu, A. 1992. Fibrinolytic and anticoagulant activities of highly sulfated fucoidan. *Biochem. Pharmacol.* 43, 1853–1858.
104. Vintila, C.D., Schneider, J., Pollack, S. and Farley, T. 2001. Heparin Anticoagulation Promotes CD34 Positive Hematopoietic Progenitor Cells (HPC) Mobilization into the Peripheral Blood (PB). *Proc. am. Soc. Clin. Oncol.*, abstr 46.
105. Frenette, P.S. and Weiss, L. 2000. Sulfated glycans induce rapid hematopoietic progenitor cell mobilization: evidence for selectin-dependent and independent mechanisms. *Blood* 96, 2460–2468.
106. Sweeney, E.A., Lortat-Jacob, H., Priestley, G.V., Nakamoto, B. and Papayannopoulou, T. 2002. Sulfated polysaccharides increase plasma levels of SDF-1 in monkeys and mice: involvement in mobilization of stem/progenitor cells. *Blood* 99, 44–51.
107. Luyt, C.E., Meddahi-Pelle, A., Ho-Tin-Noe, B., Colliec-Jouault, S., Guezennec, J., Louedec, L., Prats, H., Jacob, M.P., Osborne-Pellegrin, M., Letourneur, D. and Michel, J.B. 2003. Low-molecular-weight fucoidan promotes therapeutic revascularization in a rat model of critical hindlimb ischemia. *J. Pharmacol. Exp. Ther.* 305, 24–30.

108. Irhimeh, M.R., Fitton, J.H. and Lowenthal, R.M. 2007. Fucoidan ingestion increases the expression of CXCR4 on human CD34 + cells. (In Press) *Exp Hematol.*
109. Barthe, L., Woodley, J., Lavit, M., Przybylski, C., Philibert, C. and Houin, G. 2004. In vitro intestinal degradation and absorption of chondroitin sulfate, a glycosaminoglycan drug. *Arzneimlttelforschung* 54, 286–292.
110 Li, N., Zhang, Q. and Song, J. 2005. Toxicological evaluation of fucoidan extracted from *Laminaria japonica* in Wistar rats. *Food Chem. Toxicol.* 43, 421–426.
111. de Maar, E.F., Kleibeuker, J.H., Boersma-van Ek, W., The, T.H. and van Son, W.J. 1996. Increased intestinal permeability during cytomegalovirus infection in renal transplant recipients. *Transpl. Int.* 9, 576–580.
112. Pernet, P., Vittecoq, D., Kodjo, A., Randrianarisolo, M.H., Dumitrescu, L., Blondon, H., Bergmann, J.F., Giboudeau, J. and Aussel, C. 1999. Intestinal absorption and permeability in human immunodeficiency virus-infected patients. *Scand. J. Gastroenterol.* 34, 29-34.
113. Mutwiri, G., Bowersock, T., Kidane, A., Sanchez, M., Gerdts, V., Babiuk, L.A. and Griebel, P. 2002. Induction of mucosal immune responses following enteric immunization with antigen delivered in alginate microspheres. *Vet. Immunol. Immunopathol.* 87, 269–276.
114. Cohen, S.M. and Ito, N. 2002. A critical review of the toxicological effects of carrageenan and processed eucheuma seaweed on the gastrointestinal tract. *Crit. Rev. Toxicol.* 32, 413–444.
115. Almela, C., Algora, S., Benito, V., Clemente, M.J., Devesa, V., Suner, M.A., Velez, D. and Montoro, R. 2002. Heavy metal, total arsenic, and inorganic arsenic contents of algae food products. *J. Agric. Food Chem.* 50, 918–923.
116. Laparra, J.M., Velez, D., Montoro, R., Barbera, R. and Farre, R. 2003. Estimation of arsenic bioaccessibility in edible seaweed by an in vitro digestion method. *J. Agric. Food Chem.* 51, 6080–6085.
117. Wei, C., Li, W., Zhang, C., Van Hulle, M., Cornelis, R. and Zhang, X. 2003. Safety evaluation of organoarsenical species in edible Porphyra from the China Sea. *J. Agric. Food Chem.* 51, 5176–5182.
118. Feldmann, J., John, K. and Pengprecha, P. 2000. Arsenic metabolism in seaweed-eating sheep from Northern Scotland. *Fresenius J. Anal. Chem.* 386, 116–121.
119. Conradt, L. 2000. Use of a seaweed habitat by red deer (*Cervus elaphus* L.). *J. Zool. Lond.* 250, 541–549.
120. Van Netten, C., Cann, S.A.H., Morley, D.R. and Van Netten, J.P. 2000. Elemental and radioactive analysis of commercially available seaweed. *Sci. Total Environ.* 255, 169–175.
121. Bartle, W.R., Madorin, P. and Ferland, G. 2001. Seaweed, vitamin K and warfarin. *Am. J. Health Syst. Pharm.* 58, 2300.
122. Nishino, T., Nishioka, C., Ura, H. and Nagumo, T. 1994. Isolation and partial characterization of a novel amino sugar-containing fucan sulfate from commercial *Fucus vesiculosus* fucoidan. *Carbohydr. Res.* 255, 213–224.

15 Nutraceuticals and Functional Foods from Marine Microbes: An Introduction to a Diverse Group of Natural Products Isolated from Marine Macroalgae, Microalgae, Bacteria, Fungi, and Cyanobacteria

Adam M. Burja and Helia Radianingtyas

CONTENTS

15.1 INTRODUCTION

Microbial life has contributed much to the field of pharmacology, with an astonishing one-quarter of all biologically active secondary metabolites derived from filamentous fungi [1,2]. Marine microbial natural products, produced by microorganisms that live in the sea, on the other hand, have only recently received attention from chemists and pharmacologists [3–5]. Interest on the part of chemists has been twofold: natural product chemists have probed marine microbes as sources of new and unusual organic molecules with novel biological activities, while synthetic chemists have followed by targeting these novel structures for development of new analogs and synthetic strategies [6].

The rationale for investigating the chemistry of marine organisms has changed over the past few decades. Early investigations were largely of a "symbiotic" nature, detailing the metabolite profiles of microbes known to be associated with novel natural product producing marine organisms [7,8]. However, analogous to investigations of marine organisms themselves, more recent studies of marine microorganisms have focused on isolations of more unusual microbes from extreme environments (the "extremophilic" microorganisms), as well as recent attention shifting toward studies of cyanobacteria [9]. Pharmacological evaluations of marine microbial natural products have likewise undergone an evolution: beginning with the early investigations of toxins, followed by studies of antitumor and anti-HIV activities, to the present day, where a myriad of activities based on whole-animal models and receptor-binding assays are being pursued [10]. Indeed, even the focus of these studies has altered with time. Today, there is as much interest in the derivation of nutraceutical compounds for human consumption as there is for pharmaceutical treatments.

The term nutraceutical resulted from the association of the words nutrition and pharmaceutical. Some nutraceuticals are midway between food and medication. Nutraceutical (or nutriceutical) compounds were defined in 1994 by the U.S. Institute of Medicine's food and nutrition board as "any substance that is a food or part of a food which provides medicinal or health benefits including the prevention and treatment of disease, beyond the traditional nutrients it contains." They are not the same as pharmaceuticals, not being designed specifically for medical use under supervision. Similarly, functional foods are defined as "products derived from natural sources, whose consumption is likely to benefit human health and enhance performance." Functional foods and nutraceuticals are generally used as products to nourish the human body after physical exertion or as a preventative measure against ailments.

Consumers are progressively more concerned about debilitating illnesses such as cancer, heart disease, cholesterol, and osteoporosis. At the same time, many consumers are reluctant to rely only on prescription medications to treat or prevent these illnesses, so they are increasingly examining the link between diet and health. As their research demonstrates links between nutritional ingredients and better health, food companies are petitioning the U.S. Food and Drug Administration (FDA) to permit health claims on the labels of foods that contain these beneficial, functional ingredients. The combination of research with positive results, an increasing number of permissible health claims on functional food packaging, and media attention (concerning functional ingredients and disease prevention) has spurred growth in the nutraceutical industry that has surpassed that of the food industry as a whole.

In the United States and Japan, the homelands of these new food philosophies, business is booming with estimated nutraceutical sales of US$16.7 billion (estimated at US$28 billion by 2006) and US$4.8 billion, respectively, whereas in Europe, the current market estimate for nutraceuticals is approximately US$3.2 billion. Conservative assessments of the nutraceuticals market worldwide between now and 2010 estimate its worth to be within the vicinity of US$700 billion. In 2002, retail sales of functional foods grew by an estimated 9.2%, while the nutraceuticals market as a whole increased in value by 37.7%.

Traditionally, plant-derived nutritional supplements have predominated the market. Nevertheless, with the dramatic increase in profitability within the nutraceutical and functional food markets, new environments are being investigated. Recently, the sea is beginning to show promise as an abundant, hitherto untapped source of nutraceuticals. Marine microorganisms, in particular, represent a relatively neglected source of extracts and natural products derived from marine macroalgae, microalgae, cyanobacteria (blue-green algae), bacteria, and fungi. The health benefits of marine microorganisms have been appreciated for many years in countries such as Japan, where 21 species of macroalgae are used daily for cooking, 6 of which have been used since the eighth century. There are also reports that the cyanobacterium, *Spirulina* was first used as a food source in Mexico during the Aztec civilization, some 700 years ago. Furthermore, the medicinal qualities of cyanobacteria in continental Europe were first noted as early as 1500 BC, when *Nostoc* species were used to treat gout, fistula, and several forms of cancer. Indeed, there are many areas of the world where microalgae are harvested from the wild, such as *Nostoc* sp. (Bolivia, China, Ecuador, Fiji, Indonesia, Mongolia, Peru, Thailand, Japan, and Siberia), *Spirogyra* sp. (Burma, Indochina), *Oedogonium* sp. (India), *Apanothece sacrum* (Japan), *Chroococcus turgidus* (Mexico), *Enteromorpha* sp. (Hawaii, United States), and *Aphanizomenon flos-aquae* (Oregon, United States). Meanwhile, the culture of edible macroalgae also has a long tradition in the East, particularly the Rhodophyta, *Porphyra tenera*, which is the object of a large trade owing to its larger than normal content of proteins.

While foods with documented health benefits have been consumed for centuries (as discussed above), recognized benefits associated specifically with the ingestion of microorganisms have only become relevant, with the introduction of probiotic supplements, within the last three to four decades. Recently, the benefits of these "friendly" bacteria have become recognized by mainstream consumers, with annual sales within the manufacture market rising steadily. In this chapter, nutraceuticals and functional foods derived from marine microorganisms, including macroalgae, microalgae, bacteria, cyanobacteria, and fungi will be discussed. It is hoped that this chapter will allow the reader (scientist and lay-person alike) to gain a better understanding of the types of nutraceutical compounds that are produced by these microbes and the various beneficial properties these organisms contain either as whole-cell extracts or as their chemical derivatives.

Marine microorganisms encompass a complex and diverse assemblage of microscopic life and occur throughout the oceans, including environments of extreme pressure, salinity, and temperature. As a result marine microorganisms have developed complex and unique metabolic and physiological capabilities that not only ensure survival in highly stressful growth conditions (e.g., highly oxidative environments, competition for nutrients, symbiosis, hypersaline, and high- and low-temperature environments), but also offer the potential for a unique source of novel nutraceuticals that would not be observed from terrestrial sources. This chapter is not intended to provide a comprehensive review of marine natural products; as there are several that already give such coverage [11,12]. Instead, this chapter will focus solely upon various types of natural products that are currently under consideration as nutraceutical products and their various sources. The structure of the main groups of compounds referred to in this chapter are shown throughout, as a means of illustrating the complex chemical compounds these organisms can produce.

From a look through the scientific literature currently available on the subject, Table 15.1 lists the main groups of nutraceutical compounds currently known for marine microorganisms. For the purpose of this review, macroalgae species will also be considered owing to the current ambiguity between what is classified like microalgal and macroalgal. Recently, work has been published whereby it is possible to culture marine macroalgal species within embryonic forms much as microalgae and cyanobacteria [13,14]. Therefore, for the purpose of this review we are classifying any organism that can be cultured using standard microbiological techniques as microbial. Furthermore, within this review we will be focusing on the various polysaccharide, polyunsaturated fatty acids, and antioxidants currently derived from these marine microorganisms. Moreover, we shall also introduce several lesser-known compounds such as sterols and vitamins, as well as whole-cell extracts from several microalgal and cyanobacterial species, in which the active compound has not been identified, isolated or synthesized *in vitro* (see Table 15.1 for more detail).

15.2 NUTRACEUTICAL COMPOUNDS

15.2.1 Marine Microbial-Derived Polysaccharides

Polysaccharides are water soluble or insoluble, high-molecular-weight carbohydrates formed by the condensation of monosaccharide, or monosaccharide derivative units into linear or branched chains, and include homopolysaccharides (composed of one type of monosaccharide only) and heteropolysaccharides (composed of a mixture of different monosaccharides). Homoglycans or homopolysaccharides are exemplified by (1) glycans such as cellulose, starch, glycogen, charonin, laminaran, and dextran; (2) fructans such as inulin, levan, mannans, and xylans; (3) galacturonans such as pectin; (4) mannuronans such as alginates; and (5) *N*-acetylglucosamine polymers such as chitin. While heteroglycans or heteropolysaccharides are exemplified by (1) guaran, mannan, heparin, and chondroitin or (2) diheteroglycans including hyaluronic acid, fucoidan, and agarose. Both types of polysaccharides are yielded by the cultivation and harvesting of various macroalgal, microalgal, and cyanobacterial species (Table 15.1).

Examples of useful polysaccharides produced by macroalgae include linear, sulfated polysaccharides, collectively known as carrageenans **{1}** and chitin **{2}**. These compounds possess potent anticoagulating, blood-purifying, and antitumor activities [15,16]. Carrageenans may be isolated from selected species of brown and red macroalgae and some cyanobacteria [17] (Table 15.1). Synthesis of exocellular polysaccharides in microbes plays a major role in protecting the cells from stress in extreme habitats and from other harmful conditions, such as antibacterial agents or predation by protozoans. They have the unique ability to form an almost infinite variety of gels at room temperature, rigid or compliant, tough or tender with high or low melting points [18]. The carrageenan family has three main branches named kappa, iota, and lambda, which are well differentiated in terms of their gelling properties and protein reactivity [19]. Kappa carrageenans produce strong rigid gels while those made with iota products are flaccid and compliant [20]. Although lambda carrageenans do not gel in water, they interact strongly with proteins to stabilize a wide range of dairy products [18]. Over the past 30 years, the use of carrageenans within foods has grown by 5–7% per annum. In addition to their use as bioplastics [21] or wound defenses and stabilizers [22], sulfated polysaccharides have been demonstrated to provide nutritional support for immune health [16,23].

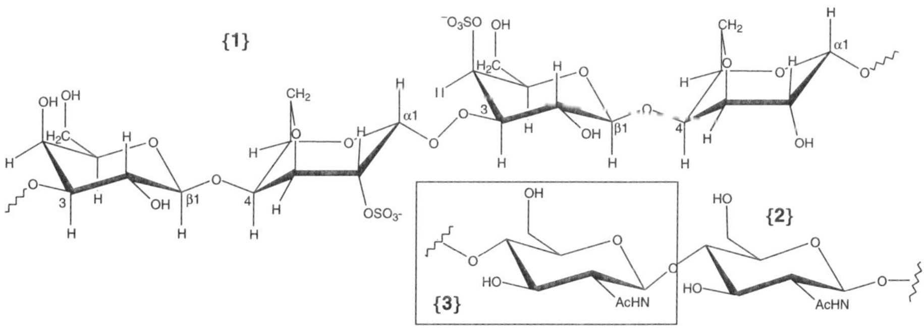

Chitin **{2}** (alpha- and beta-) is a linear polysaccharide composed of repeating *N*-acetyl-D-glucosamine units **{3}**, whose number can vary widely between producers (5000–8000 *N*-acetyl-D-glucosamine residues in crustaceans to only 100 in fungi). Traditional sources of chitin for industrial processing are crustacean shells and fungal mycelia. Chitin is the second-most abundant biopolymer on Earth, and within the marine environment is found mainly within invertebrates, marine diatoms, green macroalgae, and fungi. Annual worldwide synthesis of this polysaccharide in the marine ecosystem has been estimated at about

TABLE 15.1
Types of Nutraceuticals and Functional Foods Discussed within This Chapter

Compound	Products	Microbial Origin	References
Polysaccharides			
Sulfated, phosphated (homo-) or (hetero-) polysaccharides, chitin, glucosamine	Milk stabilizer, food and drink supplements, tablets, granules, powders, capsules	*Acanthopeltis japonica; Anabaena* sp., *A. cylindrica, A. flos-aquae*; *Anacystis nidulans*; *Aphanocapsa halophytica*; *Campyloephora hypnaeoides*; *Ceramium kondoi*; *Chlorella ellipsoidea, C. pyrenoidosa*; *Chondrus crispus*; *Chordariales nemacystus*; *Chroococcus minutus*; *Cladosiphon okamuranus*; *Cyanospira capsulate*; *Cyanothece* sp.; *Eucheuma cottonii*; *Fucus* sp.; *Gelidium amansil, G. japonicum, G. pacificum, G. subcostatum*; *Gigartina* sp.; *Gloeothece* sp.; *Gracilaria verrucosa, Gr. gigas*; *Hydrilla* sp.; *Hypeaceae* sp.; *Iridaea* sp.; *Lyngbya confervoides*; *Mastigocladus laminosus*; *Microcoleus* sp.; *Microcystis aeruginosa, M. flos-aquae*; *Monostroma nitidum*; *Nostoc* sp., *N. calcicola, N. commune, N. insulare, N. linckia f. muscorum*; *Oscillatoria* sp., *O. amphibian, O. corallinae*; *Palmaria palmat*; *Phaeophycea* sp.; *Phormidium* sp., *P. ectocarpi, P. foveolarum, P. minutum*; *Pteocladia tenuis*; *Spirulina platensis, Synechocystis* sp.	[104]
Polyunsaturated fatty acids			
n-3 and n-6 Polyunsaturated fatty acids	Food and drink supplements, tablets, granules, powders, capsules	*Alteromonas putrefaciens*; *Amphidinium* sp., *Amph. carterae*; *Conidiobolus nanodes, Crypthecodinium cohnii*; *Cyclotella cryptica*; *Entomophthora* sp., *Ent. exitalis*; *Haliphthoros philippinensis*; *Isochrysis galbana*; *Japonochytrium* sp.; *Labyrinthula* sp.; *Mortierella alpinapeyron, M. marina*; *Nannochloropsis oculata*; *Oleispira antartica*; *Pavlova lutheri*; *Phaeodactylum tricornutum*; *Photobacterium profundum*; *Prophyridium* sp.; *Psychromonas kaikoae*; *Phytium* sp., *Pyt. acanthicum*; *Saprolegnia parasitica*; *Schizochytrium* sp., *Schizo. aggregatum, Schizo. limacinum*; *Shewanella frigidimarina, S. gelidimarina, S. hanedai, S. marinintestina, S. olleyana, S. putrefaciens, S. sairae, S. schlegeliana*; *Spirulina* sp., *Sp. platensis*; *Thalassiosira pseudonana*; *Thraustochytrium* sp., *Thr. aureum, Thr. roseum*; *Trichodesma* sp.; *Ulkenia* sp.; *Vibrio* sp., *V. marinus*	[60,136–155]

Antioxidants			
Astaxanthin, beta-carotene, mycosporine amino acids, DMSP, DMS, scytonemin, (7-hydroxy)-cymopol, avrainvilleol, aragilamide	Food and drink supplements, tablets, granules, powders, capsules	*Avrainvillia* sp.; *Chondrus crispus*; *Cymopolia barbata*; *Dunaliella salina*; *Haematococcus* sp., *H. pluvialis*; *Microcoleus chthonoplastes*; *Mortensia fragilis*; *Phaffia rhodozyma*; *Prochloron* sp.; *Pseudoalteromonas* sp.; *Scytonema* sp.; *Spirulina platensis*; *Vibrio* sp., *V. harveyi*	[68]
Other nutritional compounds			
Vitamin B_{12}, sterols	Research needed	*Aphanizomenon flos-aquae, Spirulina* sp., *Sp. platensis*	[49,93,95]
Whole-cell extracts			
	Food and drink supplements, tablets, granules, powders, capsules	*Aphanizomenon flos-aquae*; *Chlorella sp., Chl. ellipsoidae*; *Dunaliella* sp., *D. salina*; *Scenedesmus* sp.; *Spirulina* sp., *Sp. maxima, Sp. platensis*	[49,92,93,95]

1600 million tons. Recent investigations confirm the suitability of chitin and its derivatives in chemistry, biotechnology, medicine, veterinary, dentistry, agriculture, nutraceutical, environmental protection, and textile production. The development of technologies based on the utilization of chitin derivatives (especially chitosan and glucosamine) is driven by their unusual polyelectrolite properties, the presence of reactive functional groups, gel-forming ability, high adsorption capacity, biodegradability, and bacteriostatic, fungistatic, and antitumor influence [24–28].

Today there are several companies, mainly within the United States and Japan, producing chitin and chitosan on a commercial scale. Within the crustacean shell waste industry, the industrial production of chitin is limited by season, limited supply in some countries, and environmental pollution considerations. Conversely, as chitin and its soluble derivative chitosan and different acidic polysaccharides are the principle components of several fungal species, attention has been drawn to fungi as an attractive, alternative source. Specifically, the mycelia of *Allomyces, Aspergillus, Penicillium, Fusarium, Mucor, Rhizopsus, Choanephora, Tamnidium, Zygorrhynchus,* and *Phycomyes* species (members of the lower Zygomycetes), as well as higher Basidiomycetes have been documented to produce industrially relevant quantities of these products (Table 15.1). Species from both orders have been documented within the marine environment, although their origin is unknown, and are included here accordingly. In addition to being sources of these compounds, many microorganisms (fungi and bacteria) are able to effectively degrade crustacean shell waste into chitin and chitosan products [29,30].

Chitin is a nontoxic, biodegradable polysaccharide that has been shown to be helpful in the prevention and healing of various diseases. Chemical modifications of this product have improved both solubility and biological activity, and include such things as macrophage activation, antithrombogenic properties, stimulation of lyzosome secretion, pathogen inhibition, immunoadjuvant, antiurucemic, and antiosteoporotic activites [31]. An alternative application of chitin is the production of the functional food supplement *N*-acetyl-D-glucosamine, with evidence of glucosamine isolation possible from marine macroalgae, as well as fungal species. Glucosamine is sold in various forms including sulfated, hydrochlorinated, iodinated, and *N*-acetylated [32]. It has been suggested that sulfated glucosamine is the most effective nutraceutical formulation because sulfate is needed in the synthesis of various cartilage compounds [33].

The Kabushiki Kaisha Yakult Honsha company, Japan,* has patented another polysaccharide derivative (which contains fucoidan and rhamnan or rhamnan sulfate polysaccharides), extracted from the marine macroalgae, *Phaeophyceae* species, *Cladosiphon okamuranus*, *Chordariales nemacystus*, *Hydrilla* sp., *Fucus* sp. and *Monostroma nitidum* [34]. This compound is used specifically as a therapeutic agent for the prevention and treatment of gastric ulcers (specifically inhibiting the adhesion of *Helicobacter pylori*) [34] and administered as tablets, granules, powders, or capsules. Furthermore, Takara Shuzo Company, Kyoto, Japan has

* Famous for the invention of the first, internationally sold, probiotic drink—Yakult™.

developed a medicinal composition exemplified by viscous polysaccharides isolated from red algae (specifically: *Gelidium amansil, G. japonicum, G. pacificum, G. subcostatum, Pteocladia tenuis, Acanthopeltis japonica, Gracilaria verrucosa, Gr. gigas, Ceramium kondoi, Campylaephora hypnaeoides,* and *Gigartinaceae, Solieriaceae,* and *Hypneaceae* species), consisting of at least one 3,6-anhydrogalactopyranose, proposed for the treatment or prevention of diabetes, rheumatism, cancer, and contains various inhibitory factors. This medicinal component is administered as either a food or drink supplement [35].

Recently, Ocean Nutrition Canada has developed several polysaccharide complexes (containing glucose and any combination of galactose, rhamnose, mannose, arabinose, *N*-acetyl glucosamide, and *N*-acetyl galactosamine) having immunomodulating properties, specifically immune stimulatory activity, from *Chlorella pyrenoidosa* (and possibly *C. ellipsoidea*) [36,37]. *Chlorella* extracts may be administered to a mammal to increase proliferation of splenocytes and increase production of cytokines and may be administered as a supplement to further stimulate the immune response. Further, these complexes were shown to inhibit the proliferation of *Listeria monocytogenes* and *Candida albicans* [37].

Since the 1950s, many cyanobacteria have been reported to be capable of synthesizing exocellular polysaccharides and, in some cases, releasing them into the surroundings. To date, approximately 70 strains, covering subsections I, III, and IV of the cyanobacterial classification scheme, have been studied. On the whole, only 10 different monosaccharides have been found to be produced by cyanobacteria: the hexoses, glucose, galactose, and mannose; the pentoses, ribose, xylose, and arabinose; the deoxyhexoses, fucose and rhamnose; and the acidic hexoses, glucuronic and galacturonic acids [38]. In a few cases, however, the presence of additional types of monosaccharides have also been reported [39,40]. For example, in benthic cyanobacteria, adhesion to sediment is modulated by cell hydrophobicity, which is determined by extracellular polymeric substances. These are polysaccharides bearing nonsugar components. For instance, *Phormidium* sp. synthesize a sulfated heteropolysaccharide, named emulcyan, which contains fatty acids and proteins that confer variable degrees of hydrophobicity [41].

Moreover, until recently, only eukaryotic cells were believed to be capable of producing sulfated polysaccharides, with sulfation occurring in the Golgi apparatus. In the past 10 years, however, an increasing number of cyanobacterial polysaccharides produced by strains isolated from both freshwater and saline or hypersaline environments have shown the presence of sulfated groups. The current opinion is that sulfated groups may be present also in some prokaryotic polysaccharides, but only those produced by cyanobacteria or archaea. For example, the marine filamentous cyanobacterium, *Spirulina platensis* has been shown to produce several antiviral sulfate- and phosphate-bound polysaccharides (comprising rhamnose, glucose, fructose, ribose, galactose, xylose, mannose, and glucuronic and galacturonic acids). These compounds have recently been patented by the Nippon Oil Company, Japan, for inclusion as additives within various foods and drinks [42]. This is the first instance where an antiviral activity has been noted within these compounds, and this patent documents a method for the prophylactic or therapeutic treatment of various viral diseases via ingestion. *Spirulina* extracts (whereby the main ingredient are these sulfate- and phosphate-bound polysaccharides) were reported to lower blood sugar level in diabetes sufferers; lower cholesterol and alleviate the symptoms of gastritis; as well as produce an antiallergic action on human test subjects [42]. Viruses to be treated include the herpes group of viruses and HIV viruses [42]. In this perspective, cyanobacterial strains of marine origin, among which the ability to synthesize and release sulfated polysaccharides seems to be prevalent, may provide a great resource for the production of new bioactive polymers.

15.2.2 Long-Chain Polyunsaturated Fatty Acids

Fatty acids are hydrocarbon chains that terminate in a carboxyl group; they are termed unsaturated if they contain at least one carbon–carbon double bond, and polyunsaturated when they contain multiple double bonds. Long-chain polyunsaturated fatty acids (PUFAs), especially n-3 and n-6 series fatty acids such as eicosapentaenoic (EPA **{4}**), docosahexaenoic (DHA **{5}**), and arachidonic (AA **{6}**), are pharmacologically significant and important in dietetics and therapeutics [43]. They are believed to have a positive effect on cardio-circulatory diseases, atherosclerosis, coronary heart diseases, cholesterols, hypertension, and cancer treatment [44–48]. They have been used for prophylactic and therapeutic treatment of chronic inflammations, for example, rheumatism, skin diseases, and inflammation of the mucosa of the gastrointestinal tract [49,50]. Currently, the main source of PUFAs is fish oils, but there are problems with flavor taint, uncontrollable fluctuations in availability, natural fish oil content variability, as well as the potential to accumulate harmful environmental pollutants [51]. Therefore, alternative sources are being sought and the ease of culture of PUFA-producing microorganisms makes them an attractive alternative (Table 15.1).

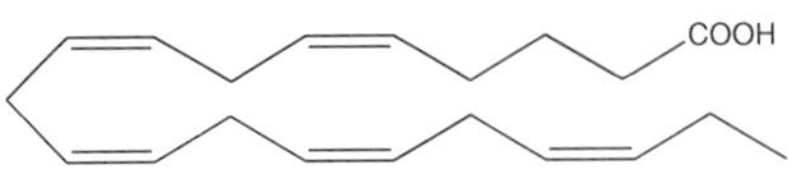

{4}

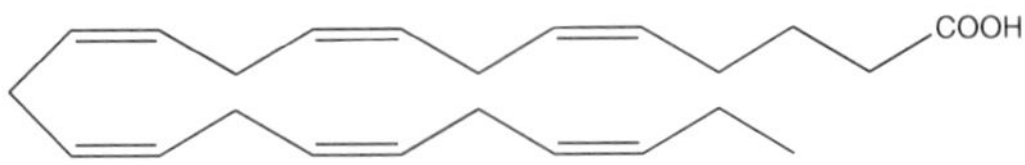

{5}

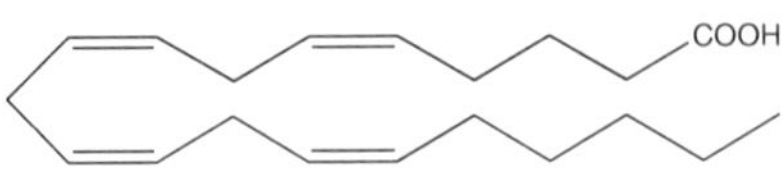

{6}

Of particular interest is the fact that microbial production is not subject to fluctuations caused by external variables such as weather and food supply. Microbially produced oil is substantially free of contamination by environmental pollutants. Originally, it was thought that apart from one group of cyanobacteria (*Spirulina* sp.), bacteria did not contain PUFAs as acyl functions of the complex lipids in their membranes [52]. Today, numerous bacterial species of marine origin have been shown to produce very long-chain PUFAs such as EPA and DHA [53]. Such isolates have been found to be particularly ubiquitous in high-pressure, low-temperature deep-sea habitats and permanently cold marine environments, and being principally members of the gamma-proteobacteria family (Table 15.2) [54]. The prevalence of PUFA-producing strains from these environments has led to speculation that PUFA synthesis is an important adaptation for countering the effects of elevated hydrostatic pressure and low temperature on membrane fluidity and indeed there is some evidence to support this hypothesis [55]. In a study by Allen and Bartlett, the gamma-proteobacterium *Photobacterium profundum* SS9 was found to exhibit enhanced proportions of both monounsaturated and polyunsaturated fatty acids when grown at a decreased temperature or elevated pressure [56]. Furthermore, transcriptional analysis indicated that elevated production of PUFAs in response to pressure resulted from posttranscriptional changes and has interesting potential for industrial manipulations [55].

Evidence of marine microorganisms (bacteria and prostids) naturally producing fatty acid has been around for at least 15 years and there are several interesting reviews on the topic, notably by Bajpai and Bajpai [53] and Ratledge [57]. Recently, fungi have been shown to be the preferred choice for microbial PUFA production having a higher PUFA content than other microbes, mainly because of the fact that these acids are components of storage oils (e.g., triacylglycerols [TAGs]), whereas in bacteria they are bound within membrane lipids. Presently, focus has been placed upon the unusual marine prostid group, phylogenetically related to microalgae, known as the *Thraustochytriaceae*, which include PUFA-producing species from the *Thraustochytrium*, *Japonochytrium*, *Ulkenia*, and *Schizochytrium* genus [58] (Table 15.2).

Microalgae are also able to vary PUFA production rates and contents with growth phases (possibly being a useful variable for future optimization of desirable PUFA content) [59]. This has been exemplified, within a study of differences in the time course of production and incorporation of DHA and EPA into TAGs, when measured in the microalgae *Nannochloropsis oculata, Phaeodactylum tricornutum,* and *Thalassiosira pseudonana*, and the macroalga *Pavlova lutheri* [60]. Differences were not only observed between species but also during the various growth phases. A much higher percentage of the total cellular EPA was partitioned to TAGs in stationary phase cells of *N. oculata* compared to *P. tricornutum* [60]. While *T. pseudonana* and *P. lutheri* produce EPA and DHA and partition these to TAGs during the stationary phase of growth [60]. Furthermore, elevation of CO_2 within photoautotrophic culture and indeed induction of heterotrophic growth in these organisms was shown to both increase and allow variations to the PUFA compositions produced by microalgae, such as *Nannochloropsis* sp. and *Crypthecodinium cohnii*, respectively (Table 15.2).

A number of sources exist for the commercial production of DHA including a variety of marine organisms, oils obtained from cold-water marine fish, and egg yolk fractions. The purification of DHA from fish sources is relatively expensive owing to technical difficulties, making DHA expensive and in short supply. In macroalgae such as *Amphidinium* sp. and marine prostids such as *Thraustochytrium* sp. and *Schizochytrium* sp., DHA may represent up to 48% of the fatty acid content of the cell. A few bacteria, usual deep-sea bacteria such as *Vibrio marinus,* also are

TABLE 15.2
PUFA Producing Ability of Prokaryotic and Eukaryotic Marine Microorganisms

Species	Habitat	Character	Growth	Activities	Ref.
Prokaryotic PUFA-producing microbes					
Moritella marina strain MP-1	Pacific Ocean sediment	Psychrophilic, halophilic	Anaerobe	Produce DHA	[149, 156]
Oleispira antarctica (DSM 14852(T))	Antarctic coastal sea water Rod Bay, Ross Sea	Psychrophilic, halophilic	Aerobe	Synthesize EPA	[155]
Photobacterium profundum strain SS9	Deep sea	Piezophilic	Aerobe	Produce EPA and DHA	[50]
Psychroflexus torquis	Antarctic Sea ice	Psychrophilic, halophilic	Aerobe	Synthesize EPA and AA	[141]
Psychromonas kaikoae (ATCC BAA-363T)	Cold-seep sediments in the Japan Trench	Psychrophilic, obligate piezophilic	Aerobe	Produce EPA and DHA in membrane	[151]
Shewanella benthica, S. frigidimarina, S. gelidimarina (ACAM 456T), *S. hanedai, S. marinintestina* (JCM 11558T), *S. schlegeliana* (JCM 11561T), *S. sairae* (JCM 11563T), *S. olleyana* (ACAM 9(T))	Antarctic Sea ice; sea-animal intestine (Japan); temperate, humic-rich river estuary in Tasmania	Halotolerant, psychrophilic, halophilic, and mesophilic	Anaerobe	Produce EPA	[141, 153, 154]
Vibrio sp. T3615, *V. marinus* (ATCC 15381)	Deep-sea fish intestines	Psychrophilic	Anaerobe	Produce EPA and DHA	[137, 142]
Eukaryotic PUFA-producing microbes					
Amphidinium sp. (CS-259); *A. carterae* (UTEX LB 1002)				Produce EPA	[145]
Crypthecodinium cohnii, Japonochytrium sp., *Ulkenia* sp. (SAM2179), *Labyrinthula* sp. (3-2)				Produce DHA	[150,157,158]

Haliphthoros philippinensis	Produce EPA and AA	[143]
Isochrysis galbana	Produce DHA	[152]
Mortierella alpina (1S-4)	Produce AA	[136]
Nannochloropsis oculata, Pavlova lutheri, Phaeodactylum tricornutum, Thalassiosira pseudonana	Produce EPA and DHA	[60,148]
Pythium acanthicum (ATCC 18660)	Produce DHA	[144]
Schizochytrium sp., *S. aggregatum* (ATCC 28209), *S. limacinum* (SR21)	Produce DHA	[146,147]
Thraustochytrium sp. (ATCC 20892), *Th. aureum* (ATCC 34304), *Th. roseum, Conidiobolus nanodes, Entomorphthora exitalis, Saprolegnia parasitica*	Produce DHA and EPA	[53,144,159–162]

Note: Definitions of terminology:

Psychrophilic: A microorganism that can survive and grow at low temperatures, but grows optimally between 15 and 20°C.

Mesophilic: A microorganism that can survive and grow at moderate temperature, but grows optimally between 20 and 45°C.

Halophilic: A microorganism that requires a high concentration of salt for optimal growth.

Halotolerant: A microorganism that is able to tolerate high concentrations of salt, but optimally grows with lesser amounts present.

Piezophilic: A microorganism that requires a high-pressure environment (either hydrostatic or atmospheric) for optimal growth.

reported to produce DHA[61]. For AA, microorganisms including the genera *Mortierella*, *Entomophthora*, *Phytium,* and *Porphyridium* can be used for commercial production (Table 15.2) [62]. Additionally, microbial oil production can be manipulated by controlling culture conditions, notably by providing particular substrates for microbially expressed enzymes, or by addition of compounds, which suppress undesired biochemical pathways. In addition to these advantages, production of fatty acids from recombinant microbes provides the ability to alter the naturally occurring microbial fatty acid profile by providing new synthetic pathways in the host or by suppressing undesired pathways, thereby increasing levels of desired PUFAs, or conjugated forms thereof, and decreasing levels of undesired PUFAs.

15.2.3 Antioxidant Compounds

Oxygen is required for metabolic functions, but it also presents challenges to cells. The human organism has a wide range of metabolic enzymes and antioxidants to rid its cells of oxygen-derived molecules. This oxidative stress is believed to be a contributing factor in conditions such as rheumatoid arthritis, ischemic heart disease and stroke, Alzheimer's dementia, cancer, and ageing [63,64]. Therefore, antioxidants have the potential to protect against a wide spectrum of diseases. Several antioxidant compounds have been isolated from marine microbial sources; these include astaxanthin **{7}**, beta-carotene and other carotenoids **{8}**, mycosporines or mycosporinelike amino acids **{9},** and dimethylsulfoniopropionate.

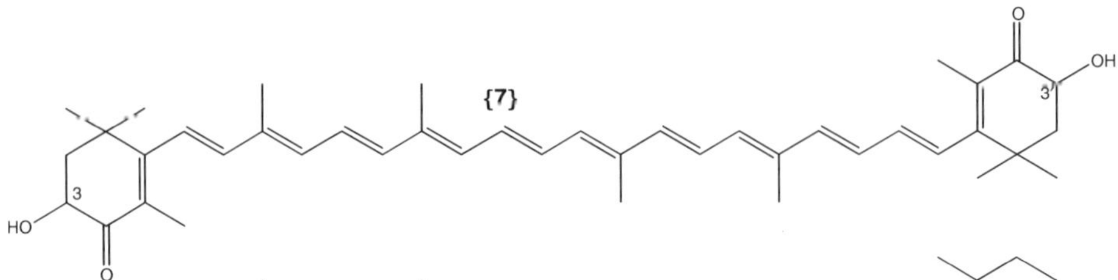

{7}

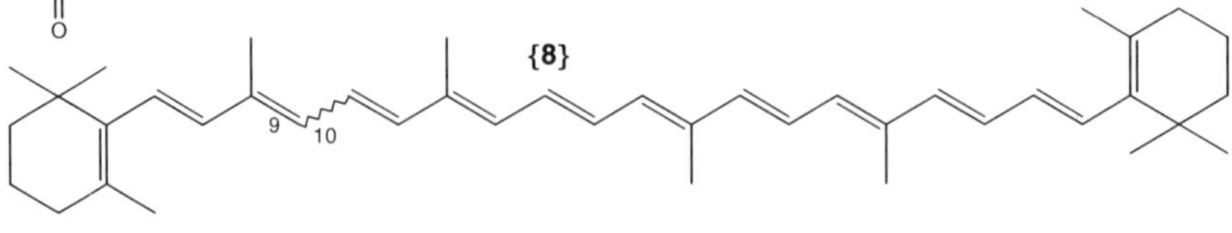

{8}

Carotenoids are important natural pigments produced by many microorganisms and plants, usually red, orange, or yellow in color. Traditionally, carotenoids have been used in the feed, food, and nutraceutical industries. They are known to be essential for plant growth and photosynthesis, and are a main dietary source of vitamin A in humans. Dietary antioxidants, such as carotenoids (beta-carotene, lycopene, astaxanthin, canthaxanthin, capsanthin, lutein, annatto, beta-apo-8-carotenal, and beta-apo-8-carotenal ester), exhibit significant anticancer activities and play an important role in the prevention of chronic diseases [65–67]. Carotenoids are potent biological antioxidants that can absorb the excited energy of singlet oxygen onto the carotenoid chain, leading to the degradation of the carotenoid molecule but preventing other molecules or tissues from being damaged. They can also prevent the chain reaction production of free radicals initiated by the degradation of PUFAs, which can dramatically accelerate the degradation of lipid membranes. In 1999, the world market for carotenoids was US$750–800 million and projections estimated around US$1 billion by 2005 [68].

Carotenoids are a widely distributed group of naturally occurring pigments. Common carotenoid sources include various fruits and vegetables. Yet, there are over 700 natural lipid-soluble pigments that are primarily produced within microalgal, macroalgal, and some bacterial and fungal species, including various unique carotenoids. For example, astaxanthin is a high-value carotenoid produced from microalgae that is achieving commercial success, being very good at protecting membranous phospholipids and other lipids against peroxidation. In the marine environment, astaxanthin is biosynthesized in the food chain by fungi and microalgae. The fungi *Phaffia rhodozyma* and the green microalgae *Haematococcus pluvialis* provide the most concentrated source of astaxanthin (Figure 15.1). *Haematococcus* sp. primarily contain monoesters of astaxanthin linked to 16:0, 18:1, and 18:2 fatty acids [69]. When nutrients become limiting, or the environment begins to dry out, these microbes produce massive amounts of astaxanthin. Cells with a high concentration of astaxanthin are more resistant to environmental hazards, such as high light and oxygen radicals, and thus

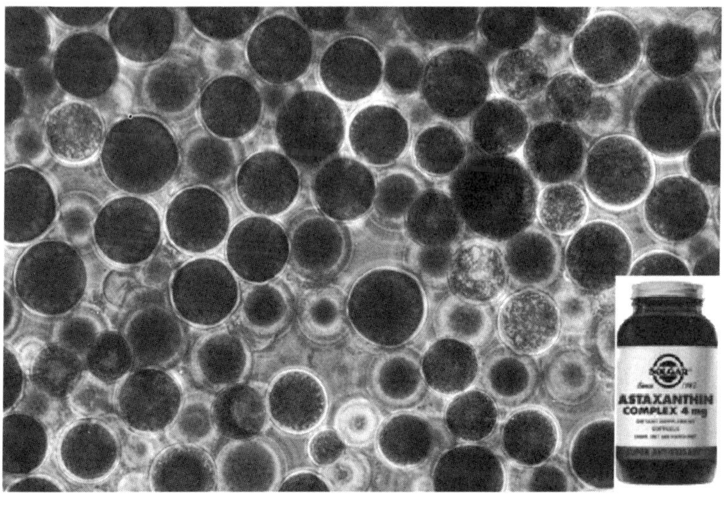

FIGURE 15.1 *Haematococcus pluvialis* hematocysts that have accumulated astaxanthin (dark colored cells) as a result of nutritional starvation and sunlight (magnification ×400). (Example of commercially available astaxanthin complex from *H. pluvialis* produced by the Solgar Vitamins and Herbs, a division of Wyeth Nutritions, Madison, NJ, USA.)

strongly suggests that astaxanthin performs a protective role for these organisms. Astaxanthin accumulates in the aquatic food chain from microalgae, macroalgae, and plankton to zooplankton and crustaceans, and then fish, such as salmon or trout, where it is deposited and concentrated within the muscle layer, probably protecting its lipid tissue from peroxidation. It is an extremely effective antioxidant protector, and is biosynthesized through the isoprenoid pathway, which is also responsible for a vast array of lipid-soluble molecules such as sterols, steroids, prostaglandins, hormones, and vitamins D, K, and E. Unlike beta-carotene, astaxanthin readily crosses the blood–brain/retina barrier, and therefore also has potential to protect from diseases of the brain and the eyes.

Preclinical studies suggest various beneficial effects of consuming astaxanthin such as: (1) inhibit cancer formation and growth in the bladder, colon, liver, mammary, and the oral cavity; (2) protect the retina of the eye from oxidative damage and thus has an effect against age-related macular disease; (3) promote increased immune activity; (4) provide protection from ultraviolet light damage; as well as (5) provide increased muscle endurance. Astaxanthin is currently available commercially in two forms, as free and esterified astaxanthin (e.g., astaxanthin dipalmitate). Free astaxanthin is available as a synthetic substance produced by *Phaffia rhodozyma*. Esterified astaxanthin, however, is available as a natural red food dye, isolated from *Haematococcus pluvialis* (approx. 1.5–3% of dry weight) [70]. *Haematococcus* algae have recently been cleared by the U.S. FDA for marketing as a dietary supplement and has been approved in several European countries for human consumption. Although currently the astaxanthin market is relatively small, it is expected that it could reach several hundred million US dollars within 5–10 years [69].

In addition to *Phaffia rhodozyma* and *Haematococcus pluvialis*, many other microorganisms inhabiting shallow water or intertidal marine environments are able to protect themselves from long-term solar damage by producing natural, UV-absorbing sunscreens of the generic class known as mycosporine amino acids [71]. Many species of marine macroalgae (e.g., *Chondrus crispus*) and cyanobacteria (e.g., *Prochloron* sp. and *Microcoleus chthonoplastes*) have been shown to be rich in imino-mycosporine or mycosporinelike amino acids **{9}** [72–74]. These compounds were also demonstrated to be biodegradable into the strong antioxidant precursor 4-deoxygadusol **{10}** by the marine bacterium *Vibrio harveyi*, *Vibrio* sp. (RML017), or by several *Pseudoalteromonas* sp. [75]. Currently, the Japanese noodle company, Toyo Suisan is investigating the commercial production of this precursor for use as a natural antioxidant and chemoprotective compound against neurodegenerative and cardiovascular diseases, in food-processing and cosmetic applications.

Other compounds produced by marine microbial sources, such as dimethylsulfoniopropionate (DMSP) and its enzymatic cleavage products (dimethylsulfide [DMS], acrylate, dimethylsulfoxide, and methane sulfinic acid), have been shown to readily scavenge hydroxyl radicals and other reactive oxygen species, and thus act as an antioxidant [76]. DMSP occurs at high cellular concentrations (100–400 mmol/L) in many marine macroalgae and dinoflagellates, and thus

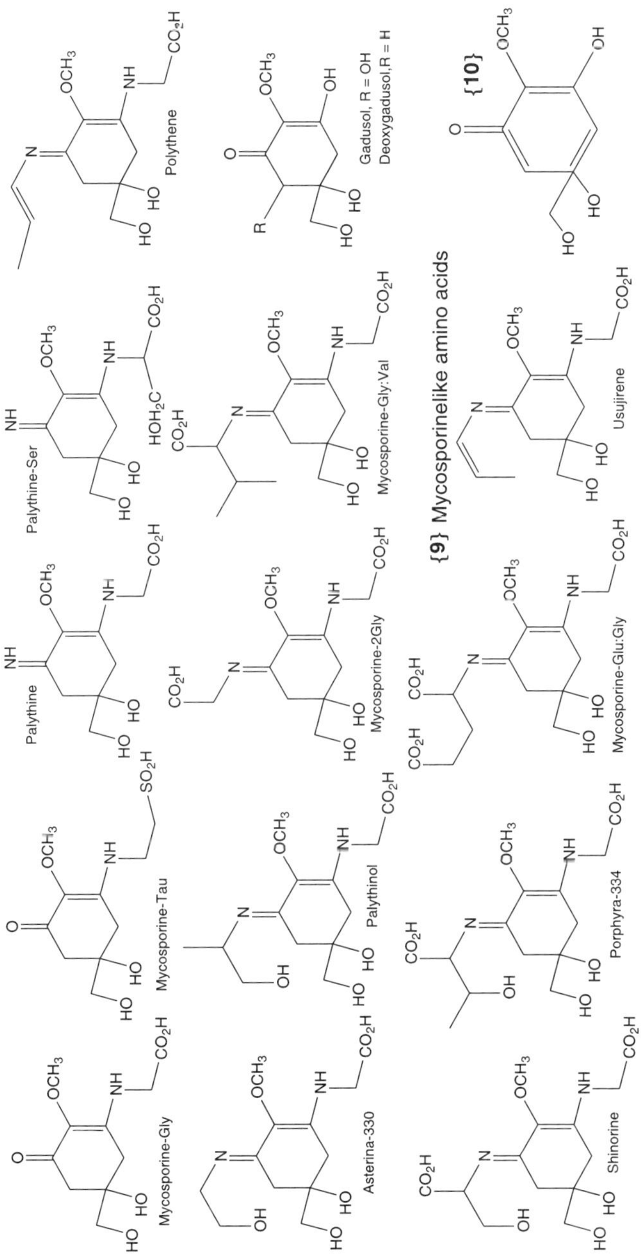

{9} Mycosporinelike amino acids

{10}

functions as an osmolyte in these algae. DMSP has also been proposed to serve as a cryoprotectant in polar macroalgae and as a grazing deterrent via its cleavage to acrylate, although its overall physiological functions remain unclear. Interestingly, its cellular concentration increases with light in many macroalgal species and may have an additional UV-protective function as mentioned above [77]. DMSP and its cleavage products have also been shown to possess antioxidant activity greater than other well-recognized antioxidants such as ascorbate. Recent literature evidence supports the hypothesis that antioxidant compounds like these, may indeed be derivatives of algicolous marine fungi, rather than the macroalgal species itself [78,79].

Until recently, marine antioxidant research has largely focused on the effects of crude extracts [80], with the resultant substances related mainly to plant-derived products. These marine antioxidants include pigments such as chlorophylls, carotenoids, tocopherol derivatives such as vitamin E and related isoprenoids and various phenolic substances (not to mention those already discussed above). Within a recent study of antioxidants by researchers at the University of Mississippi and Oregon State University, 130 structurally diverse, pure marine microbial-derived, natural products were assembled and evaluated for antioxidant activity using a combined solution- and cellular-based assay system, to develop a structure-to-function understanding of the antioxidant mode of action [81]. Compounds assayed included alkaloids, lipopeptides, furocoumarins, phenolics, peptides, and terpenoids. Results showed a number of marine cyanobacterial- and macroalgal-derived compounds, such as scytonemin **{11}** from the heterocystous cyanobacterium *Scytonema* sp. [82], cymopol **{12}** and 7-hydroxycymopol **{13}** from the green macroalga *Cymopolia barbata* [83], avrainvilleol **{14}** from the green macroalga *Avrainvillia* sp. [84], and aragilamide **{15}** from the red macroalga *Martensia fragilis* [85].

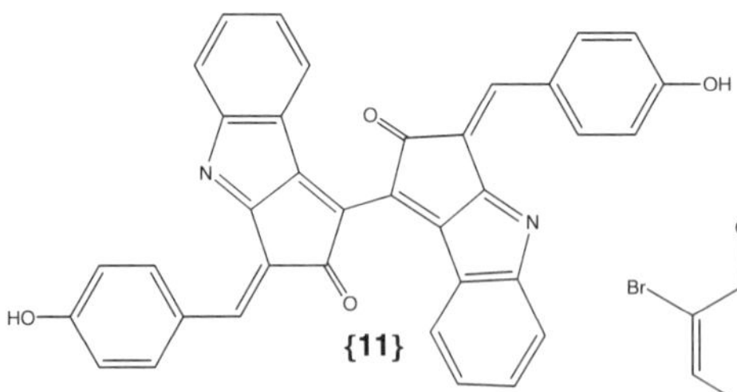

{11}

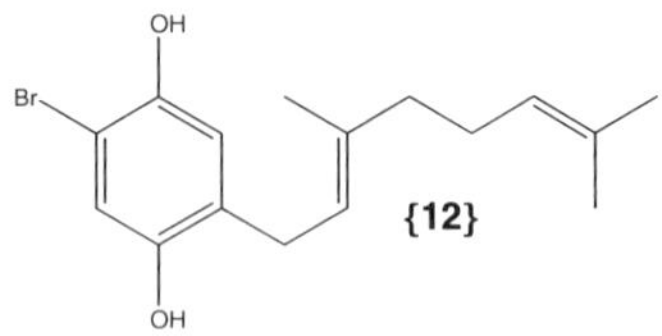

{12}

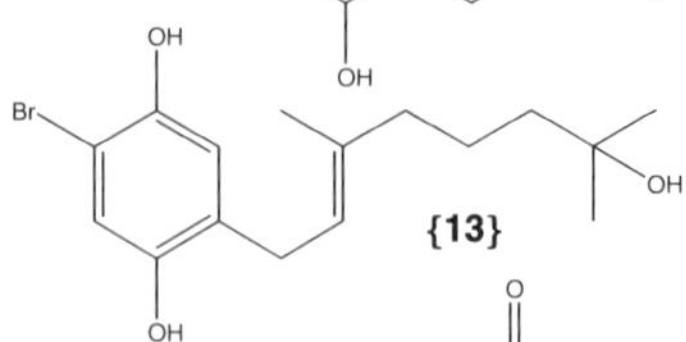

{13}

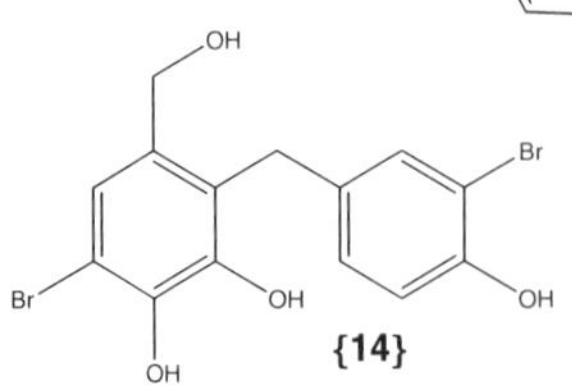

{14}

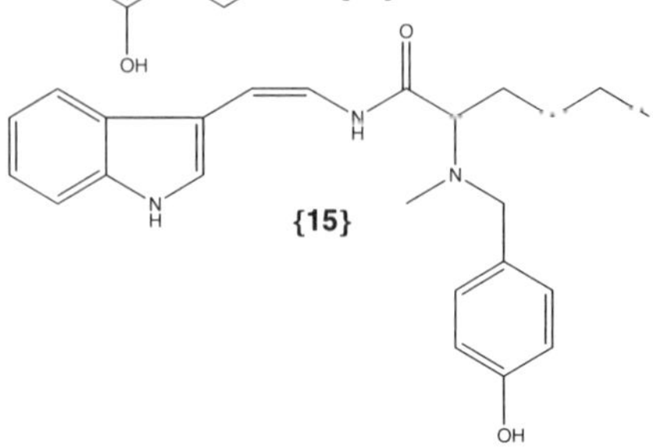

{15}

15.2.4 Other Nutritional Compounds

Initial analysis of microalgae for vitamin content frequently presented values for vitamin B_{12} (cobalamin). Recently, however, the method of analysis (typically using *Lactobacillus leichmanni* as a bioassay) was shown to be misleading [86]. Modern assay techniques, such as radioassays and clinical observations produced data that indicated much of the cobalamin values were actually vitamin B_{12} analogs [87]. Further research is necessary to adequately understand the metabolic consequences of consuming vitamin B_{12} from algal sources. The cyanobacterium *Aphanizomenon flos-aquae* (discussed in detail below) was shown to contain significant levels of cobalamin (Table 15.1). Additionally, microalgae have been found to contain several different types of sterols, including clionasterol (isolated from *Spirulina* sp.) and fucosterol (extracted from brown macroalgae), which has been shown to increase the production of plaminogen-activating factor in vascular endothelial cells and thus facilitate cardiovascular disease prevention (Table 15.1). Finally, in general, microalgae are viewed as having a protein quality value greater than other vegetable sources, for example, wheat, rice, and legumes, but poorer than animal sources, for example, milk and meat. Thus, although macro- and microalgal proteins are looked upon favorably by researchers, processing procedures still need to be optimized to extract the greatest possible content of proteins possible for nutraceutical adaptations.

15.3 WHOLE-CELL EXTRACTS

Whole-cell extracts of marine microorganisms used within the nutraceutical industry represent extracts whereby the bioactive molecular responsible for its beneficial properties is either currently unknown, unable to be isolated, or synthesized. Within the marine environment these extracts are usually isolated from microalgal and cyanobacterial species. Specifically, the freshwater and marine-derived filamentous [88], heterocystous cyanobacterium *Aphanizomenon flos-aquae* has been commercially available as a functional food source for over 20 years (Table 15.2) [49]. It is currently not cultivated but rather harvested from a wild, virtually monocultural bloom [49]. Crude extracts of which have been shown to contain glycoproteins, vitamin B_{12}, minerals, proteins (approx. 60–70% of total dry weight), polysaccharides, and PUFAs. Recent studies show that *Aphanizomenon flos-aquae* extracts have immunostimulant, anti-inflammatory, antioxidant, biomodulator, anticancer, and hypochloresterolemic properties (Figure 15.2) [89,90].

Chlorella sp. are unicellular green microalgae that are members of the phylum Chlorophyta (Figure 15.3). There are numerous references on the health benefits of consuming *Chlorella* sp., including documentary evidence that ingestion of *Chlorella* sp. extracts can lower blood sugar levels, increase hemoglobin concentrations, act as a hypocholesterolemic agent, and a hepatoprotective agent during malnutrition and ethionine intoxication [91,92]. Indeed, between 1942 and 1946, a plankton soup consisting primarily of *Chlorella* sp. cells was used as a

FIGURE 15.2 Light micrograph of *Aphanizomenon flos-aquae*, as well as a representation of the wild algae 500 mg capsules harvested from Lake Klamath, Oregon, USA.

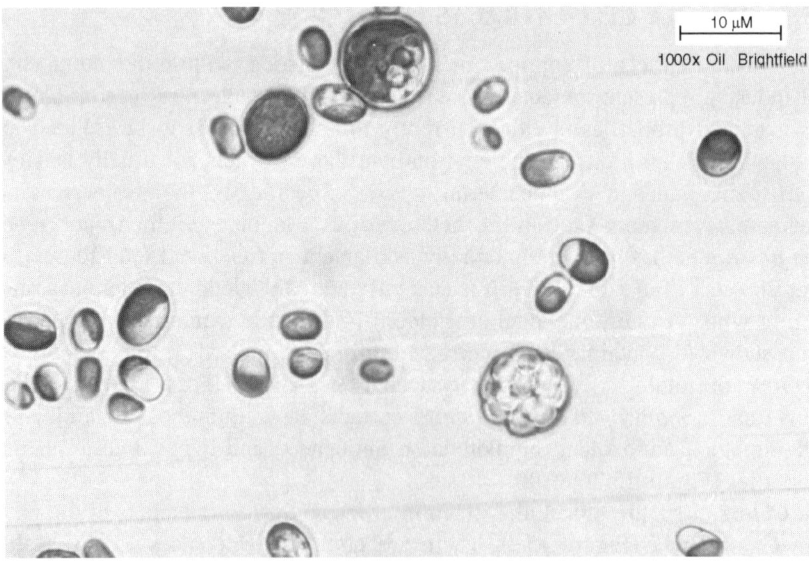

FIGURE 15.3 Bright field light micrograph of *Chlorella* sp. (in this case *C. ellipsoidea*) (×1000 magnification). (Image courtesy of The Culture Collection of Algae at The University of Texas at Austin, TX, USA.)

dietary aid for leprosy patients, resulting in dramatic increases in energy, weight, and general health of patients [91,92]. In 1954, Japanese researchers developed experimental food products using *Chlorella ellipsoidea* cells. The major focus was the enrichment of foods with this species and the overall palatability of pilot products such as powdered green tea, soups, noodles, bread and rolls, cookies, ice cream, and soy sauce. Results showed that all generated products were deemed to be acceptable as food supplements [91,92]. In 1961, human feeding studies were carried out on a combination of *Chlorella* and *Scenedesmus* sp. extracts, showing a similar acceptable level for human consumption.

Dunaliella sp. (especially *D. salina*) are rapidly becoming popular as food-grade green microalgae. In particular, their beta-carotene content (up to 4% of dry weight), glycerol concentration, protein content, and unique ability to grow under brackish conditions have been focal points. They are currently being cultivated by several companies in both Israel and Australia (Figure 15.4) as sources of the aforementioned compounds and as dietary and vitamin supplements and powders. *Scenedesmus* sp. were also previously investigated as a nutritional food source. Yet, their production on a commercial scale currently remains limited. Currently, there are no products presently marketed

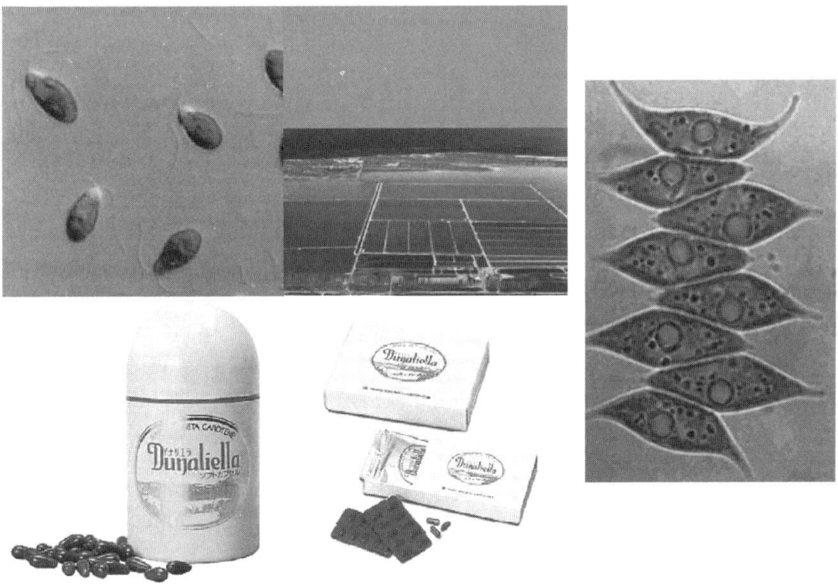

FIGURE 15.4 Left-hand side (top): phase contrast image of *Dunaliella* sp. (in this case *D. bardawill*) (×400 magnification); middle (top) this picture shows the *D. salina* production plant operated by Western Biotechnology Pty Ltd at Hutt Lagoon, Western Australia. Right-hand side (top): phase contrast image of *Scenedesmus* sp. (in this case *S. falcatus*) (×1000 magnification). Bottom: an example of two *Dunaliella* sp. products currently available on the nutraceutical market in Japan.

FIGURE 15.5 Examples of functional foods (left to right): phase contrast light micrograph image of *Spirulina platensis* (×400 magnification); Source Naturals™ Hawaiian *Spirulina* 500 mg tables; Italian Erba Vita™ *Spirulina* 400 mg tablets, Nutrex™ Hawaiian *Spirulina* 500 mg tables; Japanese Roibo™ *Spirulina*/Lime cordial; commercially produced Hawaiian *Spirulina* smoothie; *Spirulina* muffins and smoothie; and *Spirulina* cuisine cook book written by Robert Sterbene.

in North America that contain *Scenedesmus* sp. as a dietary material. In general, *Scenedesmus* sp. extracts have been found to be an acceptable candidate for incorporation into commonly used foods, such as soups, desserts, fruit puddings, noodles, and ravioli; however, it seems that at this time this market has been underdeveloped.

Spirulina sp. (the two main species being *Sp. platensis* and *Sp. maxima*) are multicellular, filamentous cyanobacteria that have the appearance of coils in cross section (Figure 15.5). *Spirulina* sp. have been used for over 1000 years as a food source, having a protein content as high as 55–70% of the total dry weight. The Aztecs harvested this cyanobacterium from Lake Texcoco (Mexico), dried it, and sold it as small cakes, which was then incorporated into bread. More recently, in 1940, the use of *Spirulina* sp., again as small cakes (this time as an ingredient in sauces used on millet meals), was used by the natives of Chad. Recently, a dietary supplement for promoting healthy hormonal balance in adult human subjects, comprising 7-ketodehydroepiandrosterone in combination with a pituitary secretagogue comprising a glycoamino acid complex of L-glutamine, L-arginine pyroglutamate, L-lysine monohydrochloride, glycine, gamma-aminobutyric acid as well as omega-3 fatty acid and antioxidant blends was developed by Reliv' International Ltd. [93]. This dietary supplement is an herbal blend of *Salsburia adiantifloria, Spirulina plantensis, Lepidium menyii, Dioscorea villosa* L., and *Chlorella* powder, to be served as a dry powder daily supplement to drinks [93] (Figure 15.5). The use of dietary microalgae has also been reported to increase the number of lactic acid bacteria in the gastrointestinal tract. Preliminary *in vitro* data support the premise that microalgae can enhance the production of lactic acid significantly [94].

Furthermore, it has been postulated that the carotenoids found in *Spirulina* sp. and *Dunaliella* sp. may be more potent anticancer agents than beta-carotene [95]. It appears that extracts from these microalgae enhance the antitumor activity of macrophages. Schwartz et al. [95] have demonstrated excellent results in the regression of induced cancers in hamsters with the use of a proprietary macro-algal extract called Phycotene™.

15.4 THE FUTURE OF MARINE MICROBIAL NUTRACEUTICALS

Interestingly, to date, the majority of nutraceutical compounds derived from marine microorganisms have been isolated from the more "primitive" microbes, those that are able to take simple carbon sources, such as CO_2 and manufacture more complex intermediates (possibly because of an associated abundance of both generation and degenerating enzymes). The presence of these proteins thus allows these microbes to generate the complex fatty acids, proteins, vitamins, and carbohydrates that nutraceutical and pharmaceutical researchers make use of.

Furthermore, it has been hypothesized recently that the ultimate source of most nutraceuticals from the marine environment are not the larger, more complex animals, but rather those species that are closer to the beginning of the food chain. For example, it has been suggested that the PUFAs that characterize oily fish are actually derived from marine algae (both macro- and micro-) species that make up the primary food source for these organisms [96]. To fully exploit these inherent abilities within marine microorganisms (especially photosynthetic microbes), it will be necessary to determine the mode of synthesis for the various complex molecules these organisms produce. Methods such as those that probe the exact nature of the biosynthetic genes present within the microorganism (exploration) and methods to increase the production of these compounds within a laboratory environment (exploitation) will need to be perfected.

15.4.1 EXPLORATION: PROBING METHODS

The probing of suitable microorganisms for cryptic genes shown to be involved in the production of a targeted compound within related organisms, has recently been demonstrated within the pharmaceutical industry [97]. These methods could easily be developed for use within the nutraceutical industry, with similar genes of interest and appropriate microorganisms being studied in each case. Specifically, DNA encoding desired polyketide synthase (PKS)-like genes can be identified in a variety of ways. In one method, a source of a desired PKS-like gene, for example, genomic libraries from a *Shewanella*, *Thraustochytrium, Schizochytrium,* or *Vibrio* spp., are screened with detectable enzymatically or chemically synthesized probes. Sources of open reading frames (ORFs) having PKS-like genes are those microorganisms that produce a desired PUFA. For example, marine microbes such as *Shewanella* sp. or *Vibrio marinus,* which produce EPA or DHA also can be used as a source of PKS-like genes. Oligonucleotide probes, based on sequences of known PKS-like genes, can also be used to screen sources and may include sequences conserved among known PKS-like genes, or on peptide sequences obtained from a desired purified protein. The probes can be made from DNA, RNA, or nonnaturally occurring nucleotides, or mixtures thereof. Furthermore, probes can be enzymatically synthesized from DNAs of known PKS-like genes for normal or reduced-stringency hybridization methods. Oligonucleotide probes based on amino acid sequences can be made degenerate to encompass the degeneracy of the genetic code, or can be biased in favor of the preferred codons of the source organism. Alternatively, a desired protein can be entirely sequenced and total synthesis of the DNA encoding that polypeptide performed [58].

15.4.2 Exploitation

15.4.2.1 Fermentation and Bioprocess Intensification

The advantage of microorganisms that can be cultured over more classical synthetic production of compounds is that a sustainable supply of a "targeted" compound may be achieved. Thus, the interest in marine microbes as a novel source of bioactives is both warranted and opportune. Although there has been some interest in the exploitation of cultured microbes to develop pharmaceutical and nutraceutical compounds, it is believed that the study of these secondary metabolites and their controlled long-term production is still in its infancy. Of the limited research that has been undertaken, there has been little concerted effort to move the biotechnological process forward beyond the characterization phase. Although a number of groups have touched on aspects dealing with bioprocess intensification (improvement in process efficiency and effectiveness) [98–103], no program has tried to increase product yield of the relevant marine microorganisms via culture and reactor configuration manipulation, through to the commercialization stage. Indeed, the application of such bioprocess intensification methods is likely to become an important strategy for improved supply and exploitation of product.

For example, a relatively new field of possible exploitation has arisen in the last decade, fueled by the growing industrial interest in polysaccharides of microbial origin [104]. Toward this end, cyanobacteria, compared to other bacteria, macro-, and microalgae, synthesize more complex polysaccharides, being composed of six or more monosaccharides, rather than the up to four monosaccharides usually noted within these species [105]. During cell growth in batch cultures, aliquots of the polysaccharide material of cyanobacterial sheaths may be released as water-soluble material into the surrounding medium, causing a progressive increase in viscosity. These water-soluble polysaccharides, being easily recoverable from liquid cultures, demonstrate the usefulness of performing fermentation and bioprocess intensification procedures for the production of these highly desirable industrial polymers.

Very few studies have been devoted to verifying if these strains produce polymers with a stable chemical composition, not changing from batch to batch and unaffected by growth conditions (a very important feature for microbial strains proposed for industrial applications) [106–109]. This feature is of great significance, not just for the production of polysaccharide polymers, but for the production of other natural products of industrial interest by microorganisms. Furthermore, many microorganisms producing these compounds of interest lose this property when cultivated under laboratory conditions, when environmental stresses have been removed [104].

Additional considerations include, whether or not the microorganism produces the compound of interest via primary or secondary metabolic pathways. These differences have a drastic effect on both the rates of production and exact makeup of the compounds produced. For example, with respect to polysaccharide production by cyanobacteria, *Anabaena halophytica*, *Spirulina platensis*, and *Cyanospira capsulata* produce polysaccharide constituents at rates that consistently parallel biomass production, so that the polymer may be considered a primary metabolite [107–110]. On the contrary, other cyanobacteria produce these polysaccharides as a typical secondary metabolite. Specifically, *Cyanothece* sp. BH68K, *Nostoc calcicola*, and *Phormidium* J-1 show a significant release of polysaccharide only starting from the later exponential growth phase [41,111–113]. In contrast, in *A. flos-aquae* A37 and *A. cylindrica* 10C, polysaccharide production occurred during all growth phases, with the highest production rates observed in the late exponential or stationary phase of microbial growth [104]. These examples demonstrate quite clearly that each strain should be carefully tested to envisage the right culture strategies aimed at optimizing production, within this and every production situation.

15.4.2.2 Gene Evolution

For newly discovered microbial sources of nutraceuticals to be introduced into the market, it is usually not enough for the organism of choice to be merely identified. Various factors need to be optimized before industrial production can begin. Various limitations to exploitation can exist, for example, the growth rate of the organism may be low, or production of the compound of interest is; this compound may be unstable or perhaps may be utilized in subsequent metabolic pathways before it can be harvested. All of these factors and many more may be responsible for the compounds not progressing to market. Today, with the advent of sophisticated cloning and directed evolution techniques these hurdles may be overcome. These technologies may be exemplified by the production of novel microbial strains that produce the compound of interest in both a high concentration and in an easy-to-harvest manner. Alternatively, the produced enzyme can be used *in vitro* as immobilized catalytic systems, independent of the microorganism. This section will highlight examples whereby these gene evolution techniques could be used to increase the production of long-chain PUFAs and carotenoids nutraceuticals.

15.4.2.2.1 Long-Chain Polyunsaturated Fatty Acids

Long-chain PUFAs are critical components of membrane lipids or storage oils in most eukaryotes and are the precursors of certain hormones and signaling molecules [43]. Known pathways in higher eukaryotes involve the processing of saturated products of fatty acid synthase (FAS) genes by elongation and aerobic desaturation reactions [96]. Lower eukaryotes and prokaryotes, however, use a nonspecific (aerobic or anaerobic) PKS-like gene cluster to synthesize PUFAs [43,58]. Currently, the exact mechanism by which these PUFAs are assembled via the PKS-like system remains unknown. Yet, the relative simplicity of these PKS-like systems makes it an attractive target for the transgenic production of PUFAs. Various gene evolution methods have been developed toward this end and include (1) the establishment of PUFA-specific PKS probes (as mentioned previously) and (2) the development and use of various non-PUFA producing microorganisms, such as *Saccharomyces cerevisiae*, *Escherichia coli,* or *Synechococcus* sp., as hosts for the heterologous expression of various PUFAs.

Specifically, polyketides are secondary metabolites, the synthesis of which involves a set of enzymatic reactions analogous to those of FAS [114,115]. It has been proposed to use PKS to produce novel antibiotics [116]. For the most part, some or all of the coding sequences for the polypeptides having PKS-like gene activity are from a natural source. In some situations, however, it is desirable to modify all or a portion of the codons, for example, to enhance expression, by employing host-preferred codons. Host-preferred codons can be determined from the codons of highest frequency in the proteins expressed within a particular host species. Thus, the coding sequence for a polypeptide having PKS-like gene activity can be synthesized in whole or in part. All or portions of the DNA may also be synthesized to remove any destabilizing sequences or regions of secondary structure, which would be present in the transcribed mRNA, or alter the base composition to one more preferable to the desired host cell.

Furthermore, *in vitro* mutagenesis, site-directed mutagenesis, or other means can be employed to obtain mutations of naturally occurring PKS-like genes to produce a polypeptide having PKS-like gene activity *in vivo* with more desirable physical and kinetic parameters for function in the host cell. For example, a longer half-life or a higher rate of production of a desired PUFA. Researchers of the Sagami Chemical Research Institute, Japan, have reported EPA production in *E. coli,* which have been transformed with a gene cluster from the marine bacterium *Shewanella putrefaciens* [117].

15.4.2.2.2 Carotenoids

More than 150 genes encoding 27 different carotenoid enzymes have been cloned from bacteria, plants, macroalgae, microalgae, cyanobacteria, and fungi [68]. Characteristically, carotenoids are derived from the isoprenoid pathway via one of two pathways: (1) the mevalonate pathway in macroalgae, microalgae, and fungi and (2) the pyruvate pathway in bacteria, cyanobacteria, and plastid-containing microorganisms [118]. Presently, various molecular engineering groups have been able to clone genes responsible for the formation of the carotenoid backbone [119,120], the formation of the acylic xanthophylls enzyme [121] and several cyclic carotenoids such as astaxanthin [122]. Most carotenoid genes and gene clusters are cloned and expressed in the genetically well-established, noncarotenogenic host *E. coli*.

Despite the recent accomplishments in metabolic engineering of *E. coli* cells for carotenoid production, production levels are not yet competitive with carotenoid levels presently produced by fermentation, synthesis, or isolation [123]. Carotenoid production levels within such a host have however been typically very low [124]. Compared to the up to several hundredfold higher levels produced by carotenogenic microalgae and microbial strains, such as *Dunaliella*, *Haematococcus,* and *Flavobacterium* [123]. In contrast to *E. coli*, yeast exhibit an efficient isoprenoid metabolism. Although metabolic engineering of recombinant yeasts for carotenoid production has not yet been explored widely, the few published examples show that this alternative is promising. For instance, the noncarotenogenic yeasts *Saccharomyces cerevisiae* and *Candida utilis* were successfully engineered to produce lycopene, beta-carotene, and astaxanthin by diverting the ergosterol pathway with genes derived from *Erwinia* [125–127].

Metabolic engineering of photosynthetic bacteria could provide an alternative route for the heterologous production of carotenoids. With some of the earliest examples of rerouting carotenoid pathways coming from *Erwinia herbicola* and *Rhodobacter sphaeroides*, resulting in the production of the *Erwinia* carotenoids lycopene, beta-carotene, and zeaxanthin in *Rhodobacter* [128,129]. More recently, carotenoid biosynthesis in recombinant *Synechocystis* sp. has been attempted, with overexpression of genes encoding various carotenoid enzyme subunits in different combinations cloned into this cyanobacterium. In a different example, overexpression of a macroalgal carotene oxygenase within *Synechocystis* sp. was also performed successfully [130,131].

The recombinant production of carotenoids in noncarotenogenic microorganisms allows engineering of new pathways for the production of diverse carotenoid structures. Combining identified genes from different organisms into pathways and altering the catalytic functions of selected enzymes by *in vitro* evolution can greatly expand the recombinant production capabilities of known and completely new carotenoid structures as a result of chemical biosynthesis [132,133]. Recombinant biosynthesis of carotenoids presently represents the best example for combinatorial biosynthesis, which involves the combination of nonmodular single enzymes into new pathways. For example, Umeno, Tobias, and Arnold have demonstrated the production of asymmetrical C_{35} carotenoids through the coexpression of *Erwinia* and *Staphylococcus aureus* carotenoid genes within *E. coli* [134,135]. While this level of metabolic engineering, directed evolution, and molecular breeding seems impressive, a systematic flux analysis of carotenoid biosynthesis, including the determination of individual enzyme activities and functional expression levels, has not yet been carried out. Understanding the biochemical mechanisms governing carotenoid pathway flux, would seem to be key to successful future metabolic engineering of commercially relevant microbial carotenoid production.

15.5 CONCLUDING REMARKS

Marine microorganisms have recently become popular as a renewed source of both nutraceutical and pharmaceutical compounds. Once isolated they are relatively easy to grow, produce novel classes and types of compounds, and have less complex biological systems compared to higher organisms. On the basis of their unique phenotypic and genotypic diversity (compared to the terrestrial environment), and in recognition of the compounds isolated from them, marine microorganisms represent a novel natural products source of high economic value. In this postgenomic era, many powerful tools for genetic evolution and metabolic engineering have been developed, in addition to bioprocess intensification, to aid in our ability to explore, understand, exploit, and even improve nature's potential for human benefit. Therefore, it is fortuitous for the development of the next generation of nutraceutical compounds to be sourced from this unique environment effectively.

REFERENCES

1. Appleyard, V.C.L., Unkles, S.E., Legg, M. and Kinghorn, J.R. 1995. Secondary metabolite production in filamentous fungi displayed. *Mol. Gen. Genet.* 247, 338–342.
2. Walton, J.D. 2000. Horizontal gene transfer and the evolution of secondary metabolite gene clusters in fungi: a hypothesis. *Fungal Genet. Biol.* 30, 167–171.
3. Pietra, F. 1990. *A Secret World: Natural Products of Marine Life*, Birkhäuser Verlag, Basel, Switzerland.
4. Fenical, W. 1993. Chemical studies of marine bacteria: developing a new resource. *Chem. Rev.* 93, 1673–1683.
5. Davidson, B. 1995. New dimensions in natural products research: cultured marine microorganisms. *Curr. Opin. Biotechnol.* 6, 284–291.
6. Faulkner, D. 2000. Highlights of marine natural products chemistry (1972–1999). *Nat. Prod. Rep.* 17, 1–6.
7. Ireland, C., Copp, B., Foster, M., McDonald, L., Radisky, D. and Swersey, J. 1993. Biomedical potential of marine natural products. In: *Pharmaceutical and Bioactive Natural Products* (Attaway, D. and Zaborsky, O., Eds.), Vol. 1, pp. 1–43. Plenum Press, New York.
8. Burja, A.M. and Hill, R.T. 2001. Microbial symbionts of the Australian Great Barrier Reef sponge, *Candidaspongia flabellata*. *Hydrobiologica* 461, 41–47.
9. Burja, A.M., Banaigs, B., Abou-Mansour, E., Burgess, J. and Wright, P.C. 2001. Marine cyanobacteria—a prolific source of natural products. *Tetrahedron* 57, 9347–9377.
10. Cragg, G.M. and Newman, D.J. 2001. Natural product drug discovery in the next millennium. *Pharm. Biol.* 39, 8–17.
11. Haefner, B. 2003. Drugs from the deep: marine natural products as drug candidates. *Drug Discovery Today* 8, 536–544.
12. Blunt, J.W., Copp, B.R., Munro, M.H.G., Northcote, P.T. and Prinsep, M.R. 2003. Marine natural products. *Nat. Prod. Rep.* 20, 1–48.
13. Wise, M.L., Rorrer, G.L., Polzin, J.J. and Croteau, R. 2002. Biosynthesis of marine natural products: solation and characterization of a myrcene synthase from cultured tissues of the marine red alga *Ochtodes secundiramea*. *Arch. Biochem. Biophys.* 400, 125–132.
14. Polzin, J.P. and Rorrer, G.L. 2003. Halogenated monoterpene production by microplantlets of the marine red alga *Ochtodes secundiramea* within an airlift photobioreactor under nutrient medium perfusion. *Biotechnol. Bioeng.* 82, 415–428.
15. Shanmugam, M., Mody, K.H., Oza, R.M. and Ramavat, B.K. 2001. Blood anticoagulant activity of a green marine alga *Codium dwarkense* (Codiaceae, Chlorophyta) in relation to its growth stages. *Indian J. Mar. Sci.* 30, 49–52.

16. Nika, K., Mulloy, B., Carpenter, B. and Gibbs, R. 2003. Specific recognition of immune cytokines by sulphated polysaccharides from marine algae. *Eur. J. Phycol.* 38, 257–264.
17. van de Velde, F., Knutsen, S.H., Usov, A.I., Rollema, H.S. and Cerezo, A.S. 2002. ^{1}H and ^{13}C high resolution NMR spectroscopy of carrageenans: application in research and industry. *Trends Food Sci. Tech.* 13, 73–92.
18. Lahaye, M. 2001. Developments on gelling algal galactans, their structure and physico-chemistry. *J. Appl. Phycol.* 13, 173–184.
19. Prado-Fernandez, J., Rodriguez-Vazquez, J.A., Tojo, E. and Andrade, J.M. 2003. Quantitation of kappa-, iota- and lambda-carrageenans by mid- infrared spectroscopy and PLS regression. *Analytica Chimica Acta* 480, 23–37.
20. Belton, P.S., Wilson, R.H. and Chenery, D.H. 1986. Interaction of group-I cations with iota-carrageenans and kappa-carrageenans studied by fourier-transform infrared-spectroscopy. *Int. J. Biol. Macromol.* 8, 247–251.
21. Aldor, A.S. and Keasling, J.D. 2003. Process design for microbial plastic factories: metabolic engineering of polyhydroxyalkanoates. *Curr. Opin. Biotechnol.* 14, 475–483.
22. Yoshida, T. 2001. Synthesis of polysaccharides having specific biological activities. *Prog. Polym. Sci.* 26, 379–441.
23. Kampf, N. 2002. The use of polymers for coating of cells. *Polym. Adv. Technol.* 13, 896–905.
24. Gracy, R.W. 2003. Chitosan and glucosamine derivatives in the treatment of osteoarthritis. *Agro Food Ind. Hi-Tech* 14, 53–56.
25. Sung, J.H., Choi, H.J., Sohn, J.I. and Jhon, M.S. 2003. Electro-rheology of chitosan polysaccharide suspensions in soybean oil. *Colloid Polym. Sci.* 281, 1196–1200.
26. Okamoto, Y., Inoue, A., Miyatake, K., Ogihara, K., Shigemasa, Y. and Minami, S. 2003. Effects of chitin/chitosan and their oligomers/monomers on migrations of macrophages. *Macromol. Biosci.* 3, 587–590.
27. Chen, A.S., Taguchi, T., Sakai, K., Kikuchi, K., Wang, M.W. and Miwa, I. 2003. Antioxidant activities of chitobiose and chitotriose. *Biol. Pharm. Bull.* 26, 1326–1330.
28. Okamoto, Y., Yano, R., Miyatake, K., Tomohiro, I., Shigemasa, Y. and Minami, S. 2003. Effects of chitin and chitosan on blood coagulation. *Carbohydr. Polym.* 53, 337–342.
29. Novikov, V.Y., Mukhin, V.A. and Kharzova, L.P. 2000. Complex processing of crustacean armor. *Russ. J. Appl. Chem.* 73, 1612–1616.
30. Arcidiacono, S. and Kaplan, D.L. 1992. Molecular-weight distribution of chitosan isolated from *Mucor rouxii* under different culture and processing conditions. *Biotech. Bioeng.* 39, 281–286.
31. Tharanathan, R.N. and Kittur, F.S. 2003. Chitin—the undisputed biomolecule of great potential. *Crit. Rev. Food Sci. Nutr.* 43, 61–87.
32. Clark, D. 1991. Current concepts in the treatment of degenerative joint disease. *Compendium Continuing Educ. Pract. Veterinarian* 13, 1439–1445.
33. Pirohamned, R. and Garris, R. 1998. Glucosamine sulfate in the treatment of osteoarthritis. *Pharmacist* August, 66–74.
34. Nagaoka, M., Shibata, H., Kimura, I. and Hashimoto, S. 2003. Oligosaccharide derivatives and process for producing the same. United States Patent: 6,645,940, Assigned to Kabushiki Kaisha Yakult Honsha (Tokyo, Japan).
35. Enoki, T., Sagawa, H., Tomono, J., Tominaga, T., Nishiyama, E., Wu, H.-K. and Kato, I. 2003. Medicinal compositions. United States Patent: 6,608,032, Assigned to Takara Shuzo Company, Ltd. (Kyoto, Japan).

36. Watanabe, S. and Seto, A. 1989. Ingredient effective for activating immunity obtained from *Chlorella minutissima*. United States Patent: 4,831,020, Assigned to The Nisshin Oil Mills, Ltd. (Tokyo, Japan).
37. Kralovec, J. 2003. Fractions of *Chlorella* extract containing polysaccharide having immunomodulating properties. United States Patent: 6,551,596, Assigned to Ocean Nutrition Canada Limited (Halifax, NS, CA).
38. de Philippis, R., Sili, C., Paperi, R. and Vincenzini, M. 2001. Exopolysaccharide-producing cyanobacteria and their possible exploitation: a review. *J. Appl. Phycol.* 13, 293–299.
39. Popper, Z.A., Sadler, I.H. and Fry, S.C. 2003. Alpha-D-glucuronosyl-(1 -> 3)-L-galactose, an unusual disaccharide from polysaccharides of the hornwort *Anthoceros caucasicus*. *Phytochemistry* 64, 325–335.
40. Bellezza, S., Paradossi, G., De Philippis, R. and Albertano, P. 2003. *Leptolyngbya* strains from *Roman hypogea*: cytochemical and physico-chemical characterisation of exopolysaccharides. *J. Appl. Phycol.* 15, 193–200.
41. Fattom, A. and Shilo, M. 1985. Production of emulcyan by *Phormidium* J-1—its activity and function. *FEMS Microbio. Ecol.* 31, 3–9.
42. Hayashi, T., Hayashi, K. and Kojima, I. 1996. Antiviral polysaccharide. United States Patent: 5,585,365, Assigned to: Nippon Oil Company, Ltd. (Tokyo, Japan).
43. Wallis, J.G., Watts, J.L. and Browse, J. 2002. Polyunsaturated fatty acid synthesis: what will they think of next? *Trends Biochem. Sci.* 27, 467–473.
44. Fisher, J., Ryu, S.Y. and Snook, J.T. 1999. Polyunsaturated fatty acid (PUFA) intake in the development of cancer in participants in the lipid research clinics coronary primary prevention trial (LRC-CPPT). *FASEB J.* 13, A933–A933.
45. Di Minno, G., Tufano, A., Garofano, T. and Di Minno, M.N.D. 2002. Polyunsaturated fatty acids, thrombosis and vascular disease. *Pathophysiol. Haemostasis Thromb.* 32, 361–364.
46. Jones, P.J.H. and Lau, V.W.Y. 2002. Effect of n-3 polyunsaturated fatty acids on risk reduction of sudden death. *Nutr. Rev.* 60, 407–409.
47. Stillwell, W. and Wassall, S.R. 2003. Docosahexaenoic acid: membrane properties of a unique fatty acid. *Chem. Phys. Lipids* 126, 1–27.
48. Wen, Z.Y. and Chen, F. 2003. Heterotrophic production of eicosapentaenoic acid by microalgae. *Biotech. Adv.* 21, 273–294.
49. Kay, R.A. 1991. Microalgae as food and supplement. *Crit. Rev. Food Sci. Nutr.* 30, 555–573.
50. Allen, E.E. and Bartlett, D.H. 2002. Structure and regulation of the omega-3 polyunsaturated fatty acid synthase genes from the deep-sea bacterium *Photobacterium profundum* strain SS9. *Microbiology* 148, 1903–1913.
51. Calder, P.C. 2001. N-3 polyunsaturated fatty acids, inflammation and immunity: pouring oil on troubled waters or another fishy tale? *Nutr. Res.* 21, 309–341.
52. Knacker, T., Harwood, J.L., Hunter, C.N. and Russell, N.J. 1985. Lipid biosynthesis in synchronized cultures of the photosynthetic bacterium *Rhodopseudomonas sphaeroides*. *Biochem. J.* 229, 701–710.
53. Bajpai, P. and Bajpai, P.K. 1993. Eicosapentaenoic acid (EPA) production from microorganisms—a review. *J. Biotech.* 30, 161–183.
54. Nichols, D., Bowman, J., Sanderson, K., Nichols, C.M., Lewis, T., McMeekin, T. and Nichols, P.D. 1999. Developments with Antarctic microorganisms: culture collections, bioactivity screening, taxonomy, PUFA production and cold-adapted enzymes. *Curr. Opin. Biotechnol.* 10, 240–246.

55. Allen, E.E. and Bartlett, D.H. 2000. FabF is required for piezoregulation of cis-vaccenic acid levels and piezophilic growth of the deep-sea bacterium *Photobacterium profundum* strain SS9. *J. Bacteriol.* 182, 1264–1271.
56. Allen, E.E., Facciotti, D. and Bartlett, D.H. 1999. Monounsaturated but not polyunsaturated fatty acids are required for growth of the deep-sea bacterium *Photobacterium profundum* SS9 at high pressure and low temperature. *Appl. Environ. Microbiol.* 65, 1710–1720.
57. Ratledge, C. 1993. Single-cell oils—have they a biotechnological future. *Trends Biotechnol.* 11, 278–284.
58. Metz, J., Roessler, P., Facciotti, D., Levering, C., Dittrich, F., Lassner, M., Valentine, R., Lardizabal, K., Domergue, F., Yamada, A., Yazawa, K., Knauf, V. and Browse, J. 2001. Production of polyunsaturated fatty acids by polyketide synthases in both prokaryotes and eukaryotes. *Science* 293, 290–293.
59. Lewis, T.E., Nichols, P.D. and McMeekin, T.A. 2001. Sterol and squalene content of a docosahexaenoic acid producing thraustochytrid: influence of culture age, temperature, and dissolved oxygen. *Mar. Biotechnol.* 3, 439–447.
60. Tonon, T., Harvey, D., Larson, T.R. and Graham, I.A. 2002. Long chain polyunsaturated fatty acid production and partitioning to triacylglycerols in four microalgae. *Phytochemistry* 61, 15–24.
61. Morita, N., Ueno, A., Tamaka, M., Ohgiya, S., Hoshino, T., Kawasaki, K., Yumoto, I., Ishizaki, K. and Okuyama, H. 1999. Cloning and sequencing of clustered genes involved in fatty acid biosynthesis from the docosahexaenoic acid producing bacterium, *Vibrio marinus* strain MP-1. *Biotechnol. Lett.* 21, 641–646.
62. Certık, M. and Shimizu, S. 1999. Biosynthesis and regulation of microbial polyunsaturated fatty acid production. *J. Biosci. Bioeng.* 87, 1–14.
63. Barnham, K.J., Masters, C.L. and Bush, A.I. 2004. Neurodegenerative diseases and oxidative stress. *Nat. Rev. Drug Discov.* 3, 205–214.
64. Bandyopadhyay, D., Chattopadhyay, A., Ghosh, G. and Datta, A.G. 2004. Oxidative stress-induced ischemic heart disease: protection by antioxidants. *Curr. Med. Chem.* 11, 369–387.
65. Edge, R., McGarvey, D.J. and Truscott, T.G. 1997. The carotenoids as antioxidants—a review. *J. Photochem. Photobiol. B-Biol.* 41, 189–200.
66. Singh, D.K. and Lippman, S.M. 1998. Cancer chemoprevention—Part 1: Retinoids and carotenoids and other classic antioxidants. *Oncol.-New York* 12, 1643–1658.
67. Smith, T.A.D. 1998. Carotenoids and cancer: prevention and potential therapy. *Brit. J. Biomed. Sci.* 55, 268–275.
68. Lee, P.C. and Schmidt-Dannert, C. 2002. Metabolic engineering towards biotechnological production of carotenoids in microorganisms. *Appl. Microbiol. Biotechnol.* 60, 1–11.
69. Guerin, M., Huntley, M.E. and Olaizola, M. 2003. *Haematococcus* astaxanthin: applications for human health and nutrition. *Trends Biotechnol.* 21, 210–216.
70. Lorenz, R.T. and Cysewski, G.R. 2000. Commercial potential for *Haematococcus* microalgae as a natural source of astaxanthin. *Trends Biotechnol.* 18, 160–167.
71. Dunlap, W., Chalker, B. and Oliver, J. 1986. Photoadaptation by reef-building corals from Davies Reef, Great Barrier Reef, Australia. III. UV-B absorbing pigments. *J. Exp. Mar. Biol. Ecol.* 104, 239–248.
72. Karsten, U. (2002) Effects of salinity and ultraviolet radiation on the concentration of mycosporine-like amino acids in various isolates of the benthic cyanobacterium *Microcoleus chthonoplastes. Phycol. Res.* 50, 129–134.

73. Franklin, L.A., Krabs, G. and Kuhlenkamp, R. 2001. Blue light and UV-A radiation control the synthesis of mycosporine-like amino acids in *Chondrus crispus* (*Florideophyceae*). *J. Phycol.* 37, 257–270.
74. Jeffrey, S.W., MacTavish, H.S., Dunlap, W.C., Vesk, M. and Groenewoud, K. 1999. Occurrence of UVA- and UVB-absorbing compounds in 152 species (206 strains) of marine microalgae. *Mar. Ecol.-Prog. Ser.* 189, 35–51.
75. Dunlap, W.C. and Shick, J.M. 1998. Ultraviolet radiation-absorbing mycosporine-like amino acids in coral reef organisms: a biochemical and environmental perspective. *J. Phycol.* 34, 418–430.
76. Sunda, W., Kieber, D., Kiene, R. and Huntsman, S. 2002. An antioxidant function for DMSP and DMS in marine algae. *Nature* 418, 317–320.
77. Karsten, U., Wiencke, C. and Kirst, G.O. 1990. The effect of light-intensity and day-length on the beta-dimethylsulphoniopropionate (DMSP) content of marine green macroalgae from Antarctica. *Plant Cell Environ.* 13, 989–993.
78. Abdel-Lateff, A., Konig, G.M., Fisch, K.M., Holler, U., Jones, P.G. and Wright, A.D. 2002. New antioxidant hydroquinone derivatives from the algicolous marine fungus *Acremonium* sp. *J. Nat. Prod.* 65, 1605–1611.
79. Abdel-Lateff, A., Fisch, K.M., Wright, A.D. and Konig, G.M. 2003. A new antioxidant isobenzofuranone derivative from the algicolous marine fungus *Epicoccum* sp. *Planta Medica* 69, 831–834.
80. Guzman, S., Gato, A. and Calleja, J.M. 2001. Antiinflammatory, analgesic and free radical scavenging activities of the marine microalgae *Chlorella stigmatophora* and *Phaeodactylum tricornutum*. *Phytother. Res.* 15, 224–230.
81. Takamatsu, S., Hodges, T.W., Rajbhandari, I., Gerwick, W.H., Hamann, M.T. and Nagle, D.G. 2003. Marine natural products as novel antioxidant prototypes. *J. Nat. Prod.* 66, 605–608.
82. Proteau, P.J., Gerwick, W.H., Garciapichel, F. and Castenholz, R. 1993. The structure of scytonemin, an ultraviolet sunscreen pigment from the sheaths of cyanobacteria. *Experientia* 49, 825–829.
83. Estrada, D.M., Martin, J.D. and Perez, C. 1987. A new brominated monoterpenoid quinol from *Cymopolia barbata*. *J. Nat. Prod.* 50, 735–737.
84. Sun, H.H., Paul, V.J. and Fenical, W. 1983. Avrainvilleol, a brominated diphenylmethane derivative with feeding deterrent properties from the tropical green-alga *Avrainvillea longicaulis*. *Phytochemistry* 22, 743–745.
85. Kirkup, M.P. and Moore, R.E. 1983. Indole alkaloids from the marine red alga *Martensia fragilis*. *Tetrahedron Lett.* 24, 2087–2090.
86. Yamada, K., Yamada, Y., Fukuda, M. and Yamada, S. 1999. Bioavailability of dried asakusanori (*Porphyra tenera*) as a source of cobalamin (vitamin B_{12}). *Int. J. Vit. Nutr. Res.* 69, 412–418.
87. Brown, M.R., Mular, M., Miller, I., Farmer, C. and Trenerry, C. 1999. The vitamin content of microalgae used in aquaculture. *J. Appl. Phycol.* 11, 247–255.
88. Laamanen, M.J., Forsstrom, L. and Sivonen, K. 2002. Diversity of *Aphanizomenon flos-aquae* (cyanobacterium) populations along a Baltic Sea salinity gradient. *Appl. Environ. Microbiol.* 68, 5296–5303.
89. Kushak, R., VanCott, E., Drapeau, C. and Winter, H. 1999. Effect of algae *Aphanizomenon flos-aquae* on digestive enzyme activity and polyunsaturated fatty acids level in blood plasma. *Gastroenterology* 116, G2455.
90. Carmichael, W.W., Drapeau, C. and Anderson, D.M. 2000. Harvesting of *Aphanizomenon flos-aquae* Ralfs ex Born. & Flah. var. *flos-aquae* (Cyanobacteria) from Klamath Lake for human dietary use. *J. Appl. Phycol.* 12, 585–595.

91. Yamaguchi, K. 1996. Recent advances in microalgal bioscience in Japan, with special reference to utilization of biomass and metabolites: a review. *J. Appl. Phycol.* 8, 487–502.
92. Watanabe, F., Takenaka, S., Kittaka-Katsura, H., Ebara, S. and Miyamoto, E. 2002. Characterization and bioavailability of vitamin B_{12}-compounds from edible algae. *J. Nutr. Sci. Vitaminol.* 48, 325–331.
93. Hastings, C.W., Barnes, D.J. and Daley, C.A. 2002. Dietary supplement. United States Patent: 6,368,617, Assigned to: Reliv' International, Inc. (Chesterfield, MO, USA).
94. Irianto, A. and Austin, B. (2002) Probiotics in aquaculture. *J. Fish Dis.* 25, 633–642.
95. Schwartz, R., Hirsch, C., Sesin, D., Flor, J., Chartrain, M., Frontling, R., Harris, G., Salvatore, M., Liesch, J. and Yudin, K. 1990. Pharmaceuticals from cultured algae. *J. Ind. Micro. Biotech.* 5, 113–124.
96. Napier, J.A. 2002. Plumbing the depths of PUFA biosynthesis: a novel polyketide synthase-like pathway from marine organisms. *Trends Plant Sci.* 7, 51–54.
97. Zazopoulos, E., Huang, K.X., Staffa, A., Liu, W., Bachmann, B.O., Nonaka, K., Ahlert, J., Thorson, J.S., Shen, B. and Farnet, C.M. 2003. A genomics-guided approach for discovering and expressing cryptic metabolic pathways. *Nat. Biotech.* 21, 187–190.
98. Chisti, Y. and Moo-Young, M. 1991. Fermentation technology, bioprocessing scale-up and manufacture in biotechnology. In: *The Science and the Business* (Noves, V. and Cape, R., Eds.), p. 167. Harwood Academic Publishers, New York.
99. Chisti, Y. and Moo-Young, M. 1996. Bioprocess intensification through bioreactor engineering. *Chem. Eng. Res. Design* 74, 575–583.
100. Marwick, J.D., Wright, P.C. and Burgess, J.G. 1999. Bioprocess intensification for production of novel marine bacterial antibiotics through bioreactor operation and design. *Mar. Biotech.* 1, 495–507.
101. Armstrong, J.E., Janda, K.E., Alvarado, B. and Wright, A.E. 1991. Cytotoxin production by a marine *Lyngbya* strain (cyanobacterium) in a large-scale laboratory bioreactor. *J. Appl. Phycol.* 3, 277–282.
102. Gerwick, W.H., Roberts, M.A., Proteau, P.J. and Chen, J.-L. 1994. Screening cultured marine microalgae for anticancer-type activity. *J. Appl. Phycol.* 6, 143–149.
103. Moore, R.E., Patterson, G.M.L. and Carmichael, W.W. 1988. New pharmaceuticals from cultured blue-green algae. In: *Biomedical Importance of Marine Organisms—Memoirs of the Californian Academy of Science* (Fautin, D.G., Ed.), Vol. 13, pp. 143–150. Californian Academy of Science, San Francisco, CA.
104. de Philippis, R. and Vincenzini, M. 1998. Exocellular polysaccharides from cyanobacteria and their possible applications. *FEMS Microbiol. Rev.* 22, 151–175.
105. Sutherland, I. 1990. *Biotechnology of Microbial Exopolysaccharides*, 171 pp. Cambridge University Press, Cambridge, UK.
106. Panoff, J., Priem, B., Morvan, H. and Joset, F. 1988. Sulfated exopolysaccharides produced by two unicellular strains of cyanobacteria, *Synechocystis* PCC 6803 and PCC 6714. *Arch. Microbiol.* 150, 558–563.
107. Vincenzini, M., Dephilippis, R., Sili, C. and Materassi, R. 1990. Studies on exopolysaccharide release by diazotrophic batch cultures of *Cyanospira capsulata*. *Appl. Microbiol. Biotech.* 34, 392–396.
108. Filalimouhim, R., Cornet, J., Fontaine, T., Fournet, B. and Dubertret, G. 1993. Production, isolation and preliminary characterization of the exopolysaccharide of the cyanobacterium *Spirulina platensis*. *Biotech. Lett.* 15, 567–572.

109. Vincenzini, M., Dephilippis, R., Sili, C. and Materassi, R. 1993. Stability of molecular and rheological properties of the exopolysaccharide produced by *Cyanospira capsulata* cultivated under different growth-conditions. *J. Appl. Phycol.* 5, 539–541.
110. Sudo, H., Burgess, J., Takemasa, H., Nakamura, N. and Matsunaga, T. 1995. Sulfated exopolysaccharide production by the halophilic cyanobacterium *Aphanocapsa halophytia*. *Curr. Microbiol.* 30, 219–222.
111. Fattom, A. and Shilo, M. 1984. *Phormidium* J-1 bioflocculant—production and activity. *Arch. Microbiol.* 139, 421–426.
112. Flaibani, A., Olsen, Y. and Painter, T. 1989. Polysaccharides in desert reclamation—compositions of exocellular proteoglycan complexes produced by filamentous blue-green and unicellular green edaphic algae. *Carbohydr. Res.* 190, 235–248.
113. Reddy, K., Soper, B., Tang, J. and Bradley, R. 1996. Phenotypic variation in exopolysaccharide production in the marine, aerobic nitrogen-fixing unicellular cyanobacterium *Cyanothece* sp. *World J. Microbiol. Biotechnol.* 12, 311–318.
114. Hopwood, D.A. 1990. Molecular-genetics of polyketides and its comparison to fattyacid biosynthesis. *Ann. Rev. Genet.* 24, 37–66.
115. Katz, L. and Donadio, S. 1993. Polyketide synthesis—prospects for hybrid antibiotics. *Ann. Rev. Microbiol.* 47, 875–912.
116. Hutchinson, C.R. and Fujii, I. 1995. Polyketide synthase gene manipulation—a structure-function approach in engineering novel antibiotics. *Ann. Rev. Microbiol.* 49, 201–238.
117. Yazawa, K. 1996. Production of eicosapentaenoic acid from marine bacteria. *Lipids* 31, S297–S300.
118. Rohmer, M. 1999. The discovery of a mevalonate-independent pathway for isoprenoid biosynthesis in bacteria, algae and higher plants. *Nat. Prod. Rep.* 16, 565–574.
119. Bartley, G., Scolnik, P. and Beyer, P. 1999. Two *Arabidopsis thaliana* carotene desaturases, phytoene desaturase and zeta-carotene desaturase, expressed in *Escherichia coli*, catalyze a poly-cis pathway to yield pro-lycopene. *Eur. J. Biochem.* 259, 396–403.
120. Raisig, A. and Sandmann, G. 2001. Functional properties of diapophytoene and related desaturases of C-30 and C-40 carotenoid biosynthetic pathways. *Biochim. Biophys. Acta— Mol. Cell Biol. Lipids* 1533, 164–170.
121. Komori, M., Ghosh, R., Takaichi, S., Hu, Y., Mizoguchi, T., Koyama, Y. and Kuki, M. 1998. A null lesion in the rhodopin 3,4-desaturase of *Rhodospirillum rubrum* unmasks a cryptic branch of the carotenoid biosynthetic pathway. *Biochemistry* 37, 8987–8994.
122. Lotan, T. and Hirschberg, J. 1995. Cloning and expression in *Escherichia coli* of the gene encoding b-C-4-oxygenase, that converts β-carotene to the ketocarotenoid canthaxanthin in *Haematococcus pluvialis*. *FEBS Lett.* 364, 125–128.
123. Johnson, E. and Schroeder, W. 1995. Astaxanthin from the yeast *Phaffia rhodozyma*. *Stud. Mycol.* 38, 81–90.
124. Ruther, A., Misawa, N., Boger, P. and Sandmann, G. 1997. Production of zeaxanthin in *Escherichia coli* transformed with different carotenogenic plasmids. *Appl. Microbiol. Biotechnol.* 48, 162–167.
125. Yamano, S., Ishii, T., Nakagawa, M., Ikenaga, H. and Misawa, N. 1994. Metabolic engineering for production of beta-carotene and lycopene in *Saccharomyces cerevisiae*. *Biosci. Biotechnol. Biochem.* 58, 1112–1114.
126. Miura, Y., Kondo, K., Shimada, H., Saito, T., Nakamura, K. and Misawa, N. 1998. Production of lycopene by the food yeast, *Candida utilis* that does not naturally synthesize carotenoid. *Biotechnol. Bioeng.* 58, 306–308.

127. Miura, Y., Kondo, K., Saito, T., Shimada, H., Fraser, P.D. and Misawa, N. 1998. Production of the carotenoid lycopene, beta-carotene, and astaxanthin in the food yeast *Candida utilis*. *Appl. Environ. Microbiol.* 64, 1226–1229.
128. Hunter, C.N., Hundle, B.S., Hearst, J.E., Lang, H.P., Gardiner, A.T., Takaichi, S. and Cogdell, R.J. 1994. Introduction of new carotenoids into the bacterial photosynthetic apparatus by combining the carotenoid biosynthetic pathways of *Erwinia herbicola* and *Rhodobacter sphaeroides*. *J. Bacteriol.* 176, 3692–3697.
129. Harker, M. and Hirschberg, J. 1998. Molecular biology of carotenoid biosynthesis in photosynthetic organisms (In: Photosynthesis: Molecular Biology of Energy Capture). *Methods Enzymol.* 297, 244–263.
130. Harker, M. and Hirschberg, J. 1997. Biosynthesis of ketocarotenoids in transgenic cyanobacteria expressing the algal gene for b-C-4-oxygenase, *crtO*. *FEBS Lett.* 404, 129–134.
131. Lagarde, D., Beuf, L. and Vermaas, M. 2000. Increased production of zeaxanthin and other pigments by application of genetic engineering techniques to *Synechocystis* sp. PCC 6803. *Appl. Environ. Microb.* 66, 64–72.
132. Schmidt-Dannert, C. 2000. Engineering novel carotenoids in microorganisms. *Curr. Opin. Biotechnol.* 11, 255–261.
133. Mijts, B. and Schmidt-Dannert, C. 2003. Engineering of secondary metabolite pathways. *Curr. Opin. Biotechnol.* 14, 597–602.
134. Umeno, D., Tobias, A. and Arnold, F. 2002. Evolution of the C_{30} carotenoid synthase CrtM for function in a C_{40} pathway. *J. Bacteriol.* 184, 6690–6699.
135. Umeno, D. and Arnold, F. 2003. A C_{35} carotenoid biosynthetic pathway. *Appl. Environ. Microb.* 69, 3573–3579.
136. Shinmen, Y., Kawashima, H., Shimizu, S. and Yamada, H. 1992. Concentration of eicosapentaenoic acid and docosahexaenoic acid in an arachidonic acid-producing fungus, *Mortierella alpina* 1S-4, grown with fish oil. *Appl. Microb. Biotechnol.* 38, 301–304.
137. Yano, Y., Nakayama, A., Saito, H. and Ishihara, K. 1994. Production of docosahexaenoic acid by marine-bacteria isolated from deep-sea fish. *Lipids* 29, 527–528.
138. Bowman, J.P., McCammon, S.A., Nichols, D.S., Skerratt, J.H., Rea, S.M., Nichols, P.D. and McMeekin, T.A. 1997. *Shewanella gelidimarina* sp. nov. and *Shewanella frigidimarina* sp. nov., novel Antarctic species with the ability to produce eicosapentaenoic acid (20:5 omega 3) and grow anaerobically by dissimilatory Fe(III) reduction. *Int. J. Syst. Bacteriol.* 47, 1040–1047.
139. Watanabe, K., Yazawa, K., Kondo, K. and Kawaguchi, A. 1997. Fatty acid synthesis of an eicosapentaenoic and producing bacterium: *De novo* synthesis, chain elongation, and desaturation systems. *J. Biochem.* 122, 467–473.
140. Bowman, J.P., Gosink, J.J., McCammon, S.A., Lewis, T.E., Nichols, D.S., Nichols, P.D., Skerratt, J.H., Staley, J.T. and McMeekin, T.A. 1998. *Colwellia demingiae* sp. nov., *Colwellia hornerae* sp. nov., *Colwellia rossensis* sp. nov. and *Colwellia psychrotropica* sp. nov.: psychrophilic Antarctic species with the ability to synthesize docosahexaenoic acid (22:6 omega 3). *Int. J. Syst. Bacteriol.* 48, 1171–1180.
141. Bowman, J.P., McCammon, S.A., Lewis, T., Skerratt, J.H., Brown, J.L., Nichols, D.S. and McMeekin, T.A. 1998. *Psychroflexus torquis* gen. nov., sp. nov., a psychrophilic species from Antarctic sea ice, and reclassification of Flavobacterium gondwanense (Dobson et al. 1993) as *Psychroflexus gondwanense* gen. nov., comb. nov. *Microbiol.-UK* 144, 1601–1609.
142. Henderson, R.J. and Millar, R.M. 1998. Characterization of lipolytic activity associated with a *Vibrio* species of bacterium isolated from fish intestines. *J. Mar. Biotechnol.* 6, 168–173.

143. Kim, H., Gandhi, S.R., Moreau, R.A. and Weete, J.D. 1998. Lipids of *Haliphthoros philippinensis*: An oomycetous marine microbe. *J. Am. Oil Chem. Soc.* 75, 1657–1665.
144. Singh, A. and Ward, O.P. 1998. Docosapentaenoic acid (C22:5, omega-3) production by *Pythium acanthicum*. *J. Ind. Microb.Biotechnol.* 20, 187–191.
145. Vazhappilly, R. and Chen, F. 1998. Heterotrophic production potential of omega-3 polyunsaturated fatty acids by microalgae and algae-like microorganisms. *Botanica Marina* 41, 553–558.
146. Yokochi, T., Honda, D., Higashihara, T. and Nakahara, T. 1998. Optimization of docosahexaenoic acid production by *Schizochytrium limacinum* SR21. *Appl. Microb. Biotechnol.* 49, 72–76.
147. Ashford, A., Barclay, W.R., Weaver, C.A., Giddings, T.H. and Zeller, S. 2000. Electron microscopy may reveal structure of docosahexaenoic acid-rich oil within *Schizochytrium* sp. *Lipids* 35, 1377–1386.
148. Fernandez, F.G.A., Perez, J.A.S., Sevilla, J.M.F., Camacho, F.G. and Grima, E.M. 2000. Modeling of eicosapentaenoic acid (EPA) production from *Phaeodactylum tricornutum* cultures in tubular photobioreactors. Effects of dilution rate, tube diameter, and solar irradiance. *Biotech. Bioeng.* 68, 173–183.
149. Morita, N., Tanaka, M. and Okuyama, H. 2000. Biosynthesis of fatty acids in the docosahexaenoic acid-producing bacterium *Moritella marina* strain MP-I. *Biochem. Soc. Trans.* 28, 943–945.
150. Ratledge, C., Kanagachandran, K., Anderson, A.J., Grantham, D.J. and Stephenson, J.C. 2001. Production of docosahexaenoic acid by *Crypthecodinium cohnii* grown in a pH-auxostat culture with acetic acid as principal carbon source. *Lipids* 36, 1241–1246.
151. Nogi, Y., Kato, C. and Horikoshi, K. 2002. *Psychromonas kaikoae* sp nov., a novel piezophilic bacterium from the deepest cold-seep sediments in the Japan Trench. *Int. J. Syst. Evol. Microb.* 52, 1527–1532.
152. Qi, B., Beaudoin, F., Fraser, T., Stobart, A., Napier, J. and Lazarus, C. 2002. Identification of a cDNA encoding a novel C18-D^9 polyunsaturated fatty acid-specific elongating activity from the docosahexaenoic acid (DHA)-producing microalga, *Isochrysis galbana*. *FEBS Lett.* 510, 159–165.
153. Skerratt, J.H., Bowman, J.P. and Nichols, P.D. 2002. *Shewanella olleyana* sp. nov., a marine species isolated from a temperate estuary which produces high levels of polyunsaturated fatty acids. *Int. J. Sys. Evol. Microb.* 52, 2101–2106.
154. Satomi, M., Oikawa, H. and Yano, Y. 2003. *Shewanella marinintestina* sp. nov., *Shewanella schlegeliana* sp. nov. and *Shewanella sairae* sp. nov., novel eicosapentaenoic acid-producing marine bacteria isolated from sea animal intestines. *Int. J. Sys. Evol. Microb.* 53, 491–499.
155. Yakimov, M.M., Giuliano, L., Gentile, G., Crisafi, E., Chernikova, T.N., Abraham, W.R., Lunsdorf, H., Timmis, K.N. and Golyshin, P.N. 2003. *Oleispira antarctica* gen. nov., sp nov., a novel hydrocarbonoclastic marine bacterium isolated from Antarctic coastal sea water. *Int. J. Sys. Evol. Microb.* 53, 779–785.
156. Morita, M., Watanabe, Y. and Saiki, H. 2000. Investigation of photobioreactor design for enhancing the photosynthetic productivity of microalgae. *Biotech. Bioeng.* 69, 693–698.
157. de Swaaf, M.E., Grobben, G.J., Eggink, G., de Rijk, T.C., van der Meer, P. and Sijtsma, L. 2001. Characterisation of extracellular polysaccharides produced by *Crypthecodinium cohnii*. *Appl. Microb. Biotechnol.* 57, 395–400.

158. de Swaaf, M.E., Sijtsma, L. and Pronk, J.T. 2003. High-cell-density fed-batch cultivation of the docosahexaenoic acid producing marine alga *Crypthecodinium cohnii*. *Biotechnol. Bioeng.* 81, 666–672.
159. Bowles, R.D., Hunt, A.E., Bremer, G.B., Duchars, M.G. and Eaton, R.A. 1999. Long-chain n-3 polyunsaturated fatty acid production by members of the marine protistan group the thraustochytrids: screening of isolates and optimisation of docosahexaenoic acid production. *J. Biotechnol.* 70, 193–202.
160. Iida, I., Nakahara, T., Yokochi, T., Kamisaka, Y., Yagi, H., Yamaoka, M. and Suzuki, O. 1996. Improvement of docosahexaenoic acid production in a culture of *Thraustochytrium aureum* by medium optimization. *J. Ferment. Bioeng.* 81, 76–78.
161. Kendrick, A. and Ratledge, C. 1992. Phospholipid fatty acyl distribution of 3 fungi indicates positional specificity for n-6 vs. n-3 fatty-acids. *Lipids* 27, 505–508.
162. Lewis, T.E., Nichols, P.D. and McMeekin, T.A. 1999. The biotechnological potential of thraustochytrids. *Mar. Biotechnol.* 1, 580–587.

16 Shark Cartilage: Potential for Therapeutic Application for Cancer—Review Article

Kenji Sato, Tsukasa Kitahashi, Chieko Itho, and Masahiro Tsutsumi

CONTENTS

16.1 INTRODUCTION

Sharks are frequently landed as a bycatch in tuna fishing. Shark meat has traditionally been utilized as a food ingredient in the forms of fillet and minced meat (surimi), among others. In addition to meat, dried collagenous fibers on both sides of fin cartilage have been used for high-grade Chinese cuisine.

However, skin and cartilages have not been used much as a food ingredient. Recently, gelatin and its enzymatically produced hydrolyzates were prepared from shark skin and used as a functional food ingredient [1]. From cartilage, the acidic polysaccharide referred to as chondroichin sulfate has been prepared after degrading the protein by exhaustive proteolytic digestion. The shark condroichin sulfate has a long history of use in cosmetics and medicines to treat surgical accretion, arthritis, and so on. In addition to chondroichin sulfate, shark cartilage has been demonstrated to contain some biologically active components such as protease inhibitors, stimulators, and suppressors of cell proliferation [2–4]. As most of the shark endoskeleton consists of cartilage, it might serve as a potential source for production of cartilage–derived bioactive components. On the basis of these findings, dried shark cartilage powder and its water extract have been prepared in industrial scale for possible use as functional food ingredients [5,6]. However, there is discouraging data on the efficacy of shark cartilage products in human trial [7]. In addition, only limited information is available on biological response to oral administration of shark cartilage–based products. In this chapter, recent findings on anticancer effects of shark cartilage extract on the basis of *in vitro* and *in vivo* inhibition against matrix metalloproteinase-9 are reported.

16.2 IMPACT OF INHIBITION OF MATRIX METALLOPROTEINASE ON CANCER PROGRESSION

Folkman [8,9] has demonstrated that malignant tumor induces new blood vessels from preexisting vessels to obtain nutrients from host and waste metabolite, which is a critical step for growth of tumor larger than 1 mm. This phenomenon is referred to as tumor-induced angiogenesis. They have also demonstrated that tumor-induced angiogenesis is a good target for cancer therapy [10,11]. They demonstrated in their early studies that cartilage implanted near the tumor inhibits tumor-induced angiogenesis and consequently tumor growth [12,13]. To identify substance(s) responsible for inhibition of tumor-induced angiogenesis, activity-directed fractionations were carried out [3], which suggested that collagenase inhibitor(s) might play a significant role in the inhibition of tumor-induced angiogenesis. Some researchers have supported this concept by demonstrating the presence of collagenase-specific inhibitors, namely, tissue inhibitor of matrix metalloproteinase (TIMP)-like proteins, in shark cartilage [2,14]. In addition to collagenase inhibition, direct inhibition of signal transmittance triggered by angiogenesis stimulators such as vascular endothelial growth factor (VEGF) has been proposed as possible mechanism of anticancer effect by shark cartilage [15]. However, a considerably high dose of shark cartilage extract is necessary to inhibit the signal transmittance even in *in vitro* cell culture system and the inhibition of the signal transmittance has not been confirmed by oral ingestion of shark cartilage–based product.

Collagenase has been defined as an endoproteinase, which can cleave triple helical domain of collagen. Collagen refers to a molecular family consisting of more than 20 types. Collagen can be divided into fibril-forming and non-fibril-forming type. Types I, II, III, V, XI collagens are classified as fibril-forming collagen. Other collagens are non-fibril-forming ones. Types I, II, and III collagens are the main constituents in skin, tendon, bone, and cartilage. Type II and XI are cartilage-specific collagen. Type V collagen is preferentially distributed in pericellular connective tissue. Type IV collagen forms a netlike structure and a main constituent of basement membrane. Other collagens are located on the surface of collagen fibrils and between collagen fibrils and basement membrane (see review by van der Rest and Garrone [16]). The tissue type collagenases belong to a metalloprotease family referred to as matrix metalloproteinases (MMP), which consists of more than 23 distinct enzymes. Two groups of MMPs can cleave triple helical domain of collagen. MMP-1 (formerly referred to as collagenase in a narrow sense) can cleave triple helical domain of type I collagen, the major constituent of interstitial connective tissues, such as bone, skin, and tendon, among others. The second group, MMPs-2 and -9 (formerly referred to as gelatinase A and B, respectively) preferentially cleave gelatin (denatured form of type I collagen) and triple helical domains of type IV and V collagens, major constituents of basement membrane and pericellular connective tissues, respectively. MMPs-2 and -9 are secreted as inactive latent forms and activated by proteolytic digestion. No significant amounts of active forms of these MMPs are observed in normal tissues except for physiological tissue remodeling and embryonic development. However, these MMPs are frequently overexpressed and activated on tumor cells with high malignancy (Figure 16.1, left). In addition, angiogenesis stimulator such as VEGF upregulates MMP-9. Therefore, consensus is reached that these MMPs play an important role in the proteolytic destruction of extracellular matrix near tumor and stimulated endothelial cell, thereby promoting tumor invasion, metastasis, and angiogenesis. See recent reviews by Curran and Murray [17]; Johansen et al. [18]; McCawley and Matrisian [19]; Foda and Zucker [20].

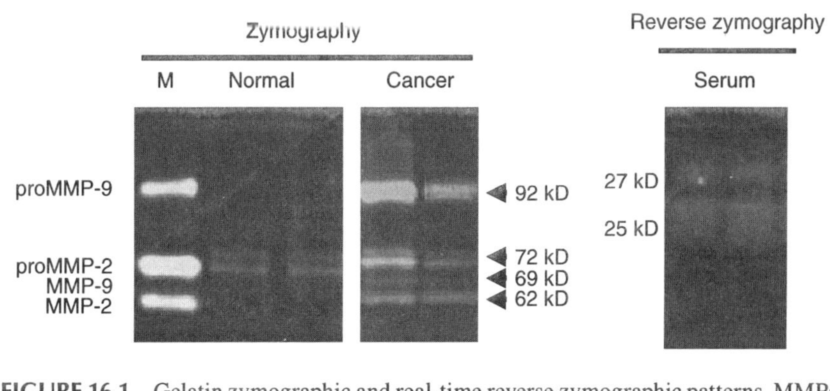

FIGURE 16.1 Gelatin zymographic and real-time reverse zymographic patterns. MMPs markers (M) and PBS-Tween 20 extract of normal and cancer legion of hamster pancreas were analyzed by gelatin zymography [22]. Serum from hamster-bearing pancreatic duct cancer was analyzed by real-time reverse zymography. proMMP represents latent form of MMP, MMP represents active form.

As described above, endogenous MMP inhibitors such as TIMPs-like proteins have been identified in shark cartilage. Recently, we found another type of MMP inhibitor in proteoglycan fraction from shark cartilage. More interestingly, we also found increased MMP-9 inhibitory activity in serum of animals and humans who ingested the proteoglycan fraction [21,22]. In the following sections, we introduce *in vitro* and *in vivo* inhibition of MMP-9 by proteoglycan fraction from shark cartilage and discuss its potential therapeutic effect for cancer therapy. The MMP-9 inhibitory activity in serum could not be evaluated by conventional method of using synthetic peptide. Thus, new approach for evaluation of MMP-9 activity in serum is introduced and advantages and disadvantages of the conventional and present approaches are discussed.

16.3 EVALUATION OF MMP AND ITS INHIBITORY ACTIVITY

16.3.1 Gelatin Zymography

MMPs-2 and 9 are secreted as latent forms and converted to active forms on the invasive front of tumor. Total content and expression of MMPs can be estimated by immunological and molecular biological approaches. It is, however, difficult to distinguish the latent and active forms of MMPs by these techniques. To solve these problems, gelatin zymography has been used extensively. Nonheated and nonreduced sample is used for zymography. The sample is loaded onto a gel copolymerized with 0.06% (w/v) gelatin. After electrophoresis, using Laemmli's buffer system, the gel is immersed in a nonionic detergent solution such as 2.5% Triton-X 100 to remove sodium dodecyl sulphate (SDS) and then incubated in a suitable buffer for a whole day. During incubation, renaturation and activation of MMPs occur in the gel. The activated MMPs degrade gelatin in the gel. By staining with Comaassie Brilliant Blue, MMP activity can be visualized as transparent bands owing to degradation of gelatin in the gel. As shown in Figure 16.1 (left), this technique can distinguish active and latent forms of MMPs-2 and -9. MMP inhibitors, if present, in the sample may be resolved from MMPs (latent and active forms) by electrophoresis. MMPs activities would then be observed even in the presence of excess amounts of MMP inhibitors. Therefore, this technique cannot evaluate inhibitory activity in the sample for the electrophoresis. Alternatively, another sample, which may contain potential MMP inhibitor can be added into the immersing buffer for the gel. If the sample contains MMP inhibitor, the inhibitor may penetrate into the gel and inhibit MMPs in the gel. By using this approach, some low-molecular-weight components from foods, such as green tea catechin, have been demonstrated to have MMPs-2 and -9 inhibitory activities [23]. However, large molecular proteinous inhibitors may not penetrate into the polyacrylamide gel matrix by just immersing. In addition, we found that some proteases and peptidases in human and animal serums can degrade gelatin, which may interfere with evaluation of MMP inhibitory activity. Therefore, this approach cannot evaluate total MMP inhibitory activity in biological samples containing proteinous inhibitors and proteases.

For estimation of high-molecular-weight inhibitors, real-time reverse gelatin zymography has been developed [24]. The sample is resolved by electrophoresis by using the gel copolymerized with fluorescein isothiocyanate (FITC)-labeled gelatin and MMP of interest. The gel after electrophoresis is treated similar to that described above. The MMP in the gel cleaves FITC labeled gelatin in gel, whereas the FITC-gelatin near MMP inhibitor cannot be cleaved. Then, MMP inhibitory activity can be visualized as band with fluorescence. As shown in Figure 16.1 (right), this technique can detect inhibitor bands at 27 and 25 kD in hamster serum, which may be TIMPs. However, this technique cannot detect the inhibitory activity of low-molecular-weight components owing to difficulty in resolution by SDS-PAGE. Therefore, balance of MMP activities and their inhibitory activities of serum and tissue extract cannot be estimated by these techniques. For this purpose, MMP activity or its inhibitory activity must be directly determined in the same solvent system.

16.3.2 MMP Activity Assay in Solvent

Type IV collagen is one of the important endogenous substrates of MMPs-2 and -9. However, the triple helical domains of type IV collagen are interrupted by globular domains, which can be degraded by proteases without collagenase activity. To solubilize type IV collagen from tissue, limited digestion by pepsin has extensively been used. Therefore, most of type IV collagen preparation consists of some triple helical fragments [16]. By SDS-PAGE analysis, several fragments appear depending on the intensity of pepsin digestion. In addition, type IV collagen has inter- and intramolecular disulfide bonds. Presence of reducing agents may affect MMP-9 activity and SDS-PAGE patterns of the degradation products. Therefore, it is difficult to quantify degradation products of type IV collagen by SDS-PAGE. In contrast, type V collagen, which is another endogenous substrate for MMPs-2 and -9, has globular domains only at each end of a triple helical domain. Therefore, type V collagen has single triple helical domain without disulfide bond even after pepsin digestion. As shown in Figure 16.2 (left), MMP-9 can cleave specific site of triple helical domain of type V collagen. The specific degradation products can be resolved and determined by SDS-PAGE followed by scanning staining intensity. Good linearity was observed between the staining intensity of degradation product bands (arrows) and MMP-9 activity (Figure 16.2, right). Therefore, the MMP-9 activity can be estimated by using intact type V collagen as substrate. MMP-2 activity can also be estimated in the same manner (data not shown). However, this assay is extensively time-consuming, as it takes a whole day for the reaction and another day for SDS-PAGE analysis.

The MMPs-2 and -9 can degrade gelatin, a denatured form of collagen. Therefore, FITC-labeled gelatin can be used as substrate for MMPs-2 and -9. After reaction with MMPs, the MMPs activity can be estimated by determining fluorescence intensity in the supernatant after precipitation of nondegraded gelatin. Now, some commercial products have been available and used to estimate the

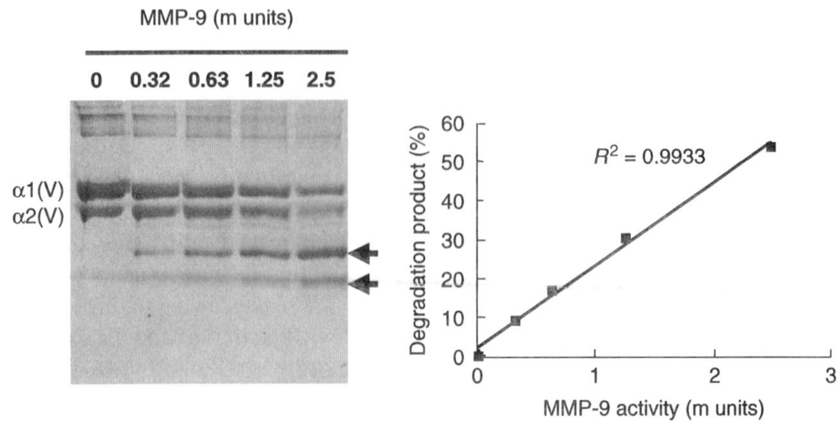

FIGURE 16.2 Degradation of native type V collagen by human active MMP-9. 10 μg of type V collagen was reacted with 0–2.5 m units of MMP-9 at 37°C for 24 h. The reaction mixture was resolved by 7.5% gel in the Laemmli's buffer system. Staining intensity of degradation products of type V collagen (arrows) was plotted against MMP-9 activity. Type V collagen consists of two subunit chains designated as α1(V) and α2(V).

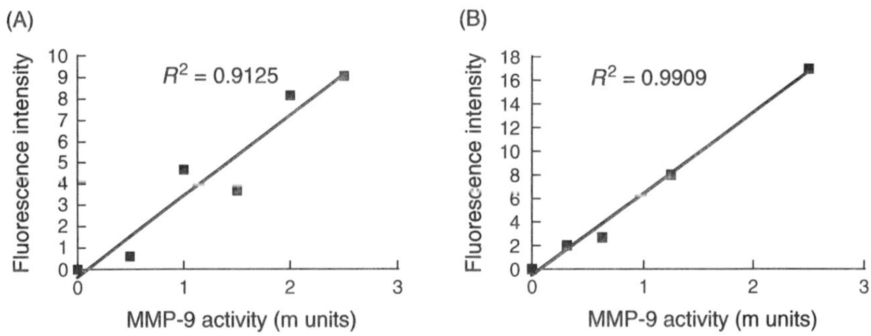

FIGURE 16.3 Standard curves between MMP-9 activity and fluorescence intensity. FITC-labeled gelatin (A) and Mca-Pro-Leu-Gly-Leu-Dap-Ala-Arg-NH_2 (B) were reacted with human active MMP-9.

MMPs activity and their inhibitory activity. As shown in Figure 16.3A, linearity can be obtained between fluorescence intensity and MMP-9 activity.

For estimation of MMP-9 activity, fluorescence-quenching synthetic peptides such as (7-methoxycoumarin-4-yl)acetyl-L-prolyl-L-leucyl-glycyl-L-leucyl-[N$^{\beta}$-(2,4-dinitrophenyl)-L-2,3-diaminopropyl]-L-alanyl-L-arginine amide (Mca-Pro-Leu-Gly-Leu-Dap-Ala-Arg-NH_2) have been used extensively [25]. As shown in Figure 16.3B, good linearity can be obtained between fluorescence intensity and MMP-9 activity. This approach is believed to allow specific, sensitive, and high-throughput analysis of the MMPs.

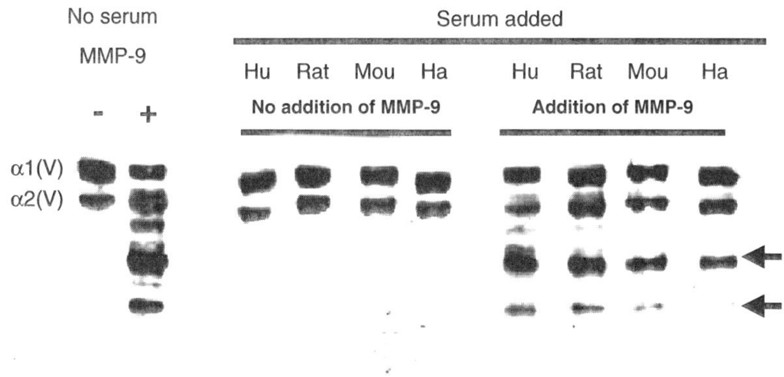

FIGURE 16.4 Effects of human and animal serums on degradation of type V collagen in the presence and absence of added MMP-9 (2.5 m units). The reaction mixtures were analyzed by 7.5% gel. The proteins in the gel were transferred to PVDF membrane and stained with antiserum against type V collagen.

16.3.3 Application for Estimation of MMP Inhibitory Activity in Animal and Human Serum

Serums from human, hamster, rat, and mouse were mixed with type V collagen and incubated for 24 h at 37°C. As shown in Figure 16.4, no degradation products were observed (Sato et al., unpublished data). However, addition of these serums into the MMP-9-type V collagen reaction mixture resulted in decreased staining intensity of degradation products. These facts clearly indicate that these serums have no collagenase activity against type V collagen, namely no MMP-2 and -9 activities but have inhibitory activity against MMP-9. The serums also have MMP-2 inhibitory activity (data not shown).

The FITC-gelatin and Mca Pro-Leu-Gly-Leu-Dap-Ala-Arg-NH_2 were mixed with the same serum samples. Unexpectedly, these substrates were extensively degraded by serum samples, which had no type V collagenase activity (Figure 16.5). These facts indicate that these substrates without triple helical structure are degraded nonspecifically by some serum peptidases that have no collagenase activity. Thus, false MMP activity appeared when such substrates were used. Therefore, collagenous substrates such as native type V collagen must be used for evaluation of collagenase activity and its inhibitory activity in biological samples.

16.4 *IN VITRO* INHIBITION OF MMPs BY SHARK CARTILAGE

It has been demonstrated that guanidine chloride and water extracts of shark cartilage have inhibitory activities against MMPs [2–4,14]. Recently, occurrence of TIMP-like proteins in these extracts has been described [2,14]. In these studies, the extracts were prepared from fresh shark cartilage. In contrast, dried shark cartilage preparations are produced as by-products from food industry. These products suffered from heat

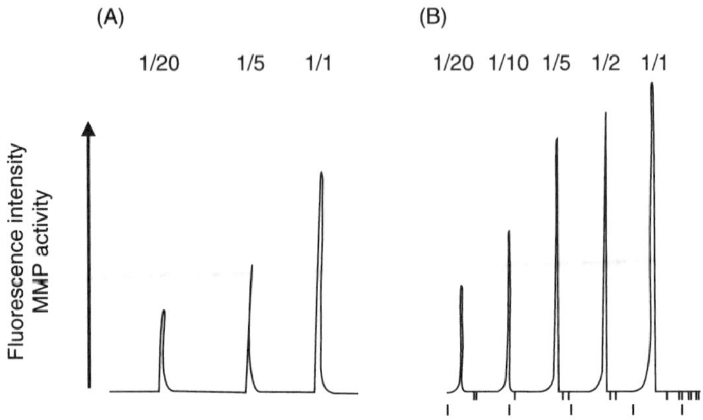

FIGURE 16.5 Degradation of FITC-labeled gelatin (A) and Mca-Pro-Leu-Gly-Leu-Dap-Ala-Arg-NH_2 (B) by human serum. These substrates were reacted with diluted serum. Fluorescence intensity was estimated by flow-injection method.

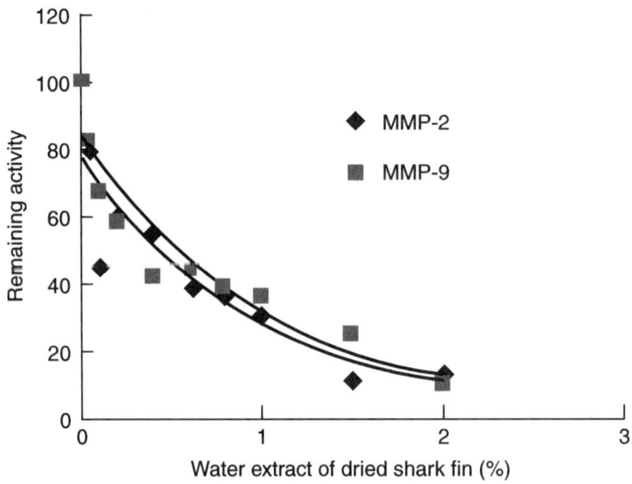

FIGURE 16.6 Inhibition of human fibrosarcoma MMPs-2 and -9 by water extract of shark cartilage.

treatment at 50–60°C to remove adherent tissues and subsequent sun drying. As shown in Figure 16.6, water extract of such dried cartilage has inhibitory activity, at least against MMPs-2 and -9. These inhibitory activities were stable after heating at 90°C for 10 min. Activity-directed fractionation of the water extract revealed that MMP-9 inhibitory activity was recovered in acidic (p$I<3.0$ by isoelectrofocusing), hydrophilic (nonadsorbed fraction by reversed phase high performance liquid chrmatography (HPLC)) and high-molecular-weight fractions (>450 kD by Superedex 200 size exclusion chromatography). Together with compositional features, we concluded that proteoglycan fraction is responsible for the observed MMPs inhibitory activity [22]. However, purified chondroichin sulfate C, glucosaminoglycan of shark cartilage proteoglycan, has no inhibitory activity. Then, it can be assumed that peptide or peptide–sugar conjugate is MMP-9 inhibitory domain of the shark cartilage proteoglycan. However, we could not detect MMP inhibitory activity in fraction with neutral pI and a molecular weight of less than 40 kD, which may contain TIMP. TIMP-like proteins in shark cartilage might lose the MMP-9 inhibitory activity during heat and dry treatments. The proteoglycan fraction from raw shark cartilage also has the MMP-9 inhibitory activity (data not shown).

The inhibitory activity against MMPs by shark proteoglycan has not been observed in previous studies, which can be explained by difference in preparation procedures. Lee et al. [4] described that most of proteoglycan remained in the residue after their extraction. Recent findings also indicate that water extract of shark cartilage is free from proteoglycan [6].

16.5 EFFECT OF ORAL ADMINISTRATION OF SHARK CARTILAGE ON SERUM MMP INHIBITORY ACTIVITY

16.5.1 Animal Experiment

There are some studies demonstrating that oral administration of shark cartilage or its extract can delay progression of cancer [26–28]. However, in these studies, the effects of oral ingestion of these products on MMPs inhibition in tissue were not examined. By using native type V collagen as substrate, we examined the inhibitory activity against MMPs-2 and -9 in serum of cancer-bearing animals fed basal and experimental diet containing proteoglycan fraction of shark cartilage. The proteoglycan fraction was prepared as described elsewhere [22]. The dried shark fin cartilage was cut into small pieces (approximately 5 × 5 mm) and milled with dry ice. The fine powder was mixed with 5 volumes of cold deionized water and stirred for 30 min and then pressed between cheesecloth. The filtrate was centrifuged at 5000 g for 20 min. The supernatant was mixed with 3 volumes of ethanol. The resultant precipitate was harvested by centrifugation and dried. The dried material was milled and used as shark cartilage proteoglycan fraction (SCPG). Hamsters bearing chemo-induced pancreatic duct cancer were given with basal and experimental diet containing 0.4% SCPG diet for 50 days. As shown in Figure 16.7, administration of 0.4% SCPG diet increased inhibitory activity against MMP-9 in the serum [22]. Although no significant change in the inhibitory activity against MMP-2 was observed, the same amount of serum could inhibit approximately 10 times higher activity of MMP-2 than that of MMP-9 [22]. Thus it could be anticipated that the increased MMP-9 inhibitory activity in serum could be linked to suppression of degradation of basement membrane by cancer. Oral administration of the SCPG also suppresses progress of carcinogenesis [21,22]. A similar result was obtained by using mice implanted with Ehrich ascetic sarcoma (unpublished data). In this animal model, an increased MMP-9 inhibitory activity with suppression of growth of tumor was also observed.

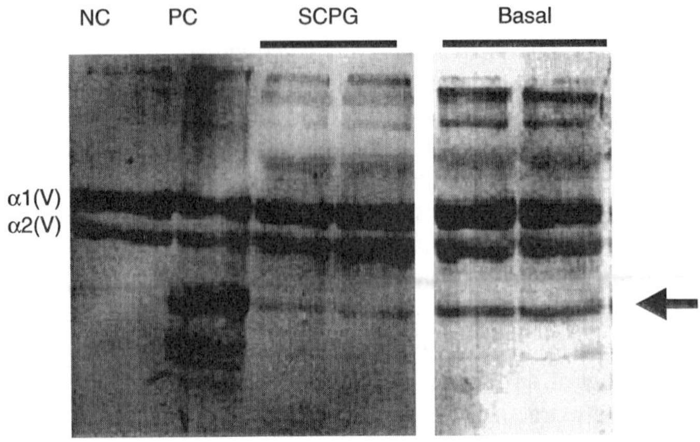

FIGURE 16.7 Increased MMP-9 inhibitory activity in serum from hamster bearing pancreatic duct cancer. The animals were fed diet containing 0.4% SEPG and basal diet. NC, negative control consists of substrate; PC, positive control substrate and MMP-9. Serums diluted to 1/5 were added to the reaction mixture.

16.5.2 Human Studies

Recent phase II clinical trial suggested that high dose (240 mL/day) of oral administration of a formulated water extract of shark cartilage shows a survival benefit in renal cell carcinoma [29]. In those studies, effects of oral ingestion of shark cartilage–based products on MMPs inhibitory activity in human blood and tissues were not examined. Then, we examined serum MMP-9 inhibitory activity of patients suffering from pancreatic cancer after oral administration of the SCPG (2–3 g/day) for 40 days. As shown in Figure 16.8, the inhibitory activity increased after 40 days of administration (Sato et al., unpublished data). The serum type IV collagen degradation product was also determined by ELISA. As shown in Figure 16.9, a higher level of degradation product was observed in the serum of cancer patients in comparison with healthy volunteers, indicating degradation of type IV collagen, possibly by MMPs, was enhanced in the cancer patients. After 40 days of SCPG administration, the degradation product decreased. This result is based on only one case owing to difficulty in preparation of sufficient amounts of SCPG in the laboratory for clinical trial. Therefore, further studies are necessary to prove *in vivo* inhibition of the MMPs in cancer patients by oral administration of shark cartilage–based products.

16.6 CONCLUSIONS AND FUTURE PROSPECTS

FITC-labeled gelatin and fluorescence-quenching synthetic peptide, which are designed for evaluation of MMP-9 activity, are nonspecifically degraded by serum peptidase. Therefore, false MMP-9 activity may be detected in biological samples when such substrates are used. To overcome these problems, native

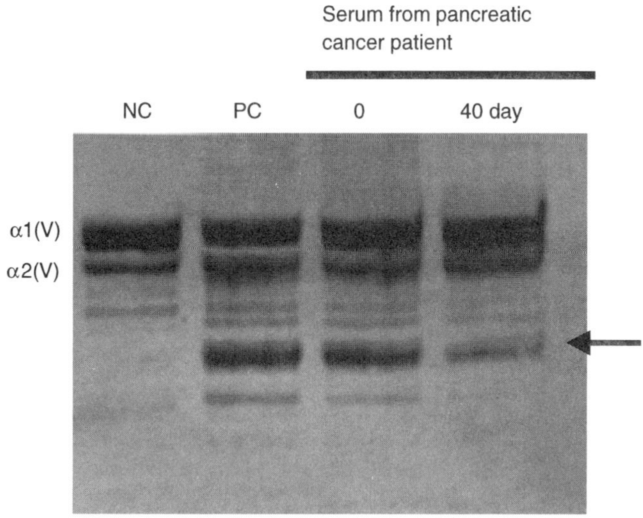

FIGURE 16.8 Increased MMP-9 inhibitory activity in serum from a patient suffering from pancreatic cancer. NC, negative control consists of substrate; PC, positive control substrate and MMP-9. Serums before and after administration of SCPG (2 g/day) for 40 days were added to the reaction mixture.

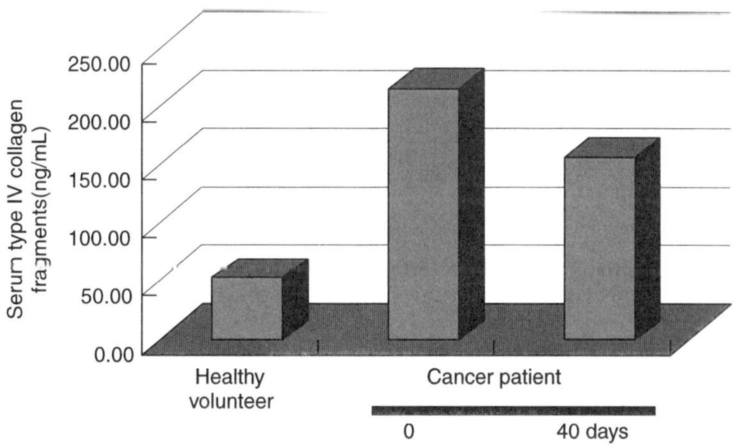

FIGURE 16.9 Serum level of type IV collagen fragments in healthy volunteers and a cancer patient (same person in Figure 16.8).

type V collagen can be used as substrate. By using this approach, we demonstrated that the protein moiety of proteoglycan fraction is responsible for the *in vitro* MMP-9 inhibitory activity in the dried shark cartilage and that oral administration of SCPG can increase MMP-9 inhibitory activity in the serum of some cancer-bearing animals. A similar positive result was observed for a cancer patient. This preliminary observation has encouraged us to examine the effect of oral administration of SCPG on MMP-9 inhibitory activity and degradation of type IV collagen in other cancer patients. The techniques presented in this chapter for evaluation of inhibitory activity of MMPs-2 and -9 would be a good tool in further clinical studies because it requires only blood as sample, which can be easily collected from humans and animals.

REFERENCES

1. Nomura , Y. 2004. Properties and utilization of shark collagen. In: *More Efficient Utilization of Fish and Fisheries Products, Developments in Food Science 42*, Sakagachi, M. Ed., Elsevier, Oxford, UK, pp. 147–158.
2. Mose, M.A. and Langer, R. 1991. A metalloproteinase inhibitor as an inhibitor of neovascularization. *J. Cell. Biochem.* 47, 230–235.
3. Oikawa, T., Ashino-Fuse, H., Shimamura, M., Koide, U. and Iwaguchi, T. 1990. A novel angiogenetic inhibitor derived from Japanese shark cartilage (I). Extraction and estimation of inhibitory activities toward tumor and embryonic angiogenesis. *Cancer Lett.* 51, 181–186.
4. Lee, A.K., van Beuzekom, M., Glowacki, J. and Langer, R. 1984, Inhibitors enzymes and growth factors from shark cartilage. *Comp. Biochem. Physiol.* 78B, 609–616.
5. Lane, I.W. and Contreras, Jr. E. 1992. High rate of bioactivity (reduction in gross tumor size) observed in advanced cancer patients treated with shark cartilage material. *J. Naturopathic Med.* 3, 8–88.
6. Dupont, E., Brazeau, P. and Juneau, C. 1995. Extracts of shark cartilage having an angiogenenic activity and an effect on tumor progression: process of making thereof. United State Patent #5,618,925.
7. Miller, D.R., Anderson, G.T., Stark, J.J., Granick, J.L. and Richardson, D. 1998. Phase I/II trial of the safety and efficiency of shark cartilage in the treatment of advanced cancer. *J. Clin. Oncol.* 16, 3649–3655.
8. Folkman, J. 1989. What is the evidence that tumors are angiogenesis dependent? *J. Natl. Cancer Inst.* 82, 4–6.
9. Folkman, J. 2001. Angiogenesis-dependent diseases. *Semin Oncol.* 28, 536–542.
10. O'Reilly, M.S., Holmgren, L., Shing, Y., Chen, C., Rosenthal, R.A., Moses, M., Lane, W.S., Cao, Y., Sage, E.H. and Folkman, J. 1994. Angiostatin: a novel angiogenesis inhibitor that mediates the suppression of metastasis by Lewis lung carcinoma. *Cell* 79, 315–328.
11. O'Reilly, M.S., Boehm, T., Shing, Y., Fukai, N., Vasios, G., Lane, W.S., Flynn, E., Birkhead, J.R., Olsen, B.R. and Folkman, J. 1997. Endostatin: an endogenous inhibitor of angiogenesis and tumor growth. *Cell* 88, 277–285.
12. Berm, H. and Folkman, J. 1975. Inhibition of tumor angiogenesis mediated by cartilage. *J. Exp. Med.* 141, 427–439.
13. Lee, A. and Langer, R. 1983. Shark cartilage contains inhibitors of tumor angiogenesis. *Science* 221, 1185–1187.
14. Gingras, D., Renaud, A., Mousseau, N., Beaulieu, E., Kachra, Z. and Beliveau, R. 2001. Matrix protease inhibition by AE-941, a multifunctional angiogenetic compound. *Anticancer Res.* 21, 145–156.
15. Beliveau, R., Gingras, D., Kruger, E.A., Lamy, S., Sirois, P., Simard, B., Sirois, M.G., Tranqui, L., Baffert, F., Beaulieu, E., Dimitriadou, V., Pepin, M.-C., Courjal, F., Ricard, I., Poyet, P., Falardeau, P., Figg, W.D. and Dupont E. 2002. The antiangiogenic agent neovastat (AE-941) inhibits vascular endothelial growth factor-mediated biological effects. *Clin. Cancer Res.* 8, 1242–1250.

16. van der Rest, M. and Garrone, R. 1991. Collagen family of proteins. *FASEB J.* 5, 2814–2823.
17. Curran, S. and Murray, G.I. 1999. Matrix metalloproteinases in tumor invasion and metastasis. *J. Pathol.* 189, 300–308.
18. Johansen, N., Ahonen, M. and Kahari, V.M. 2000. Matrix metalloproteinases in tumor invasion. *Cell Mol. Life Sci.* 20, 5–15.
19. McCawley, L.J. and Matrisian, L.M. 2000. Matrix metalloproteinase: multifunctional contributors to tumor progression. *Mol. Med. Today* 6, 149–156.
20. Foda, H.D. and Zucker, S. 2001. Matrix metalloproteinases in cancer invasion, metastasis and angiogenesis. *Drug Discov. Today* 6, 478–482.
21. Murata, N. 2002. Effects of bovine and shark cartilage water extracts on pancreatic ductal carcinogenesis in hamster. *J. Nara Med. Assoc.* 53, 241–252.
22. Sato, K., Murata, N., Tstsumi, M., Shimizu-Suganuma, M., Shichinohe, K., Kitahashi, T., Nishimura, K., Nakamura, Y. and Ohtsuki, K. 2003. Moderation of chemo-induced cancer by water extract of dried shark fin: anti-cancer effect of shark cartilage. In: *More Efficient Utilization of Fish and Fisheries Products, Developments in Food Science 42*, Sakagachi, M. Ed., Elsevier, Oxford, UK, pp. 159–168.
23. Maeda-Yamamoto, M., Kawahara, H., Tahara, N., Tsuji, K., Hara, Y. and Isemura, M. 1999. Effects of tea polyphenols on the invasion and matrix metalloproteinases activities of human fibrosarcoma HT1080 cells. *J. Agric. Food Chem.* 47, 2350–2354.
24. Hattori, S., Fujisaki, H., Kiriyama, T., Yokoyama, T. and Irie, S. 2002, Real-time zymography: A method for detecting activities of matrix metalloproteinases and their inhibitors using FITC-labeled collagen and casein as substrates. *Anal. Biochem.* 301, 27–34.
25. Barber, R., Delahunt, B., Grebe, S.K.G., Davis, P.F., Thornton, A. and Slim, G. 2001. Oral shark cartilage does not abolish carcinogenosis but delays tumor progression in a murine model. *Anticancer Res.* 21, 1065–1070.
26. Knight, C.G., Willenbrock, F. and Murphy, G. 1992. A novel coumarine-labelled peptide for sensitive continuous assays of the matrix metalloproteinases. *FEBS Lett.* 27, 263–266.
27. Dupont, E., Falardeau, P., Mousa, S.A., Dimitriadou, V., Pepin, M.C., Wang, T. and Alaoui-Jamali, M.A. 2002. Antiangiogenic and antimetastatic properties of Neovastat (AE-941), an orally active extract derived from cartilage tissue. *Clin. Exp. Metastasis* 19, 145–153.
28. Weber, M.H., Lee, J. and Orr, F.W. 2002. The effect of Neobastat (AE-941) on an experimental metastatic bone tumor model. *Int. J. Oncol.* 20, 299–303.
29. Batist, G., Patenaude, F., Champagne, P., Croteasu, D., Levinton, C., Harton, C., Escudier, B. and Dupont, E. 2002. Neovastat (AE-941) in refractory renal cell carcinoma patients: report of a phase II trial with two dose levels. *Ann. Oncol.* 13, 1259–1263.

17 Calcium from Fish Bone and Other Marine Resources

Won-Kyo Jung, Fereidoon Shahidi, and Se-Kwon Kim

CONTENTS

17.1 INTRODUCTION

Calcium is known to be an essential element required for numerous functions in our body including the strengthening of teeth and bones, nerve function, and many enzymatic reactions that require calcium as a cofactor. It is also necessary for muscle contraction and regulation of the permeability of sodium ion across cell membranes including those of nerve cells. The concentration of calcium in the blood plasma remains almost constant and varies only slightly over time for a given individual [1,2]. In marine ecosystem, there is a large amount of calcium in the major form of calcite ($CaCO_3$) or calcium divalent cation (Ca^{2+}). Moreover, calcium with diverse physiological roles such as fertilization of egg, formation and growth of skeleton and shell, and nutritional metabolite is widely distributed in various marine phytoorganisms, microorganisms, invertebrates, and vertebrates. The ubiquitous occurrence and plentiful mass of calcium sources in marine ecosystem have provided a motive for study on utilization in bioavailable calcium supplements and fortifiers. However, studies on application of marine bioresources for bioavailable calcium [3–8] are scarce. This chapter focuses on current status of knowledge on the utilization of marine organisms for use as calcium supplements or fortifiers.

17.2 NEEDS FOR ALTERNATIVE CALCIUM SUPPLEMENTS OTHER THAN DAIRY PRODUCTS

Except for nutritional and physiological applications, calcium that originates from dolomite, bone meal, and oyster shell are also utilized as important ingredients in various industries such as food industry, electronic industry, and leather industry, among others. For example, calcium is used to produce acryl resin, make emulsion coagulants in the rubber industry, to produce additives in pulp and paper industry, and to make early strengthening agents (concrete strengthening agent and coating material coagulating agent, and others) in the construction industry. Especially in food and agricultural industries, calcium is utilized as a foodstuff antiseptic to prevent putrefaction of fruits and vegetables and help the process of cheese making. Although most people are aware of calcium as an important element in their bodies, it is still severely deficient in most diets. Calcium deficiency in the United States has been considered as a major cause of osteoporosis, affecting approximately 26 million people annually [9]. In 1994, the National Institute of Health (NIH) Consensus panel revised the recommendations for calcium intake [10].

As shown in Table 17.1, the optimal calcium intake has been recommended to be 800 mg/day during childhood below 5 years of age, 800–1200 mg/day for children from age 6 to 10, 1200–1500 mg/day for adolescence or young adults from age 12 to 24 and pregnant or lactating women, 1000 mg/day from age 25 to the time of estrogen deprivation or age 65, and 1500 mg/day for elderly people.

Generally, most common and trusted source of calcium (Table 17.2) is milk and other dairy products [2] However, some people, especially Asians, do not prefer to take milk because of lactose indigestion and intolerance, which make them allergic to milk. Thus, as an alternative, these people prefer to take calcium-fortified fruit juice, calcium-rich foods, and calcium salt supplements,

TABLE 17.1
Recommended Calcium Intake for Various Population Groups (NIH Consensus Development)

Age (Years)	Calcium Needs (mg/day)
Children	800–1200
Adolescents	1200–1500
Adults	1000
Elderly	1500
<65 on hormonal replacement therapy	1000

TABLE 17.2
High-Bioavailable Calcium Sources in Foods

Food Source	Serving Size	Calcium (mg)
Milk and Yogurt	8 oz or 1 cup	300–450
Cheese	3 oz	300–450
Bones in canned sardines and salmon	3 oz	181–325
Calcium-fortified foods (i.e., orange juice, soy milk, tofu)	8 oz	200–300
Dark green, leafy vegetables	1/2 cup cooked, 1 cup raw	50–100
Nuts and seeds	1 oz	25–75

Source: http//:ag.arizona.edu/pubs/health/az1296.pdf

such as calcium fumarate, citrate, lactate, carbonate, di- and tribasic phosphate, and gluconate. These salts are available in ingredient forms, each with their own calcium content, solubility, taste, and cost issues. Especially, the solubility and bioavailability of calcium-containing ingredients are important. Although the low pH condition in the stomach renders all calcium into its ionic form, precipitation as insoluble calcium phosphate, depending on the amount of phosphate present, can occur in the intestine, where the pH range is 6–7.

The human body cannot absorb the calcium present in precipitated calcium phosphate. To improve solubility and bioavailability of calcium, various proprietary blends of calcium salts have been developed with milk protein, food acids and sugar, polysaccharides and calcium/amino acid chelate complexes like Ca-casein phosphopeptides, as specific end-use products, depending on their final pH [1]. Casein phosphopeptides (CPPs) derived from the intestinal digestion of casein have been shown to enhance bone calcification in rats [11,12]. Calcium fortifiers like CPPs, egg yolk phosphopeptides (phosvitin), and some organic ingredients (citrate, malate, acetate, etc.) have the capacity to chelate Ca ion and to prevent precipitation of Ca-phosphate salts at neutral intestinal pH [13], thereby increasing the amount of soluble Ca available for absorption across the mucosa [14,15].

In the following parts, we introduce dairy foodlike calcium supplements and CPP-like fortifiers derived from marine organisms, and survey current researches for bioavailable calcium.

17.3 CALCIUM FROM MARINE ORGANISMS

17.3.1 Utilization of Fish Skeletal Frame in Bioactive Calcium Substances

Marine capture fisheries contribute over 50% of total world fish production, and more than 70% of this production has been utilized for processing [16]. As a result, every year a considerable amount of total catch is discarded as processing leftovers and these include trimmings, fins, frames, heads, skin, and viscera. In addition to fish processing, a large quantity of processing by-products are accumulated as shells of crustaceans and shellfish from marine bioprocessing plants. Recent estimates revealed that current discards from the world's fisheries exceed 20 million tons equivalent to 25% of the total production of marine capture fisheries [16]. Therefore, there is a great potential in marine bioprocess industry to convert and utilize more of these by-products as valuable products. Majority of fisheries by-products are presently employed to produce fish oil, fishmeal, fertilizer, pet food, and fish silage [17,18]. However, most of these recycled products possess low economic value. Recently, bioactive compounds from remaining fish muscle proteins, collagen and gelatin, fish oil, fish bone, internal organs, and shellfish and crustacean shells were reviewed by Kim and Mendis [6]. Among fish by-products, fish bone or skeleton is considered as a potential source to obtain calcium, which is an essential element for the human health. However, only few studies have been carried out to identify bioavailability of fish bone calcium and its potential applications.

Generally, calcium is obtained from the diet and it is severely deficient in most of the regular diets. Therefore, to improve calcium intake, several calcium-fortified products are in the market and demand for these products is growing continuously. It is well documented that consumption of whole small fish is nutritionally beneficial providing with a rich source of calcium. Calcium in fish could be absorbed to the body as tested *in vivo* [3]. However, very little information is available on the beneficial effects of larger fish bone and few attempts have been made to test their usage for benefits of human health. Fish bone material derived from processing of large fish is a useful calcium source where the quantity of calcium is concerned. To incorporate fish bone into calcium-fortified food, it should be converted into an edible form by softening its structure. This can be achieved utilizing different methods including hot water treatment and hot acetic acid solutions. In addition, Ishikawa et al. [19] used superheated steam to reduce the loss of soluble components from fish tissue and that enabled better recovery of bone within a shorter period. Jung et al. [20] performed enzymatic degradation in acetic acid solution (pH 2.0) by pepsin to easily dissolve both mineral and organic parts of fish bone. Pepsinolytic degradation of Alaska pollack bone in acidic condition could lead to the highest degree of hydrolysis in comparison with those of other enzymes. Moreover, Jung et al. [7,8] isolated high calcium-binding oligophosphopeptide from hoki fish skeletons by an enzymatic degradation method with carnivorous intestinal crude enzymes in an ultrafiltration membrane bioreactor system. In this study, it has been observed that calcium-binding activity of the fish bone peptides (FBP) II was similar to that of CPP. Further, the pH of the reaction system was maintained at 7.8, because low pH could increase the solubility of the insoluble calcium salt. As reported by Jiang and Mine [21], the solubility of 36.3 mg/L of Ca could be obtained at 200 mg/L of the oligophosphopeptide from egg yolk phosvitin with 35% phosphate retention, and the solubility was higher than that of commercial

CPP II (Meiji Seika Co., Ltd., Tokyo, Japan). Furthermore, *in vivo* studies with osteoporosis modeling rat have elucidated that the FBP II fractionated in the molecular weight range of 5.0–1.0 kDa increased Ca solubility and bioavailability. Menopause is a time when estrogen deficiency leads to accelerated bone resorption and negative bone balance. Another study has been undertaken to evaluate the beneficial effects of FBP as a Ca fortifier in osteoporosis induced by ovariectomy and a concurrent low-Ca diet. During the experimental period corresponding to the menopause with osteoporosis disease, the loss of bone mineral (Ca) was decreased by FBP II supplementation in the ovariectomized rats. After the low-Ca diet, the FBP II diet, including both normal levels of Ca and vitamin D, significantly decreased Ca loss in feces and increased Ca retention as compared with the control. The levels of femoral total Ca, bone mineral density, and breaking strength were also significantly increased by FBP II diet to a level similar to those of the CPP diet group (no difference; $P < 0.05$). Based on these data, it was suggested that increased Ca retention by FBP II intake may lead to the prevention of mineral loss in the osteoporosis-modeling rats. As reported by Larsen et al. [3,4], the intake of small fish with bones can increase Ca bioavailability, and the small fish may be an important source of Ca, especially in population groups with low intake of milk and dairy products. Thus, these results prove the beneficial effects of fish meal in preventing Ca deficiency owing to increased Ca bioavailability by FBP intake. Furthermore, there is a potential to these fish peptides to provide a novel nutraceutical with a high bioavailability for Ca to Oriental people with lactose indigestion and intolerance and Ca-fortified supplements, such as fruit juice or Ca-rich foods, as alternatives to dairy products.

In the case of marine calcium in medicinal application, attempts have been taken to isolate fish bone–derived hydroxyapatite and use them as an alternate for synthetic hydroxyapatite [22,23]. Recently, hydroxyapatite [$Ca_{10}(PO_4)_6(OH)_2$] has been introduced as a bone graft material in a range of medical and dental applications because of their similar chemical composition. Generally, bone substitution materials such as autografts, allografts, and xenografts are used to solve problems related to bone fractures and damages. However, none of these materials provides a perfect bone healing owing to mechanical instability and incompatibility. Currently, calcium phosphate bioceramics such as tetracalcium phosphate, amorphous calcium phosphate, tricalcium phosphate, and hydroxyapatite are identified as most suitable bone substitution materials to address the demand. Fish bone material may serve as an important source for biomedical applications owing to the presence of hydroxyapatite as the major inorganic constituent. Unlike other calcium phosphates, hydroxyapatite does not break under physiological conditions. In fact, it is thermodynamically stable at physiological pH and actively takes part in bone bonding. This property has been exploited for rapid bone repair after major trauma or surgery. Hydroxyapatite is derived from natural materials such as coral and fish bone [24]. Generally, very high heat treatment is used for isolation of hydroxyapatite from fish bone and this temperature gives a higher strength to hydroxyapatite structure [25] and results an excellent biocompatible inorganic substance [26–28].

17.3.2 High Absorbable Coral Calcium

Ayurveda, an ancient system of Indian medicine has mentioned several calcium preparations in the correction of bone metabolic disorders such as osteoporosis. *Praval bhasma* (PB, *Coral calx*) a natural source of rich calcium in marine ecosystem, predominantly consists of $CaCO_3$ and is widely used in ayurveda as calcium supplement from time immemorial. Moreover, because of appropriate ayurvedic processing of PB, it has the advantage of easy absorption from the intestine [29].

Reddy et al. [30] evaluated the efficacy of PB on the progress of bone loss in calcium-deficient diet fed ovariectomized (OVX) rats. According to the results, calcium-deficient OVX rats developed bone changes similar to those seen in osteoporotic women as indicated by a decrease in femur weight, density, and bone mineral content. Treatment with PB significantly prevented the reduction in bone density and bone mineral content, especially in calcium and phosphorus levels despite ovariectomy and calcium deficiency. The unchanged levels of calcium and phosphorus in plasma of sham and Ca-deficient/OVX (CD-OVX) group indicates that homeostatic mechanisms were able to maintain plasma levels of these minerals despite ovariectomy. Treatment with PB significantly increased and maintained the serum calcium level, thus indicating good absorption from the intestine. Fasting urinary calcium excretion is also a useful variable for estimating net bone resorption. PB-treated animals showed decreased urinary calcium excretion despite elevated serum calcium levels, suggesting that more calcium might be deposited in bones. The assumption could be true as it is evident from the increased ash weight, percent ash, and mineral content in femurs of PB-treated group compared to CD-OVX group. In addition, high-resolution radiography such as CT-scanning technique employing magnification to assess cortical bone loss has indicated a decreased cortical area and increased medullary width and cross-sectional area suggesting increased bone loss in CD-OVX animals. In conclusion, the authors have discussed that treatment with PB in the adult rat model of osteoporosis exerted desired beneficial effects on the inhibition of bone resorption, thereby justifying its continued use. In addition, a group fed with normal calcium diet as a positive control have strengthened the claim that the efficacy of PB in rats is through improvement of calcium absorption. Nevertheless, the results of this study clearly indicate the beneficial effects of PB in preventing bone loss as indicated by various parameters in the PB-treated group compared to the CD-OVX group. Furthermore, it is widely believed that calcium supplements of natural origin such as PB probably could have trace amounts of lead or other toxic substances that may cause concern for safety.

17.3.3 Calcium Bioavailability of Marine Algal Powder

Marine algae have been a valued food in Asia for thousands of years because of its highly nutritious qualities. They provide the body with a full array of nutrients including complete protein, complex carbohydrates, essential fatty acids, fiber, vitamins, minerals, enzymes, and trace elements. Marine algae are thought to exert medicinal properties to prevent a number of disease complications in the human body.

However, marine algal–derived valuable proteins, minerals, and isoflavons on bone metabolism has not yet been clarified. Recently, the effect of various algae on bone calcification in the femoral-metaphyseal tissues of rats was investigated by Yamaguchi et al. [31]. The study was undertaken to determine the effect of various marine algae on bone calcification in the femoral metaphyseal tissue of rats *in vivo* and *in vitro*. Marine algae (*Undaria pinnatifida*, *Sargassum horneri*, *Eisenia bicyclis*, *Cryptonemia scmitzana*, *Gelidium amansii*, and *Ulva pertusa* Kjellman), which are utilized in food, have been used in this study. Water suspensions of marine algae powder were orally administered to rats and it was observed that bone calcium content was significantly increased by the administration of *U. pinnatifida*, *S. horneri*, *E. bicyclis*, or *C. scmitziana*. Bone alkaline phosphatase activity, which is an enzyme for calcification, was significantly enhanced by the administration of *S. horneri* or *G. amansii*. Moreover, bone calcium content was significantly elevated in the presence of *S. horneri* extract (25 and 50 μg/mL). It was elucidated that *S. horneri* extract had an anabolic effect on bone calcification *in vivo* and *in vitro*. In this study, it was suggested that prolonged intake of *S. horneri* extract may play a role in the prevention of bone loss with increasing age. These data suggest that much more remains to be elucidated in animal models of osteoporosis. Thus, marine algae (*S. horneri*) extract has an anabolic effect on bone calcification in animal models of osteoporosis and hence may play a role in the prevention of osteoporosis.

17.3.4 Low Molecular Phosphorylated Chitooligosaccharides Derived from Crab Shell as a Calcium Fortifier

Chitosan is a deacetylated polymer of *N*-acetylglucosamine, which is obtained after alkaline deacetylation of the chitin derived from the exoskeletons of crustaceans and arthropods. Recently, considerable attention has been given for its commercial applications in biomedical, food, and chemical industries. In addition, chitosan has been widely used in vastly diverse fields such as pharmaceuticals, medicine, and biotechnology. However, increasing attention has recently been paid to convert chitosan into its oligosaccharides because of their biological activities, such as antitumor activity [32,33], immunostimulating effects [34,35], enhancing protective effects against infection with some pathogens in mice [36], antifungal activity [37], antimicrobial activity [38–40], angiotensin I converting enzyme (ACE)-inhibitory activity [41], and radical scavenging activity [42,43]. Kim and Mendis [6], reported that phosphorylated chitooligosaccharides (P-COSs) exhibited inhibitory activity against the formation of insoluble calcium phosphate at neutral pH. Furthermore, P-COS with low molecular weight exhibited the highest inhibitory activity of calcium phosphate precipitation. Its inhibitory activity, especially at concentrations of more than 4 mg/mL, was similar to that of CPP, which is widely used as a calcium-fortifying agent that improves calcium absorbability. Therefore, this study illustrated that phosphorylated chitooligosaccharides can be considered as potential inhibitors of calcium phosphate precipitation. Furthermore, *in vivo* effect of COSs was also elucidated by Jung et al. [8]. In this study, low-molecular-weight COSs obtained using an ultrafiltration (UF) membrane reactor system, inhibited the formation of insoluble calcium salts in the neutral pH. *In vivo* effects of COSs on Ca bioavailability were further studied in the osteoporosis rats model induced by ovariectomy and concurrent low calcium intake. During the experimental period corresponding to the menopause with the osteoporosis disease, calcium retention was increased and bone turnover was decreased by COS IV (molecular mass of <5.0 kDa) supplementation in the OVX rats. After a low-Ca diet, COS IV diet including both normal level of calcium and vitamin D significantly decreased calcium loss in feces and increased calcium retention compared to the control diet. The levels of femoral total calcium, bone mineral density (BMD), and femoral strength were also significantly increased by the COS IV diet in a similar level to those of CPP diet group. In the study, the results proved the beneficial effects of low-molecular-weight COS IV in preventing negative mineral balance.

In addition, Kim et al. [44] previously reported that COSs do not exert any toxic effects in experimental diet groups of Sprague–Dawley rats fed with 500, 1000, and 2000 mg/kg/day. This suggestion was made observing weight change, general symptoms, food consumption, urinalysis, hematology, blood biochemistry, and relative organ weights of COS-treated rats. Further, this study states that further studies are needed to confirm the bioavailability of the chitosan derivative, P-COS using *in vivo* studies and it may serve as potential inhibitors of calcium phosphate precipitation.

ACKNOWLEDGMENTS

This research was funded by a grant (p-2004-01) from Marine Bioprocess Research of the Marine Bio 21 Center funded by the Ministry of Maritime Affairs and Fisheries, Republic of Korea.

REFERENCES

1. Allen, L.H. 1982. Calcium bioavailability and absorption: a review. Am. J. Clin. Nutr. 35, 738–808.
2. Anderson, J.J.B. and Garner, S.C. 1996. Calcium and phosphorous nutrition in health and disease. In J.J.B. Anderson and S.C. Garner, (Eds.), *Calcium and Phosphorous in Health and Disease* (pp. 1–5). New York: CRC Press.
3. Larsen, T., Thilsted, S.H., Kongsbak, K. and Hansen, M. 2000. Whole small fish as a rich calcium source. Br. J. Nutr. 83, 191–196.
4. Larsen, T., Thilsted, S.H., Biswas, S.K. and Tetens, I. 2003. The leafy vegetable amaranth (*Amaranthus gangeticus*) is a potent inhibitor of calcium bioavailability and retention in rice-based diets. Br. J. Nutr. 90, 521–527.
5. Kim, S.K., Park, P.J., Byun, H.G., Je, J.Y., Moon, S.H. and Kim, S-.H. 2003. Recovery of fish bone from hoki (*Johnius belengerii*) frame using a proteolytic enzyme isolated from mackerel intestine. J. Food Biochem. 27 (3), 255–266.

6. Kim, S.K. and Mendis, E. 2006. Bioactive compounds from marine processing byproducts—a review. Food Res. Int. 39, 383–393.
7. Jung, W.K., Park, P.J., Byun, H.G., Moon, S.H. and Kim, S.K. 2005. Preparation of hoki (*Johnius belengerii*) bone oligophosphopeptide with a high affinity to calcium by carnivorous intestine crude proteinase. Food Chem. 91, 333–340.
8. Jung, W.K., Lee, B.J. and Kim, S.K. 2006. Fish bone peptide increases Ca solubility and bioavailability in ovariectomized rats. Brit. J. Nutr. 95, 124–128.
9. Melton, L.J. 1995. How many women have osteoporosis now? J. Bone Min. Res. 10, 175–177.
10. NIH Consensus Development Panel on Optimal Calcium Intake. 1994. NIH Consensus conference. Optimal calcium intake. JAMA 272(24), 1942–1948. http://jama.ama-assn.org/content/vol272/issue24/index.dtl.
11. Lee, Y.S., Noguchi, T. and Naito, H. 1980. Phosphopeptides and soluble calcium in the small intestine of rat given a casein diet. Br. J. Nutr. 43, 457–467.
12. Tsuchita, H., Sekiguchi, I., Kuwata, T., Igarashi, T. and Ezawa, I. 1993. The effect of casein phosphopeptides on calcium utilization in young ovariectomized rats. Z. Ernährungswiss 32, 121–130.
13. Berrocal, R., Chanton, S., Juillerat, M.A., Pavillard, B., Scherz, J.C. and Jost, R. 1989. Tryptic phosphopeptides from whole casein. II. Physiochemical properties related to the solubilization of calcium. J. Dairy Res. 56, 335–341.
14. Yuan, Y.V. and Kitts, D.D. 1991. Conformation of calcium absorption and femoral utilization in spontaneously hypertensive rats fed casein phosphopeptide supplemented diets. Nutr. Res. 11, 1257–1272.
15. Yuan, Y.V. and Kitts, D.D. 1994. Calcium absorption and bone utilization in spontaneously hypertensive rats fed on native and heat-damaged casein and soyabean protein. Br. J. Nutr. 71, 583–603.
16. FAOSTAT, FAO statistical databases, fisheries data (2001). Food and Agriculture Organization of the United Nations, Rome, Italy. Available from http://www.fao.org.
17. Choudhury, G.S. and Bublitz, C.G. 1996. Computer-based controls in fish processing industry. In G.S. Mittal (Ed.), Computerized Control Systems in the Food Industry (pp. 513–538). New York: Marcel Dekker Inc.
18. Choudhury, G.S. and Gogoi, B.K. 1995. Extrusion processing of fish muscle. J. Aqu. Food Prod. Tech. 4, 37–67.
19. Ishikawa, M., Kato, M., Mihori, T., Watanabe, H. and Sakai, Y. 1990. Effect of vapor pressure on the rate of softening of fish bone by super-heated steam cooking. Nippon Suisan Gakkaishi 56, 1687–1691.
20. Jung, W.K., Jeon, Y.J., Karawita, R., Heo, S.J., Siriwardhana, N., Lee, B.J. and Kim, S.K. 2006. Recovery of a novel Ca-binding peptide from Alaska pollack (*Theragra chalcogramma*) backbone by pepsinolytic hydrolysis. Process Biochem. 41, 2097–2100.
21. Jiang, B. and Mine, Y. 2000. Preparation of novel functional oligophosphopeptides from hen egg yolk phosvitin. J. Agric. Food Chem. 48, 990–994.
22. Kim, S.K., Lee, C.K., Byun, H.G., Jeon, Y.J., Lee, E.H. and Choi, J.S. 1997. Synthesis and biocompatibility of the hydroxyapatite ceramic composites from tuna bone (I)—the sintering properties of hydroxyapatite and hydroxyapatite-containing wollastonite crushed with dry milling process. J. Korean Ind. Eng. Chem. 8(6), 994–999.
23. Ozawa, M. and Suguru, S. 2002. Microstructural development of natural hydroxyapatite originated from fish-bone waste through heat treatment. J. Am. Ceram. Soc. 85, 1315–1317.

24. Jensen, S.S., Aaboe, M., Pinhold, E.M., Hjrting-Hansen, Z., Melsen, F. and Ruyter, I.E. 1996. Tissue reaction and material characteristics of four bone substitutes. Int. J. Oral Maxillofacial Implants 11, 55–66.
25. Choi, J.S., Lee, C.K., Jeon, Y.J., Byun, H.G. and Kim, S.K. 1999. Properties of the ceramic composites and glass ceramics prepared by using the natural hydroxyapatite derived from tuna bone. J. Korean Ind. Eng. Chem. 10, 394–399.
26. Kim, G.H., Jeon, Y.J., Byun, H.G., Lee, C.K., Lee, E.H. and Kim, S.K. 1998. Effect of calcium compounds from oyster shell bound fish skin gelatin peptide in calcium deficient rats. J. Korean Fish. Soc. 31(2), 149–159.
27. Kim, S.K., Choi, J.S., Lee, C.K., Byun, H.G., Jeon, Y.J., Lee, E.H. and Park, I. Y. 1998. Synthesis and biocompatibility of the hydroxyapatite ceramic composites from tuna bone(III)-SEM photographs of bonding properties between hydroxyapatite ceramics composites in the simulated body fluid. J. Korean Ind. Eng. Chem. 9(3), 322–329.
28. Kim, S.K., Choi, J.S., Lee, C.K., Byun, H.G., Jeon, Y.J. and Lee, E. H. 1997. Synthesis and biocompatibility of the hydroxyapatite ceramic composites from tuna bone (II) —the sintering properties of hydroxyapatite treated with wet milling process. J. Korean Ind. Eng. Chem. 8(6), 1000–1005.
29. Mitra, S.K., Venkataranganna, M.V., Udupa, U.V., Gopumadhavan, S., Sheshadri, S.J., Rafiq, M., Anturlikar, S.D., Sundaram, R. and Tripathi, M. 2001. The beneficial effect of OST-6 (Osteocare), a herbomineral preparation, in experimental osteoporosis. Phytomedicine 8(3), 195–201.
30. Reddy, P.N., Lakshmana, M. and Udupac, U.V. 2003. Effect of Praval bhasma (*Coral calx*), a natural source of rich calcium on bone mineralization in rats. Pharmacol. Res. 48 (2003) 593–599.
31. Yamaguchi, M., Hachiya, S., Hiratuka, S. and Suzuki, T. 2001. Effect of marine algae extract on bone calcification in the femoral-metaphyseal tissues of rats: anabolic effect of *Sargassum horneri*. J. Health Sci. 47(6), 533–538.
32. Suzuki, K., Mikami, T., Okawa, Y., Tokorom, A., Suzuki, S. and Suzuki, M. 1986. Antitumor effect of hexa-*N*-acetylchitohexaose and chitohexaose. Carbohydr. Res. 151, 403–408.
33. Jeon, Y.J. and Kim, S.K. 2002. Antitumor activity of chitosan oligosaccharides produced in ultrafiltration membrane reactor system. J. Microbiol. Biotech. 12, 503–507.
34. Suzuki, S., Watanabe, T., Mikami, T., Matsumoto, T. and Suzuki, M. 1992. Immunoenhancing effects of *N*-acetylchitohexaose. In: C.J. Brine, P.A. Sanford, and J.P. Zikakis (Eds.), *Advances in Chitin and Chitosan* (pp. 277–316). Barking: Elsevier.
35. Jeon, Y.J. and Kim, S.K. 2001. Potential immuno-stimulating effect of antitumoral fraction of chitosan oligosaccharides. J. Chitin Chitosan 6, 163–167.
36. Yamada, A., Shibuya, N., Kodama, O. and Akatsuka, T. 1993. Induction of phytoalexin formation in suspension-cultured rice cells by *N*-acetylchitooligosaccharides. Biosci. Biotechnol. Biochem. 57, 405–409.
37. Kendra, D.F., Christian, D. and Hadwiger, L.A. 1989. Chitosan oligomers from *Fusarium solani*/pea interactions, chitinase/(b-glucanase digestion of sporelings and from fungal wall chitin actively inhibit fungal growth and enhance disease resistance. Physiol. Mol. Plant Pathol. 35, 215–230.
38. Jeon, Y.J., Park, P.J. and Kim, S.K. 2001. Antimicrobial effect of chitoligosaccharides produced by bioreactor. Carbohydr. Polym. 44, 71–76.

39. Park, P.J., Je, J.Y., Byun, H.G., Moon, S.H. and Kim, S.K. 2004. Antimicrobial activity of hetero-chitosans and their oligosaccharides with different molecular weights. J. Microbiol. Biotechnol. 14, 317–323.
40. Park, P.J., Lee, H.K. and Kim, S.K. 2004. Preparation of heterochitooligosaccharides and their antimicrobial activity on *Vibrio parahaemolyticus*. J. Microbiol. Biotechnol. 14, 41–47.
41. Park, P.J., Je, J.Y. and Kim, S.K. 2003. Angiotensin I converting enzyme (ACE) inhibitory activity of hetero-chitooligosaccharides prepared from partially different deacetylated chitosans. J. Agric. Food Chem. 51, 4930–4934.
42. Park, P.J., Je, J.Y. and Kim, S.K. 2003. Free radical scavenging activity of chitooligosaccharides by electron spin resonance spectrometry. J. Agric. Food Chem. 51, 4624–4627.
43. Park, P.J., Je, J.Y. and Kim, S.K. 2004. Free radical scavenging activities of differently deacetylated chitosans using an ESR spectrometer. Carbohydr. Polym. 55, 17–22.
44. Kim, S.K., Park, P.J., Yang, H.P. and Han, S.S. 2001. Subacute toxicity of chitosan oligosaccharide in Sprague–Dawley rats. Arzneim-Forsch/Drug Res. 51, 769–774.

18 Immunoenhancing Preparations of Marine Origin

Jaroslav A. Kralovec and Colin J. Barrow

CONTENTS

18.1 INTRODUCTION

It has become increasingly accepted that nutrition and health are closely interconnected. There has also been a shift in the concept of optimal nutrition from simply preventing diseases stemming from malnutrition, to reducing the risk of chronic disease [1,2]. Intense research has focused on the effect that nutrition has on the immune system and its proper functioning, and also on the changes in nutrient utilization caused by immune response [3–6]. Consequently, a great deal of attention has been paid to the management of chronic disease involving the immune system and naturally, to agents capable of modifying immune response, also known as immunomodulators.

Immunomodulators drive the host's biological response and thus provide therapeutic effects by stimulating the immune system, or in some cases suppressing aspects of immunity such as allergic response. These agents can induce the body's own defense mechanisms in both a specific and nonspecific manner, but only have a weak antigenic effect. They generally influence the proliferation of immunocompetent cells without leaving behind any memory reaction. The primary targets of the immunomodulatory substances are macrophages, granulocytes, as well as T and B lymphocytes. Their effects may be direct or indirect, for example, via the complement system or the lymphocytes; via the production of interferon or lysosomal enzymes; as well as via an increase in macrophagocytosis and microphagocytosis.

The agents that enhance host resistance could potentially prove effective in the prevention of opportunistic infections in patients at risk, in treatment of chronic, persistent, chemotherapy-resistant bacterial and viral infections, cancer, immunodeficiency disease, generalized immunosuppression, and as adjuvants for vaccines. The agents that suppress immune reactions are used in managing autoimmune disease, gastrointestinal tract diseases, and transplantation surgery. The complex interplay between specific and nonspecific defense mechanisms produces cascade effects and simultaneous influences on defense mechanisms of a different nature. Immunomodulators are, along with vaccines, part of a fast growing category of new bioactives.

18.2 IMMUNOSTIMULANTS OF MARINE ORIGIN

The marine environment is a rich source of materials with incredible structural diversity [7–9]. In the past few years, more than 3000 compounds have been identified and some have entered clinical trials [10]. These chemical entities are prime candidates for the development of new drugs and nutraceuticals. Much attention has been paid to seaweed and algae, traditionally food sources, since it was observed that some have not only nutritional effects, but are also important for maintaining good health and even curing disease [11–14]. Indeed there is much scientific evidence that certain materials of marine origin contain substances that exhibit major physiological effects, including immunomodulatory activity.

Emerging scientific evidence that supports the benefit of nutraceuticals for the prevention of disease and improvement of health is a strong factor in the growth of this industry. The immune system by its very nature holds the key to optimum health, and therefore considerable attention has been paid to immunomodulators. This chapter will discuss sources and chemical nature of marine-based immunomodulators, preparations that enhance host resistance or decrease aspects of immunity such as inflammatory or allergic response. Two of Ocean Nutrition Canada's products, ONC-106, a material isolated from shark cartilage and Respondin™ (ONC-107), a proprietary potent immune response modifier extracted from an edible green microalga *Chlorella pyrenoidosa* will be discussed in some detail later in the chapter.

18.2.1 Low Molecular Mass Immunostimulants

18.2.1.1 Omega-3 Polyunsaturated Fatty Acids and Alkylglycerols

There is some evidence that fish and plant seed fatty acids have anti-inflammatory and immunomodulatory properties [15]. However, a great deal of work is needed to develop a better understanding of their effects on the immune system. Some data suggest that they could serve as a base for the development of more benign therapy for rheumatoid arthritis. For instance, studies show that fish oil preparations (1.7 g-EPA-dose spiked with DHA) have the capacity to reduce nonsteroidal anti-inflammatory drug (NSAID) or disease modifying antirheumatic drugs (DMARD) requirements [16].

Marine- and plant-derived omega-3 polyunsaturated fatty acids (PUFA) (Figure 18.1) also had an effect on T cell-mediated immune response in zynomolgus monkeys. Animals were first fed a 14-week baseline diet; then for two consecutive 14-week periods, two groups of animals were fed diets containing eicosapentaenoic acid (EPA) and docosahexaenoic acid (DHA) and two groups were fed a diet containing α-linolenic acid (ALA). Both types of diet significantly decreased the percentage of T cells, T helper cells, and T suppressor cells. In both,

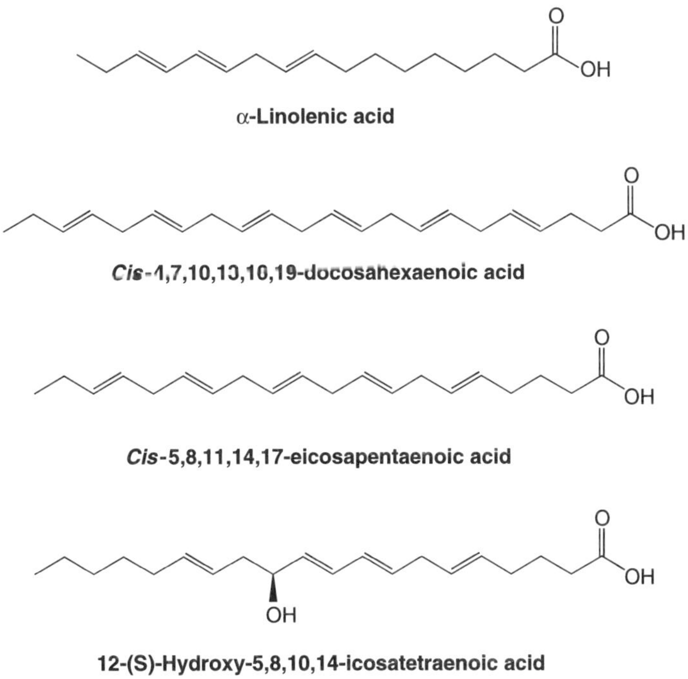

FIGURE 18.1 Structures of ALA, DHA, EPA, and ITA.

proliferative response of lymphocytes to T-cell mitogens and levels of IL-2 were significantly increased in the group of animals receiving diet containing a higher dose of EPA and DHA. No significant changes in mitogenic response or IL-2 production were found among animals receiving ALA diets [17]. There is emerging evidence that PUFA control the normal inflammatory process by acting as signal modulators, which regulate specific enzymatic activities typically elevated during inflammation at the inflammation site. For instance, E-SAR 94010, a marine fish extract, lowered triglyceride plasma levels of mice pups and their mothers. Lymphocyte cell studies revealed a clear immune activation as measured by the increased levels of CD25, CD28, CD54, CD56, and CD62 markers on lymphocytes derived from both E-SAR 94010–treated mothers and their newborns. The study concluded that E-SAR 94010 dietary supplementation improved the quality of immune surveillance [18].

Mohgaddam et al. [19] reported the first isolation of 12-(S)-hydroxy-5,8,10,14-eicosatetraenoic acid (ITA) (Figure 18.1) from a plant, the tropical marine alga *Platysiphonia miniata* (*C. agardh*) Borgesen. This potent mammalian immune hormone is a 12-lipoxygenase metabolite of arachidonic acid that is widely distributed in animal tissues. It is produced and secreted by platelet cells and elicits both chemotactic and degranulatory responses in target neutrophils. This immune hormone has previously been found in animal tissues only, but this is the first time it has been found in plants.

The oils from the ratfish (*Chimera monstrosa*) liver have been used as a folk remedy by Scandinavian fishermen for centuries. They were used for maintaining general health as well as for specific purposes, including wound healing and the treatment of irritation. These oils are particularly rich in O-alkylglycerols. O-alkylglycerols were also isolated from other sources, including oil from shark and flat fish liver [20]. There is a great deal of evidence in the scientific literature demonstrating the immunomodulatory activity of O-alkylglycerols and their derivatives.

O-alkyldiacylglycerols and O-alkylglycerols obtained from bone marrow and shark liver oil, respectively, were effective in the treatment of leukopenia caused after irradiation and were also beneficial in the treatment of thrombocytopenia. For instance, in one trial, 100 patients suffering from irradiation leukopenia were examined and a selected cohort received O-alkylglycerols. Seventy-five percent of these patients showed a significant increase in leukocyte count. The leukocyte count was also boosted in patients with cervical cancer receiving radiotherapy in another trial [21]. In another study, patients with cervical cancer were vaccinated against *typhus-paratyphus* (TABC) on the day before and after implantation of radium. The treatment group (O-alkylglycerol + radiation) was receiving 0.3 g of O-alkylglycerols immediately after the first vaccination. The serum samples were taken before the vaccination and after 3 weeks, that is, before the second implantation of radium. The formation of antibodies was more pronounced for the alkoxyglycerol group than for the group subjected to the standard treatment of radiation only [22].

Complex injuries due to the cancer growth and radiation were reduced to about one-third in a group receiving daily dose of 0.6 g of alkylglycerols prophylactically [23]. In a trial in which patients with cancer of the uterine cervix were treated with preparations from Greenland shark liver oil in combination with radiation, mortality was reduced to 50% after 36 months of treatment, compared to the group that received radiation only. The general immune response of the alkylglycerol-treated group was enhanced [21]. A functional response in human neutrophils was stimulated with O-alkylglycerols as well [24]. The data on immune boosting potential of O-alkylglycerols should be interpreted with caution since apart from a few studies conducted by other groups [25–28], most of the data come from studies carried out by one group led by The Brohults. In addition, The Brohults has attempted to commercialize an O-alkylglycerol product through AB Astra Pharmaceuticals. Between 1958 and 1975, The Brohults published data on 841 patients who have received radiotherapy together with O-alkylglycerols. The results showed significant statistic improvement with a 12.1% reduction in advanced cancers, compared to the group not receiving prophylactic O-alkylglycerol treatment. However, rigorous evaluation of their methodologies and results' interpretation leaves many unanswered questions and perhaps undermines to some extent evidence for the significance of alkylglycerol treatment [29].

In relation to the work of The Brohults, it should be pointed out that others found immunomodulating activity in structurally relevant materials. For instance, Homma and Yamamoto [30] found that a brief *in vitro* treatment of peritoneal cells with 50 ng/mL of dodecylglycerol significantly enhanced Fc-receptor-mediated ingestion activity of macrophages. Dodecylglycerol alone in combination with lysophosphatidylcholine enhanced antibody production in mice and may have practical use as an adjuvant [31]. Interestingly enough, alkyl-lysophospholipids that have been developed for combating cancer are synthetic, and coincidentally, platelet activating factor (PAF) is a phospholipid with almost identical structure to that of the most active synthetic preparations [32]. There is also scientific evidence that biosynthesis and metabolism of O-alkylglycerols is strongly associated with oncogenic transformation of cells [33]. For instance, cells with high levels of diacyl alkylglycerols generate more tumors when transplanted to mice than neoplastic cells with low levels of these molecules [34].

18.2.1.2 Other Low-Molecular-Weight Immunostimulants

Bryostatin 1, a member of a family of structurally related macrocyclic lactones, has been shown to impact the steady-state expression of the human immunodeficiency virus (HIV) receptor, CD4, produced by normal peripheral blood T lymphocytes. Incubation of the cells with 5 nM of bryostatin 1 caused a substantial loss of CD4 from the cell surface and the modulation of CD4 expression was not due to a cytotoxic effect. In accordance with previous findings by May et al. [35] and Hess et al. [36] showing that bryostatin 1 promotes the development of murine bone marrow progenitor stem cells and secretion by T lymphocytes, bryostatin 1 also elevated the expression of the IL-2 gene. The data also indicate that bryostatin 1 inhibits the glycosylation and expression of CD4 similar

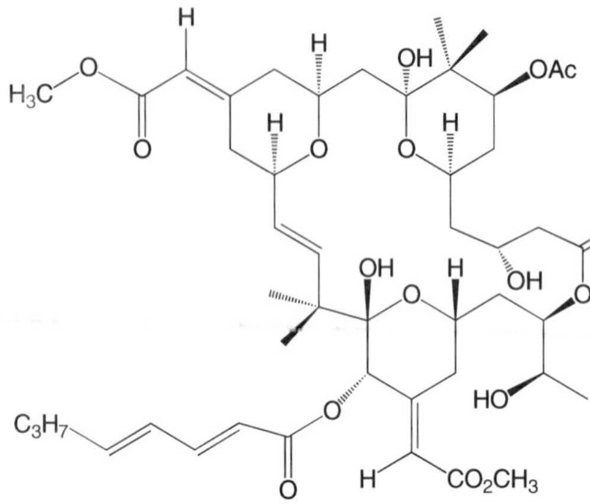

FIGURE 18.2 Structure of bryostatin 1.

to tunicamycin [37], and the study indicates that the compound could be useful in restorative immunotherapy. Bryostatin 1 also induces cytokine release and expansion of tumor-specific lymphocyte populations [38,39] (Figure 18.2).

Two novel steroids with an unusual carboxyl group at the C-14 site, penasterone, and acetylpenasterol, isolated from the Okinawan marine sponge *Penares incrustans* inhibited histamine release from rat peritoneal mast cells induced by anti-IgE in a dose-dependent manner. Penasterone was markedly more potent than cromolyn (Aventis, Bridgewater, NJ), a popular antiallergy medicine [40].

Another Okinawan sponge *Agelas mauritianus* contains antitumor immunostimulants called agelaspins. These marine natural products are glycolipids with a characteristic α-galactosylceramide structure [41,42], and exhibit potent immunostimulatory activity in a mixed lymphocyte reaction assay. This discovery led to the development of KRN7000, a new type of antitumor immunostimulant [43] (Figure 18.3). The drug is a dendritic cell activator, it first stimulates antigen-presenting cells and natural killer T (NKT) cells and then activates natural killer (NK) cells, macrophages, and T cells.

18.2.2 Biopolymer Immunostimulants

18.2.2.1 Peptides and Proteins

Eisenin (L-pyroGlu-L-Gln-L-Ala) is a tripeptide extracted from a brown marine algae known as arame (*Eisenia bicyclis* Setchell) (Figure 18.4). This algae has been used as a folk remedy for human cancers, and eisenin together with fucoidan are the known principles responsible for its bioactivity. Kojima et al. [44] demonstrated in a ^{51}Cr release assay that this peptide enhanced natural cytotoxicity of peripheral blood lymphocytes (PBLs) in humans, and the effect was attributed to NK cells activity. Eisenin made K-562 target cells resistant to lysis by PBL. Interestingly, both L-pyroglutamic acid and L-alanine, or a mixture of all

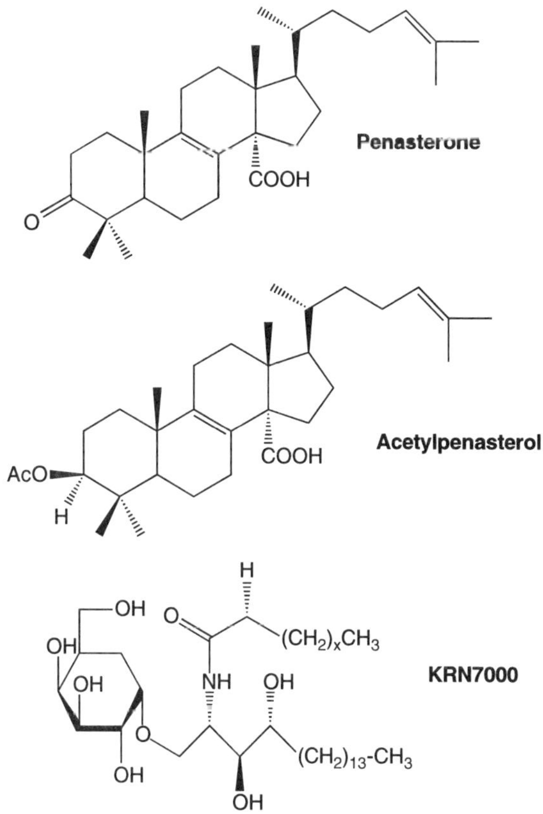

FIGURE 18.3 Structures of penasterone, acetylpenasterol, and KRN7000.

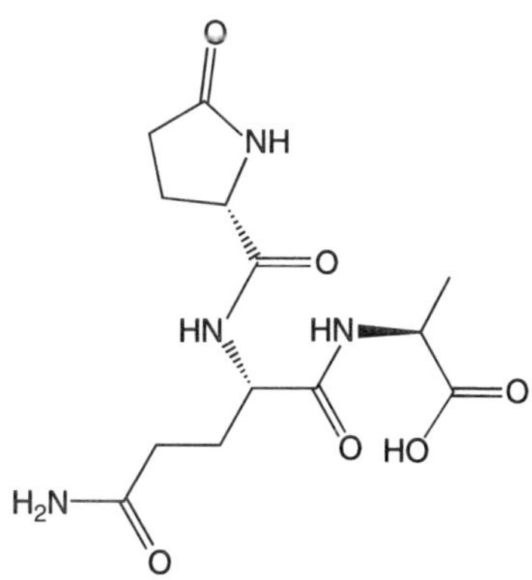

FIGURE 18.4 Structure of eisenin.

three building blocks of eisenin could augment the NK cell activity in the assay as well. Eisenin is similar in composition and its behavior in certain biological assays to a familiar immunostimulator *N*-acetylmuramyl-L-alanyl-D-isoglutamine, muramyldipeptide (MDP). Eisenin also resembles in size and bioactivity the immunostimulatory tetrapeptide tuftsin [45]. However, contrary to tuftsin, which stimulates NK cells directly, eisenin needs assistance from accessory cells for augmentation of natural cytotoxicity of PBL.

In 1976, a paper appeared [46] reporting on a preparation, sea star factor, isolated from the echinoderm *Asterias forbesi celomycetes*. The study demonstrated that the factor was a potent inhibitor of the primary immune response to T-dependent antigens and a suppressor of concavalin A (Con A)–induced mitogenesis. A later study revealed that macrophages obtained from rats receiving i.p. injection of 0.2 mg of the factor suppressed proliferation of tumor cells *in vitro* [47]. The cytostatic effect of these macrophages was demonstrated against syngenic, allogenic, or xenogenic (murine origin, P815-X2 of DBA/2 strain) targets and was not immune specific. The data indicate that the cytostatic effect is manifested in three major stages: cyto-adhesion, macrophage-dependent state, and macrophage-independent state. Suppression of tumor cell division by the sea star factor–activated macrophages required direct contact between viable effector macrophages and tumor target cells for a relatively prolonged period to achieve maximum cytostatic effect. The main active component of the sea star factor is a 38 kDa protein consisting of a heavy chain (25.3 kDa) and a light chain (12.7 kDa) [46].

Lectins are well known for their immunostimulatory effects, and they have been in use as specific lymphocyte stimulators [48]. They have also been recently isolated from algae [49]. Amansin, a lectin from *Amansia multifida* stimulated proliferation of human peripheral mononuclear cells at concentrations as low as 3.12 μg/mL. The activity was significantly reduced at the concentration of 12.5 μg/mL and particularly at 50.0 μg/mL. Amansin stimulated production of IFN-γ, particularly at concentrations as low as 6.25 mg/μL [50].

Several studies support claims that keyhole limpet hemocyanin (KLH), a protein widely used as hapten carrier, is also a nonspecific immune stimulant. The first clinical use of KLH in immunotherapy was for posttreatment of bladder cancer after radiotherapy and was reported by Olsson 30 years ago. This study showed a decreased rate of recurrence of the disease in the group using KLH [51]. In a later clinical trial, patients were first given an injection to evoke an immune response, followed by 18 monthly intravesical instillations in saline solution. The trial included 101 patients, and significantly higher numbers of patients receiving KLH had complete response or some degree of improvement, compared to a number of patients in the study arm receiving Mitomycin C, indicating response rates comparable to Bacille Calmette-Guérin (BCG), though those using KLH reported fewer side effects. No systemic toxicity or other adverse side effects were observed in subjects using KLH [52]. KLH was also shown to produce a more predictable behavior than BCG, with similar results and fewer side effects [52,53]. The advantage of KLH is its apparent lack of toxicity. Lamm et al. compared KLH and Immucothel®, a form of KLH modified for clinical use (Biosyn Arzneimittel GmbH, Fellbach, FRG) and concluded that both crude KLH and Immucothel appear to be effective immunomodulators in the treatment of transitional cell carcinoma [54]. Immucothel is primarily used for treatment of bladder cancer and has received marketing approval in the Netherlands, Austria, and Korea.

The best responders to KLH immunotherapy are patients with refractory carcinoma *in situ* (CIS), which is a more difficult form of transitional cell carcinoma (TCC) to treat. For instance, in a study by Lamm et al. [55], 11 of 19 CIS patients, 10 of 20 patients with papillary TCC, and 4 of 12 patients with both forms of bladder cancer showed a positive response. KLH has been shown to be effective against tumor recurrence in bilharzial bladder with papillary TCC. The immunotherapy resulted in the reduction of the recurrence rate to 15.4% compared to 76.9% not receiving KLH therapy [56]. KLH enhances the host's immune response by interacting with T cells, monocytes, macrophages, and polymorphonuclear macrophages. It induces both a cell-mediated and humoral response in both animal and man. Other multicenter clinical trials have confirmed the beneficial effect of KLH given intravesically for 6 weeks to patients with various stages of bladder cancer and a variety of genitourinary tumors [57,58].

In addition, multiple cancer cell lines were tested for their responsiveness, including estrogen-dependent breast (MCF-7), estrogen-independent breast (ZR75-1), pancreas (PANC-1, MIA-PaCa), and prostate (DU145). Growth of cancer line ZR75-1 was slowed by 43% at 72 h, whereas treated MCF-7 cells had an average of 40% growth inhibition at the equivalent dosing. Treated PANC-1 cells had a mean growth inhibition of 19% at 72 h, and the DU145 prostate cancer cell line averaged a 6% growth inhibition over the concentrations tested [59]. The authors found in previous experiments that KLH induces changes in IL-2 and IL-12 produced by cancer cells.

Use of KLH as an immunomodulator is interesting since it is very popular as a hapten carrier to help trigger production of antihapten antibodies, that is, it is an antigenic material. It is not a typical immunomodulator since immunomodulators are predominantly preparations displaying very low degree of antigenicity. KLH are blue copper proteins, which serve as oxygen carriers in the blood of the keyhole limpet *Megathura crenulata*, a marine gastropod. KLH occurs as two distinct isoforms. Each of these molecules is based on a very large polypeptide chain, the subunit (molecular mass ca 400 kDa), which is folded into a series of eight globular functional units (molecular mass ca 50 kDa each). Twenty copies of this subunit form a cylindrical quaternary structure with molecular mass of about 8 MDa [60].

18.2.2.2 Protein Polysaccharide Complexes

Cho et al. [14] discovered an antitumor activity of the protein–polysaccharide fractions from sea staghorn and laver toward sarcoma-180 cells. The polysaccharide contents of these extracts were above 60% and fructose was the most common sugar. Similar to the immunomodulators ONC-106 and Respondin™ developed by

Ocean Nutrition Canada, which also contain protein and polysaccharide active components (see Sections 18.2.4 and 18.2.5), the amino acids of the protein portion of the extracts were mostly acidic or neutral; the major amino acids were Asp, Glu, Gly, and Cys. Dose of 50 mg/kg of sea staghorn extract inhibited the tumor growth by more than 50% and increased life span (ILS) after laver extract administration was extended by about 17% at that dose. Also, the number of total peritoneal exudate cells in the group with administered sea staghorn extract was increased significantly in comparison with that of the control group.

18.2.2.3 Nonglucan Polysaccharides

More than 3600 varieties of marine macroalgae and seaweed are known. Some of the bioactivities are clearly attributed to the polysaccharides they contain, particularly to those that are sulfated [8]. For instance, the effects of carrageenan on the immune system have been known for many years; the polysaccharide has been used as an immunosuppressant in renal transplant management [61]. Immunoadjuvant, immunostimulant, and immunosuppressant activity of carrageenans were observed in multiple studies [62,63]. It was discovered that carrageenans act as T-cell mitogens [64], induce production of granulocyte monocyte colony stimulating factor (GM-CSF) [65], and stimulate proliferation of mouse B cells [66].

Chitosan oligosaccharides are also known for their immunostimulatory activity. Chitosan is a deacetylated form of chitin, a naturally occurring linear polysaccharide component of the shells of crustaceans, invertebrates, and fungi (Figure 18.5). Chitin is a heteropolymer composed of *N*-acetyl-D-glucosamine units, typically with less than 15% being deacetylated, whereas chitosan is typically deacetylated chitin containing usually not more than a 15% of acetyl groups. Although most chitosans are manufactured from chitin, naturally occurring chitosans exist as well. Cellulase digestion of chitosan followed by acetylation of the hydrolysate resulted in a mixture of chitosan oligomers that stimulated proliferation of humanlike macrophages. A sample with 100% acetylation level stimulated

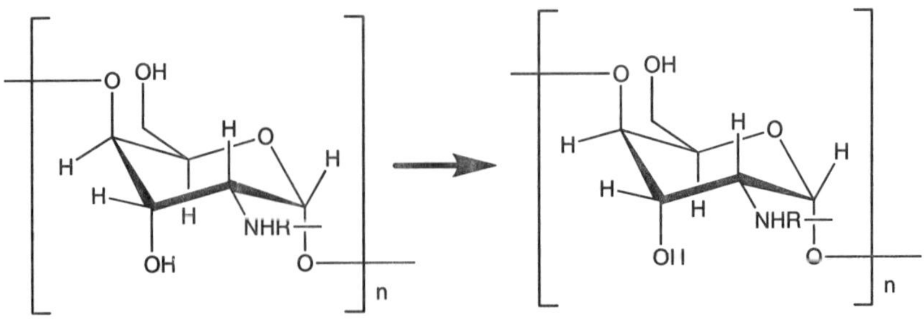

R = $H_2^+ X^-$ ~40%, Ac ~60% n > 500 (typically) R = $H_2^+ X^-$ ~60%, Ac ~40%

FIGURE 18.5 Structures of chitin (left) and chitosan (right).

cell proliferation most effectively. Also the levels of serum immunoglobulin G (IgG) and immunoglobulin M (IgM) were significantly higher with the totally acetylated preparation. The product also enhanced the Con A- and lipopolysaccharide (LPS)-induced proliferation of mouse lymphocytes [67]. It was also discovered that intranasal application of chitin microparticles down-regulates symptoms of allergic hypersensitivity to *Dermatophagoides pteronyssinus* and *Aspergillus fumigatus* in murine models. Previous studies have demonstrated that chitin in the form of microparticles that can be phagocytosed is a potent macrophage stimulator and promotes a Th1 cytokine response. It has been shown that oral administration of chitin microparticles is effective in down-regulating serum IgE and lung eosinophilia in a mouse model of ragweed allergy. The intranasal application of chitin microparticles effectively reduces serum IgE and peripheral blood eosinophilia, airway hyperresponsiveness and lung inflammation in both allergy models. It results in elevation in Th1 cytokines IL-12, IFN-γ, and TNF-α and reduction in IL-4 production during allergen challenge. The stimulation of the nasal-associated lymphoid tissue with chitin microparticles could offer a natural approach to treating human allergies [68].

Okai et al. [11,69] observed stimulation of murine spleen cells with polysaccharide fractions from seaweed, an edible marine algae *Hizikia fusiformis*. The extract stimulated the release of TNF-α from macrophages of endotoxin-non-responder C3H/HeJ mice. The activity was not only detected in both the polysaccharide and nonpolysaccharide fractions but was associated primarily with the polysaccharide components. The extract displayed stimulation of B cells and it was not effective in T-cell stimulation. The active components in the carbohydrate fractions are polysaccharides with molecular masses of about 2000 and 70 kDa. The smaller polysaccharide was much more stimulatory with regard to the release of TNF-α from macrophages. The authors did not specify the compositions of the polysaccharides, but indicated that they primarily contained galactose and mannose but did not contain significant amount of glucuronic and mannuronic acid. These polysaccharides, particularly the smaller-sized one, also enhanced macrophage-dependent suppression against the growth of EL-4 tumor cells in an *in vitro* culture experiment. The immune boosting properties of the polysaccharides may play a significant role in the prevention of and protection from cancer. The polysaccharide fraction was further refined using an anion exchanger and the purified fraction stimulated the production of IgG and IgM, and release of IL-1α and TNF-α from the macrophages of endotoxin nonresponder mice.

Yoshizawa et al. [70,71] demonstrated in both *in vivo* and *in vitro* models, macrophage-stimulating activity of fractions predominantly composed of polysaccharides isolated from edible marine algae *Porphyra yezoensis*. The average molecular mass of these polysaccharides was 400 kDa and 1.73 MDa, respectively. The polysaccharides were made of a significant number of largely sulfated 3,6-anhydrogalactose units and porphyran was the main component. Both polysaccharides stimulated production of TNF-α and nitric oxide by macrophages but the larger one was a better stimulator. The larger polysaccharide also stimulated secretion of IL-1, and the smaller polysaccharide fraction had a stronger effect on carbon clearance activity of phagocytes of the injected mice.

Liu et al. [72] observed that certain seaweed stimulate proliferation of mouse B cells *in vitro*, including those of LPS nonresponder C3H/HeJ mice strain. The production of immunoglobulins by B cells was also enhanced, and the extracts also stimulated production of TNF-α by mouse macrophages. The most potent extracts were those of *Hizikia fusiformis* and *Meristotheca papulosa*. A glycoprotein from *Meristotheca papulosa* with molecular mass of 100 kDa was responsible for the activity of the extract, but no information on the composition of polysaccharide portion was given. The effect of the extracts on human lymphocytes *in vitro* was also studied [73]. The extracts of *Hizikia fusiformis* and *Meristotheca papulosa* (green) greatly stimulated proliferation of human lymphocytes, whereas *Eucheuma muricatum* and *Meristotheca papulosa* (red) were not found to be strong stimulators. T cells purified by sheep red blood cell (SRBC) rosette-formation were significantly stimulated, but B cells were not. The extracts also enhanced the induction of cytotoxic T lymphocyte activity, but they did not enhance NK cell activity. They also had a stimulatory effect on immunoglobulin production by B cells and TNF-α production by monocytes. There was no detailed discussion on the composition of the immune principles but the activity of *Hizikia fusiformis* was found to be associated with polysaccharides of average molecular weight of 100 kDa.

Itoh et al. [74] discovered that marine algal polysaccharide, GIV-A from *Sargassum thunbergii*, significantly inhibited the growth of Ehrlich ascites carcinoma at the dose of 20 mg/kg/day with no sign of toxicity in mice. GIV-A is a fucan sulfate (fucoidan) with a molecular mass of 19 kDa. A carbon-clearance test demonstrated that the fucoidan activated the reticuloendothelial system (RES), enhanced the phagocytosis and chemiluminescence of macrophages. Binding of the complement cleavage product C3 to macrophages and the proportion of C3 positive cells were stimulated.

Fucoidans isolated from marine algae such as Laminariales, Chondariales, Fucales, Echinodermata, Echnoidea, and Asterozoa are described as therapeutic or prophylactic agents for diseases requiring regulation of cytokine production, nitric oxide production, and allergic diseases in which they act as IgE production suppressors. The molecular mass of the fucoidans is around 200 kDa. For instance, the fucoidans from *Kjellmaniella crassifolia* are mixtures of U-fucoidans (20% sulfated polysaccharides) containing mostly fucose, mannose, galactose, and glucuronic acid, and F-fucoidans (50% sulfated polysaccharides) containing mostly fucose and galactose [75].

Funoran, a polysaccharide isolated from *Gloiopeltis tenax*, significantly inhibited the growth of Ehrlich ascites carcinoma, solid Ehrlich, Meth-A fibrosarcoma, and Sarcoma-180 tumors. It also significantly induced the enhancement of delayed type hypersensitivity (DTH) response to SRBC in tumor-bearing mice. When administered i.p., funoran increased the spleen weight of mice. In addition, changes in the T-cell subsets in the spleen, thymus, and peripheral blood were monitored and it was found that the percentages of L3T4 and Lyt 2 T-cells were markedly increased in the peripheral blood. The percentages of asialo CM1 cells in the thymus and peripheral blood were significantly increased as well. The results suggest that the antitumor effect of funoran is associated with the augmentation of T-helper, T-cytotoxic, and NK cells [76].

A novel sulfated polysaccharide (p-G603) was isolated from red-tide microalga *Gyrodinium impudicum*. The activities of peritoneal macrophages and NK cells isolated from mice and treated with a single dose of 100 mg or 200 mg/kg of p-G603 were increased. In addition, plaque forming cell response to SRBC, as well as the levels of IgM and IgG were enhanced. The material also stimulated production of IL-1β, IL-6 and TNF-α. [77]. The average molecular mass is 187 kDa, with galactose as a main sugar and the degree of sulfation of 10.3% [78].

A polysaccharide fraction corresponding to a sulphated xylomannan obtained from water extracts of the red seaweed *Nothogenia fastigiata* efficiently inhibited the replication of Herpes simplex virus (HSV)-1. The preparation also selectively inhibited the replication of several other enveloped viruses including HSV-2 (human cytomegalovirus [HCMV]), respiratory syncytial virus, influenza A and B virus, Junin and Tacaribe virus, and simian immunodeficiency virus, but was only marginally effective against HIV-1 and HIV-2. The preparation inhibited HSV-1 and HCMV viruses via adsorption [79].

Viva-Natural, is a cold water extract from *Undaria pinnantifida*, a dietary seaweed from the *Phaeophycophyta* algae family. The extract contains a macrophage-activating polysaccharide (80% by weight), it is noncytotoxic in KB cell cultures, enhances the cytolytic activity of peritoneal macrophages and activates immune system [80] in a nonspecific manner. The material was also shown to be active against Lewis lung carcinoma and spontaneous AKR T-cell leukemia. The anti-Lewis lung carcinoma (LLC) activity of Viva-Natural was stronger than that of isoprinosine and weaker than that of pyran copolymer, but a crude polysaccharide fraction of the extract demonstrated curative activity similar to that of pyran copolymer. Viva-Natural also reversed the amplification of cyclosporin effect on the development of leukemia in AKR mice at preleukemic stage [81]. Antileukemic activity of Viva-Natural was also confirmed against spontaneous AKR T-cell leukemia in mice and it was the only immunomodulator active against AKR T-cell leukemia, while all standard immunomodulators such as pyran copolymer, tilorone, isoprinosine, and levamisole were not active [82]. Viva-Natural was found to be therapeutically very effective (347% ILS) against mice in advanced stage of leukemia, and induced complete clinical remissions.

18.2.2.4 Glucans

Significant immunomodulating activity was found in an aqueous extract of the most popular edible seaweed in Japan, *Laminaria japonica* (Makonbu). The extracts exhibited a stimulation of DNA synthesis of spleen cells from C3H/HeJ mice. The polysaccharide fraction of the extract stimulated activity of mouse phagocytes against *Staphylococcus aureus* and the release of IL-1α and TNF-α from the same cells, and also exhibited enhancing effects on polyclonal antibody

(IgM and IgG) production in spleen cells. No details on the composition of the polysaccharides were given, except for the note that they did not contain a significant amount of galacturonic or mannuronic acid [83]. The paper cites work of Kato and Owen [84] who discussed possibilities of immunological regulation of epithelial mucosa of the intestinal tract and suggested, that M cells in the Peyer's patches capture macromolecules, which were transported to the enfolded macrophages and lymphocytes. It is believed that some β-glucans work in a similar fashion [85]. Another study concluded that *Laminaria* is an important factor contributing to the relatively low breast cancer rates in Japan [86]. *Laminaria* may either prevent the commencement of breast cancer or inhibit cancer development by endogenous physiological factors. Several factors could contribute to this anticancer effect. *Laminaria* is a source of nondigestible fiber, thereby increasing fecal bulk and decreasing bowel transit time. It also changes the posthepatic metabolism of sterols and contains an antibiotic substance that may influence fecal ecology. *Laminaria* contains β-(1-3)-D-glucan (laminarin) and ((1-4)-D-mannuronan (alginate). Okai employed gentle extraction methodology that left most of the alginate unextracted [83] possibly leaving β-(1-3)-D-glucan as the main component. β-(1-3)-D-Glucans form the backbone of a number of immunomodulators, including Lentinan (Yamamouchi/Ajinomoto, Japan) and SPG (schizophylan) Sonfilan® (Kaken/Taito, Japan), two of the most successful anticancer immunostimulants currently used in clinics in Japan and Korea. It is also known that β-(1-3)-D-glucans alter enzymatic activity of fecal flora and stimulate the host-mediated immune response.

A Russian group isolated bioglycans from marine invertebrates that had an effect on the humoral and cellular immunity when administered parenterally in a single dose. They increased the level of the rosette-forming cells (primarily T2- and B-subpopulations) in the spleen of the mice and enhanced titers of serum antibodies. They were administered at different periods with respect to immunization and influenced the DTH induced by SRBC [87]. Later on they isolated three immunostimulatory bioglycans: mitilan (isolated from mussel *Crenomytilus grayanu*), a glycoprotein, containing α-(1-4)-D-glucan, (mol. mass 3 MDa); translam β-D-(1-3); β-1-6-) -D-glucan, isolated from *Laminaria cichorioides* (mol. mass 8–10 kDa), and zosterin, a low-methoxylated pectin isolated from marine plant of genera *Zosteraceae* [88] (Figure 18.6). The immunomodulatory activity of the bioglycans was based on a complex interaction of immune cells involving cytokines. All bioglycans produced changes in immune system, such as enlargement of spleen mass, redistribution of lymphocyte subpopulations, enhancement of nonspecific T-suppressor activity, an increase of interferon content in blood serum in mice. The immunomodulatory effects of the investigated biopolymers are because of the polysaccharide nature and are in relation to their stereochemistry and size. Translam also demonstrated potent healing effect in experimental radiation disease, exhibited preventing effect at experimental bacterial infections, stimulated hematopoiesis, influenced humoral and cell immunity, and factors of nonspecific resistance [89]. The same research group found that fucoidan was able to markedly inhibit the growth of Gram-negative and Gram-positive organisms,

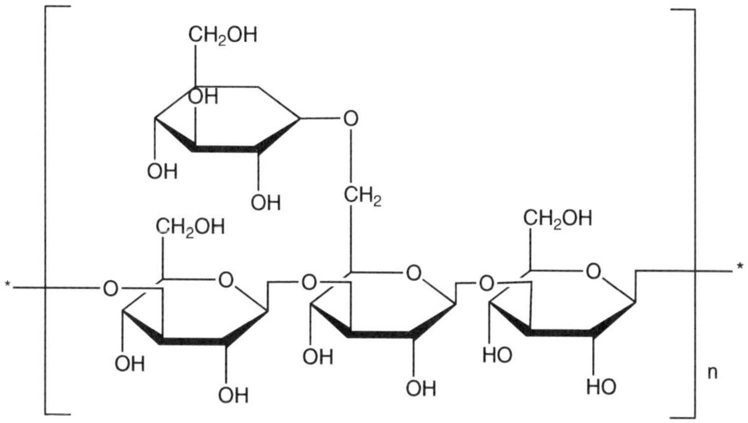

FIGURE 18.6 Structure of translam.

while stimulating immune response of the humoral and cellular types, as well as phagocytosis. Fucoidan seems to be a good candidate for a development of a drug combining antibacterial and immunomodulating activity [90].

Phycarine, β-(1-3)-D-glucan, isolated from sporophytes of *Laminaria digitata* was found to enhance phagocytic activity and stimulated production of IL-1, IL-6 and TNF-α. It also increased NK cell-mediated killing of tumor cells both *in vitro* and *in vivo* while acting via complement receptor type 3 (CR3) receptors [91]. Phycarine also amplified potency of cyclophosphamide in the treatment of LLC and increased the speed recovery form leucopenia caused by chemotherapy or irradiation. The glucan stimulates both cellular and humoral immunity [92]. Laminarin and its hydrolysates also stimulated proliferation of thymocytes. A mouse cDNA microarray showed induction of genes, coding for proteins involved in immune response [93].

18.2.3 Nonchemically Defined Immunostimulants

Just a few examples will be given in this review to illustrate this category. An unidentified extract derived from the seaweed *Ascophyllum nodusum* significantly improved growth performance of pigs challenged with *Salmonella typhimurium*. This was demonstrated in a 28-day study involving 95 pigs. However, except for activation of alveolar macrophages *in vitro* leading to secretion of prostaglandin E_2 (PGE_2) by high dose of the extract, no influence on immune response was observed [94]. Wakame (*U. pinnatifida*) seaweed is known to suppress the proliferation of 7,12-dimethylbenz(a)-anthracene-induced mammary tumors in rats [95]. Filtrates of marine bacterium *Vibrio anguillarum* culture are also known to exhibit immunostimulatory effects [96]. There are likely many examples of products of this subcategory, however, owing to their lack of structural characterization they are not well described in the scientific literature and more extensive review of these products is out of the scope of this review.

18.2.4 *Spirulina* and *Chlorella*

18.2.4.1 *Spirulina*

Microalgae are a promising source of leads for new drugs. Many isolated compounds have novel structures and exhibit potent physiological activity. In terms of drug discovery, most efforts in the past revolved around the investigation of blue-green algae [97]. *Spirulina* and *Chlorella* are edible algae that have been widely studied both in terms of their component molecules and their biological activities. Owing to the traditional role they have played in nutrition in Far East for millennia, they have been primarily investigated for their potential as nutritional supplements and adjuvants to drug therapy [98,99].

Spirulina, some blue-green algae, and *Chlorella* were all shown to contain large polysaccharides with immunostimulatory activity. *Spirulina* is a marine plant that grows in salt water. It is a genus of the phylum Cyanobacteria and exists as several species. The most popular, in terms of human consumption, are *Spirulina platensis* and *Spirulina maxima*. These species have been used as food supplement in Asia for many years and are becoming popular in North America. However, in addition to its nutritional value, *Spirulina* boosts the immune system, helping to prevent both viral infection and cancer. For instance, the NK cells isolated from the blood of volunteers after oral administration of *Spirulina* produced IL-12, IL-18 and increased level of IFN-γ. These cells were also more responsive to BCG stimulation *in vitro*, as indicated by an increased production of IL-12. *Spirulina* may be involved in the signaling responses through Toll in blood cells when orally administered and it acts in humans directly on myeloid linkages and either directly or indirectly on NK cells [100]. Other *Spirulina* preparations were found to inhibit mast cell–mediated immediate type allergic reactions in rats. *Spirulina* may amplify mast cell degranulation, possibly by affecting the mast-cell membrane [101]. In addition, extracts of *Spirulina platensis* enhance macrophage function in cats [102] and enhanced humoral and cell-mediated immune functions in chickens [103]. An extract of *Spirulina* inhibited *in vitro* replication of HSV-1 [9]. A recent study showed a beneficial effect of *Spirulina* on asthma indicating Th2 suppression [104]. The best-characterized biopolymer component of *Spirulina* is a polysaccharide called spirulan [105]. It consists of two types of sulfated disaccharide repeating units, O-hexuronosyl-rhamnose and -rhamnosyl-3-O-methyl-rhamnose. The known bioactivity of this polysaccharide is associated with inhibition of cultured bovine aortic endothelial cell proliferation and anti-HSV and anti-HIV activity [106]. C-phycocyanin, a protein pigment isolated from *Spirulina platensis* is another component of the alga with a significant medical and pharmacological importance [107,108].

18.2.4.2 *Chlorella*

18.2.4.2.1 Summary of the Key Findings

Chlorella is a unicellular plant with size similar to that of human erythrocytes, growing in both fresh and marine waters. It is one of the most widely cultivated species of microalgae and widely used as a dietary supplement. It shows excellent utilization efficiency of sunlight and other light energy, shows more rapid proliferation than other conventional plants and it is a valuable protein source and contains all the essential amino acids [109]. *Chlorella* has a strong cell wall that prevents its native form being adequately digested. However, the cell wall can be broken down using processes such as Dyno-Mill processing, so that the cellular components can be digested by humans [110].

Original scientific reports on immunostimulatory properties of *Chlorella* preparations came from Kojima et al. [111] isolated from *Chlorella* cells, a substance exhibiting accelerating effect to the phagocytic function of RES. Later, a paper appeared describing antibody modulation effect by *Chlorella* [112]. Several years ago, a polysaccharide demonstrating immunostimulatory activity in a transcription factor–based bioassay for nuclear factor kappa B (NF-kappa B) activation in THP-1 human monocytes was isolated. In this assay, the *Chlorella*-originating polysaccharide was 1000 times more active than schizophyllan, lentinan, krestin and acemannan, the major commercial immunostimulatory polysaccharides [113]. Most biological studies focused on antitumor effects stemming from the immunostimulating activity of *Chlorella*. An acidic polysaccharide isolated from *Chlorella pyrenoidosa* showed life-prolongation effects in mice with sarcoma 180, IMC carcinoma, Meth-A fibrosarcoma, B 16 melanoma and LLC. It stimulated proliferation of lymphocytes, had carbon clearance activity, and no direct cytotoxic effect [114]. Chlorella cells or relevant extracts also exhibited a tumor-suppressing effect on mammary carcinoma (MM-2, CA-755, Ehrlich) and mouse leukemia (EL-4), and again there was no direct cytotoxicity against the tumor cells *in vitro* and the effect was mediated by host immune response [115]. Tanaka et al. [116] found significant suppression of Meth A tumor in mice fed with a *Chlorella* extract and concluded that its antitumour resistance was mediated through cytostatic T cells and not through NK cell activity. A novel glycoprotein exhibiting an antimetastatic effect was isolated by the same group [117]. There is ample additional evidence from animal models that immunostimulating activity of *Chlorella* is effectual in suppressing various forms of cancer.

Immunomodulatory effects of *Chlorella* extracts were demonstrated in other models as well. Enhanced resistance to pathogen invasion was demonstrated in a number of animal models. For instance, *Chlorella* extracts had no direct antibacterial activity in *E. coli* model, but the protective effect was mediated through the acceleration of superoxide generation and chemokinesis in polymorphonuclear leukocytes [118]. Also, resistance to *E. coli* of mice having undergone cyclophosphamide treatment was restored when treated with a *Chlorella* extract [119]. In another study, a beneficial effect on mice with mouse AIDS (MAIDS) infected with *Listeria monocytogenes* was observed. The study suggests that the treatment could be beneficial as a prophylactic therapy of patients with a high risk of septicemia [120].

Hasegawa et al. [121] demonstrated that hot water extract of *C. vulgaris* enhanced resistance to *Listeria monocytogenes* via augmentation of Th1 responses producing IFN-γ. Oral administration of the extract resulted in suppressing the production of IgE against casein antigen, accompanied by increased

IFN-γ, IL-12, and mRNA expression. Oral administration of the extract enhanced Thl response to casein in the spleen of immunized mice and it may be useful for prevention of allergies with a predominant Th2 response. An effect on bone marrow progenitor cells of mice infected with this pathogen was also demonstrated [122]. Hot water extracts of *C. vulgaris* increased physical stamina in mice [123].

Price tested methanolic extracts of taxonomically diverse algae for their ability to influence release of histamine by mast cells. A *Chlorella* extract was one of the extracts tested and it exhibited potent anti-inflammatory activity. LH-20 chromatography yielded two active fractions [124]. Recently, Guzman et al. [125] isolated polysaccharides exhibiting anti-inflammatory and immunomodulatory properties from marine algae *Chlorella stigmatophora* and *Phaeodactylum tricornutum*. The anti-inflammatory effect was demonstrated in the carrageenan-induced paw edema test and was superior to that of indomethacin. *C. stigmatophora* was immunosuppressive in DTH tests and stimulated phagocytocic activity *in vitro* and *in vivo* whereas *P. tricornutum* did not. The *C. stigamtophora* polysaccharide was of 22 kDa size, whereas the *C. tricornutum* contained three polysaccharide fractions with molecular mass of 27, 408, and 449 kDa, respectively. The *C. stigmatophora* polysaccharide contained glucose, glucuronic acid, xylose, and ribose/fructose, whereas *P. tricornutum* contained glucose glucuronic acid and mannose.

Several clinical trials were conducted to study the beneficial effects of *Chlorella* supplementation on humans. For instance, Merchant et al. [126] demonstrated that glioma patients may have benefited from adding *C. pyrenoidosa* to their diet. However, the researchers cautioned that it was a limited study, so further work was required to confirm that longer survival and fewer infections related to an improved immune status. Another study performed by this group suggests that *Chlorella* supplementation may help relieve the symptoms of fibromyalgia in some patients [127].

18.2.4.2.2 Respondin™

18.2.4.2.2.1 In Vitro *Studies and Animal Models*

Respondin™ is a proprietary preparation discovered and developed by Ocean Nutrition Canada Ltd. It is composed of a mixture of biopolymers, predominantly polysaccharide-based complexes with molecular mass of the main immunoactive principles in excess of 100 kDa.

Initial *in vitro* studies demonstrated that Respondin™ stimulated proliferation of mouse spleen B cells and did not have a stimulatory effect on T cells (Figure 18.7). Figure 18.8 shows a comparison of Respondin™ with other *Chlorella* preparations and the popular immune herbal supplement *Echinacea* in their capabilities to stimulate proliferation of B cells *in vitro*. The data demonstrate superior performance of Respondin™ over the other tested materials. Respondin™ is also a powerful stimulator of peritoneal mouse macrophages and macrophage cell lines. It stimulates production of several cytokines, including IL-6, IL-12, and nitric oxide by peritoneal macrophages [128–130]. Comparison of

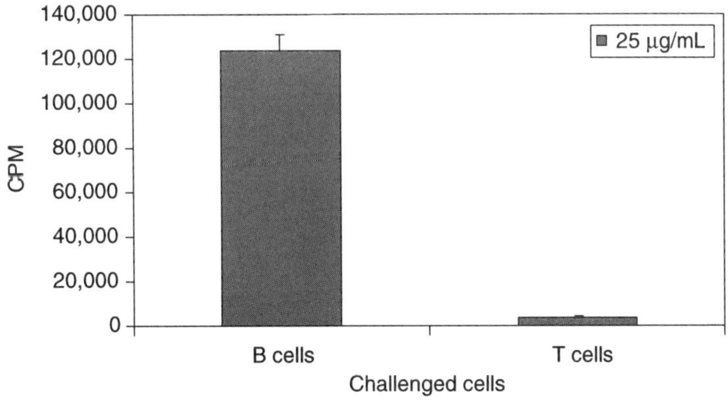

FIGURE 18.7 Stimulation of B cells and T cells by Respondin™.

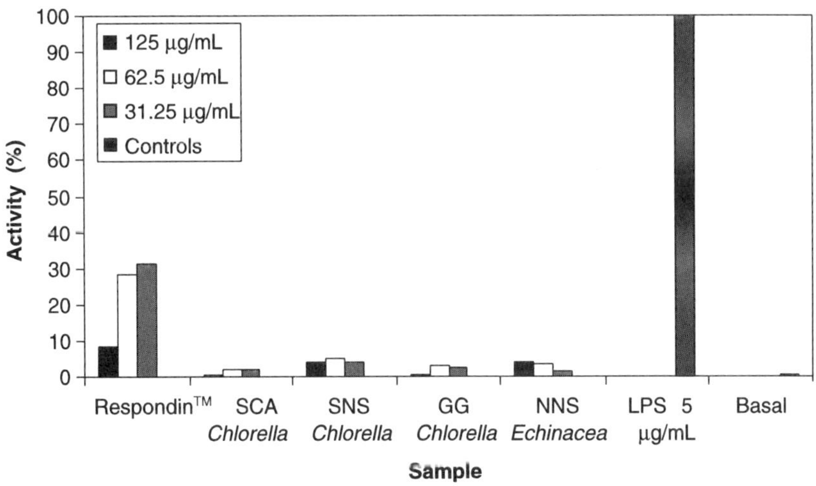

FIGURE 18.8 Comparison of Respondin™ with other *Chlorella* preparations and *Echinacea* in their capabilities to stimulate proliferation of mouse splenic B cells *in vitro.*

Respondin™ with selected products in stimulation of mouse macrophages measured as NO production is illustrated in Figure 18.9. Respondin™ was also examined *in vivo* in mouse models and was found to significantly augment resistance toward infection with bacterium *Listeria monocytogenes*, as well as with the fungus *Candida albicans.* Figure 18.10 illustrates the effect of Respondin™ on the development of *Listeria monocytogenes* in mice.

Allergies are a widespread problem particularly in the industrialized world, and immunomodulators may be beneficial in their prevention or treatment. However, compounds that impact immunity could potentially also trigger or promote allergies. To confirm whether Respondin™ has any antiallergy activity, it was

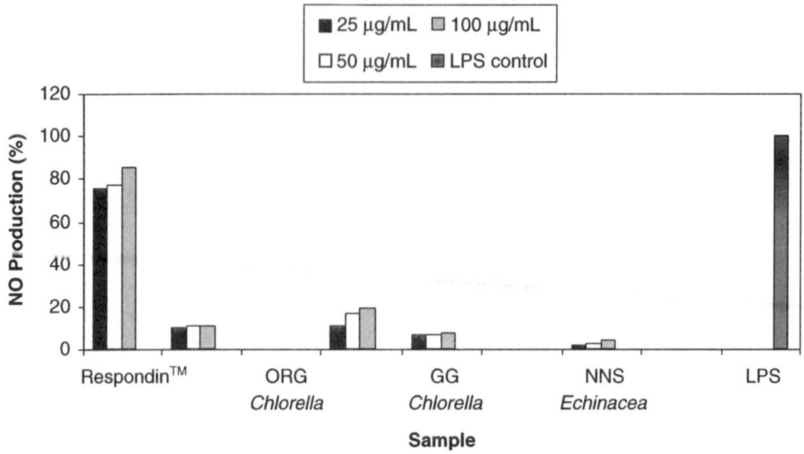

FIGURE 18.9 Comparison of Respondin™ with selected products in stimulation of mouse macrophages measured as NO production.

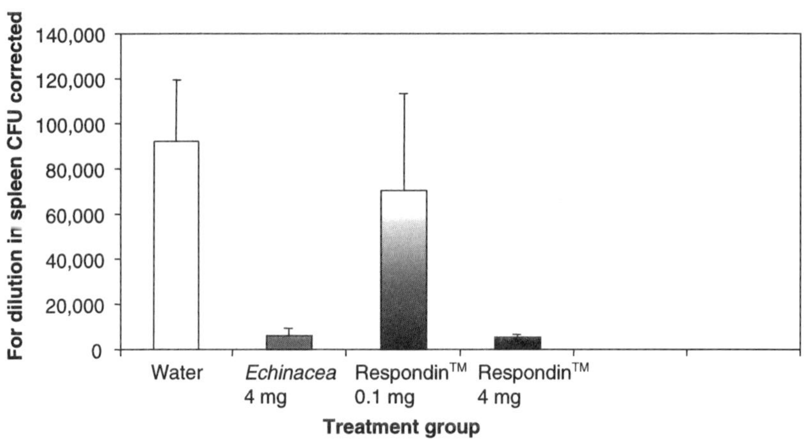

FIGURE 18.10 Effect of Respondin™ on the development of *Listeria monocytogenes* in mice.

tested in an *in vivo* allergy model. This model involved inducing experimental allergy with mercuric chloride in the lupus mouse. However, in this model no effect on disease progression was observed, as demonstrated in the immune complex deposits and immunoglobulin levels in the mercuric chloride treated group, and in terms of immune complex deposits and mortality in the lupus model. We also investigated effect of Respondin™ on development of allergic airway disease. We initially evaluated the effect of Respondin™ on degranulation of mast cells and production of cytokines [131] and later also challenged mice with airway stimuli and measured cytokine levels in bronchoalveolar lavage fluid

(BALF) and lung tissues as well as RNA expression of various cytokines and chemokines in the lung. Respondin™ exhibited significant effects on reduction of IL-5 levels. Its mechanism of action is similar to Mepolizumab (GSK), a monoclonal anti-IL-5 antibody; an effective antiallergy drug. However, Mepolizumab therapy might induce partial maturational arrest of the eosinophil lineage in the bone marrow. The effect of Respondin™ on airway inflammation was assessed as well and the data showed that both myeloperoxidase (MPO) and eosinophil peroxidase (EPO) activity in the BALF from Respondin™-treated mice was lower than that from the control mice, suggesting that Respondin™ treatment reduced allergen (OVA)-induced neutrophil and eosinophil infiltration into the airway. Thus, Respondin™ demonstrated a potential for treatment of asthma and relevant airway diseases [123].

18.2.4.2.2.2 In Vitro *Experiments with Human Cells; Clinical Trial*

Series of *in vitro* experiments with human blood cells, examining production of cytokines revealed a stimulatory effect, similar to that seen in the mouse model. Figure 18.11 demonstrates effect of Respondin™ on the stimulation of peripheral blood mononuclear cells measured as production of IFN-γ.

Effect of Respondin™ on antibody response to influenza vaccination was studied recently in a randomized, placebo-controlled, double-blind clinical trial. Adults over 50 years of age received a 200 mg or 400 mg dose of Respondin™, or a placebo capsule. The medication was taken for 28 days; after 21 days, participants were immunized with a commercially available, inactivated, split virion influenza vaccine. Measuring antibodies to the three influenza virus strains before, 7 and 21 days after immunization assessed antibody response to the immunization.

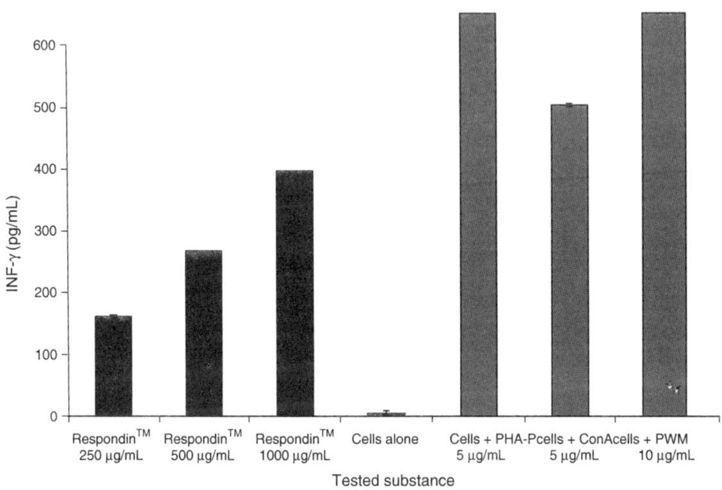

FIGURE 18.11 Effect of Respondin™ on the stimulation of human peripheral blood mononuclear cells measured as the production of IFN-γ.

Respondin™ significantly increased the antibody response to influenza vaccine as measured at 7 days in participants of 55 years of age or younger. There was a consistent trend for all three viruses, and this could be seen with both the 200 and 400 mg dose, but was more consistently apparent with the higher dose [129].

18.2.4.2.2.3 Chemical Makeup of Respondin™

The main active principles of Respondin™ are polysaccharide-based complexes. Although the basic extract contains almost 50% glucose, other monosaccharides comprise major immunoactive polysaccharide fractions. GC-MS and GC-FID of the most active fractions showed that the major monosaccharides are Ara, Gal, and Rha, suggesting that arabinogalactans- and rhamnose-rich biopolymers, possibly rhamnogalacturonans and glucorhamnans are major polysaccharides comprising the immunoactive principles. More comprehensive studies, involving ethanol and Cetavlon precipitation, size exclusion chromatography (SEC), anion exchange chromatography (AEC), and hydrophobic interaction chromatography (HIC) revealed that the larger polysaccharide/protein complexes were of arabinogalactan nature whereas the smaller ones were polysaccharides containing high content of rhamnose. A unique arabinogalactan with –2)-α-L-Araf-(1-3)-D-L-araf-(1-4)-β-D-Galp-1- repeating unit with molecular mass of about 47 kDa was isolated but was not immunoactive [132]. Later 1→2-linked α-D-galactofuranan with molecular weight of about 15 kDa as well as arabinogalactans were isolated but only larger arabinogalactans (188 ± 109 kDa and 1020 ± 370 kDa) were immunoactive [133]. *Chlorella* cells are known to contain LPS, which have been shown to be structurally and functionally similar to LPS of Gram-positive bacteria or endotoxins. Like LPS from gram-negative bacteria, the algal molecule stimulates exocytosis of the Limulus blood cell and the clotting of coagulin [134]. Unlike bacterial LPS they inhibit release of TNF-α from macrophages of the infected host and thereby inhibit sepsis [135–137]. Thus, it is important to realize while testing for immunoactivity that if the tests are not executed with caution, presence of LPS-like materials in *Chlorella* preparations could lead to false positives suggesting bacterial contamination.

18.2.5 Shark Cartilage

Shark cartilage has traditionally been associated with a number of medical benefits and particularly with anticancer effects [138–142]. The basis of shark cartilage cancerostatic activity is antiangiogenesis [143–145] and several clinical trials have been conducted or are underway [146–149].

There has been speculation, rather than scientific evidence, that the cartilage may stimulate both the cellular and humoral branches of the immune system, presumably making it effective against tumor growth, bacterial, viral, and fungal infections. In this context, a Canadian patent application was issued describing the use of shark cartilage as a component of a medicine that could be used to treat, alleviate, prevent, restore, or preserve health from various diseases, such as hepatitis *C.*, chronic fatigue syndrome, arthritis, bacteria, viruses, and fungal infections, but unfortunately the document lacks any data [150]. An interesting work examining the immunosupportive properties of shark cartilage and rice bran supplements using a rat model was presented at a Society for Neuroscience meeting several years ago [151]. Male Lewis rats were divided into three groups, a control group fed only oatmeal, a shark cartilage group fed oatmeal mixed with a commercially available shark cartilage supplement, and a third group fed oatmeal mixed with a commercially available rice bran supplement. Rice bran and shark cartilage combo increased splenic NK-cell activity and lymphocyte proliferation challenged with Con-A, phytohemagglutinin (PHA), and toxic shock syndrome toxin (TSST). However, we were the first to publish scientific evidence on immunostimulatory properties of shark cartilage extracts in a peer-reviewed journal [152]. Soon after our findings were published, a paper appeared supplying additional evidence for the immunostimulatory activity of shark cartilage, that is, modulation of antitumor immunity through infiltration by CD4+ and CD8+ lymphocytes [153].

The same group published a paper on immunoactivity of a shark cartilage fraction composed of two low molecular weight proteins. This fraction enhances DTH response against SRBC in mice, and decreases the cytotoxic activity of NK cells. It also increases T cell infiltration into the tumor, and decreases the tumor lesion size. Additionally, this fraction has strong inhibitory effect on HBMEC proliferation and migration in fibrin matrix [154].

18.2.5.1 ONC-106

We have tried a number of methods for extracting immunostimulatory principles from shark cartilage, and for some time extraction using buffered 2 M guanidine was our method of choice. The guanidine extraction was later replaced with a simple water extraction at 80°C that led to producing an immunostimulatory preparation known as ONC-106. Both ONC-106 and guanidine preparations equally stimulated proliferation of splenocytes *in vitro*, but hot water extractions delivered products in higher yields (7.4% versus 4.9%). Although ONC-106 has a lower content of proteoglycans than guanidine extracts (in average about 43% of that of guanidine extracts), it displayed comparable immunostimulatory activity. In our *in vitro* tests, both types of shark cartilage extracts behaved similar to Respondin™. They are B-cell proliferation stimulators and macrophage activators.

The patterns of production of a number of cytokines induced by activation with our shark cartilage extracts were examined, and an increase in production of IL-1, IL-6, IL-10, IL-12 and IFN-γ was seen [152]. Stimulation of macrophage activity was also demonstrated by generating elevated levels of nitric oxide upon the macrophage challenge with ONC-106.

ONC-106 also stimulated activity of the NK cells isolated from spleen cells (Figure 18.12). The activity was measured in terms of killing YAC-1 target cells.

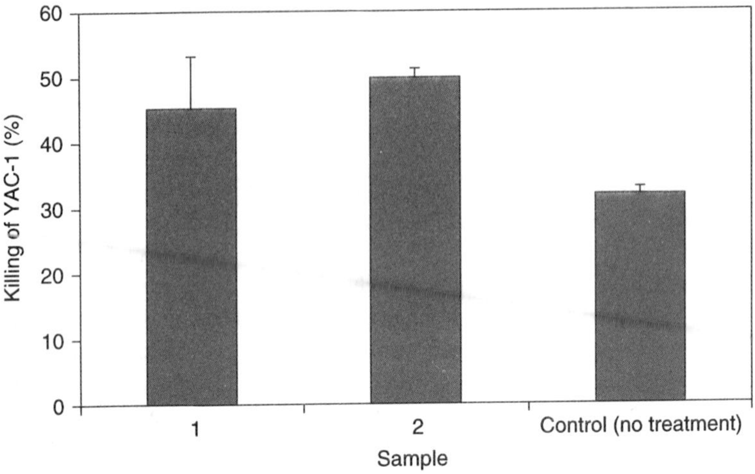

FIGURE 18.12 Stimulation of the activity of mouse spleen NK cells by two different preparations of ONC-106 (YAC-1 as the target cells).

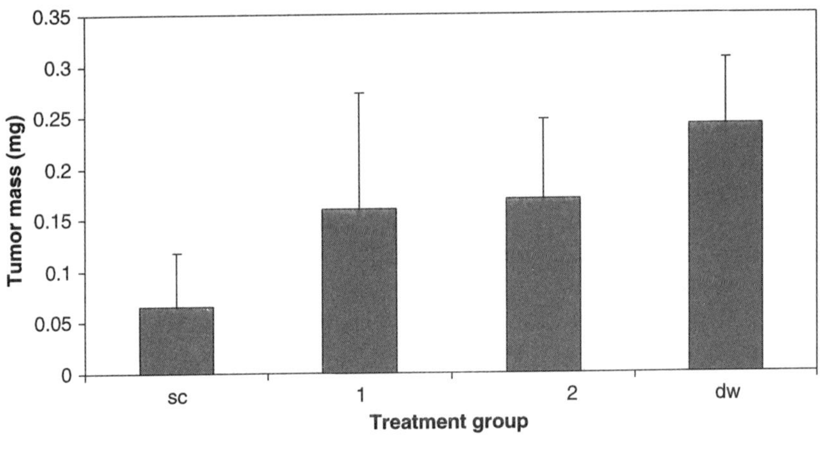

FIGURE 18.13 Effect of shark cartilage (sc) and two ONC-106 preparations on the growth of MCA-38 mouse carcinoma (dw stands for the untreated control).

Although no significant increase in killing was seen using the whole spleen cells, the separated NK cells showed a significant increase in killing when incubated with either water- or guanidine-based shark cartilage extracts. In addition to *in vitro* studies, our shark cartilage preparations were also tested in various models *in vivo* and it was discovered that ONC-106 and selected extracts had a significant inhibitory effect on proliferation of *Listeria monocytogenes* in mouse cells [155]. The material was also effective in reducing tumor mass of MCA-38 mouse colon adenocarcinoma (Figure 18.13).

Shark cartilage extracts are complex mixtures of molecules; the main immune principles of ONC-106 and our guanidine-based extracts are proteoglycans and collagen with molecular mass greater than 100 kDa. Owing to a very complex interplay among the molecules, it is premature to speculate on molecular mechanisms of these immunomodulators.

18.3 DISCUSSION

The rationale behind the idea of immunomodulators is to assure a proper response to a series of stress factors, which continuously attack human body, weaken the function of the immune system (immunosuppression), and consequently cause illness. Immunosuppression can be generated by severe bacterial and viral infections, cancer, environmental agents such as pesticides or allergens, long-term chemo- or radiotherapy, malnutrition, psychological stress, or endogenic autoimmune reaction. There has been dramatic increase in a number of new antibiotic-resistant microorganisms, proliferation of AIDS, and return of old afflictions. Many chronic diseases are the consequence of an unbalanced or impaired immune system. Among them, recurrent opportunistic infections, and skin and intestine inflammations are the most important ones, and it is hypothesized that the incidence of some severe infections such as AIDS might be positively influenced by restoration of the chronically suppressed immune system. Although Japanese have used clinical immunostimulants for a number of years, this has not been the case in North America. Therapeutic BCG (ImmmunoCysts©, Connaught Laboratories Ltd., Willowdale, ON), was the first immunomodulator licensed in the United States and Canada. Similar to PSK, OK-432, lentinan and SPG Immmunocysts is intended for cancer therapy, namely for treatment of carcinoma of the urinary bladder. The product license was granted following the conduct of a well-designed, multicentered, independent, and controlled efficacy study with well-defined end points and long follow-up period [156]. All new immunomodulator products should be standardized and have adequate postmarketing surveillance to determine their toxicity.

Knowledge of which specific cell types were being stimulated is important for designing potential combination therapies. Combining substances that give complementary stimulation, rather than those that simply stimulate the same cells, are more likely to succeed. Activated macrophages produce nitric oxide, in the form of a free radical with antimicrobial activity, and since some of the products, including ONC-106 and Respondin™, stimulated production of nitric oxide by macrophages, they could potentially play an important role in enhancing the body's defense against bacteria. The capacity to stimulate production of certain cytokines by macrophages brings important insights with respect to the underlying mechanisms involved in the immunostimulating activities of some of the mentioned materials including Respondin™ and ONC-106. IL-6 is an important effector molecule involved in B-lymphocyte activation and stimulation of antibody production, while IL-12 plays a major role in stimulation of NK cells and in affecting the balance of Th1 and Th2 cells [157]. This is important because Th1 cells are primarily involved in cellular immunity and Th2 cells appear primarily involved in humoral immunity. Immunomodulators stimulate at one dose and inhibit at another, usually at a higher dose. However, any substance at high enough dose can be inhibitory, thus it is important to know whether the inhibition represents activation of suppressor mechanisms, antiproliferative effects on lymphocytes, specific immunotoxic effects, or nonspecific toxicity. Persistent effects on the immune system by immunomodulators are very rare, which is contrary to many cases of conventional pharmaceuticals.

Some of the nutraceuticals, including shark cartilage products and *Chlorella* have been introduced as dietary supplements or even tested in clinical practice; however, many of the claims of purity, bioactivity, and clinical efficacy are not backed by rigorous scientific investigation. A variety of marine-based pharmaceutical products and nutraceuticals have been developed as marine by-products. Chitosan, glucosamine, PUFA concentrates, and protamines are good examples [10]. Except for a few preparations, they are sometimes semipurified preparations or mostly crude extracts. However, more sophisticated products, particularly those intended for pharmaceutical use, are now under development. The area has been greatly advanced by improvements in technologies involved in deep-sea sample collection and large-scale drug and nutraceutical production through aquaculture. Immunomodulators appear to play a very important role in future medicine since the objective of their use is not only to enhance the immune response, but also to help manage the immunosuppressive consequence of stress [158].

Naturally, the fact that the majority of marine-based immunomodulators have large molecular masses raises questions regarding efficacy after oral administration. Some studies demonstrated evidence that certain proteins modulate intestinal epithelium permeability and thus enhance the intestinal absorption of orally administered macromolecules through the paracellular pathway [159]. The fact that some large biopolymer immunostimulants are bioactive after oral administration is supported by a number of examples. Perhaps the most compelling evidence for oral effectiveness of large immunotherapeutics comes from two of the most clinically successful immunostimulants, PSK and SPG [160,161]. High-molecular-mass substances may be absorbed from the lymphoid nodules and PP of the gut via the lymphatic pathway [162,163]. It is assumed that pinocytotic cells or follicle-associated epithelial cells of PP present in the intestinal epithelium engulf large molecular mass molecules, bacterial cells, and viruses that have entered the intestinal cells [164]. Interestingly, glucans of high molecular mass seem to be more immunoactive than those with low molecular mass and they are more active at lower dosage and vice versa. This dose–effect relationship bears some relationship to the immunobiological reactions of the host [165].

Crude extracts of complex natural products are generally associated with a degree of compositional variability, depending on the differences in the genetic makeup, seasonal, and environmental factors. It is clear from our study that a complex mixture of molecules is responsible for the immunostimulatory activity of the above-described crude extracts and more refined fractions. Isolation, purification, analysis, and standardization of these complex materials are still extremely challenging, and therefore it is of the utmost importance to identify markers associated with a particular bioactivity. That would serve as a basis of effective quality control of marketed nutraceuticals and alternative medicine products, and ensure consistent performance. Therefore, one of our current research priorities is a detailed characterization and a quantitative determination of the representative biomarkers for immunostimulatory activity of marine-based nutraceuticals.

REFERENCES

1. Levander, O.A. and Whanger, P.D. 1996. Deliberation and evaluation of the approaches, endpoints and paradigms for selenium and iodine dietary recommendations. *J. Nutr.* 126, 2427S–2434S.
2. Shils, M.E. and Rude, R.K. 1996. Deliberation and evaluation of the approaches, endpoints and paradigms for magnesium dietary recommendations. *J. Nutr.* 126, 2398S–2403S.
3. Feigin, R.D. 1977. Interaction of nutrition and infection: plans for future research. *Am. J. Clin. Nutr.* 30, 1553–1563.
4. Scrimshaw, N.S. 1977. Effect of infection on nutrient requirements. *Am. J. Clin. Nutr.* 30, 1536–1544.
5. Cunningham-Rundles, S. 1982. Effects of nutritional status on immunological function. *Am. J. Clin. Nutr.* 35, 1202–1210.
6. Grimble, R.F. 1995. Interactions between nutrients and the immune system. *Nutr. Health* 10, 191–200.
7. Fenical, W. 2003. D. John Faulkner (1942–2002): marine natural products chemistry and marine chemical ecology. *Angew. Chem. Int. Ed. Engl.* 42, 1438–1439.
8. Renn, D.W. 1993. Medical and biotechnological applications of marine macroalgal polysaccharides. In *Marine Biotechnology, Volume 1: Pharmaceutical and Bioactive Natural Products*, D.H. Attaway and O.R. Zaborsky (ed.). Plenum Press, New York, pp.181–196.
9. Ayehunie, S., Belay, A., Baba, T.W. and Ruprecht, R.M. 1998. Inhibition of HIV-1 replication by an aqueous extract of *Spirulina platensis* (*Arthrospira platensis*). *J. Acquir. Immune Defic. Syndr. Hum. Retrovirol.* 18, 7–12.
10. Oshima, T. 1998. Recovery and use of nutraceutical products from marine resources. *Food Technol.* 52, 50–55.
11. Okai, Y., Higashi-Okai, K., Ishizaka, S. and Yamashita, U. 1997. Enhancing effect of polysaccharides from an edible brown alga *Hijikia fusiforme* (Hijiki) on release of tumor necrosis factor-α from macrophages of endotoxin-nonresponder C3J HeJ mice. *Nutr. Cancer* 27, 74–79.
12. Schaeffer, D.J. and Krylov, V.S. 2000. Anti-HIV activity of extracts and compounds from algae and cyanobacteria. *Ecotoxicol. Environ. Safety* 45, 208–227.
13. Carper, J. 1989. Seaweed or kelp. In *The Food Pharmacy*, Bantam Books, New York, pp. 264–268.
14. Cho, K.J., Lee, Y.S. and Ryu, B.H. 1990. Antitumor and immunological effects of seaweeds against sarcoma-180. *Han'guk Susan Hakhoechi* 23, 345–352.
15. Volgarev, M.N., Levachev, M.M., Trushchina, E.N., Sergeeva, K.V., Kulakova, S.N., Aseeva, L.E., Mustafina, O.K. and Tokareva, T.V. 1993. The modulation of the function of the immunocompetent system and of nonspecific body resistance in rats with different ratios of essential fatty acids in the diet. *Biull. Eksp. Biol. Med.* 116, 607–609.
16. DeLuca, P., Rothman, D. and Zurier, R.B. 1995. Marine and botanical lipids as immunomodulatory and therapeutic agents in the treatment of rheumatoid arthritis. *Rheum. Dis. Clin. North Am.* 21, 759–777.

17. Wu, D., Meydani, S.N., Meydani, M., Hayek, M.G., Huth, P. and Nicolosi, R.J. 1996. Immunologic effects of marine- and plant-derived n-3 polyunsaturated fatty acids in nonhuman primates. *Am. J. Clin. Nutr.* 63, 273–280.
18. Lombardi, V.R.M., Cagiao, A., Fernandez Novoa, L., Alvarez, X.A., Corzo, M.D., Zas, R., Sampedro, C. and Cacabelos, R. 2001. Short term food supplementation effects of a fish derived extract on the immunological status of pregnant rats and their sucking pups. *Nutr. Res.* 21, 1425–1435.
19. Moghaddam, M.F., Gerwick, W.H. and Ballantine, D.L. 1989. Discovery of 12-(S)-hydroxy-5,8,10,14-icosatetraenoic acid [12-(S)-HETE] in the tropical red alga *Platysiphonia miniata*. *Prostaglandins* 37, 303–308.
20. Tsujimoto, M. and Toyama, Y. 1922. About saponification of components (higher alcohols) of shark liver- and flat fish oils. *Chem. Umschau.* 29, 27–29.
21. Brohult, A., Brohult, J. and Brohult, S. 1970. Biochemical effects of alkylglycerols and their use in cancer therapy. *Acta. Chem. Scand.* 24, 730–732.
22. Brohult, A., Brohult, J. and Brohult, S. 1972. Effect of irradiation and alkoxyglycerol treatment on the formation of antibodies after *Salmonella* vaccination. *Experientia* 28, 954–955.
23. Brohult, A., Brohult, J., Brohult, S. and Joelson, I. 1977. Effect of alkoxyglycerols on the frequency of injuries following radiation therapy for carcinoma of the uterine cervix. *Acta. Obster. Gynecol. Scand.* 56, 441–448.
24. Palmblad, J., Samuelsson, J. and Brohult, J. 1990. Interactions between alkylglycerols and human neutrophil granulocytes. *Scand. J. Clin. Lab. Invest.* 50, 363–370.
25. Buford, R.G. and Gowdey, C.W. 1968. Anti-inflammatory activity of alkoxyglycerol in rats. *Arch. Int. Pharmacodyn.* 173, 56–70.
26. Boeryd, B., Nilsson, T., Lindholm, L., Lange, S., Hallgren, B. and Stallberg, G. 1978. Stimulation of immune reactivity by methoxy-substituted glycerol ethers incorporate into the feed. *Eur. J. Immunol.* 8, 678–680.
27. Oh, S.Y. and Jadhaw, L.S. 1994. Effects of dietary alkylglycerols in lactating rats on immune responses in pups. *Pediatric. Res.* 36, 300–305.
28. Werbach, M.R. 2001. Alkylglycerols. In *Research Booklet BRG*, T. Fujita (ed.), Lane labs http://www.lanelabs.com/published%20clinical%20Research.pdf.
29. Chorostowska-Wynimko, J., Krotkiewski, M., Radomska-Lesniewska, D., Sokolnicka, I. and Skopinska-Rozewska, E. 2001. The synergistic effect of lactic acid bacteria and alkylglycerols on humoral immunity in mice. *Int. J. Tissue React.* 23, 81–87.
30. Homma, S. and Yamamoto, N. 1990. Activation process of macrophages after *in vitro* treatment of mouse lymphocytes with dodecylglycerol. *Clin. Exp. Immunol.* 79, 307–313.
31. Ngwenya, B. and Foster, D.M. 1991. Enhancement of antibody production by lysophosphatidylcholine and alkylglycerol. *Proc. Soc. Exp. Biol. Med.* 196, 69–75.
32. Demopoulos, C.A., Pinckard, R.N. and Hanahan, D.J. 1979. Platelet-activating factor. Evidence for 1-O-alkyl-2-acetyl-sn-glyceryl-3-phosphorylcholine as the active component (a new class of lipid chemical mediators). *J. Biol. Chem.* 254, 9355–9358.
33. Snyder, F. 1972. Ether-linked lipids and fatty alcohol precursors in neoplasma. In *Ether Lipids. Chemistry and Biology*, Academic Press, New York, pp. 273–295.
34. Roos, D.S. and Choppin, P.W. 1984. Tumorigenicity of cell lines with altered lipid composition. *Proc. Nat. Acad. Sci. USA* 81, 7622–7626.
35. May, S.W., Sharkis, S.J., Esa, A.H., Gebbia, V., Clapham, V., Kraft, A.S., Pettit, G.R. and Sensenbrenner, L.L. 1997. Antineoplastic bryostatins are multipotential stimulators of human hematopoietic progenitor cells. *Proc. Natl. Acad. Sci. USA* 84, 8483–8487.

36. Hess, A.A., Silanskis, A.H., Esa, A.H., Pettit, G.R. and Sitkovsky, M.V. 1988. Activation of human T lymphocytes by bryostatin 1. *J. Immunol.* 140, 433–439.
37. Boto, W.M., Brown, L., Chrest, J. and Adler, W.H. 1991. Distinct modulatory effects of bryostatin 1 and staurosporine on the biosynthesis and expression of the HIV receptor protein (CD4) by T cells. *Cell Regul.* 2, 95–103.
38. Trenn, G., Pettit, G.R., Takayama, H., Hu-Li, J., Sitkovsky, G.R., Pettit, H. and Sitkovsky, M.V. 1988. Immunomodulating properties of a novel series of protein kinase C activators. The bryostatins. *J. Immunol.* 140, 433–439.
39. Hornung, R.L., Pearson, J.W., Beckwith, M. and Longo, D.L. 1992. Preclinical evaluation of bryostatin as an anticancer agent against several murine tumor cell lines: *in vitro* and *in vivo* activity. *Cancer Res.* 52, 101–107.
40. Shoji, N., Umeyama, A., Motoki, S., Arihara, S., Ishida, T., Nomoto, K., Kobayashi, J. and Takei, M. 1992. Potent inhibitors of histamine release, two novel triterpenoids from the Okinawan marine sponge *Penares incrustans*. *J. Nat. Prod.* 55, 1682–1685.
41. Natori, T., Morita, M., Akimoto, K. and Koezuka, Y. 1994. Agelaspins, novel antitumor and immunostimulatory cerebrosides from marine sponge *Agelas Mauritinius*. *Tetrahedron* 50, 2771–2784
42. Sakai, T., Ueno, H., Natori, T., Uchimura, A., Motoki, K. and Koezuka, Y. 1998. Effects of alpha-, and beta-galactolysated C2-ceramides on the immune system. *J. Med. Chem.* 41, 650–652.
43. Natori, T., Motoki, K., Higa, T. and Koezuka, Y. 2000. KRN7000 as a new type of antitumor and immunostimulatory drug. In *Drugs from the Sea*, N. Fusetani (ed.), Basel, Karger, pp. 86–97.
44. Kojima, T., Koike, A., Yamamoto, S., Kanemitsu, T., Miwa, M., Kamei, H., Kondo, T. and Iwata, T. 1993. Eisenin (L-pyroGlu-L-Gln-L-Ala), a new biological response modifier. *J. Immunother.* 13, 36–42.
45. Kralovec, J. 1984. Tuftsin. *Drugs Future* 9, 541–547.
46. Prendergast, R.A. and Liu, S.H. 1976. Isolation and characterization of sea star factor. *Scand. J. Immunol.* 5, 873–880.
47. Liu, S.H., McChesney, M.B. and Prendergast, R.A. 1983. Kinetics of tumor cell cytostasis by sea star factor-activated macrophages. *Develop. Compar. Immunol.* 7, 545–554.
48. Lis, H. and Sharon, N. 1973. The biochemistry of plant lectins (*Phytohemagglutinins*). *Ann. Rev. Biochem.* 42, 541–574.
49. Rogers, D.J. and Hori, K. 1993. Marine algal lectins: new developments. *Hydrobiologia* 260/261, 589–593.
50. Lima, C.H., Costa, F.H.F., Sampaio, A.H., Neves, S.A., Benevides, N.M.B., Teixeira, D.I.A., Rogers, D.J. and Freitas, A.L.P. 1998. Induction and inhibition of human lymphocyte transformation by the lectin from the red marine alga *Amansia multifida*. *J. Appl. Phycol.* 10, 153–162.
51. Olsson, C.A., Chute, R. and Rao, C.N. 1974. Immunologic reduction for bladder cancer recurrence rates. *J. Urol.* 111, 173–176.
52. Jurincic, C.D., Engelmann, U., Gasch, J. and Klippel, K.F. 1988. Immunotherapy in bladder cancer with keyhole limpet hemocyanin: a randomized study. *J. Urol.* 139, 723–726.
53. Lamm, D.L. 1990. Recurrent superficial transition cell carcinoma of the bladder: adjuvant topical chemotherapy versus immunotherapy a prospective randomized trial. *J. Urol.* 144, 260–263.
54. Lamm, D.L., DeHaven, J.I., Riggs, D.R., Delgra, C. and Burrell, R. 1993. Keyhole limpet hemocyanin immunotherapy of murine bladder cancer. *Urol. Res.* 21, 33–37.

55. Lamm, L.D., Morales, A, Grossman, H.B, Lowe, B. and Swerdlow, R. 1996. Keyhole limpet hemocyanin (KLH) immunontherapy of papillary and in situ transitional cell carcinoma of the bladder. A multicenter phase I-II clinical trial. *J. Urol.* 155, A1405.
56. Wishahi, M.M., Ismail, I.M., Rubben, H. and Otto, T. 1995. Keyhole-limpet hemocyanin immunotherapy in the bilharzial bladder: a new treatment modality? Phase II trial: superficial bladder cancer. *J. Urol.* 153, 926–928.
57. Jurincic-Winkler, C.D., Metz, K.A., Beuth, J. and Klippel, K.F. 2000. Keyhole limpet hemocyanin for carcinoma *in situ* of the bladder: a long-term follow-up study. *Eur. Urol.* 37 (Suppl. 3),45–49.
58. Lamm, D.L., DeHaven, J.I. and Riggs, D.R. 2000. Keyhole limpet hemocyanin immunotherapy of bladder cancer: laboratory and clinical studies. *Eur. Urol.* 37 (Suppl.3), 41–44.
59. Riggs, D.R., Jackson, B., Vona-Davis, L. and McFadden, D. 2002. *In vitro* anticancer effects of a novel immunostimulant: keyhole limpet hemocyanin. *J. Surg. Res.* 108, 279–284.
60. Markl, J., Lieb, B., Gebauer, W., Altenhein, B., Meissner, U. and Harris, J.R. 2001. Marine tumor vaccine carriers: structure of the molluscan hemocyanins KLH and HtH. *J. Cancer Res. Clin. Oncol.* 27 (Suppl 2), R3–R9.
61. Thomson, A.W. and Fowler, E.F. 1981. Carrageenan. A review of its effects on the immune system. *Agents Actions* 11, 265–272.
62. Hanazawa, S., Ishikawa, T. and Yamaura, K. 1982. Comparison of the adjuvant effect of the antibody response of three types of carrageenan and the cellular events in the induction of the effect. *Int. J. Immunopharmacol.* 4, 521–527.
63. Manicino, D. and Minucci, M. 1983. Adjuvant effects of ι, κ and λ carrageenans on antibody production in BALB/c mice. *Int. Arch. Allergy Appl. Immunol.* 72, 359–361.
64. Sugawara, I., Ishizuka, S. and Muller, G. 1982. Carrageenans, highly sulfated polysaccharides and macrophage-toxic agents: newly found human T lymphocyte activator. *Immunobiology* 163, 527–538.
65. Shikita, M., Tsuneoka, K., Hagiwara, S. and Tsurufuji, S. 1981. A granulocyte-macrophage colony-stimulating factor (GM-CSF) produced by carrageenin-induced inflammatory cells of mice. *J. Cell Physiol.* 109, 161–169.
66. Kolb, J-P.B., Quan, P.C., Poupon, M-F. and Desaymard, C. 1981. Carrageenan stimulates population of mouse B cells mostly overlapping with those stimulated with LPS or dextran sulfate. *Cell Immunol.* 57, 348–360.
67. Wu, G-J., Lin, H-T. and Tsai, G-J. 2003. Production of chitooligosaccharides from shrimp chitosan with immune-enhancing activity. In *Advance in Chitin Science Vol. V*, K. Suchiva, S. Chandrkrachang, P. Methacanon, M.G. Peter (eds.), National Metal and Materials Technology Center (MTEC), Bangkok, pp. 77–80.
68. Strong, P., Clark, H. and Reid, K. 2002. Intranasal application of chitin microparticles down-regulates symptoms of allergic hypersensitivity to *Dermatophagoides pteronyssinus* and *Aspergillus fumigatus* in murine models of allergy. *Clin. Exp. Allerg.* 32, 1794–1800.
69. Okai, Y., Higashi, Okai K., Ishizaka, S., Ohtani, K., Matsui-Yuasa, I. and Yamashita, U. 1998. Possible immunomodulating activities in an extract of edible brown alga, *Hijikia fusiforme* (Hijiki). *J. Sci. Food Agric.* 76, 56–62.
70. Yoshizawa, Y., Enomoto, A., Todho, H., Ametani, A. and Kaminogawa, S. 1993. Activation of murine macrophages by polysaccharide fractions from marine alga (*Porphyra yerzensis*). *Biosci. Biotech. Biochem.* 57, 1862–1866.

71. Yoshizawa, Y., Enomoto, A., Todho, H., Ametani, A. and Kaminogawa, S. 1993. Macrophage stimulation activity of polysaccharide fractions from a marine alga (*Porphyra yerzensis*): structure-function relationships and improved solubility. *Biosci. Biotech. Biochem.* 59, 1933–1937.
72. Liu, J.N., Yoshida, Y., Wang, M.Q., Okai, Y. and Yamashita, U. 1997. B cell stimulating activity of seaweed extracts. *Int. J. Immunopharmacol.* 19, 135–142.
73. Shan, B.E., Yoshida, Y., Kuroda, E. and Yamashita, U. 1999. Immunomodulating activity of seaweed extract on human lymphocytes *in vitro. Int. J. Immunopharmacol.* 21, 59–70.
74. Itoh, H., Noda, H., Amano, H., Zhuaug, C., Mizuno, T. and Ito, H. 1993. Antitumor activity and immunological properties of marine algal polysaccharides, especially fucoidan, prepared from *Sargassum thunbergii* of *Phaeophyceae. Anticancer Res.* 13, 2045–2052.
75. Tominaga, T., Yamashita, S., Mizutani, S., Sagawa, H. and Kato, I. 2002. Remedies. European patent EP 1,226,826 A1.
76. Ren, D.L., Wang, J.Z., Noda, H., Amano, H. and Ogawa, S. 1995. The effects of an algal polysaccharide from *Gloiopeltis tenax* on transplantable tumors and immune activities in mice. *Planta Med.* 61, 120–125.
77. Yim, J.H., Son, E., Pyo, S. and Lee, H.K. 2005. Novel sulfated polysaccharide derived from red-tide microalga *Gyrodinium impudicum* strain KG03 with immunostimulating activity in vivo. *Mar. Biotechnol.* 7, 331–338.
78. Yim, J.H., Kim, S.J, Ahn, S.H. and Lee, H.K. 2006. Characterization of a novel bioflocculant, p-KG03, from a marine dinoflagellate, *Gyrodinium impudicum* KG03. *Bioresour. Technol.* 98, 361–367.
79. Damonte, E., Neyts, J., Pujol, C.A., Snoeck, R., Andrei, G., Ikeda, S., Witvrouw, M., Reymen, D., Haines, H., Matulewicz, M.C., Cerezo, A., Coto, C.E. and De Clercq, E. 1994. Antiviral activity of a sulphated polysaccharide from the red seaweed *Nothogenia fastigiata. Biochem. Pharmacol.* 47, 2187–2192.
80. Furusawa, E. and Furusawa, S. 1985. Anticancer activity of a natural product, Viva-Natural, extracted from *Undaria pinnantifida* or intraperitoneally implanted: Lewis lung carcinoma. *Oncology* 42, 364–369.
81. Furusawa, E. and Furusawa, S. 1989. Anticancer potential of Viva-Natural, a dietary seaweed extract, on Lewis lung carcinoma in comparison with chemical immunomodulators and on cyclosporine-accelerated AKR leukemia. *Oncology* 46, 343–348.
82. Furusawa, E. and Furusawa, S. 1988. Effect of pretazettine and Viva-Natural, a dietary seaweed extract, on spontaneous AKR leukemia in comparison with standard drugs. *Oncology* 45, 180–186.
83. Okai Y., Ishizaka. S., Higashi-Okai, K. and Yamashita, U. 1996. Detection of immunomodulating activities in an extract of Japanese edible seaweed, *Laminaria japonica* (Makonbu). *J. Sci. Food Agric.* 72, 455–460.
84. Kato, T. and Owen, R.L. 1994. M cell functions. In *Handbook of Mucosal Immunology*, P.L. Ogra (ed.), Academic Press, NY, pp. 20–24.
85. Suzuki, I., Hashimoto, K., Ohno, N., Tanaka, H. and Yadomae, T. 1989. Immunomodulation by orally administered β-glucan in mice. *Int. J. Immunopharmacol* 11, 761–769.
86. Teas, J. 1983. The dietary intake of *Laminaria*, a brown seaweed, and breast cancer prevention. *Nutr. Cancer* 4, 217–222.
87. Zaporozhets, T.S., Ovodova, R.G., Besednova, N.N., Molchanova, V.I. and Nikitin, A.V. 1985. Effect of polysaccharides isolated from marine invertebrates on various functions of T- and B lymphocytes. *Antibiot. Med. Biotekhnol.* 30, 43–47.

88. Zaporozhets, T.S., Besednova, N.N., Molchanova, V.N. and Zvyagintseva, T.N. 2001. Comparative immunomodulating activity of marine origin bioglycans. *Antibiot. Khimiother.* 46, 6–10.
89. Ivanushko, L.A., Besednova, N.N., Zaporozhets, T.S. and Zviagintseva, T.N. 2001. 1,3:1,6-D-glucan translam: results of studying and prospects for application. *Antibiot. Khimiother.* 46, 14–18.
90. Zapopozhets, T.S., Besednova, N.N. and Loenko, I.uN. 1995. Antibacterial and immunomodulating activity of fucoidan. *Antibiot. Khimioter.* 40, 9–13.
91. Vetvicka, V. and Yvin, J.C. 2004. Effects of marine beta-1,3 glucan on immune reactions. *Int. Immunopharmacol.* 4, 721–730.
92. Vetvicka, V., Dvorak, B., Vetvickova, J., Richter, J., Krizan, J., Sima, P. and Yvin, J.C. 2006. Orally administered marine (1-3)-beta-d-glucan phycarine stimulates both humoral and cellular immunity. *Int. J. Biol. Macromol.* (Epub ahead of print).
93. Kim, K.H., Kim, Y.W., Kim, H.B., Lee, B.J. and Lee, D.S. 2006. Anti-apoptotic activity of laminarin polysaccharides and their enzymatically hydrolyzed oligosaccharides from *Laminaria japonica. Biotechnol. Lett.* 28, 439–446.
94. Turner, J.L., Dritz, S.S., Higgins, J.J. and Minton, J.E. 2002. Effects of *Ascophyllum nodosum* extract on growth performance and immune function of young pigs challenged with *Salmonella typhimurium. J. Anim. Sci.* 80, 1947–1953.
95. Funahashi, H., Imai, T., Tanaka, Y., Tsukamura, K., Hayakawa, Y., Kikumori, T., Mase, T., Itoh, T., Nishikawa, M., Hayashi, H., Shibata, A., Hibi, Y., Takahashi, M. and Narita, T. 1999. Wakame seaweed suppresses the proliferation of 7,12-dimethylbenz(a)-anthracene-induced mammary tumors in rats. *Jpn. J. Cancer Res.* 90, 922–927.
96. Shimizu, T. and Mifuchi, I. 1983. Antitumor activity and immunostimulating effect of culture filtrate of a marine bacterium, *Vibrio anguillarum*, in mice. *Yakugaku Zasshi* 103, 761–765.
97. Shimizu, Y. 2000. Microalgae as a drug source. In *Drugs from the Sea*, N. Fusetani (ed.), Karger, Basel, pp. 30–45.
98. Kay, R.A. 1991. Microalgae as food and supplement. *Crit. Rev. Food Sci. Nutr.* 30, 555–573.
99. Chamorro, G., Salazar, M., Araujo, K.G., dos Santos, C.P., Ceballos, G. and Castillo, L.F. 2002. Update on the pharmacology of *Spirulina* (Arthrospira), an unconventional food. *Arch. Latinoam Nutr.* 52, 232–240.
100. Hirahashi, T., Matsumoto, M., Hazeki, K., Saeki, Y., Ui, M. and Seya, T. 2002. Activation of the human innate immune system by *Spirulina*—augmentation of interferon production and NK cytotoxicity by oral administration of hot-water-extract of *Spirulina platensis. DIC Tech. Rev.* 8, 43–49.
101. Kim, H.M., Lee, E.H., Cho, H.H. and Moon, Y.H. 1998. Inhibitory effect of mast cell-mediated immediate-type allergic reactions in rats by *Spirulina. Biochem. Pharmacol.* 55, 1071–1076.
102. Quereshi, M.A. and Ali, R.A. 1996. *Spirulina platensis* exposure enhances macrophage phagocytic function in cats. *Immunopharmacol. Immunotoxicol.* 18, 457–463.
103. Quereshi, M.A., Garlich, J.D. and Kidd, M.T. 1996. Dietary *Spirulina platensis* enhances humoral and cell-mediated function in chickens. *Immunopharmacol. Immunotoxicol.* 18, 465–476.
104. Labhe, R.U., Mani, U.V., Iyer, U.M., Mishra, M., Jani, K. and Bhattacharya, A. 2001. The effect of *Spirulina* in the treatment of bronchial asthma. *J. Nutr. Funct. Med. Foods* 3, 53–59.

105. Kaji, T., Fujiwara, Y., Hamada, C., Yamamoto, C., Shimada, S., Lee, J-B. and Hayashi, T. 2002. Inhibition of cultured bovine aortic endothelial cell proliferation by sodium spirulan, a new sulfated polysaccharide isolated from *Spirulina platensis. Planta Med.* 68, 505–509.
106. Hayashi, K., Hayashi, T. and Kojima, I. 1996. A natural sulfated polysaccharide, calcium spirulan, isolated from *Spirulina platensis*: *in vitro* and *ex vivo* evaluation of anti-herpes simplex virus and anti-human immunodeficiency virus activities. *AIDS Res. Hum. Retroviruses* 12, 1463–1471.
107. Li, B., Zhang, X., Gao, M. and Chu, X. 2005. Effects of CD59 on antitumoral activities of phycocyanin from *Spirulina platensis. Biomed. Pharmacother.* 59, 551–560.
108. Li, B., Gao, M.H., Zhang, X.C. and Chu, X.M. 2006. Molecular immune mechanism of C-phytocyanin from *Spirulina platennsis* induces apoptosis in HeLa cells in vitro. *Biotechnol. Appl. Biochem.* 43, 155–164.
109. Yalcin, I., Hicsazmas, Z., Boz, B. and Bozoglu, F. 1994. Characterization of the extracellular polysaccharide from freshwater microalgae *Chlorella* sp. *Lebens.-Wiss. U. Technol.* 27, 154–165.
110. Mitsuda, H., Nishikawa, Y., Higuchi, M., Nakajima, K. and Kawai, F. 1977. Effect of the breaking of Chlorella cells on the digestibility of *Chlorella* protein. *J. Jpn. Soc. Food Nutr.* 30, 93–98.
111. Kojima, M., Kasajima, T., Imai, Y., Kobayashi, S., Dobashi, M. and Uemura, T. 1973. A new *Chlorella* polysaccharide and its accelerating effect on the phagocytic activity of the reticuloendothelial system. *Recent Adv. RES Res.* 13, 101–122.
112. Neveu, P.J., Morin, O., Miegeville, M., LeMevel, B.P. and Vermeil, C. 1978. Modulation of antibody synthesis by an anti-tumor alga. *Experientia* 34, 1644–1645.
113. Pugh, N., Ross, S.A., ElSohly, H.N., ElSohly, M.A. and Pasco, D.S. 2001. Isolation of three high molecular weight polysaccharide preparations with potent immunostimulatory activity from *Spirulina platensis, Aphanizomenon flos-aque* and *Chlorela pyreinodosa. Planta Med.* 67, 737–742.
114. Komiyama, K., Hirokawa, Y., Morota, T. and Umezawa, I. 1986. An acidic polysaccharide, Chlon A, from *Chlorella pyrenoidosa*. II. Antitumor activity and immunological response. *Chemotherapy* 34, 302–307.
115. Miyazawa, Y., Murayama, T., Ooya, N., Wang, L.F., Tung, Y.C. and Yamaguchi, N. 1988. Immunomodulation by a unicellular green algae (*Chlorella pyrenoidosa*) in tumor-bearing mice. *J. Ethnopharmacol.* 24, 135–146.
116. Tanaka, K., Tomita, Y., Tsuruta, M., Konishi, F., Okuda, M., Himeno, K. and Nomoto, K. 1990. Oral administration of *Chlorella vulgaris* augments concomitant antitumor immunity. *Immunopharmacol. Immunotoxicol.* 12, 277–291.
117. Tanaka, K., Yamada, A., Noda, K., Hasegawa, T., Okuda, M., Shoyama, Y. and Nomoto, K. 1998. A novel glycoprotein obtained from *Chlorella vulgaris* strain CK22 shows antimetastatic immunopotentiation. *Cancer Immunol. Immunother.* 45, 313–320.
118. Tanaka, K., Koga, T., Konishi, F., Nakamura, M., Mitsuyama, M., Himeno, K. and Nomoto, K. 1986. Augmentation of host defense by unicellular green alga, *Chlorella vulgaris*, to *Escherichia coli* infection. *Infect. Immunol.* 53, 267–271.
119. Konishi, F., Tanaka, K., Kumamoto, S., Hasegawa, T., Okuda, M., Yano, I., Yoshikai, Y. and Nomoto, K. 1990. Enhanced resistance against *Escherichia coli* infection by subcutaneous administration of the hot-water extract of *Chlorella vulgaris* in cyclophosphamide-treated mice. *Cancer Immunol. Immunother.* 32, 1–7.

120. Hasegawa, T., Kimura, Y., Hiromatsu, K., Kobayashi, N., Yamada, A., Makino, M., Okuda, M., Sano, T., Nomoto, K. and Yoshikai, Y. 1997. Effect of hot water extract of *Chlorella vulgaris* on cytokine expression patterns in mice with murine acquired immunodeficiency syndrome after infection with *Listeria monocytogenes. Immunopharmacology* 35, 273–282.
121. Hasegawa, T., Ito, K., Ueno, S., Kumato, S., Ando, Y., Yamada, A., Nomoto, K. and Yasunobu, Y. 1999. Oral administration of hot water extracts of *Chlorella vulgaris* reduces IgE production against milk casein in mice. *Int. J. Immunopharmacol.* 21, 311–323.
122. Dantas, D.C.M. and Queiroz, M.L.S. 1999. Effects of *Chlorella vulgaris* on bone marrow progenitor cells of mice infected with *Listeria monocytogenes. Int. J. Immunopharmacol.* 21, 499–508.
123. An, H.J., Choi, H.M., Park, H.S., Han, J.G., Lee, E.H., Park, Y.S., Um, J.Y., Hong, S.H. and Kim, H.M. 2006. Oral administration of hot water extracts of *Chlorella vulgaris* increases physical stamina in mice. *Ann. Nutr. Metab.* 50, 380–386.
124. Price, III J.A., Sanny, C. and Shevlin, D. 2002. Inhibition of mast cells by algae. *J. Med. Food* 5, 205–210.
125. Guzman, S., Gato, A., Lamela, M., Freire-Garabal, M. and Calleja, J.M. 2003. Anti-inflammatory and immunomodulatory activities of polysaccharide from *Chlorella stigmatophora* and *Phaeodactylum tricornutum. Phytother. Res.* 17, 665–670.
126. Merchant, R.E., Rice, C.D. and Young, H.F. 1990. Dietary *Chlorella pyrenoidosa* for patients with malignant glioma: effects on immunocompetence, quality of life, and survival. *Phytother. Res.* 4, 220–231.
127. Merchant, R.E., Carmack, C.A. and Wise, C.M. 2000. Nutritional supplementation with *Chlorella pyrenoidosa* for patients with fibromyalgia syndrome: a pilot study. *Phytother. Res.* 14, 167–173.
128. Kralovec, J.A. 2003. Fractions of *Chlorella* extract containing polysaccharide having immunomodulating properties, US patent 6,551,596 B2.
129. Halperin, S.A., Smith, B.S., Nolan, C., Shay, J. and Kralovec, J. 2003. Randomized, double-blind, placebo-controlled trial of the safety and immunostimulatory effect of a *Chlorella*-derived food supplement in healthy adults undergoing influenza immunization. *Can. Med. Assoc. J.* 169, 111–117.
130. Kralovec, J.A., Metera, K.L., Kumar, J.R., Watson, L.V., Girouard, G.S., Guan, Y., Carr, R.I., Barrow, C.J. and Ewart, H.S. 2007. Immunomodulating principles from edible green microalga *Chlorella pyrenoidosa*. Part 1. Isolation and biological assessment *in vitro. Phytomedicine* 14, 57–64.
131. Kralovec, J.A., Power, M.R., Liu, F., Maydanski, E., Ewart, H.S., Watson, L.V., Barrow, C.J. and Lin, T-J. 2005. An aqueous *Chlorella* extract inhibits IL-5 production by mast cells *in vitro* and reduces ovalbumin-induced eosinophil infiltration in the airway in mice *in vivo. Int. Immunopharmacol.* 5, 689–698.
132. Reyes Suárez, E., Kralovec, J.A., Noseda, M.D., Ewart, H.S., Barrow, C.J., Lumsden, M.D. and Grindley, T.B. 2005. Isolation, characterization, and structural determination of a unique type of arabinogalactan from an immunostimulatory extract of *Chlorella pyrenoidosa. Carbohydrate Res.* 340, 1489–1498.
133. Reyes Suárez, E., Kralovec, J.A., Noseda, M.D., Barrow, C.J., Ewart, H.S., Syvitski, R., Lumsden, M.D. and Grindley, T.B. 2006. Application of DOSY experiments for the determination of structures and sizes o polysaccharides from immunostimulatory extract of *Chlorella pyrenoidosa*: the effect of molecular weight on immunostimulation. *Biomacromolecules* 7, 2368–2376.

134. Conrad, M.L., Pardy, R.L., Wainwright, N., Child, A. and Armstrong, P.B. 2006. Response of the blood clotting system of the American horseshoe crab, *Limulus polyphemus*, to a novel form of lipopolysaccharide from a green alga. *Comp. Biochem. Physiol. A Mol. Integr. Physiol.* 144, 423–428.
135. Royce, C.L. and Pardy, R.L. 1996. Endotoxin-like properties of an extract from a symbiotic, eukaryotic *chlorella*-like green alga. *J. Endotoxin. Res.* 3, 437–444.
136. Armstrong, P.B., Armstrong, M.T., Pardy, R.L., Child, A. and Wainwright, N. 2002. Immunohistochemical demonstration of a lipopolysaccharide in the cell wall of a eukaryote, the green alga, *Chlorella. Biol. Bull.* 203, 203–204.
137. Pardy, R.L., Morrison, D.C., Shnyra, A. and Royce, C.L. 2003. Isolated algal lipopolysaccharides and use of same to inhibit endotoxin-initiated sepsis. US patent 6,534,648.
138. Green, S. 1997. Shark cartilage therapy against cancer. *Nutr. Health Forum* 14, 1–5.
139. Hunt, T.J. and Connelly, J.F. 1995. Shark cartilage for cancer treatment. *Am. J. Health Syst. Pharm.* 52, 1756–1760.
140. Mathews, J. 1993. Media feeds frenzy over shark cartilage as cancer treatment. *J. Natl. Cancer Inst.* 85, 1190–1191.
141. McGuire, T.R., Kazakoff, Hoie, E.B. and Fienhold, M.A. 1996. Antiproliferative activity of shark cartilage with and without tumor necrosis factor alpha I human umbilical vein endothelium. *Pharmacotherapy* 2, 237–244.
142. Lane, J.W. 1991. Method and dosage unit for inhibiting angiogenesis or vascularization in animal using shark cartilage. US patent 5,075,112.
143. Folkman, J. and Klagsbrun 1987. Angiogenic factors. *Science* 235, 442–447.
144. Lee, A. and Langer, R. 1983. Shark cartilage contains inhibitors of tumor angiogenesis. *Science* 221, 1185–1187.
145. Dupont, E., Brazeau, P. and Juneau, C. 1997. Extracts of shark cartilage having an anti-angiogenic activity and an effect on tumor regression; process of making thereof. US patent 5,618,925.
146. McCutcheon, L. 1997. Taking a bite out of shark cartilage. *Skept. Inq.* 21, 44–48.
147. Couzin, J. 1998. Beefed-up NIH center probes unconventional therapies. *Science* 282, 2175–2176.
148. Miller, D.R., Anderson, G.T., Stark, J.J., Granick, J.L. and Richardson, D. 1998. Phase I/II trial of the safety and efficacy of shark cartilage in the treatment of advanced cancer. *J. Clin. Oncol.* 16, 2649–2655.
149. Batist, G., Patenaude, F., Champagne, P., Crotcau, D., Levinton, C., Hariton, C., Escudier, B. and Dupont, E. 2002. Neovastat (AE-941) in refractory renal cell carcinoma patients: report of a phase II trial with two dose levels. *Ann. Oncol.* 13, 1259–1263.
150. Brundritt, L.W. 1998. New medicine comprising sharks cartilage/flesh and kelp – useful for treating viruses, bacteria or fungal diseases, especially hepatitis C. Can. Pat. Appl. 2,201,025.
151. Giese, S., SaBell, G.R. and Coussons-Read, M.E. 2001. *Influence of the Oral Ingestion of Shark Cartilage and Rice Bran on the Immune System in a Rat Model.* Society of Neuroscience meeting, San Diego.
152. Kralovec, J.A., Guan, Y., Metera, K. and Carr, R.I. 2003. Immunomodulating principles from shark cartilage. Part 1. Isolation and biological assessment *in vitro. Int. Immunol.* 3, 657–669.
153. Feyzi, R., Hassan, Z.M. and Mostafaie, A. 2003. Modulation of $CD4^+$ and $CD8^+$ tumor infiltrating lymphocytes by a fraction isolated from shark cartilage: shark cartilage modulates anti-tumor immunity. *Int. Immunopharmacol.* 3, 921–926.

154. Hassan, Z.M., Feyzi, R., Sheikhian, A., Bargahi, A., Mostafaie, A., Mansouri, K., Shahrokhi, S., Ghazanfari, T. and Shahabi, S. 2005. Low molecular weight fraction of shark cartilage can modulate immune responses and abolish angiogenesis. *Int. Immunopharmacol.* 5, 961–970.
155. Kralovec, J.A., Watson, L.V., Metera, K.L., Guan, Y., Henneberry, K.P. and Carr, R.I. 2006. Isolation, characterization and effect after oral administration of shark cartilage preparations on mice infected with *Listeria monocytogenes. J. Alt. Compl. Med.* (submitted).
156. Barreto, L., Csizer, Z. and Sparkes, J.D. 1992. Evaluation of clinical data in bladder cancer immunotherapy with Connaught BCG (ImmuCyst). *Dev. Biol. Stand.* 77, 229–231.
157. Morel, P.A. and Oriss, T.B. 1998. Crossregulation between Th1 and Th2 cells. *Crit. Rev. Immunol.* 18, 275–303.
158. Anderson, D.P. 1992. Immunostimulants, adjuvants, and vaccine carriers in fish: applications to aquaculture. *Ann. Rev. Fish Dis.* 2, 281–307.
159. Fasano, A. and Uzzau, S. 1997. Modulation of intestinal tight junctions by *Zonula occludens* toxin permits enteral administration of insulin and other macromolecules in an animal model. *J. Clin. Invest.* 99, 1158–1164.
160. Nio, Y., Tsubono, M., Tseng, C-C., Morimoto, H., Kawabata, K., Masai, Y., Shirashi, T., Imai, S., Ohgaki, K. and Tobe, T. 1992. Immunomodulation by orally administered protein-bound polysaccharide PSK in patients with gastrointestinal cancer. *Biotherapy* 4, 117–128.
161. Furue, H. 1987. Biological characteristic and clinical effect of Sizofilan. *Drugs Today* 23, 335–334.
162. Kikumoto, S., Miyajima, T., Yoshizumi, T. Fujimoto, S. and Kimura, K. 1970. Study on polysaccharide produced by *Schizophyllum commune fries*. Production of polysaccharide and its properties. *Nogei Kagaku Kaishi* 44, 337–342.
163. Tabata, K., Ito, W., Kojima, T., Kawabata, S. and Misaki, A. 1981. Ultrasonic degradation of schizophyllan, an antitumor polysaccharide produced by *Schizophyllum commune fries. Carbohydr. Res.* 89, 121–135.
164. Yamaki, T., Ito, W., Tabata, K., Kojima, T., Norisuye, T., Takano, N. and Fujita, H. 1983. Correlation between the antitumor activity of polysaccharide schizophyllan and its triple-helical conformation in dilute aqueous solution. *Biophys. Chem.* 17, 337–342.
165. Suzuki, I., Hashimoto, K., Ohno, N., Tanaka, H. and Yadomae, T. 1989. Immunomodulation by orally administered β-glucan in mice. *Int. J. Immunopharmacol.* 11, 761–769.

Index

A

B

E

H

I

J

M

N

O

Q

R

U

V